Lecture Notes in Computer Science   14505

Founding Editors

Gerhard Goos
Juris Hartmanis

Editorial Board Members

Elisa Bertino, *Purdue University, West Lafayette, IN, USA*
Wen Gao, *Peking University, Beijing, China*
Bernhard Steffen⬤, *TU Dortmund University, Dortmund, Germany*
Moti Yung⬤, *Columbia University, New York, NY, USA*

The series Lecture Notes in Computer Science (LNCS), including its subseries Lecture Notes in Artificial Intelligence (LNAI) and Lecture Notes in Bioinformatics (LNBI), has established itself as a medium for the publication of new developments in computer science and information technology research, teaching, and education.

LNCS enjoys close cooperation with the computer science R & D community, the series counts many renowned academics among its volume editors and paper authors, and collaborates with prestigious societies. Its mission is to serve this international community by providing an invaluable service, mainly focused on the publication of conference and workshop proceedings and postproceedings. LNCS commenced publication in 1973.

Giuseppe Nicosia · Varun Ojha ·
Emanuele La Malfa · Gabriele La Malfa ·
Panos M. Pardalos · Renato Umeton
Editors

Machine Learning, Optimization, and Data Science

9th International Conference, LOD 2023
Grasmere, UK, September 22–26, 2023
Revised Selected Papers, Part I

 Springer

Editors
Giuseppe Nicosia (iD)
University of Catania
Catania, Catania, Italy

Varun Ojha (iD)
Newcastle University
Newcastle upon Tyne, UK

Emanuele La Malfa (iD)
University of Oxford
Oxford, UK

Gabriele La Malfa (iD)
University of Cambridge
Cambridge, UK

Panos M. Pardalos (iD)
University of Florida
Gainesville, FL, USA

Renato Umeton (iD)
Dana-Farber Cancer Institute
Boston, MA, USA

ISSN 0302-9743   ISSN 1611-3349 (electronic)
Lecture Notes in Computer Science
ISBN 978-3-031-53968-8   ISBN 978-3-031-53969-5 (eBook)
https://doi.org/10.1007/978-3-031-53969-5

This Springer imprint is published by the registered company Springer Nature Switzerland AG
The registered company address is: Gewerbestrasse 11, 6330 Cham, Switzerland

Paper in this product is recyclable.

Preface

LOD is the international conference embracing the fields of machine learning, deep learning, optimization, and data science. The ninth edition, LOD 2023, was organized on September 22–26, 2023, in Grasmere, Lake District, England Like the previous edition, LOD 2023 hosted the Advanced Course and Symposium on Artificial Intelligence & Neuroscience – ACAIN 2023. In fact, this year, in the LOD Proceedings we decided to also include the papers of the third edition of the Symposium on Artificial Intelligence and Neuroscience (ACAIN 2023). The ACAIN 2023 chairs were:

Giuseppe Nicosia, University of Catania, Italy
Panos Pardalos, University of Florida, USA

The review process of 18 submissions to ACAIN 2023 was double-blind, performed rigorously by an international program committee consisting of leading experts in the field. Therefore, the last nine articles in the LOD Table of Contents (volume 14506) are the articles accepted to ACAIN 2023.

Since 2015, the LOD conference has brought together academics, researchers and industrial researchers in a unique *pandisciplinary community* to discuss the state of the art and the latest advances in the integration of machine learning, deep learning, nonlinear optimization and data science to provide and support the scientific and technological foundations for interpretable, explainable, and trustworthy AI. Since 2017, LOD adopted the *Asilomar AI Principles.*

The annual conference on machine Learning, Optimization and Data science (LOD) is an international conference on machine learning, deep learning, AI, computational optimization and big data that includes invited talks, tutorial talks, special sessions, industrial tracks, demonstrations, and oral and poster presentations of refereed papers.

LOD has established itself as a premier interdisciplinary conference in machine learning, computational optimization, and data science. It provides an international forum for presentation of original multidisciplinary research results, as well as exchange and dissemination of innovative and practical development experiences.

LOD 2023 attracted leading experts from industry and the academic world with the aim of strengthening the connection between these institutions. The 2023 edition of LOD represented a great opportunity for professors, scientists, industry experts, and research students to learn about recent developments in their own research areas and to learn about research in contiguous research areas, with the aim of creating an environment to share ideas and trigger new collaborations.

As chairs, it was an honour to organize a premier conference in these areas and to have received a large variety of innovative and original scientific contributions.

During LOD 2023, 3 plenary talks were presented by leading experts:

LOD 2023 Keynote Speakers:

Gabriel Barth-Maron DeepMind, London, UK
Anthony G. Cohn University of Leeds, UK; The Alan Turing Institute, UK
Sven Giesselbach Fraunhofer Institute - IAIS, Germany

ACAIN 2023 Keynote Lecturers:

Karl Friston, University College London, UK
Kenneth Harris University College London, UK
Rosalyn Moran King's College London, UK
Panos Pardalos, University of Florida, USA
Edmund T. Rolls University of Warwick, UK

LOD 2022 received 119 submissions from 64 countries in five continents, and each manuscript was independently double-blind reviewed by a committee formed by at least five members. These proceedings contain 71 research articles written by leading scientists in the fields of machine learning, artificial intelligence, reinforcement learning, computational optimization, neuroscience, and data science presenting a substantial array of ideas, technologies, algorithms, methods, and applications.

At LOD 2023, Springer LNCS generously sponsored the LOD Best Paper Award. This year, the paper by *Moritz Lange, Noah Krystiniak, Raphael C. Engelhardt, Wolfgang Konen, and Laurenz Wiskott* titled *"Improving Reinforcement Learning Efficiency with Auxiliary Tasks in Non-Visual Environments: A Comparison"*, received the LOD 2023 Best Paper Award.

This conference could not have been organized without the contributions of exceptional researchers and visionary industry experts, so we thank them all for participating. A sincere thank you goes also to the 20 sub-reviewers and to the Program Committee of more than 210 scientists from academia and industry, for their valuable and essential work of selecting the scientific contributions.

Finally, we would like to express our appreciation to the keynote speakers who accepted our invitation, and to all the authors who submitted their research papers to LOD 2023.

October 2023

Giuseppe Nicosia
Varun Ojha
Emanuele La Malfa
Gabriele La Malfa
Panos M. Pardalos
Renato Umeton

Organization

General Chairs

Renato Umeton   Dana-Farber Cancer Institute, MIT, Harvard T.H. Chan School of Public Health & Weill Cornell Medicine, USA

Conference and Technical Program Committee Co-chairs

Giuseppe Nicosia   University of Catania, Italy
Varun Ojha   Newcastle University, UK
Panos Pardalos   University of Florida, USA

Special Sessions Chairs

Gabriele La Malfa   University of Cambridge, UK
Emanuele La Malfa   University of Oxford, UK

Steering Committee

Giuseppe Nicosia   University of Catania, Italy
Panos Pardalos   University of Florida, USA

Program Committee Members

Jason Adair   University of Stirling, UK
Agostinho Agra   University of Aveiro, Portugal
Massimiliano Altieri   University of Bari, Italy
Vincenzo Arceri   University of Parma, Italy
Roberto Aringhieri   University of Turin, Italy
Akhila Atmakuru   University of Reading, UK
Roberto Bagnara   University of Parma, Italy
Artem Baklanov   Int Inst for Applied Systems Analysis, Austria
Avner Bar-Hen   CNAM, France

Bernhard Bauer	University of Augsburg, Germany
Peter Baumann	Constructor University, Germany
Sven Beckmann	University of Augsburg, Germany
Roman Belavkin	Middlesex University London, UK
Nicolo Bellarmino	Politecnico di Torino, Italy
Heder Bernardino	Universidade Federal de Juiz de Fora, Brazil
Daniel Berrar	Tokyo Institute of Technology, Japan
Martin Berzins	University of Utah, USA
Hans-Georg Beyer	Vorarlberg University of Applied Sciences, Austria
Sandjai Bhulai	Vrije Universiteit Amsterdam, The Netherlands
Francesco Biancalani	IMT Lucca, Italy
Martin Boldt	Blekinge Institute of Technology, Sweden
Vincenzo Bonnici	University of Parma, Italy
Anton Borg	Blekinge Institute of Technology, Sweden
Jose Borges	University of Porto, Portugal
Matteo Borrotti	University of Milano-Bicocca, Italy
Alfio Borzi	Universität Würzburg, Germany
Goetz Botterweck	Trinity College Dublin, Ireland
Will Browne	Queensland University of Technology, Australia
Kevin Bui	University of California, Irvine, USA
Luca Cagliero	Politecnico di Torino, Italy
Antonio Candelieri	University of Milano-Bicocca, Italy
Paolo Cazzaniga	University of Bergamo, Italy
Adelaide Cerveira	UTAD and INESC-TEC, Portugal
Sara Ceschia	University of Udine, Italy
Keke Chen	Marquette University, USA
Ying-Ping Chen	National Yang Ming Chiao Tung University, Taiwan
Kiran Chhatre	KTH Royal Institute of Technolgy, Sweden
John Chinneck	Carleton University, Canada
Miroslav Chlebik	University of Sussex, UK
Eva Chondrodima	University of Piraeus, Greece
Pedro H. Costa Avelar	King's College London, UK
Chiara Damiani	University of Milano-Bicocca, Italy
Thomas Dandekar	University of Würzburg, Germany
Renato De Leone	University of Camerino, Italy
Roy de Winter	Leiden Institue of Advanced Computer Science, The Netherlands
Nicoletta Del Buono	University of Bari, Italy
Mauro Dell'Amico	University of Modena and Reggio Emilia, Italy
Clarisse Dhaenens	University of Lille, France

Michael Hellwig	Vorarlberg University of Applied Sciences, Austria
Carlos Henggeler Antunes	University of Coimbra, Portugal
Alfredo G. Hernndez-Daz	Pablo de Olavide University, Spain
J. Michael Herrmann	University of Edinburgh, UK
Vinh Thanh Ho	University of Limoges, France
Colin Johnson	University of Nottingham, UK
Sahib Julka	University of Passau, Germany
Robert Jungnickel	RWTH Aachen University, Germany
Vera Kalinichenko	UCLA, USA
George Karakostas	McMaster University, Canada
Emil Karlsson	Saab AB, Sweden
Branko Kavek	University of Primorska, Slovenia
Marco Kemmerling	RWTH Aachen, Germany
Aditya Khant	Harvey Mudd College, USA
Zeynep Kiziltan	University of Bologna, Italy
Wolfgang Konen	Cologne University of Applied Sciences, Germany
Jan Kronqvist	KTH Royal Institute of Technology, Sweden
T. K. Satish Kumar	University of Southern California, USA
Nikolaos A. Kyriakakis	Technical University of Crete, Greece
Alessio La Bella	Politecnico di Milano, Italy
Gabriele Lagani	University of Pisa, Italy
Dario Landa-Silva	University of Nottingham, UK
Cecilia Latotzke	RWTH Aachen University, Germany
Ang Li	University of Southern California, USA
Zhijian Li	University of California, Irvine, USA
Johnson Loh	RWTH Aachen University, Germany
Gianfranco Lombardo	University of Parma, Italy
Enrico Longato	University of Padua, Italy
Angelo Lucia	University of Rhode Island, USA
Hoang Phuc Hau Luu	University of Lorraine, France
Eliane Maalouf	University of Neuchatel, Switzerland
Hichem Maaref	Université d'Evry Val d'Essonne, France
Antonio Macaluso	German Research Center for Artificial Intelligence, Germany
Francesca Maggioni	University of Bergamo, Italy
Silviu Marc	Middlesex University London, UK
Nuno Marques	FEUP, Portugal
Rafael Martins de Moraes	New York University, USA
Moreno Marzolla	University of Bologna, Italy
Shun Matsuura	Keio University, Japan

Jose Gilvan Rodrigues Maia	Universidade Federal do Ceará, Brazil
Jan Rolfes	KTH Royal Institute of Technology, Sweden
Roberto Maria Rosati	University of Udine, Italy
Hang Ruan	University of Edinburgh, UK
Chafik Samir	University of Clermont, France
Marcello Sanguineti	Universita di Genova, Italy
Giorgio Sartor	SINTEF, Norway
Claudio Sartori	University of Bologna, Italy
Frederic Saubion	University of Angers, France
Robert Schaefer	AGH University of Science and Technology, Poland
Robin Schiewer	Ruhr-University Bochum, Germany
Joerg Schlatterer	University of Mannheim, Germany
Bryan Scotney	University of Ulster, UK
Natalya Selitskaya	IEEE, USA
Stanislav Selitskiy	University of Bedfordshire, UK
Marc Sevaux	Université Bretagne-Sud, France
Hyunjung Shin	Ajou University, South Korea
Zeren Shui	University of Minnesota, USA
Surabhi Sinha	University of Southern California, USA
Cole Smith	New York University, USA
Andrew Starkey	University of Aberdeen, UK
Julian Stier	University of Passau, Germany
Chi Wan Sung	City University of Hong Kong, China
Johan Suykens	Katholieke Universiteit Leuven, Belgium
Sotiris Tasoulis	University of Thessaly, Greece
Tatiana Tchemisova	University of Aveiro, Portugal
Michele Tomaiuolo	University of Parma, Italy
Bruce Kwong-Bun Tong	Hong Kong Metropolitan University, China
Elisa Tosetti	Ca Foscari University of Venice, Italy
Sophia Tsoka	King's College London, UK
Gabriel Turinici	Université Paris Dauphine - PSL, France
Gregor Ulm	Fraunhofer-Chalmers Research Centre for Industrial Mathematics, Sweden
Joy Upton-Azzam	Susquehanna University, USA
Werner Van Geit	EPFL, Switzerland
Johannes Varga	TU Wien, Austria
Filippo Velardocchia	Politecnico di Torino, Italy
Vincent Vigneron	Université d'Evry, France
Herna Viktor	University of Ottawa, Canada
Marco Villani	University of Modena and Reggio Emilia, Italy
Dean Vucinic	Vrije Universiteit Brussel, Belgium

Yasmen Wahba	Western University, Canada
Ralf Werner	Augsburg University, Germany
Dachuan Xu	Beijing University of Technology, China
Shiu Yin Yuen	City University of Hong Kong, China
Riccardo Zese	University of Ferrara, Italy
Takfarinas Saber	University of Galway, Ireland

Contents – Part I

Contents – Part II

Artificial Intelligence and Neuroscience (ACAIN 2023)

Consensus-Based Participatory Budgeting for Legitimacy: Decision Support via Multi-agent Reinforcement Learning

Srijoni Majumdar$^{(\boxtimes)}$ and Evangelos Pournaras

School of Computing, University of Leeds, Leeds, UK
s.majumdar@leeds.ac.uk

Abstract. The legitimacy of bottom-up democratic processes for the distribution of public funds by policy-makers is challenging and complex. Participatory budgeting is such a process, where voting outcomes may not always be fair or inclusive. Deliberation for which project ideas to put for voting and choose for implementation lack systematization and do not scale. This paper addresses these grand challenges by introducing a novel and legitimate iterative consensus-based participatory budgeting process. Consensus is designed to be a result of decision support via an innovative multi-agent reinforcement learning approach. Voters are assisted to interact with each other to make viable compromises. Extensive experimental evaluation with real-world participatory budgeting data from Poland reveal striking findings: Consensus is reachable, efficient and robust. Compromise is required, which is though comparable to the one of existing voting aggregation methods that promote fairness and inclusion without though attaining consensus.

Keywords: participatory budgeting · reinforcement learning · consensus · legitimacy · social choice · decision support · collective decision making · digital democracy

1 Introduction

Participatory budgeting (PB) is a bottom-up collective decision-making process with which citizens decide how to spend a budget of the local municipality [4, 16]. Citizens initially submit proposals for implementation of various project ideas, i.e. public welfare amenities. These are evaluated by the city officials and finally, a subset is put for voting. Citizens then express their preferences using different input voting methods such as approval or score voting [9]. Finally, voting aggregation methods are applied to select the winner projects [4].

The selection of the winner projects depends on both input and aggregation methods [2]. As preferences via approvals or scores are based on self-interest, voting outcomes may yield different satisfaction levels, under-representation, and poor legitimacy. For a more stable, conclusive, shared and legitimate voting outcome, a form of systematic and scalable deliberation is missing among citizens

© The Author(s), under exclusive license to Springer Nature Switzerland AG 2024
G. Nicosia et al. (Eds.): LOD 2023, LNCS 14505, pp. 1–14, 2024.
https://doi.org/10.1007/978-3-031-53969-5_1

so that individual preferences are exchanged, debated and compromised in a viable way to reach consensus [6]. This challenge is addressed in this paper.

A new multi-agent reinforcement learning approach (MARL-PB) is introduced to model a novel iterative consensus-based PB process. In the proposed approach, consensus emerges as a result of (i) reward-based learning based on project ideas proposed and selected in the past and (ii) decentralized voter communication that supports information exchange and deliberation.

MARL-PB is implemented as a decision-support system that finds applicability in three use cases by three beneficiaries as shown in Fig. 1: (i) *Citizens*: digital assistance to communicate, deliberate and reach a common ground for which projects to implement. This is expected to increase the participation, satisfaction and legitimacy in participatory budgeting. (ii) *Policy-makers*: digital assistance to filter out projects during the project ideation phase with the aim to put for voting a reasonable and legitimate number of projects that results in informed and expressive choices during voting without informational overload. (iii) *Researcher*: digital assistance for the assessment of fair and inclusive voting aggregation methods (e.g. equal shares, Phragmen) via comparisons with a fine-grained consensus-based model such as the one of MARL-PB.

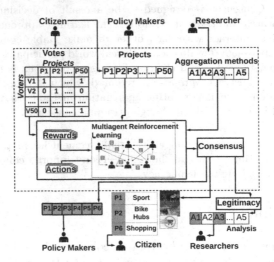

Fig. 1. Consensus-based participatory budgeting using a multi-agent reinforcement learning (MARL-PB). A decision-support framework is designed for three different use cases and beneficiaries: citizens, policy-makers and researchers.

MARL-PB is extensively assessed using state-of-the-art real-world participatory budgeting datasets [15] from Poland. The following three research questions are addressed: (i) *How effective multi-agent reinforcement learning is to assist voters reach consensus in participatory budgeting?* (ii) *What level of flexibility is required by voters to compromise and reach consensus in participatory budgeting?* (iii) *How efficient and robust a consensus-based participatory budgeting is by using multi-agent reinforcement learning?.* The quality of consensus, its efficiency and robustness are studied, along with how they are influenced by factors such as the following: (i) number of possible consensus bundles, (ii) in-degree of the communication network, (iii) number of voters, (iv) districts, (v) voting aggregation methods and (vi) project attributes.

The contributions of this paper are summarized as follows: (i) The multi-agent reinforcement learning approach of MARL-PB to model and implement an iterative consensus-based PB process. (ii) A decision-support framework based on MARL-PB to digitally assist three use cases by three beneficiaries. (iii) An extension of the reward-based learning strategy with a gossip-based agents communication protocol for decentralized information exchange and consensus building. (iv) A compilation of metrics that characterize and assess the legitimacy of the consensus-based PB process. (v) Practical and revealing insights about the nature of the achieved consensus: requires compromises comparable to the ones of the voting aggregation methods that promote fairness and inclusion. (vi) An open-source software artifact of MARL-PB for reproducibility and encouraging further research in this niche research area[1].

This paper is outlined as follows: Sect. 2 reviews related work. Section 3 introduces the consensus-based approach. Section 4 illustrates the empirical results and findings. Section 5 concludes this paper and outlines future work.

2   Related Literature Review

This section provides an overview of related literature, with a focus on iterative reward-based learning for collective decision-making processes.

Social dilemma games such as Prisoners Dilemma has been studied in the context of reward-based learning agents [12] with two agents and discrete rewards, i.e. punishment (0) or no punishment (1), to explore the compromises that two agents make to reach consensus. Using multiple agents, this experiment provides insights on how learning can stabilize using limited voters and deterministic rewards. This provides a relevant direction for dealing with voting for social choice and collective preferences.

Airiau et al. [1] model an iterative single-winner voting process in a reinforcement learning setup to analyze the learning capabilities of voting agents to obtain more legitimate collective decisions. The proposed framework provides a new variant of iterative voting that allows agents to change their choices at the same time if they wish. The rank of the winner at every stage in the preference order of voters is used as a reward for the agents to learn and re-select. The proposed work by Liekah et al. [11] additionally calculates the average satisfaction among voters in every iteration based on the winner and individual preferences.

Prediction of complete PB ballots using machine learning classification is recently studied as a way to decrease information overload of voters using partial ballots [10]. This approach could complement MARL-PB to speed up the consensus process.

Existing approaches (see Table 1) do not incorporate inter-agent communication for large-scale information exchange in multi-winner voting systems. The

---

[1] https://github.com/DISC-Systems-Lab/MARL-PB (last accessed: July 2023).

Table 1. Comparison of this work with earlier multi-agent reinforcement learning approaches for collective decision making.

Aspect	Macy et al. [12]	Airiau et al. [1]	Liekah et al. [11]	Proposed Approach (MARL-PB)
Outcome	Single Winner	Single Winner	Single Winner	Multiple Winners
Rewards	Deterministic	Stochastic*	Stochastic*	Deterministic (project attributes)
	4 discrete values			Stochastic (from communication)
Execution	Centralized	Centralized	Centralized	Shared aggregate rewards, decentralized
Action Space	4	5	5	till 100

*Rank of winner in the preference order of the voter (within an iteration).

rewards are fixed in centralized settings and do not model the preferences of the voters. Moreover, the feasibility of reaching a consensus via communication with other voters has not been studied. This is relevant to the problem of scaling up and automating deliberation in collective decision-making to reach more legitimate voting outcomes. These are some of the gaps addressed in this paper.

3   Consensus-Based Iterative Participatory Budgeting

In this section, an iterative participatory budgeting process is introduced modeled by a multi-agent reinforcement learning approach. The voters (agents) maximize their self-interest but also compromise to reach a consensus in a multi-agent system, where the choices of others are initially only partially known.

3.1   Multi-armed Bandit Formulation

In a participatory budgeting process, voters collectively choose multiple projects subject to a constraint that the total cost of the projects is within the total budget. To incorporate this knapsack constraint, a combinatorial model [2] is designed to formulate bundles from the available list of projects. So for three projects, there are seven possible bundles, out of which a subset fulfills the budget constraints. These are referred to as *valid knapsack bundles* and they constitute the possible actions in a multi-arm bandit formulation (see Fig. 2).

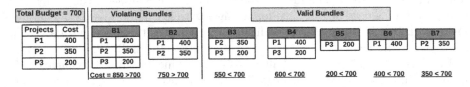

Fig. 2. Calculation of bundles: They represent all possible combinations from the listed projects where the total cost of all the projects in the bundle is within the budget. In the example, three projects with their corresponding costs are listed for voting in a participatory budgeting process. The total budget is 700. Five out of the seven possible bundles are valid and satisfy the budget constraint.

The bundles encode all possible multi-winner preferences the voters can collectively have. Learning valid knapsack bundles instead of individual project selections prevents early terminations by budget violations [3].

The iterative version of the PB process introduces a partial voters' communications at every iteration to exchange preferences with the aim they converge (compromise) to the same bundle (same preferences). This process models a large-scale automated deliberation process. The selection of an action by a voter depends only on the rewards associated with the bundles, hence, the problem is modeled as a multi-armed bandit reinforcement learning approach [14].

The multi-arm bandits are defined in the form of a tuple $<A; R>$, where A represents actions that are the possible bundles and $R_A = P[r|A]$ is the probability distribution of rewards over the bundles (actions).

Actions: For a participatory budgeting process with a set of projects and a total budget, the actions comprise of the *valid knapsack bundles* formed from the projects and their associated costs.

Rewards: The preference modelling in the form of rewards plays an important role in reaching consensus in a large action space with multiple agents. The aggregate preferences of the voters over past and the current year are encoded for a region in the form of rewards that signify how the needs for public amenities evolve over the years and thus can help to predict the collective preference for the current participatory budgeting process. These are modeled and calculated as deterministic rewards for each bundle.

To reach a consensus, voters explore the action space of each other via information exchange. This exchange models a large-scale and automated deliberation process, which voters use to learn, compromise and adjust their choices.

Deterministic Rewards: A project is related to a type of public welfare amenities[2] such as urban greenery, sports, culture, education, environmental protection etc., or a population group that benefits such as elderly, families with children, etc. The preference of citizens are mostly associated with these attributes and can be used to estimate collective preferences for the population of a region.

The number of occurrence of such project attributes, which are put for voting and selected in the past years of a region is used as reward utilities:

$$R^a = \Sigma_{y \in Y}(\mathbb{C}(a) + \mathcal{C}(a)),$$

where a is a specific project attribute, $\mathbb{C}$ and $\mathcal{C}$ signify the normalized total count of occurrence of the project attribute across listed and selected projects respectively over Y years of participatory budgeting processes in a region.

The reward for a project is determined as follows:

$$R_p = \sigma(\Sigma_{i=1}^{\mathcal{A}}(R_i^a)) + \tanh(\frac{c_p}{\mathcal{B}}),$$

---

[2] It is assumed that preferences for such projects persist over the passage of time, in contrast to infrastructure projects that once they are implemented, they may not be preferred anymore.

where $\mathcal{A}$ is the total number of attributes associated with a project, c_p is the individual project cost and $\mathcal{B}$ is the total budget of the PB process. The rewards for a bundle is the sum of the rewards of each of its projects.

Rewards from Inter-agent Communication: At every iteration, we update a dynamic random bidirectional graph using a decentralized process such as the gossip-based peer sampling [8] for peer-to-peer communication. At each iteration, the connected agents send the bundle that has received the highest rewards (other randomized schemes are supported by the code), together with the reward itself. As the neighbors are randomly decided, the accumulated rewards from information exchange are stochastic. The stateless variant of the Q learning approach is augmented to incorporate rewards obtained from information exchange:

$$Q(b_t) \leftarrow Q(b_t) + \alpha(r + \delta(\max_{b_t^c}(Q(b_t^c) - Q(b_t)))),$$

where $Q(b_t^c)$ is the rewards obtained via agent communication for a bundle b at time t. The introduced learning rate for rewards from information exchange -δ is set empirically to 0.1. The discount factor γ in the Q learning is set to zero as future rewards are not considered. Algorithm 1 outlines the learning process.

Algorithm 1. Augmented Q-Learning for consensus in participatory budgeting.

1: Populate project list and cost for the current participatory budgeting process
2: Initialize the fixed rewards for projects
3: Calculate rewards for all valid bundles
4: **for** each iteration $i \geq 1$ **do**
5:     **for** each voter $v \in$ V **do**
6:         **if** $i == 1$ **then**
7:             Assign the bundle with highest overlap to original individual preferences
8:         Update random graph via the peer sampling service
9:         Aggregate rewards of bundles from neighbors
10:        Update total rewards for a bundle in the Q table
11:        Select action (bundle) according to ϵ-greedy policy, $\epsilon \in [0,1]$

Initially, the agents select a bundle according to their preference (first iteration) and then they start communication with other agents during which they adjust their preferred bundle. The selection of the bundle at each iteration is based on the cumulative sum of both rewards, which the agents maximize using an ϵ greedy exploration strategy. For a low number of projects and voters size, the initial preferences from the multi-winner approvals of the voters may result in a reduced action space for exploration.

3.2   Assessment Model for Consensus

The following metrics are designed to assess the quality of the consensus (legitimacy) based on popularity, representation and budget utilization that can increase the satisfaction of the citizens, increase participation and improve the quality of the overall PB process [2,13]. The level of compromise made by voters is assessed. These metrics also characterize how difficult it is to reach a consensus in an iterative voting process for participatory budgeting. The metrics that model the legitimacy are outlined as follows:

- *Compromise Cost*: The mean non-overlap (1 - mean overlap) of projects between the preferred bundle of the voters and the consensus bundle, calculated using the Jaccard Index [5].
- *Unfairness*: The coefficient of variation of the *compromise cost* over all agents.
- *Popularity* (fitness of consensus): The normalized ranking score of the projects in the consensus bundle, calculated using the number of votes of each project from the original voters' preferences.
- *Budget Utilization*: The cost of the projects in the consensus bundle divided by the total available budget in the participatory budgeting process.

4   Experimental Evaluation

This section illustrates the results obtained from the evaluation of the consensus-based participatory budgeting process, using real-world data. These results shed light on the efficiency of reward models, the communication protocol and the exploration strategy to reach consensus.

Dataset: The *pabulib* PB dataset (http://pabulib.org/) is used for the evaluation. It contains the metadata related to projects and voters along with the voting records for multiple participatory budgeting instances, for various districts and cities of Poland. Each project is associated with multiple attributes such as urban greenery, education, relavance to children etc., along with information about the project costs. There are multiple participatory budgeting instances for every district or city for multiple years and different ballot designs such as k-approval, cumulative and score voting. Furthermore, the winners are calculated using various aggregation methods such as the method equal shares, phragmen, and utilitarian greedy [4] to assess the quality of the consensus bundles compared to the ones calculated by methods that promote fairness and inclusion. Three districts are selected - Ruda, Ursynow and Rembertow, whose valid bundles vary from a smaller set (12 for Ruda) to a larger one (90 for Rembertow).

Table 2. Parameters for experimentation with each dataset. The ranges signify that experiments are performed incrementally, for instance, 5, 6, 7,....90 for the # of bundles selected randomly in *Rembertow*. The maximum number of valid bundles extracted from 20 projects is 90. The available data is for 4 years for each district. The experiments are performed for the latest year and the aggregate preference (rewards) are calculated using all years. The decay rate and learning rates are set to 0.1 after empirical investigation.

Dataset	# of Projects	# of Bundles	In-degree	# of Agents
Rembertow	20	5 to 90	2 to 26	50 to 100
Ursynow	18	5 to 75	2 to 26	50 to 100
Ruda	10	3 to 12	2 to 10	50 to 100

Design: The framework is tested using various settings such as the numbers of combinations (bundles), number of agents, learning rate, decay rate, and the in-degree of the random graph updated at every iteration (see Table 2). For each of these settings, the projects selected in the consensus are analyzed and compared with winners selected using other aggregation methods such as the method of equal shares and greedy [4].

4.1   RQ1: Effectiveness to Reach Consensus

Quality of Consensus: The convergence to the consensus bundle depends on the in-degree of the random graph and the number of bundles (action space). Figure 3 shows the budget utilization as a function of the number of bundles and in-degree. Budget utilization of the consensus bundles (Fig. 3) for Rembertow is low (0.40 to 0.45) in most cases, which could be attributed to considerable high costs for popular projects and fewer projects in the consensus bundle.

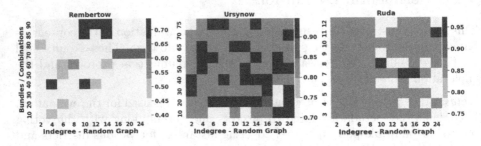

Fig. 3. Total budget utilization of the consensus bundles for different number of valid bundle combinations and in-degree. The budget utilization for the consensus bundles for Ursynow is the highest.

Figure 4 shows the popularity index (fitness of the consensus) for different number of bundles and in-degrees. In case of Ursynow, the consensus bundles have more projects and a higher percentage of popular projects (0.65 to 0.70), that also have medium costs, which results in higher overall budget utilization. The percentage of popular projects is low for Rembertow (0.40 to 0.45) and selected popular projects have considerably high costs too, as the overall budget utilization is also low (see Fig. 3).

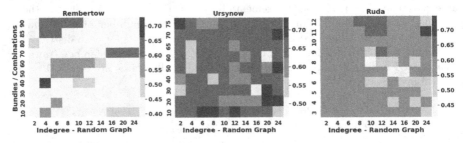

Fig. 4. Popularity index (fitness of consensus) for different number of valid bundle combinations and in-degrees.

Comparison with Other Aggregation Methods: The overlap of projects between the consensus bundle (using the maximum number of bundles for action space) and the winners from the aggregation methods are calculated (see Table 3). When a larger number of projects are listed, e.g. Rembertow, the overlap with consensus bundle is higher with equal shares (0.62) and Phragmen (0.61). Hence, these methods maximize fairness and representation and also produce more legitimate winners. For a lower number of projects, e.g. n Ruda, greedy has higher overlap (0.72) with the consensus bundle. The reward based iterative learning with communication can reach a consensus, which has a higher overlap with aggregation methods that promote fairness for a higher number of listed projects.

Table 3. The overlap between the projects in the consensus bundle and three aggregation methods: greedy, equal shares and phragmen. The highest possible number of valid knapsack bundles are used. *G: Greedy, PG: Phragmen, ES*: Equal Shares*, MARL-PB: Proposed approach*

Dataset	Size of Consensus Bundle				MARL-PB Overlap		
	MARL-PB	ES	PG	G	ES	PG	G
Rembertow	8	9	9	6	0.62	0.61	0.49
Ursynow	8	8	8	6	0.72	0.75	0.73
Ruda	7	8	8	5	0.62	0.66	0.72

* phragmen completion method used for the method of equal shares.

Analysis of the Reward Modelling: The top-3 amenities associated with all projects (listed and selected) over all years in Ursynow are public space (22%), education(17%), environmental protection (12%), impact on children (22%) adults (21.1%) and seniors (19%). Similarly, for Rembertow and Ruda the most popular ones are public space (24%) and education (22.7%). These project attributes affect a large proportion of the population. Figure 5 shows the amenities selected via MARL-PB and the aggregation methods, as well as how they compare with the original aggregate preferences based on the past data. The projects selected using equal shares and MARL-PB correspond to similar public amenities (e.g. for Rembertow, projects related to public space, education and culture are selected in higher proportion). This also signifies that consensus projects in MARL-PB prioritize fairness and better representation. The public amenities selected in the greedy method do not correlate with the ones selected in the consensus for Rembertow and Ruda. It can be observed for Ursynow that the selected projects for any aggregation method and with consensus mostly conform. Collective preferences for these projects remain stable over time in this region.

4.2   RQ2: Level of Flexibility to Compromise and Reach Consensus

Figure 6 compares the mean voters' compromise of MARL-PB with the one of different aggregation methods. The mean cost of compromise among the three districts for greedy is 0.56, while for MARL-PB is 0.68, which is close to the one of equal shares and Phragmen with 0.66 and 0.62 respectively. These results show the following: Consensus requires compromise that is not observed in the standard greedy voting aggregation method, however, this compromise is attainable and comparable to the one observed with consensus-oriented voting aggregation methods.

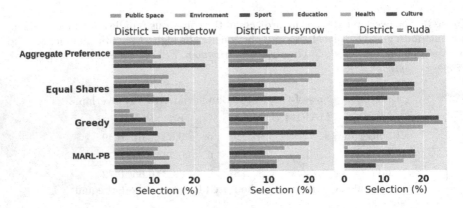

Fig. 5. The attributes of the project selections with the different methods.

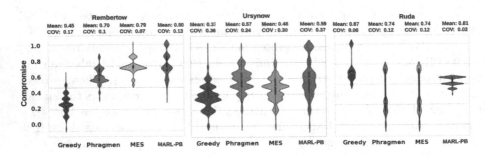

Fig. 6. Mean voters' compromise cost of MARL-PB and the different voting aggregation methods. The coefficient of variation (COV) measures unfairness, which is how compromises spread within the voters' population.

The mean unfairness of MARL-PB is 0.17, while greedy is 0.19. The equal shares and Phragmen have an unfairness of 0.16 and 0.15 respectively. The case of Ruda does not align with the results of the other districts and this is likely an artifact of the lower number of projects.

4.3   RQ3: Efficiency and Robustness

Convergence: An increase in the number of agents increase the converge time for any combination of in-degree or action (bundle) space (see Fig. 7). The iterations increase by 5% on average for an increase of 50 voters. Although this increase is only based on the data from the three districts, the system demonstrates to be scalable by converging within finite time as the number of voters increases.

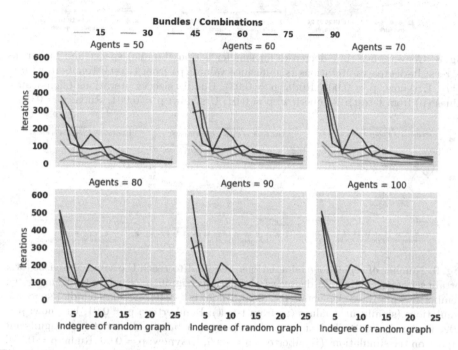

Fig. 7. Convergence time increases as the number of agents increases. Significance values (p) from t-test (iterations, agents): Rembertow: $p = 0.04$, Ursynow: $p = 0.04$, Ruda: $p = 0.04$. The values here are averaged over the three districts.

Figure 8 shows for each district the convergence time as function of in-degree and bundles size. Smaller bundles with higher in-degrees improve the speed.

Robustness: The influence of the randomness in the dynamic communication network on the stability of convergence is assessed by repeating the learning process multiple times, with different size of bundles. Figure 9 shows the required number of repetitions for a stable convergence speed. More repetitions are required for lower in-degrees due to limited information exchange for deliberation. A higher action space (number of bundles) results in a larger number of alternatives to explore and thus stability requires a higher number of repetitions.

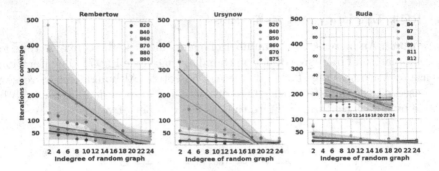

Fig. 8. Convergence time decreases for smaller action (bundle) spaces and higher in-degrees: In-degree vs. iterations (significance values (p) from t-test): Rembertow: p = 0.001, Ursynow: p = 0.002, Ruda: p = 0.03). Bundle size vs. iterations (significance values (p) from t-test): Rembertow: p = 0.01, Ursynow: p = 0.001, Ruda: p = 0.04.

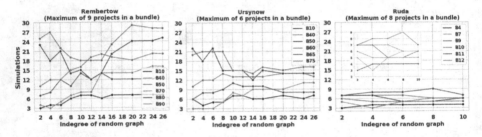

Fig. 9. Number of repetitions (simulations) required to reach the same consensus for a certain set of parameters. In-degree vs. simulations (significance values (p) from t-test): Rembertow: p = 0.03, Ursynow: p = 0.03, Ruda: p = 0.04. Action (bundle) space vs. simulations (significance values (p) from t-test): Rembertow: p = 0.01, Ursynow: p = 0.02, Ruda: p = 0.04. The number of projects in a bundle does not have significant impact on the simulations (Rembertow: p = 0.09, Ursynow: p = 0.06, Ruda: p = 0.13).

5   Conclusion and Future Work

This paper concludes that a consensus-based participatory budgeting process, with three use cases introduced, is feasible via a novel multi-agent reinforcement learning approach. The consensus process actually models a more systematic, large-scale and automated deliberation process, which has so far remained decoupled from the collective choice of voting. The experimental evaluation with real-world data confirms that the studied consensus is reachable, efficient and robust. The results also demonstrate that the consensus in MARL-PB requires compromises from voters, which are though comparable to the ones of existing voting aggregation methods that promote fairness and inclusion.

This is a key result with impact and significant implications: voters may not need in the future to rely anymore on a top-down arbitrary selection of the aggregation method. Instead, communities will be empowered to institutionalize and directly apply independently their own consensus-based decision-making processes. Moreover, city authorities may use the proposed method to filter out projects during the project ideation phase, which usually relies on subjective criteria with risks on legitimacy.

As part of future work, the agent communication may expand to different dynamic topologies that represent more closely social networks and proximity. The expansion of the multi-agent reinforcement learning approach with other preferential elicitation methods [7], beyond approval voting, is expected to further strengthen the accuracy and legitimacy of consensus-based participatory budgeting. A more advanced design of the rewards scheme will further expand the applicability of this ambitious approach.

Acknowledgment. This work is supported by a UKRI Future Leaders Fellowship (MR/W009560/1): '*Digitally Assisted Collective Governance of Smart City Commons– ARTIO*', and the SNF NRP77 'Digital Transformation' project "Digital Democracy: Innovations in Decision-making Processes", #407740_187249.

References

1. Airiau, S., Grandi, U., Perotto, F.Ś.: Learning agents for iterative voting. In: Rothe, J. (ed.) ADT 2017, pp. 139–152. Springer, Cham (2017). https://doi.org/10.1007/978-3-319-67504-6_10
2. Aziz, H., Shah, N.: Participatory budgeting: models and approaches. In: Pathways Between Social Science and Computational Social Science: Theories, Methods, and Interpretations, pp. 215–236 (2021)
3. Badanidiyuru, A., Kleinberg, R., Slivkins, A.: Bandits with knapsacks. J. ACM (JACM) **65**(3), 1–55 (2018)
4. Faliszewski, P., et al.: Participatory budgeting: data, tools, and analysis. arXiv preprint arXiv:2305.11035 (2023)
5. Fletcher, S., Islam, M.Z., et al.: Comparing sets of patterns with the jaccard index. Australas. J. Inf. Syst. **22** (2018)
6. Ganuza, E., Francés, F.: The deliberative turn in participation: the problem of inclusion and deliberative opportunities in participatory budgeting. Eur. Polit. Sci. Rev. **4**(2), 283–302 (2012)
7. Hausladen, C.I., Hänggli, R., Helbing, D., Kunz, R., Wang, J., Pournaras, E.: On the legitimacy of voting methods. Available at SSRN 4372245 (2023)
8. Jelasity, M., Voulgaris, S., Guerraoui, R., Kermarrec, A., Steen, M.: Gossip-based peer sampling. ACM Trans. Comput. Syst. (TOCS) **25**(3) (2007)
9. Kilgour, D.M.: Approval balloting for multi-winner elections. In: Laslier, J.F., Sanver, M. (eds.) Handbook on Approval Voting, pp. 105–124. Springer, Heidelberg (2010). https://doi.org/10.1007/978-3-642-02839-7_6
10. Leibiker, G., Talmon, N.: A recommendation system for participatory budgeting. In: International Conference on Autonomous Agents and Multiagent Systems (AAMAS). Workshop on Optimization and Learning in Multiagent Systems (2023)
11. Liekah, L.: Multiagent reinforcement learning for iterative voting. Ph.D. thesis, Master's thesis, Toulouse Capitole University (2019)
12. Macy, M.W., Flache, A.: Learning dynamics in social dilemmas. Proc. Natl. Acad. Sci. **99**(suppl_3), 7229–7236 (2002)

13. Miller, S.A., Hildreth, R., Stewart, L.M.: The modes of participation: a revised frame for identifying and analyzing participatory budgeting practices. Adm. Soc. **51**(8), 1254–1281 (2019)
14. Slivkins, A., et al.: Introduction to multi-armed bandits. Found. Trends Mach. Learn. **12**(1–2), 1–286 (2019)
15. Stolicki, D., Szufa, S., Talmon, N.: Pabulib: a participatory budgeting library. arXiv preprint arXiv:2012.06539 (2020)
16. Wellings, T.S., Majumdar, S., Haenggli Fricker, R., Pournaras, E.: Improving city life via legitimate and participatory policy-making: A data-driven approach in Switzerland. In: Annual International Conference on Digital Government Research, pp. 23–35 (2023)

Leverage Mathematics' Capability to Compress and Generalize as Application of ML Embedding Extraction from LLMs and Its Adaptation in the Automotive Space

Vera Kalinichenko[✉]

UCLA, Los Angeles, USA
kalin.vera@gmail.com

1 Introduction

Lately, with the public release of Chat GPT-2,3,4 many people agree that not enough attention given to the power and importance of the machinery that Mathematics brings, especially Discrete and Pure Mathematics. From simple ability to capture the essence, capacity to compress the information to its expertise in grasping the representational structure of entities, objects to projection of text blobs into the unified vector space. Mathematics mechanism to generalize can cover as a foundation in making significant progress in the area of AGI (artificial general intelligence), applications of embedding can lead quickly to very short roll out to production environment the varies of recommendation systems. Here, we present a relatively new approach to ML models in inference space when we successfully make an analog translation of business problem such as generating an offer price on the used vehicle, answering a consumer enquiry to its equivalent problem in math. We explore and exploit what does it truly mean to adapt to unknowns, to extract the skeleton representation when pricing used retail vehicles in the real time given VIN (vehicle identification number), make, model, body style, mileage and enriched data set with n-dimensional capabilities setting when n is going from hundreds to thousands of attributes extracted from the feature embedding. We also have experimented with different measures such as standard Euclidean, Manhattan, Minkowski and Cosine Similarity. We have used the above to find the closest inventory item given the user query provided a competitive advantage to scale at Shift (http://shift. com) with the recommendation systems built on the large inventory. We have selected the most optimal measure for the automotive domain space. First, we have stored the pre-computed embedding on our inventory, the retrieval of the encoding becomes instance even in the large datasets. Secondly, the encoding of the user query can certain be optimized based on pre-seeding token encoding. Finally, we use distance measure to rank the user results and experiment with weights assignments to emphasize certain token (keywords, brand names important to our industry domain). 1 Introduction Major automotive 3P sites like cargurus.com, autotrader.com and cars.com have been working with dealerships to facilitate buying and selling vehicles in the United States. Majority of startups like Carvana, Shift and Vroom have started to invest resources in the advanced

G. Nicosia et al. (Eds.): LOD 2023, LNCS 14505, pp. 15–23, 2024.
https://doi.org/10.1007/978-3-031-53969-5_2

data analytics and machine learning to recommend products, price the used vehicle in near real time settings to pick up on market trends. Shift has invested in Data Science enabling buying vehicles from consumer based on the external market conditions. We have started with standard and quick to develop set of ML models for our use cases. We had to create a dollar offers for used cars the company wanted to buy from the consumer to refurbish and sell back digitally. We had linear regressor models that used the following car level features (number of previous owners, accidents, mileage, year of the car, make, model, body style, trim, etc.) and external market data partners to compute the offers. Quickly, we had move to umbrella of machine learning models (Fig. 1).

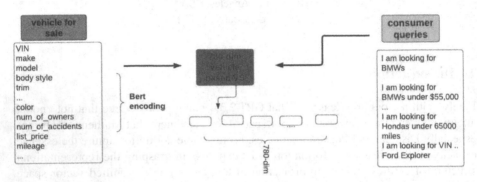

Fig. 1. Car Inventory and Consumer queries translations into Numerical VS.

2 Vector Space and Rich Data

There are many formulas and techniques available for ML use from simple algebra to advance calculus and probability theory. At Shift, we buy and sell used vehicles. Our inventory is not huge; however, we do contain thousands of cars for sale. Look at the sample summary of our inventory. It contains basic information about a vehicle such as VIN, make, model, body style, trim of vehicle, engine type, exterior and interior color, mileage, number of previous owners and accidents. We are not going to use all the available attributes we have in our data warehouse just to demonstrate how impressive the data enrichment coming from the high dimensional space works. Another very important side of this paper is the emphasis on the importance to truly go for high dimensional space and supply the inventory as a text blob for the encoding then without aligning the customer question to the similar semantic structure as the inventory simply translate/encode the user questions under the same vector space. The essence in this paper is that existence of the formula can be used to compress the information, creating a model that truly generalizes to the future settings. For instance, the simple arithmetic progression formula provides not only the way to compute fast any next number in the arithmetic sequence given its position but instantly generalize, extract the pattern at any given time. The analogous approach can be developed and leverage in the more complex scenarios. an = an − 1 + d where d is a common difference that is the same between any two consecutive terms. Of course, the simple toy example of arithmetic sequence seems very naive approach; however, we believe that tons can be adapted from mathematics. For instance, Category Theory [6] abstraction machinery can be used to pick the most suitable metric measure for automotive vector space, we have just started to apply and experiment with that answering consumer inquiries.

3 Measures

We have leveraged notion of the distance when constructed search results back to the user on her query. We best metrics for the automotive inventory retrieval was based on Euclidean distance measure. Please see examples in Table 2 in Embedding Examples out of the Box section below.

$$d(a, b) = \sqrt{\sum_{i=1}^{n} (a_i - b_i)^2} \tag{1}$$

$$d(a, b) = \left(\sum_{i=1}^{n} (a_i - b_i)^p \right)^{\frac{1}{p}} \tag{2}$$

$$d(a, b) = \sum_{i=1}^{n} |a_i - b_i| \tag{3}$$

```
def create_inv_emb_with_emp(df, m_name, token, freq):
    ### please pass in dataframe and model name
    ### m_name 'paraphrase-MiniLM-L6-v2'
    if (m_name != ''):
        model = SentenceTransformer(m_name)
    else:
        model = SentenceTransformer('paraphrase-MiniLM-L6-v2')
    # Sample inventory descriptions
    inventory_descriptions = []

    # Create an empty array to store the embeddings
    embeddings = []
    m = model.get_sentence_embedding_dimension()
    matrix = np.zeros((df.shape[0],m))

    # Loop through the data frame and generate embeddings for each row
    for i, row in df.iterrows():
        # Generate embedding for the car_index, make, model_c, year_c, list_price, mileage_c, exterior_color
        text = row['vin_c'] + ' ' + row['make_c'] + ' ' + row['model_c'] + ' ' + str(row['year_c']) + ' ' +
        # Repeat the token to emphasize its importance
        token = extract_vin(text)
        if (token != None):
            token = row['vin_c'] + ' '
            text_emphasized = text.replace(token, token * freq)
            embedding = model.encode([text_emphasized])[0]
        else:
            embedding = model.encode([text])[0]
        embeddings.append(embedding)
        matrix[i,:] = embedding

    # Create a new DataFrame from the embeddings
    embeddings_df = pd.DataFrame(embeddings)
    # Concatenate the original DataFrame with the embeddings DataFrame
    enriched_df = pd.concat([df, embeddings_df], axis=1)
    return enriched_df, matrix, model, m
```

Example of Encoding Inventory Data

We have adapted the following algorithm:

1. Translated our inventory into a 780-dimensional VS.
2. Translated user query as 780-dimensional VS.
3. Competed Euclidean distance from the query to every vehicle from the inventory.
4. Sorted the distances in the increasing order.
5. Displayed top k-results back to the user.

4   Embedding Examples Out of the Box

Please take a look at the Python code snippet below. The example demonstrates how easy is to start encoding the specific data leveraging open source libraries. We will now examine the examples provided, which were generated using an open-source transformer model, enhanced by the application of transfer learning within the automotive domain. The subsequent results showcase answers derived from various metrics. The capability to call APIs, load the model, encode data, and promptly respond to user queries is noteworthy. It is imperative to highlight the importance of adapting embeddings, not only as a measure to optimize engineering time and cost but also as an avenue to bolster business support [2] (Table 1).

Table 1. Embedding Results

*** Question –I am looking for BMW under $55,000

vin_c	make_c	model_c	list_price	mileage_c	body_style	year_c
WBA73AP0XMCF20695	BMW	4 Series	40100	18237.0	Midsize Car	2021.0
WBAPH57549NL79555	BMW	3 Series	9485	94845.0	Small Car	2009.0
WBAWB7C57AP048668	BMW	3 Series	16375	87418.0	Small Car	2010.0
WBXYJ3C30JEJ84113	BMW	X2	25675	24406.0	Small SUV	2018.0
WBA8E1C56JA167523	BMW	3 Series	25200	33606.0	Small Car	2018.0
5YMKT6C57G0R78216	BMW	X5 M	47600	34524.0	Midsize SUV	2016.0
WBAXH5C56CDW03119	BMW	5 Series	11950	79052.0	Midsize Car	2012.0
WBA8D9G55JNU70071	BMW	3 Series	23450	47263.0	Small Car	2018.0
WBAJA9C59KB392817	BMW	5 Series	31200	31282.0	Midsize Car	2019.0

*** Question –I am looking for BMWs

vin_c	make_c	model_c	list_price	mileage_c	body_style	year_c
5UXKR0C57F0K65284	BMW	X5	25200	62228.0	Small SUV	2015.0
WBAXA5C54ED690963	BMW	5 Series	17275	84843.0	Midsize Car	2014.0
WBXYJ3C30JEJ84113	BMW	X2	25675	24406.0	Small SUV	2018.0
5UXKS4C51F0J98015	BMW	X5	25700	78990.0	Small SUV	2015.0
5YMKT6C57G0R78216	BMW	X5 M	47600	34524.0	Midsize SUV	2016.0
5UXKR2C57E0C01906	BMW	X5	20615	83963.0	Small SUV	2014.0
WBA73AP0XMCF20695	BMW	4 Series	40100	18237.0	Midsize Car	2021.0
WBA3N3C59FK232310	BMW	4 Series	19025	64444.0	Midsize Car	2015.0
WBAWB7C57AP048668	BMW	3 Series	16375	87418.0	Small Car	2010.0

*** Question –Need this car VIN is JC3CFFBR5DT742742

vin_c	make_c	model_c	list_price	mileage_c	body_style	year_c
WDDHF5KB9CA526581	Mercedes-Benz	E 350	13725	97862.0	Midsize Car	2012.0
WDDHF5GB7BA340769	Mercedes-Benz	E 350	13725	94001.0	Midsize Car	2011.0
WDDKK5KF4DF204424	Mercedes-Benz	E 350	18325	76484.0	Midsize Car	2013.0
WDDHF2EB5DA722772	Mercedes-Benz	E 350	14255	93040.0	Midsize Car	2013.0
3C3CFFBR5DT742742	FIAT	500	500	61848.0	Subcompact Car	2013.0
WDD6J4GB2LN033662	Mercedes-Benz	CLA 250	32600	23890.0	Small Car	2020.0
WDDKK5KF1EF274707	Mercedes-Benz	E 350	20615	59548.0	Midsize Car	2014.0
WDDKK5KF7DF199798	Mercedes-Benz	E 350	22200	53613.0	Midsize Car	2013.0
3C3CFFHH0DT608624	FIAT	500	10135	61771.0	Subcompact Car	2013.0

** Question –I am looking for BMWs under $55,000 with low mileag

vin_c	make_c	model_c	list_price	mileage_c	body_style	year_c
WBA73AP0XMCF20695	BMW	4 Series	40100	18237.0	Midsize Car	2021.0
5YMKT6C57G0R78216	BMW	X5 M	47600	34524.0	Midsize SUV	2016.0
5UXKR0C57H0V66723	BMW	X5	25150	69830.0	Small SUV	2017.0
WBAPH57549NL79555	BMW	3 Series	9485	94845.0	Small Car	2009.0
WBXYJ3C30JEJ84113	BMW	X2	25675	24406.0	Small SUV	2018.0
WBAWB7C57AP048668	BMW	3 Series	16375	87418.0	Small Car	2010.0
WBA4A9C50FGL85609	BMW	4 Series	19025	70572.0	Midsize Car	2015.0
WBA8E1C56JA167523	BMW	3 Series	25200	33606.0	Small Car	2018.0
5UXKS4C51F0J98015	BMW	X5	25700	78990.0	Small SUV	2015.0

*** Question –Looking for 1FM5K7D84FGA08946   Ford Explorer

vin_c	make_c	model_c	list_price	mileage_c	body_style	year_c
1FM5K7D84FGA08946	Ford	Explorer	19555	86600.0	Midsize SUV	2015.0
1FM5K7D85EGB95306	Ford	Explorer	17800	88973.0	Midsize SUV	2014.0
1FM5K8B86JGC45035	Ford	Explorer	20950	77758.0	Midsize SUV	2018.0
1FM5K7D81GGC87709	Ford	Explorer	18850	101683.0	Midsize SUV	2016.0
1FM5K7DH7HGE00175	Ford	Explorer	26200	52896.0	Midsize SUV	2017.0
1FM5K7D8XJGC21862	Ford	Explorer	24200	89984.0	Midsize SUV	2018.0
1FM5K7FH0JGC94324	Ford	Explorer	25700	39344.0	Midsize SUV	2018.0
1FADP3J28DL324600	Ford	Focus	9170	95143.0	Compact Car	2013.0
1FTER4FH3MLE04964	Ford	Ranger	38200	7363.0	Small Truck	2021.0

*** Question –HONDA SUV 2017 year

vin_c	make_c	model_c	list_price	mileage_c	body_style	year_c
3CZRU5H79HM723176	Honda	HR-V	20950	67612.0	Subcompact SUV	2017.0
3CZRU6H17LM727305	Honda	HR-V	24650	58138.0	Subcompact SUV	2020.0
3CZRM3H39FG700108	Honda	CR-V	17995	73280.0	Compact SUV	2015.0
3CZRU5H7XGM728940	Honda	HR-V	20615	51215.0	Subcompact SUV	2016.0
2HKRW2H85JH652032	Honda	CR-V	27600	39024.0	Compact SUV	2018.0
1HGCP26408A121777	Honda	Accord	12135	86634.0	Midsize Car	2008.0
3CZRU5G52GM731668	Honda	HR-V	17995	77095.0	Subcompact SUV	2016.0
5J6RM4H55FL007594	Honda	CR-V	16450	96877.0	Compact SUV	2015.0
1HGCR2F79GA206068	Honda	Accord	19025	61826.0	Midsize Car	2016.0

** Question –I am looking for BMW under $25000

vin_c	make_c	model_c	list_price	mileage_c	body_style	year_c
WBADW7C52BE544213	BMW	3 Series	12665	84355.0	Small Car	2011.0
WBAXH5C56CDW03119	BMW	5 Series	11950	79052.0	Midsize Car	2012.0
WBA2J1C04L7E63570	BMW	2 Series	28700	21416.0	Compact Car	2020.0
WBA8G5C54GK752721	BMW	3 Series	25200	60183.0	Small Car	2016.0
WBAXG5C56DDY34199	BMW	5 Series	13725	98494.0	Midsize Car	2013.0
WBA8B9G57JNU96564	BMW	3 Series	27200	35783.0	Small Car	2018.0
WBAJA9C59KB392817	BMW	5 Series	31200	31282.0	Midsize Car	2019.0
WBA5A5C50ED505251	BMW	5 Series	14470	95840.0	Midsize Car	2014.0

We looked at Euclidean, Minkowski and Manhattan measures side by side. Please see below.

Table 2. Embedding Results

5　Quality of the Results

Let's discuss the results and list out some concerns we have seen below.

- When a user looks for BMW below $55,000 how good is the return results? Does it make sense to retrieve as a top response BMW 4 series with the price of $40,100 following by BMW 3 series with list $9485
- When a user looks for specific VIN number (a unique identifier of the car in the United States) the closest car is a match (VIN ending on 8946)
- When the customer looks for any BMWs the response is fairly nicely distributed in terms of mileage, list price and car size.

```
filtered_df = df[(df['make_c'] == 'BMW') & (df['list_price'] < 55000)]

print(filtered_df.shape)

(64, 17)
```

6　Applications in Pricing Models

We have historical data on our customers that sold and bought used vehicles from us. We used that data to create an inference pricing model that generated a price tag for a given used car. We have noticed improvements in our linear regression model when we have enriched the data with feature embedding coming from transformers. The enriched data has inherited some representational structure that surely helped in the price inference. Comparisons are made with baselines that use word level transfer learning via pretrained word embeddings as well as baselines do not use any transfer learning. We find that transfer learning using sentence embeddings tends to outperform word level transfer [4] or in the previous ML cases uses of one-hot encoders and custom mapper of categorical data such as vehicle model, make or trim (Table 3).

Table 3. ML models table

Model Name	Attribute Size	r2(on test)	MAE(on test)
Linear Regressor	about 15	0.9679	1362.9
xgboost, Gurobi optimization	about 15	0.9735	1380.7
xgboost	about 780-dim+	available upon request	available upon request

We have noticed that certain tokens can be emphasized simply by replacing it with itself multiple times. We have tried that with VIN token. Also, we we wanted to sort results in the most aligned to human common sense order we needed to adjust the results with cosine similarity measure plus creating a bound user enquiry such as "Mileage < = 50, 000" (Figs. 2 and 3).

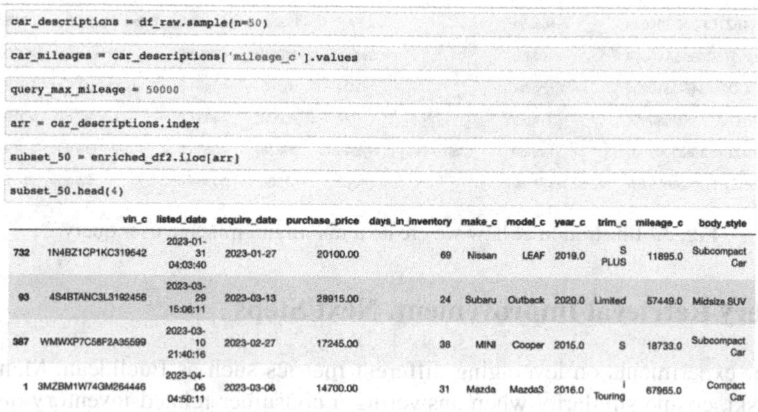

```
car_descriptions = df_raw.sample(n=50)
car_mileages = car_descriptions['mileage_c'].values
query_max_mileage = 50000
arr = car_descriptions.index
subset_50 = enriched_df2.iloc[arr]
subset_50.head(4)
```

	vin_c	listed_date	acquire_date	purchase_price	days_in_inventory	make_c	model_c	year_c	trim_c	mileage_c	body_style
732	1N4BZ1CP1KC319642	2023-01-31 04:03:40	2023-01-27	20100.00	69	Nissan	LEAF	2019.0	S PLUS	11895.0	Subcompact Car
93	4S4BTANC3L3192456	2023-03-29 15:06:11	2023-03-13	28915.00	24	Subaru	Outback	2020.0	Limited	57449.0	Midsize SUV
387	WMWXP7C58F2A35599	2023-03-10 21:40:16	2023-02-27	17245.00	38	MINI	Cooper	2015.0	S	18733.0	Subcompact Car
1	3MZBM1W74GM264446	2023-04-06 04:50:11	2023-03-06	14700.00	31	Mazda	Mazda3	2016.0	i Touring	67965.0	Compact Car

Fig. 2. Snippet on creating a sample from our inventory and encoding the data leveraging the BERT mini model.

vin_c	make_c	model_c	list_price	mileage_c	body_style	year_c
KM8R7DHE1MU277082	Hyundai	Palisade	48200	9191.0	Large SUV	2021.0
WA1ANAFY2L2067405	Audi	Q5	35200	10682.0	Midsize SUV	2020.0
1N4BZ1CP1KC319542	Nissan	LEAF	23700	11895.0	Subcompact Car	2019.0
WMWXP7C58F2A35599	MINI	Cooper	18495	18733.0	Subcompact Car	2015.0
SALCL2FX0MH885191	Land Rover	Discovery Sport	44200	19366.0	Small SUV	2021.0
WMWJZ9C0XM2P25672	MINI	Clubman	40200	21149.0	Subcompact Car	2021.0
WBA2J1C04L7E63570	BMW	2 Series	28700	21416.0	Compact Car	2020.0
1G1RC6S59JU152646	Chevrolet	Volt	22700	22510.0	Electric	2018.0
1C4HJXEG0KW514624	Jeep	Wrangler Unlimited	46200	23376.0	Small SUV	2019.0
WVWPP7AU0FW903016	Volkswagen	e-Golf	15845	26105.0	Electric	2015.0
JTDKARFP6L3134867	Toyota	Prius Prime	26700	28038.0	Subcompact Car	2020.0
1C3CCCAB4FN691894	Chrysler	200	15700	29951.0	Compact Car	2015.0
JTDKARFP7L3150625	Toyota	Prius Prime	27700	33421.0	Subcompact Car	2020.0
1N4AZ1CP2JC308639	Nissan	LEAF	19555	41747.0	Subcompact Car	2018.0
WBA3T3C59F5A40974	BMW	4 Series	28600	42962.0	Midsize Car	2015.0
1C3CCCCG4FN510205	Chrysler	200	16375	43054.0	Compact Car	2015.0
WA1LHAF71KD030601	Audi	Q7	35600	44997.0	Large SUV	2019.0
JHMZC5F36KC003501	Honda	Clarity Plug-In Hybrid	25675	46834.0	Electric	2019.0
WDDEJ7EB0CA029427	Mercedes-Benz	CL 63	41200	50364.0	Coupe	2012.0

Fig. 3. Illustration of how we create a maximum mileage user query.

7   Query Retrieval Improvement. Next Steps

From our experiments on leveraging different metrics such as Euclidean, Manhattan, Minkowski, cosine similarity when answering a consumer related inventory question, we realized that each metric will produced some errors, did better in correctly picking up and surfacing make, model or year then the other metric, please see Table 2 (Embedding Results). However, none worked perfectly in all the scenarios captured in the bank of asked questions. It demonstrated that we should create weights, proving more weight to the specific dimensions, guiding the answers based on the prior knowledge, human common sense. Humans pose/bring into the problem a common sense of course, we have so many teachers, in the similar fashion we should provide a guidance to the machine learning models. Each not just a common sense, it's human active biases that we learn to use and apply. At Shift, we wanted to try existing models, we have implemented experiments to run and select the best model embedding for our use cases. A very natural use case for Shift is to provide relevant results back when customers are shopping on our side We looked very carefully at the Supervised Learning of Universal Sentence Representation from Natural Language Inference Data [5] since we ourselves had in house models that translated text into numerical features, we started to lean more and more into higher dimensional VS; however, all the features that we had used were still hand crafted. We needed ability to scale fast, ability to experiments with dozens of different embedding models and choose the best in order to move faster than our competitors. By passing weights to either an Euclidean distance formula, based on even naive definition of probability. The next step would be to set up a domain specific priors.

We were thinking on including our based view vehicles when we leveraged historical digital customer footprint when browsing vehicle details pages, adding a specific vehicle to favorites, selecting a car for a test drive. Let's remind readers in here. Even naive definition of probability can be written as a formula (it is a number between 0 and 1, $P(A) = m\,n$ where n - count of all possible events and m is count of positive events. We call probability of the event, A, a fraction of positive occurrences (favorable outcomes) over all possible outcomes (counts of all events) but in the nutshell any formula is a level of abstraction that generalizes, pick up and exploits the pattern, is able to generate next term so to speak predicts with 100 We are in the process to use Poisson Distribution in Days to Sell to create weights and add them to the Euclidean measure as the next steps to fine grain the results given the user query. We have cluster our data and then computed basic statistics for each cluster. Certain cluster exhibited statistics of the 8 Poisson distribution since the mean was equal to its variance in days to sell distribution. To reminder the reader about Poisson distribution.

$$P(X = k) = \frac{e^{-\lambda}\lambda^k}{k!}$$

$$E(X) = e^{-\lambda} \sum_{k=0}^{\infty} k\frac{\lambda^k}{k!}$$

$$E(X) = e^{-\lambda} \sum_{k=1}^{\infty} \lambda\frac{\lambda^{k-1}}{(k-1)!} = \lambda e^{-k} e^k = \lambda$$

References

1. Sutskever, I., Vinyals, O., Le, Q.V.: Generating sequences with recurrent neural networks. In: Proceedings of the International Conference on Learning Representations (ICLR), pp. 1–14 (2013). https://doi.org/10.1162/ICLR.2013.14
2. Karpukhin, V., et al.: Dense Passage Retrieval for Open-Domain Question Answering. CoRR, abs/2004.04906 (2020). https://arxiv.org/abs/2004.04906
3. Chang, W.-C., Yu, F.X., Chang, Y.-W., Yang, Y., Kumar, S.: Pre-training tasks for embedding-based large-scale retrieval. In: Proceedings of the International Conference on Learning Representations (ICLR) (2020). Published as a conference paper at ICLR 2020. https://openreview.net/forum?id=BJeDY0NFwS
4. Cer, D., et al.: Universal Sentence Encoder. arXiv preprint arXiv:1803.11175 (2018). https://doi.org/10.48550/arXiv.1803.11175
5. Conneau, A., Kiela, D., Schwenk, H., Barrault, L., Bordes, A.: Supervised Learning of Universal Sentence Representations from Natural Language Inference Data. arXiv preprint arXiv:1705.02364 (2017). https://doi.org/10.48550/arXiv.1705.02364. arXiv: https://arxiv.org/abs/1705.02364
6. Mac Lane, S.: Categories for the Working Mathematician. https://en.wikipedia.org/wiki/Category_theory.10

Speeding Up Logic-Based Benders Decomposition by Strengthening Cuts with Graph Neural Networks

Johannes Varga[1](✉) ⓘ, Emil Karlsson[2,3] ⓘ, Günther R. Raidl[1] ⓘ,
Elina Rönnberg[2] ⓘ, Fredrik Lindsten[4] ⓘ, and Tobias Rodemann[5] ⓘ

[1] Institute of Logic and Computation, TU Wien, Vienna, Austria
{jvarga,raidl}@ac.tuwien.ac.at
[2] Department of Mathematics, Linköping University, Linköping, Sweden
elina.ronnberg@liu.se
[3] Saab AB, 581 88 Linköping, Sweden
emil.karlsson1@saabgroup.com
[4] Department of Computer and Information Science, Linköping University,
Linköping, Sweden
fredrik.lindsten@liu.se
[5] Honda Research Institute Europe, Offenbach, Germany
tobias.rodemann@honda-ri.de

Abstract. Logic-based Benders decomposition is a technique to solve
optimization problems to optimality. It works by splitting the problem
into a master problem, which neglects some aspects of the problem, and
a subproblem, which is used to iteratively produce cuts for the master
problem to account for those aspects. It is critical for the computational
performance that these cuts are strengthened, but the strengthening of
cuts comes at the cost of solving additional subproblems. In this work
we apply a graph neural network in an autoregressive fashion to approximate the compilation of an irreducible cut, which then only requires few
postprocessing steps to ensure its validity. We test the approach on a job
scheduling problem with a single machine and multiple time windows per
job and compare to approaches from the literature. Results show that our
approach is capable of considerably reducing the number of subproblems
that need to be solved and hence the total computational effort.

Keywords: Logic-based Benders Decomposition · Cut Strengthening ·
Graph Neural Networks · Job Scheduling

1 Introduction

Many different ways have been investigated lately to improve traditional combinatorial optimization methods by means of modern machine learning (ML)
techniques. For a general survey see [2]. Besides more traditional approaches

J. Varga acknowledges the financial support from Honda Research Institute Europe.

G. Nicosia et al. (Eds.): LOD 2023, LNCS 14505, pp. 24–38, 2024.
https://doi.org/10.1007/978-3-031-53969-5_3

where ML techniques are used to tune parameters or configurations of an optimization algorithm or to extract problem features then utilized by an optimization algorithm [5,6], diverse approaches have been proposed where handcrafted heuristics for guiding a search framework are replaced by learned models. The advantages of successful approaches are apparent: Significant human effort may be saved by learning components of an optimization algorithm in an automated way, and these may potentially also make better decisions, leading to better performing solvers and ultimately better or faster obtained solutions. Given historical data from a practical application scenario, a more specialized model may also be trained to more effectively solve similar future instances of this problem. Moreover, if characteristics of the problem change, a learned model may also be rather easily retrained to accommodate the changes, while handcrafted methods require manual effort.

In the context of mathematical programming techniques, in particular mixed integer linear programming, learned models are used, for example, as primal heuristics for diving, to approximate the well-working but computational expensive strong branching [1], or to select valid inequalities to add as cuts by means of a faster ML model that approximates a computationally expensive look-ahead approach [14]. We focus here in particular on logic-based Benders decomposition (LBBD) [7], which is a well-known mathematical programming decomposition technique. The basic idea is that an original problem is split into a master problem and a subproblem that are iteratively solved in an alternating manner to obtain a proven optimal solution to the original problem. The master problem considers only subsets of the decision variables and constraints of the original problem and is used to calculate a solution with respect to this subset of variables. The subproblem is obtained from the original problem by using the master problem solution to fix the values of the corresponding variables. If the original objective function includes master problem variables only, the purpose of the subproblem is to augment the solution in a feasible way. If this is not possible, a so-called feasibility cut, i.e., a valid inequality that is violated by the current master problem solution, is derived and added to the master problem. Used as is, this cut would only remove a small number of master problem solutions from the master problem's solution space and as such it is not likely to yield much progress. It is therefore crucial for the practical performance of LBBD schemes to *strengthen* cuts before adding them to the master problem. Typically, such strengthening involves to solve additional subproblems.

As a first step towards speeding up the LBBD scheme through cut strengthening, we propose a learning-based approach that reduces the number of subproblems to be solved. For the learning part, subproblems are represented by graphs, and consequently we use a graph neural network (GNN) as ML model. An autoregressive approach inspired by [11] is applied to compile a promising inequality by means of iteratively selecting and adding variables under the guidance of the GNN. This GNN is trained offline on a set of representative problem instances. As test bed we use a single-machine scheduling problem with time-windows from [8]. Results clearly indicate the benefits of the GNN-guided cut strengthening over classical strengthening strategies from the literature. In particular the number of solved subproblem instances is significantly lower, which leads to better overall runtimes to obtain proven optimal solutions.

Compared to classical Benders decomposition, which is used to solve mixed integer linear programs decomposed so that the subproblem has continuous variables, LBBD is a generalization that allows also discrete variables in the subproblem. This generalization is however done at the cost of no longer having a general strategy for deriving strong cuts from subproblem. Several different cut-strengthening techniques have been proposed in the literature, and typically they are described for the specific problem at hand. In Karlsson and Rönnberg [9], techniques for strengthening of feasibility cuts have been described within a common framework and compared on a variety of different scheduling problems to provide an overview of this area. An extension of this work to also include optimality cuts is found in [16].

It is common to apply LBBD to problems with both allocation and scheduling decisions. For example, LBBD has been applied to a single facility scheduling problem with segmented timeline [4] and an electric vehicle charging and assignment scheduling problem [17]. Other examples are [10,15].

The outline of this paper is as follows. Section 2 specifies the scheduling problem and our LBBD scheme. Section 3 describes what we aim to approximate with a GNN, how we make use of its predictions to strengthen cuts, and how we collect training data, while Sect. 4 discusses the specific structure of the GNN, the features used as its input, and the training procedure. Finally, Sect. 5 presents experimental results, and Sect. 6 concludes the paper.

2   Single-Machine Scheduling Problem and LBBD

In the single-machine scheduling problem we will use as test bed for our LBBD approach we are given a set of tasks I. Each task $i \in I$ may be scheduled in a non-preemptive way within one of its possibly multiple time windows $w \in W_i$ given by a release time r_{iw} and a deadline $d_{iw} \geq r_{iw}$. Task i has a processing time p_i, for which it exclusively requires the single available machine. Moreover, if task i runs before another task $j \in I$, the time difference between the start of task j and the end of task i has to be at least a given setup time s_{ij}. For each task $i \in I$ that is scheduled, a prize q_i is collected and the objective is to maximize these collected prizes.

An overview of the LBBD framework is shown in Algorithm 1. To apply LBBD to our scheduling problem, we split it into a master problem (MP), which determines the subset of tasks to be performed together with selected time windows, and a subproblem (SP) for determining an actual schedule. More specifically, the SP gets a set of task-time window pairs (TTWPs) $X \subseteq \{(i,w) : i \in I, w \in W_i\}$ as input and checks, whether the tasks can be scheduled without overlap within the selected time windows, also obeying the setup times. If such a schedule exists, it is an optimal solution to the original problem. Otherwise a feasibility cut, i.e., inequality that excludes this set of TTWPs from the MP's

---

Algorithm 1: LBBD scheme with cut strengthening

Data: A problem instance
Result: An optimal solution or a proof of infeasibility
1   $\mathcal{F} = \emptyset$ /* Set of feasibility cuts */;
2   $k \leftarrow 1$ /* The current LBBD iteration */;
3   $X \leftarrow \emptyset$ /* The current TTWPs */;
4   **while** *true* **do**
5     $X^k \leftarrow$ solve MP with feasibility cuts $\mathcal{F}$;
6   **if** X^k *is a feasible solution* **then**
7 | $X \leftarrow X^k$;
8   **else**
9 | **return** *problem has no feasible solution*;
10   **end**
11    $Y^k \leftarrow$ solve SP for TTWPs X;
12   **if** Y^k *is a feasible solution* **then**
13 | **return** *optimal solution* (X^k, Y^k);
14   **else**
15 | $\hat{\mathcal{X}} \leftarrow$ CutStrengthening(X);
16 | $\mathcal{F} \leftarrow \mathcal{F} \cup \hat{\mathcal{X}}$;
17   **end**
18    $k = k + 1$;
19 **end**

---

space of feasible solutions, is derived and added to the MP to prevent the same configuration from further consideration. Thus, in our case a feasibility cut is represented by a set of TTWPs.

2.1 Master Problem

We state the MP by the following binary linear program with variables $x_{iw} \in \{0, 1\}$ indicating with value one that task $i \in I$ is to be performed in its time window $w \in W_i$.

$$[\text{MP}(I, \mathcal{F})] \quad \max \sum_{i \in I} \sum_{q \in W_i} q_i x_{iw} \tag{1}$$

$$\text{s.t.} \sum_{w \in W_i} x_{iw} \leq 1, \qquad\qquad i \in I \tag{2}$$

$$\sum_{(i,w) \in X} (1 - x_{iw}) \geq 1, \qquad\qquad X \in \mathcal{F} \tag{3}$$

$$\sum_{i \in I} \sum_{w \in W_i(t_1, t_2)} p_i x_{iw} \leq t_2 - t_1, \qquad (t_1, t_2) \in T \tag{4}$$

$$x_{iw} \in \{0, 1\}, \qquad\qquad i \in I, \; w \in W_i. \tag{5}$$

The objective (1) is to maximize the total prize collected by assigning tasks to time-windows. Constraints (2) ensure that each task is scheduled in at most one of its time windows. All so far added feasibility cuts are represented by inequality (3), where the elements in $\mathcal{F}$ are sets of TTWPs that must not appear together in a feasible solution. Inequalities (4) are used to strengthen the MP. Set $T = \{(r_{iw}, d_{i'w'}) \mid (w, w') \in W_i \times W_{i'}, (i, i') \in I \times I, d_{i'w'} > r_{iw}\}$ contains release time and deadline pairs for which a segment relaxation is used and $W_i(t_1, t_2) = \{w \in W_i \mid t_1 \leq r_{iw}, d_{iw} \leq t_2\}$ is the set of time windows for task $i \in I$ that starts after t_1 and ends before t_2. The inequality ensures that the total processing time of all tasks within a time interval $[t_1, t_2)$ does not exceed the interval's duration.

2.2 Subproblem

To simplify the notation, let $I(X) = \{i : (i, w) \in X\}$ denote the set of tasks to be scheduled. In the SP, the decision variables $y_i \geq 0$ represent the start time of task $i \in I(X)$. We state the SP as the following decision problem.

$$[SP(X)] \quad \text{find } y_i, \forall i \in I \tag{6}$$
$$\text{s.t. DISJUNCTIVE}((y_i \mid i \in I(X)), (p_i \mid i \in I(X)),$$
$$(s_{ij} \mid i, j \in I(X)), i \neq j) \tag{7}$$
$$r_{iw} \leq y_i \leq d_{iw} - p_i \qquad (i, w) \in X \tag{8}$$
$$y_i \geq 0 \qquad i \in I \tag{9}$$

Constraints (7) ensure that no two tasks overlap and setup times are obeyed. Inequalities (8) guarantee that all tasks are scheduled within their time windows.

2.3 Cut Strengthening

Strengthening of feasibility cuts obtained from an infeasible subproblem is crucial for good performance of the LBBD framework. In our case, a cut is represented by a set of TTWPs X and is *strengthened* when one or more TTWPs are removed from X and the newly obtained inequality is still a valid cut in the sense that the respective SP remains infeasible. The infeasibility of the SP implies that no feasible solutions are cut away when adding the inequality to the MP. Ideally, a cut-strengthening procedure should provide one or multiple *irreducible cuts*, which are cuts that cannot be strengthened further by removing TTWPs. Should the MP become infeasible at some point, the problem instance is proven infeasible.

The deletion filter cut-strengthening algorithm is based on the deletion filter for finding an irreducible inconsistent set of constraints [3] and is used to form an irreducible cut $\hat{X}$ from a given X, see Algorithm 2. The method starts with $\hat{X} = X$ and proceeds through all TTWPs $(i, w) \in X$ one-by-one, checking the feasibility of the subproblem with (i, w) excluded. If a subproblem is infeasible, TTWP (i, w) will be permanently removed from $\hat{X}$. For example, this form of cut strengthening was already used in [12,15]. MARCO [13] is another procedure to strengthen cuts. While it is more expensive in terms of subproblems to check, it systematically enumerates *all* irreducible feasibility cuts that can be derived from X. We will use MARCO to collect training data for learning, while our cut strengthening approach is based on the faster deletion filter.

Algorithm 2: Deletion filter cut strengthening

Data: A set of TTWPs X representing a feasibility cut
Result: A set of TTWPs $\hat{X}$ representing an irreducible feasibility cut

1 $\hat{X} \leftarrow X$;
2 **for** $(i, w) \in X$ **do**
3 $\quad$ $X' \leftarrow \hat{X} \setminus \{(i, w)\}$;
4 $\quad$ **if** SP(X') *is infeasible* **then**
5 $\quad$ $\quad$ $\hat{X} \leftarrow X'$;
6 $\quad$ **end**
7 **end**
8 **return** $\hat{X}$;

---

3 Learning to Strengthen Cuts

We now introduce our main contribution, the graph neural network (GNN) based cut strengthening approach to speed up the LBBD scheme. The idea is to train a GNN offline on representative problem instances and apply it when solving new instances as guidance in the construction of a promising initial subset of TTWPs $\hat{X} \subseteq X$ that ideally already forms a strong irreducible cut without the need of any subproblem solving. Still, our approach then has to check if the respective SP($\hat{X}$) is feasible or not. In case the SP is feasible, further TTWPs are added one-by-one until the SP becomes infeasible. In any case, the deletion filter is finally also applied to obtain a guaranteed irreducible cut. The hope is that by this approach, only few subproblems need to be solved to "repair" the initially constructed TTWP subset in order to come up with a proven irreducible cut.

Algorithm 3 shows our GNN-based cut strengthening in more detail, which again receives a TTWP set X representing an infeasible SP and thus an initial feasibility cut as input. In its first part up to line 5, the algorithm selects a sequence of TTWPs S in an autoregressive manner steered by the GNN-based function $f_\Theta(I, X, S)$ with trainable parameters Θ. Here, I refers to the set of all tasks of the original problem, which are provided as global information, and S is the initially empty TTWP sequence. Function f_Θ returns a TTWP (i, w) that is appended to S as well as a Boolean indicator τ for terminating the construction. The way how function f_Θ is realized will be explained in Sect. 3.1 and how it is trained in Sect. 3.2. In the following while-loop, the algorithm adds further TTWPs that are again selected by function f_Θ as long as the corresponding SP remains feasible, which needs to be checked here by trying to solve the SP. Finally, in line 10, the deletion filter is applied, where the TTWPs are considered in reverse order as they have been previously selected by f_Θ.

3.1 Function $f_\Theta(I, X, S)$

Our trainable function f_Θ shall predict a best suited TTWP from X for appending it to the given TTWP sequence S as well as the termination indicator τ, so

---

Algorithm 3: GNN-based cut strengthening

Data: A set of TTWPs X representing a feasibility cut
Result: A set of TTWPs $\hat{X}$ representing an irreducible feasibility cut

1 $S \leftarrow []$ /* sequence of selected TTWPs */;
2 **repeat**
3 | $(i, w), \tau \leftarrow f_\Theta(I, X, S)$;
4 | append (i, w) to S;
5 **until** τ;
6 **while** $SP(S)$ *is feasible* **do**
7 | $(i, w), \tau \leftarrow f_\Theta(I, X, S)$;
8 | append (i, w) to S;
9 **end**
10 $\hat{X} \leftarrow$ Deletion Filter(reverse(S));
11 **return** $\hat{X}$;

---

that in the ideal case the final S represents an irreducible feasibility cut that is also *strong* in the sense that it is likely binding in the final MP iteration. In a first step of $f_\Theta(I, X, S)$, a complete directed graph $G = (X, A)$ with nodes corresponding to the TTWPs in X is set up, and node as well as arc features are derived from I, X, and S, cf. Sect. 4. Following in parts the autoregressive approach in [11], we then apply a GNN to graph G, which returns for each node and thus each TTWP $(i, w) \in X \setminus S$ a value $p_{iw} \in [0, 1]$ that shall approximate the probability that, under the assumption that S does not contain a feasibility cut yet, when appending (i, w) to S, this set is getting one step closer to containing some strong cut while including as few TTWPs not being part of this cut as possible. More precisely, let $\mathcal{X}$ be the set of all irreducible strong cuts being subsets of X. Moreover, let $\delta(S, \mathcal{X}) = \min_{\hat{X} \in \mathcal{X}} |S \setminus \hat{X}|$ be the minimum number of TTWPs any complete extension of S towards a feasibility cut must contain that will not be part of a strong cut. Subset $\mathcal{X}' = \{\hat{X} \in \mathcal{X} : |S \setminus \hat{X}| = \delta(S, \mathcal{X})\}$ then contains those strong cuts that might be created by extending S without increasing $\delta(S, \mathcal{X})$. Values p_{iw} should therefore approximate

$$\hat{p}_{iw} = \begin{cases} 1 & \text{if } (i, w) \in \bigcup_{\hat{X} \in \mathcal{X}'} \hat{X} \\ 0 & \text{else.} \end{cases} \tag{10}$$

We intentionally leave p_{iw} undefined for the case that S already contains an irreducible cut, although in practice f_Θ might indeed be called for such cases if the termination prediction is not precise enough. These predicted values of p_{iw} are, however, not critical as the corresponding unnecessarily appended TTWPs in S will get removed by the finally applied deletion filter anyway.

In addition to the above p_{iw} value, the GNN further returns a value $\tau_{iw} \in [0, 1]$ for each TTWP $(i, w) \in X \setminus S$ approximating the probability that when appending (i, w) to S, this augmented set contains already an irreducible feasibility cut. Function $f_\Theta(I, X, S)$ evaluates the described GNN, selects a TTWP $(i, w) \in X \setminus S$ with maximum value p_{iw}, and returns (i, w) together with $\lceil \tau_{iw} + 0.5 \rceil$ as Boolean valued τ. Ties among p_{iw} are broken in favor of higher values τ_{iw} and finally at random.

3.2 Collecting Training Data

The GNN is trained in an offline manner by supervised learning on a substantial set of representative problem instances. To obtain labeled training data, the LBBD framework is applied to each of these problem instances with the following extensions in order to identify strong irreducible feasibility cuts. In the first step, the LBBD scheme is performed with MARCO as cut-strengthening approach to enumerate all possible irreducible cuts for each infeasible subproblem. Let us denote with $\mathcal{F}$ the set of cuts obtained in this way over a full LBBD run. The second step finds an irreducible subset of *strong* cuts of the form $\mathcal{X} \subseteq \mathcal{F}$ that is actually required to get a feasible optimal solution in the master problem in the sense that the master problem will give an infeasible solution when removing one of those cuts. This is done by applying a deletion filter on $\mathcal{F}$ using a random order, iteratively removing cuts and checking whether the master problem still gives a feasible solution. Note that there might be other minimal sets of irreducible cuts, which we do not encounter with this approach.

For each training instance I we perform the above procedure and collect for each Benders iteration a tuple $(I, X, \mathcal{F}, \mathcal{X})$ where X denotes the TTWPs of the SP and thus original feasibility cut and $\mathcal{X} = \{\hat{X} \in \mathcal{F} : \hat{X} \subseteq X \wedge \hat{X} \text{is strong}\}$ is the set of strong cuts that can be derived from X. These tuples are then used to derive training samples for the GNN.

This is done by performing for each tuple $(I, X, \mathcal{F}, \mathcal{X})$ rollouts that simulate the derivation of feasibility cuts without needing to actually solve any further subproblems. Algorithm 4 details this rollout procedure. In each iteration of the loop, step 7 selects a TTWP that brings S one step closer to containing a strong cut from $\mathcal{X}$ and adds it to S. This is repeated until S is a superset of an irreducible cut. Note that in this procedure, the order of the elements in S does not matter, and we therefore realize it as a classical set in contrast to Algorithm 3, where S needs to be an ordered sequence. Additionally, in each iteration the training, labels $\hat{p} = (\hat{p}_{iw})_{iw \in X}$ and $\hat{\tau} = (\tau_{iw})_{iw \in X}$ are calculated based on the definitions of p_{iw} and τ_{iw} from Sect. 3.1 and a training sample is added. The notation $\llbracket \cdot \rrbracket$ denotes Iverson brackets, i.e., $\llbracket C \rrbracket$ is 1 if the condition C is true and 0 otherwise.

4 Graph Neural Network

The architecture of our GNN is shown in Fig. 1. Remember that the nodes of our graph correspond to the TTWPs of an original feasibility cut X. For each node $(i, w) \in X$, vector x_{iw} shall be the vector containing all node features and x the respective matrix over all nodes. An initial node embedding h_0 of size 64 per node is derived by a linear projection with trainable weights. Moreover, for

---

Algorithm 4: The rollout procedure to construct training samples from a tuple $(I, X, \mathcal{F}, \mathcal{X})$.

Data: Original feasibility cut X, all irreducible cuts $\mathcal{F}$, strong cuts $\mathcal{X}$.
Result: A set T of training samples, a sample being a tuple $(I, X, S, \hat{p}, \hat{f})$, where I is the set of all tasks with their attributes and $\hat{p}$ and $\hat{f}$ are the labels.

1   $S \leftarrow \emptyset$;
2   $T \leftarrow \emptyset$;
3 **repeat**
4     $\hat{p}_{iw} \leftarrow [\![(i, w) \in \bigcup_{\hat{X} \in \mathcal{X}'} \hat{X}]\!] \quad \forall (i, w) \in X \setminus S$;
5     $\hat{f}_{iw} \leftarrow [\![\exists \hat{X} \in \mathcal{F} : \hat{X} \subseteq S \cup \{(i, w)\}]\!] \quad \forall (i, w) \in X \setminus S$;
6     $T \leftarrow T \cup \{(I, X, S, \hat{p}, \hat{f})\}$;
7   select $(i, w) \in \bigcup_{\hat{X} \in \mathcal{X}'} \hat{X}$ uniformly at random;
8     $S \leftarrow S \cup \{(i, w)\}$;
9 **until** $\exists \hat{X} \in \mathcal{F} : \hat{X} \subseteq S$;
10 **return** T;

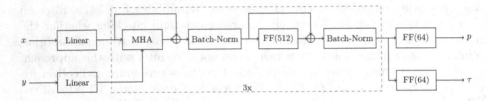

Fig. 1. Architecture of the GNN.

all arcs in A and their features, we also derive an arc embedding h^A of size 32 per arc by a trainable linear projection.

We apply a transformer-based convolution layer that also considers arc embeddings, implemented in Pytorch Geometric, with 8 heads, skip connections, and batch normalization, followed by a feed-forward layer with a hidden sublayer of dimensionality 512, parametric rectified linear units (PReLUs) as nonlinearity, and again with skip connections and batch normalization, similar as in the encoder part of [11] and usually done in transformers [18]. This multi-head attention layer plus feed-forward layer is repeated $k = 3$ times. We chose hyper-parameters by hand based on preliminary experiments and guided by [11].

In contrast to transformers and [11], we do not follow an encoder/decoder architecture, but keep things simpler. To obtain output values p_{iw} and τ_{iw} for each node (i, w), we further process the so far obtained node embeddings h_{iw}^k by two independent feed-forward neural networks with one hidden layer of dimensionality 64 and single final output nodes with sigmoid activation functions, respectively. In the autoregressive approach, the whole GNN is completely evaluated in each step with updated features, in particular concerning the already selected TTWPs. Output nodes for already selected TTWPs are masked out.

As we do binary classification, we use binary cross-entropy as loss function:

$$\mathcal{L}(p, \hat{p}, \tau, \hat{\tau}) = \frac{1}{2}(\mathcal{L}(p, \hat{p}) + \mathcal{L}(\tau, \hat{\tau})) \qquad (11)$$

with

$$\mathcal{L}(p, \hat{p}) = \frac{1}{|V \setminus S|} \sum_{v \in V \setminus S} -(\hat{p}_v \log p_v + (1 - \hat{p}_v) \log(1 - p_v)) \qquad (12)$$

and

$$\mathcal{L}(\tau, \hat{\tau}) = \frac{1}{|V \setminus S|} \sum_{v \in V \setminus S} -(\hat{\tau}_v \log \tau_v + (1 - \hat{\tau}_v) \log(1 - \tau_v)). \qquad (13)$$

The list of node and arc features we use is given in Table 1. There are three additional features with the values from the last three arc features from the reverse arc. Features corresponding to points in time are scaled with function

$$\text{time-scaling}(t) = (t - \min_{i,w}(r_{iw}))/(\max_{i,w}(d_{iw}) - \min_{i,w}(r_{iw})), \qquad (14)$$

over all tasks i and their time windows w in the whole set of training instances. All other features are either normalized across the training data to values within [0,1] or kept if they are already reasonably distributed. We further compute a basic schedule from the TTWPs in X with a variation of the earliest deadline first (EDF) rule. This greedy heuristic selects the task with the earliest deadline of only those tasks, whose release time is before the ending time of the previously scheduled task. If all unscheduled tasks have a later release time, the unscheduled task with the next release time is scheduled next. Some of the features are based on the start times s_i^{EDF} and end times e_i^{EDF} of the tasks $i \in I$ in this schedule.

5 Experimental Evaluation

We implemented our approach in Python 3.10.8 using `pytorch_geometric`[1] for the GNN and used the MILP solver Gurobi[2] 9.5 and the CP solver CPOptimizer[3] 20.1 to solve the master and the subproblem instances, respectively. To evaluate our approach we use benchmark instances that are generated as described in [8] for the avionics application with $m = 3$ resources. Note that the problem formulation in [8] differs in having secondary resources, but we are able to model this aspect with setup times. We consider instance sizes $n \in \{10, 15, 20, 25\}$, for which we are able to find optimal solutions by the LBBD approach in most cases in reasonable time.

For each instance size between 300 000 and 600 000 training samples were collected from 2 000 to 200 000 independent instances, see Table 2 for details on

---

[1] https://pytorch-geometric.readthedocs.io/.
[2] https://www.gurobi.com/.
[3] https://www.ibm.com/products/ilog-cplex-optimization-studio/cplex-cp-optimizer.

Table 1. Node and arc features. For a node feature (i, w) denotes the TTWP associated with the node, for an arc feature (i_1, w_1) represents the TTWP of the source node and (i_2, w_2) the TTWP of the target node.

Formula	Type	Normalization	Meaning
$[\![(i, w) \in S]\!]$	node	-	Whether the TTWP has already been selected
$\lvert S \rvert / \lvert X \rvert$	node	-	Fraction of selected TTWPs
p_i	node	normalized	Processing time
r_{iw}	node	time-scaling	Release time
d_{iw}	node	time-scaling	Deadline
$d_{iw} - r_{iw}$	node	normalized	Length of the time window
$q_i / \max_{j \in I}(q_j)$	node	-	Relative prize
$d_{iw} - r_{iw} - p_i + 1$	node	normalized	Absolute flexibility of scheduling the job
$1 - p_i/(d_{iw} - r_{iw})$	node	-	Relative flexibility of scheduling the job
$[\![d_{iw} - r_{iw} > p_i]\!]$	node	-	Whether there is any flexibility in scheduling the job
s_i^{EDF}	node	time-scaling	EDF start time
e_i^{EDF}	node	time-scaling	EDF end time
$[\![s_i^{\mathrm{EDF}} = r_i]\!]$	node	-	Whether the EDF start time is at the release time
$[\![e_i^{\mathrm{EDF}} = d_i]\!]$	node	-	Whether the EDF end time is at the deadline
$[\![e_i^{\mathrm{EDF}} > d_i]\!]$	node	-	Whether the task is late in the EDF schedule
$\max(e_i^{\mathrm{EDF}} - d_i, 0)$	node	time-scaling	Lateness of the task in the EDF schedule
$s_{i_1 i_2}$	arc	normalized	Setup time
$[\![s_{i_2}^{\mathrm{EDF}} = e_{i_1}^{\mathrm{EDF}} + s_{i_1 i_2}]\!]$	arc	-	Whether the two tasks cannot be any closer than in the EDF schedule
$\max(s_{i_2}^{\mathrm{EDF}} - e_{i_1}^{\mathrm{EDF}}, 0)$	arc	normalized	Time difference of the EDF schedule
-	arc	normalized	Overlap of the time windows
$[\![r_{i_1 w_1} \leq r_{i_2 w_2} < d_{i_1 w_1}]\!]$	arc	-	Whether time window w_2 starts within time window w_1
$[\![s_{i_1 i_2} > 0]\!]$	arc	-	Whether there is a non-zero setup time
-	arc	-	Whether task i_2 is the direct successor of task i_1 in the EDF schedule

Table 2. Characteristics of the training.

n	10	15	20	25
# of training instances	200 000	60 000	20 000	2 000
# of training samples	547 893	354 030	339 263	322 955
# of epochs	40	40	200	400
# of test instances	20 000	6 000	2 000	200
# of test samples	54 723	35 178	33 493	37 832
Recall of p-values [%]	92.00	82.92	80.91	81.15
Precision of p-values [%]	91.44	86.44	80.32	80.64
Recall of τ-values [%]	98.07	94.66	86.17	74.55
Precision of τ-values [%]	93.00	90.21	86.75	68.81

the training. In each of 40 to 300 epochs we train with minibatches of 64 samples, using the ADAM optimizer with a learning rate of $3 \cdot 10^{-4}$. For regularization we use dropout before each transformer-based graph convolution layer, before each feed-forward layer, and on the normalized attention coefficients with a dropout probability of 25% for $n \in \{10, 15, 20\}$ and 10% for $n = 25$.

First, we compare the performance of the LBBD with our approach as cut-strengthening procedure, denoted by *GNN*, to the deletion filter where the order of the TTWPs is (a) random (*Random*), (b) dictated by the instance (*Sorted*), or (c) sorted by decreasing slack as done by Coban and Hooker [4] (*Hooker*). For this comparison we ran each approach for each benchmark instance on a single core of an AMD Ryzen 9 5900X without using a GPU. The results for $n = 20$ and $n = 25$ are shown in Fig. 2 in the form of cumulative distribution plots. Each plot shows how many instances could be solved to optimality out of 100, with which number of subproblems to be solved, in which time, and within how many major LBBD iterations. As can be seen, our GNN-based cut-strengthening procedures significantly reduce the number of subproblems that have to be solved as well as the total runtimes. For example, for $n = 20$ *GNN* on average only needs to solve $\approx 30.4\%$ of the number of subproblems *Hooker* has to solve and requires only $\approx 54\%$ of *Hooker*'s time. The number of LBBD iterations is similar for all methods, which means that the time savings are only achieved by reducing the number of subproblems that had to be solved.

Progress of the Bounds. In each iteration of the LBBD we obtain a dual bound from the solution to the master problem. Concerning primal bounds, note that any set of TTWPs for which the subproblem is feasible represents a feasible solution to the overall problem. We use as primal bound the value of the best such solution encountered during the cut strengthening procedure. The average development of those primal and dual bounds for $n = 20$ and $n = 25$ tasks are shown in Fig. 3. Observe that the GNN-based approach performs significantly better in respect to the dual bound with an average percentage gap to the optimal solution of 4.85% after four seconds for $n = 25$ compared to 10.6% for *Hooker*.

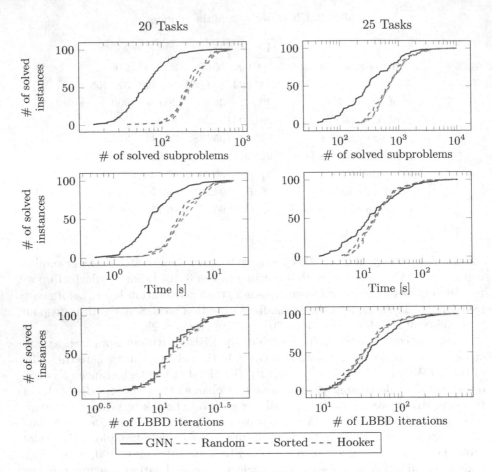

20 Tasks   25 Tasks

Fig. 2. Cumulative distribution diagrams with the number of solved instances over the number of subproblems solved, the running time and the number of Benders iterations performed.

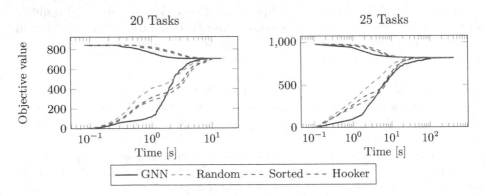

Fig. 3. Primal and dual bounds of the approaches over time.

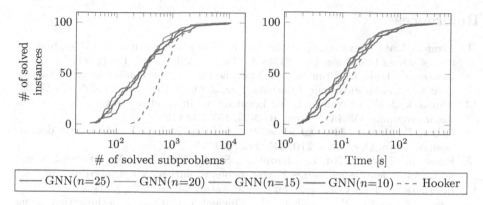

Fig. 4. Cumulative distribution diagrams showing the performance of the models trained on instances of size $n = 10$, $n = 15$, $n = 20$ and $n = 25$, respectively, on instances of size $n = 25$ and comparing to the performance of *Hooker*.

However the primal bounds perform mostly worse. This can be explained by the fact that the subproblem for the GNN-based cut strengthening is only called for smaller sets of TTWPs, which have a worse objective value. We remark that we do not focus here on getting good heuristic solutions and thus primal bounds early. To improve on this aspect, it would be natural to include some primal heuristics, such as a local search or more advanced metaheuristic for strengthening intermediate heuristic solutions.

Out-of-Distribution Generalization. So far we applied *GNN* only to instances of the same size as the neural network model has been trained with. Now, we investigate the out-of-distribution generalization capabilities in the sense that the models trained on instances with $n \in \{10, 15, 20\}$ tasks, respectively, are applied to larger instances with 25 tasks. Figure 4 shows the corresponding cumulative distribution plots for *GNN* and compares them to *Hooker*. The generalization works surprisingly well and *GNN* trained with $n = 15$ works even slightly better than the model trained with $n = 25$, requiring a geometric mean time that is only $\approx 76\%$ of *Hooker*'s.

6   Conclusion

We trained GNNs to guide the cut-strengthening in logic-based Benders decomposition. An autoregressive approach is used, where the GNN first constructs a preliminary inequality, which is postprocessed to ensure that it is indeed an irreducible cut. The approach is tested on a single machine scheduling problem with time windows. For this problem the number of Benders subproblems solved can be reduced down to one third and up to half of the runtime can be saved on average. It is up to future work to scale up the training of the GNN to larger instance sizes. Results suggest that generalization of the GNN to larger instance sizes works quite well and thus curriculum learning seems to be promising.

References

1. Alvarez, A.M., Louveaux, Q., Wehenkel, L.: A machine learning-based approximation of strong branching. INFORMS J. Comput. **29**(1), 185–195 (2017)
2. Bengio, Y., Lodi, A., Prouvost, A.: Machine learning for combinatorial optimization: a methodological tour d'Horizon. Eur. J. Oper. Res. **290**(2), 405–421 (2021)
3. Chinneck, J.W., Dravnieks, E.W.: Locating minimal infeasible constraint sets in linear programs. ORSA J. Comput. **3**(2), 157–168 (1991)
4. Coban, E., Hooker, J.N.: Single-facility scheduling by logic-based Benders decomposition. Ann. Oper. Res. **210**, 245–272 (2013)
5. Friess, S., Tiňo, P., Xu, Z., Menzel, S., Sendhoff, B., Yao, X.: Artificial neural networks as feature extractors in continuous evolutionary optimization. In: 2021 International Joint Conference on Neural Networks, pp. 1–9 (2021)
6. Gräning, L., Jin, Y., Sendhoff, B.: Efficient evolutionary optimization using individual-based evolution control and neural networks: a comparative study. In: ESANN, pp. 273–278 (2005)
7. Hooker, J.N., Ottosson, G.: Logic-based Benders decomposition. Math. Program. **96**, 33–60 (2003)
8. Horn, M., Raidl, G.R., Rönnberg, E.: A* search for prize-collecting job sequencing with one common and multiple secondary resources. Ann. Oper. Res. **307**, 477–505 (2021)
9. Karlsson, E., Rönnberg, E.: Strengthening of feasibility cuts in logic-based Benders decomposition. In: Stuckey, P.J. (ed.) Integration of Constraint Programming, Artificial Intelligence, and Operations Research, pp. 45–61. Springer, Cham (2021). https://doi.org/10.1007/978-3-030-78230-6_3
10. Karlsson, E., Rönnberg, E.: Logic-based Benders decomposition with a partial assignment acceleration technique for avionics scheduling. Comput. Oper. Res. **146**, 105916 (2022)
11. Kool, W., van Hoof, H., Welling, M.: Attention, learn to solve routing problems! In: International Conference on Learning Representations (2019)
12. Lam, E., Gange, G., Stuckey, P.J., Van Hentenryck, P., Dekker, J.J.: Nutmeg: a MIP and CP hybrid solver using branch-and-check. SN Oper. Res. Forum **1**(3), 22 (2020)
13. Liffiton, M.H., Malik, A.: Enumerating infeasibility: finding multiple MUSes quickly. In: Gomes, C., Sellmann, M. (eds.) Integration of AI and OR Techniques in Constraint Programming for Combinatorial Optimization Problems. LNCS, vol. 7874, pp. 160–175. Springer, Heidelberg (2013). https://doi.org/10.1007/978-3-642-38171-3_11
14. Paulus, M.B., Zarpellon, G., Krause, A., Charlin, L., Maddison, C.: Learning to cut by looking ahead: cutting plane selection via imitation learning. In: Proceedings of the 39th International Conference on Machine Learning, pp. 17584–17600. PMLR (2022)
15. Riedler, M., Raidl, G.R.: Solving a selective dial-a-ride problem with logic-based benders decomposition. Comput. Oper. Res. **96**, 30–54 (2018)
16. Saken, A., Karlsson, E., Maher, S.J., Rönnberg, E.: Computational evaluation of cut-strengthening techniques in logic-based benders' decomposition. Oper. Res. Forum **4**, 62 (2023)
17. Varga, J., Raidl, G.R., Limmer, S.: Computational methods for scheduling the charging and assignment of an on-site shared electric vehicle fleet. IEEE Access **10**, 105786–105806 (2022)
18. Vaswani, A., et al.: Attention is all you need. In: Advances in Neural Information Processing Systems, vol. 30. Curran Associates, Inc. (2017)

Flocking Method for Identifying of Neural Circuits in Optogenetic Datasets

Margarita Zaleshina[1](✉) ⓘ and Alexander Zaleshin[2] ⓘ

[1] Moscow Institute of Physics and Technology, Moscow, Russia
zaleshina@gmail.com
[2] Institute of Higher Nervous Activity and Neurophysiology, Moscow, Russia

Abstract. This work introduces a new approach to spatial analysis of brain activity in optogenetics datasets based on application of flocking method for an identification of stable neuronal activity locations. Our method uses a multiple local directivity and interaction in neuronal activity paths. It can be seen as a flocking behaviour that promotes sustainable structuration because they use collective information to move. We processed sets of mouse brain images obtained by light-sheet fluorescence microscopy method. Location variations of neural activity patterns were calculated on the basis of flocking algorithm. An important advantage of using this method is the identification of locations where a pronounced directionality of neuronal activity trajectories can be observed in a sequence of several adjacent slices, as well as the identification of areas of through intersection of activities. The trace activity of neural circuits can affect parameters of subsequent activation of neurons occurring in the same locations. We analyzed neuronal activity based on its distributions from slice to slice obtained with a time delay. We used GDAL Tools and LF Tools in QGIS for geometric and topological analysis of multi-page TIFF files with optogenetics datasets. As a result, we were able to identify localizations of sites with small movements of group neuronal activity passing in the same locations (with retaining localization) from slice to slice over time.

Keywords: Brain Imaging · Pattern Recognition · Optogenetics · Mouse Brain

1 Introduction

This paper presents a new approach in spatial analysis of optogenetic data using a flocking method.

Optogenetics is a widely used method to study neuronal activity in living organisms at the cellular level. Genetically encoded indicators enable high spatiotemporal resolution optical recording of neuronal dynamics in behaving mice [1]. These recordings further makes it possible to collect and process brain images, revealing important indicators of activities of sets of neurons in various behavioral tasks, as well as in the study of spontaneous activity.

Optogenetics uses sets of images of registered neuronal activity in the form of 2D slices. Each pixel in these images represents an activity of neurons at a specific point,

G. Nicosia et al. (Eds.): LOD 2023, LNCS 14505, pp. 39–52, 2024.
https://doi.org/10.1007/978-3-031-53969-5_4

which is linked to a relative coordinate system. Taking into account the fact that the time of 2D recordings is nonzero, the change in activity from slice to slice can also be used in computational operations to determine the dynamics of activity.

When detecting neuronal activity, the main problem faced by researchers is noises. A noise in images, on the one hand, leads to the detection of a "false activity" (false positives), but on the other hand, makes it difficult to identify the existing activity (false negatives), lowering the overall recognition quality.

In areas with redundant information the influence of noise is higher. As a result, those zones that are in the middle range of activity are of interest for analysis, and allow to distinguish and highlight weak effects.

The main goal of this work is to develop and apply computational methods and tools that help reduce the influence of noise on recognition of neuronal activity and increase the predictability of dynamic optogenetic (neuronal) activity through brain slices.

To eliminate the problem of noise in optogenetic images, in this paper we propose a new approach based on the principles of spatial analysis of flock trajectories (flocking method). The flocking method is based on keeping the distances and co-direction of movements of elements in the flock and can be used in analysis of dynamic changes in brain activity. The principles of flocking are already applied in brain imaging analysis [2]. In our paper, their scope is expanded to analyze the neuronal activity of fluorescently activated mouse brain cells.

Spatial relationships and neighborhood in neural networks in fluorescence microscopy datasets enrich the possibilities of processing connectivity. The fact that the activity of neurons is related not to a single element but to a set of elements makes it possible to process data by methods of spatial analysis for flocks, taking into account the joint distribution of "neuronal ensemble" activity. The presence of a spatiotemporal sweep between slices during scanning makes it possible to take into account the direction of movement of neuronal activity in multi-page TIFF files.

The usage of spatial analysis methods made it possible to reveal data from pixels of optogenetic images and conduct inter-slice spatiotemporal analysis using flocking method.

The main scheme of our work is presented in Fig. 1:

A. Typical multi-TIFF image. Schematically shows that multi-TIFF image includes slices. Each subsequent slice is recorded with a shift along the brain and with a time delay relative to the previous one.
B. Image analysis: B1. Primary image in grayscale mode. B2. Identification of activity points (shown as orange dots) by activity on a pair of neighboring slices. B3. Identification of ensemble locations (shown as lilac dots). B4. Identification of circuit tube between neighboring slices (shown as blue tube), which connect of circuit points (shown as green dots).
C. Cross-slice projection between slices 1_2 and N-1_N when identifying circuit locations, taking into account the buffer zones (shown with solid blue lines on the upper slices and dashed lines on the lower slices).

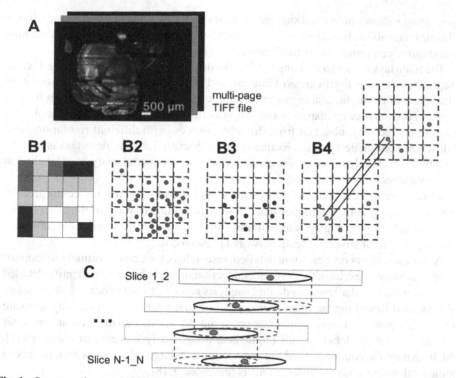

Fig. 1. Sources, elementary processed units and tools in the processing of optogenetic images (see text for details).

2   Background and Related Works

2.1   Brain Imaging Methods and Tools

The purpose of brain imaging analysis is usually to process data on brain structure, neuronal activity and their interrelationships. The possibilities and ways of analyzing the obtained data expand with the development of medical and research equipment used to obtain images of the brain. Thus, as image spatial resolution and the accuracy of localization of individual elements in the image increase, the ability to identify the topology of structures and individual brain areas improves. Also, with appropriate resolution, the level of detail is improved for describing processes in a healthy brain.

The possibility of separating activity of different neuronal populations in the brain tissue was investigated in experiments using various injections to detect activity. Thus, in [3] it was shown that two channel rhodopsins can detect two-color neural activation of spiking and downstream synaptic transmission in independent optical excitation of distinct neural populations.

Topology of structures is considered in certain size ranges typical for these structures. Within small areas of interest, an influence of topology from other scales will reduce. Curved surfaces of the brain directly affect the overall measurement of activity of ensembles from different segments. Spatial analysis in the recognition of images of brain tissue images allows individual smaller elements on a curved surface to become similar to linear elements. A formation of dynamically stable ensembles with a self-sustaining configuration can remain in its localization for a prolonged time.

The main task for understanding the functioning of both healthy and damaged brains is segmentation, selection of areas of interest, and identifying connectivity between individual parts of the brain. Linking experimental results to spatial and temporal reference points is necessary for comparative analysis of multiple heterogeneous data sets of brain structure and activity, obtained from different sources, with different resolutions, and in different coordinate systems. Evaluation of automatic labeling detection is investigated by Papp et al. [4], who propose a new workflow for spatial analysis of labeling in microscopic sections.

Modern methods of brain imaging analysis apply a transition from selecting areas of interest to tractography techniques that allow visualizing pathways of the brain (white matter tracts) using tractography algorithms. Comparison of tractography algorithms for detecting abnormal structural brain networks presented in [5].

As measurements become more detailed, researchers have an opportunity to monitor not only summary results in the form of connection or tracts but also to identify detailed elements at the cellular level. To do this, analysis algorithms are enhanced. Athey et al. [6] presented BrainLine, an open source pipeline that interacts with existing software to provide registration, axon segmentation, soma detection, visualization and analysis of results. A set of global metrics biomedical image analysis were introduced in [7], which take into account the resulting connectivity. A large overview with rankings of biomedical image analysis competitions is presented in [8].

2.2 Artificial Neural Networks for Processing of Spatial Data from High-Resolution Images in Optogenetics

The main issues when processing spatial images are data distortion due to the presence of noise artifacts and the loss of part of the data during measurements. If the extent of the defined area is comparable to the typical size of the noise that occurs in the original image, it becomes necessary to remove incorrectly recognized areas and fill in the voids where the signal was not detected by an error.

Artificial neural networks are frequently used in segmentation of biomedical images. To solve the problem of image processing in differentiated zoom levels of images, mixed sized biomedical image segmentation based on training U-Net (generic deep-learning solution for frequently occurring quantification tasks such as cell detection and shape measurements in biomedical image data) [9, 10] and DeepLabV3, DeepLabV3Plus (multi-class semantic segmentation algorithm) [11, 12] are used. Time overlap strategy used in U-Net [13] allows for seamless segmentation of images of arbitrary size, and the missing input data is extrapolated by mirroring. However, U-Net performance can be influenced by many factors, including the size of training dataset, the performance metrics used, the quality of the images and, in particular, specifics of brain functional areas to be segmented.

Despite the development of convolutional neural networks (CNNs) and their frequent use in image analysis, these solutions are not suitable for any tasks because both efforts required to prepare training dataset, and time spent on data recognition in trained neural networks are still too great. As a result, it is more convenient to solve this type of tasks using spatial analysis methods that allow performing multi-operations and selecting not all objects, but only those that are of interest for further study. These approaches can be employed either individually or in conjunction with CNNs during pre-processing or post-processing stages of data processing. In addition, a well-chosen segmentation labeling algorithm [14] helps to optimize work with neural networks.

2.3 Optogenetics as a Tool for Studying the Activity of Neurons and Controlling Brain Activity

In 2005, Boyden and Deisseroth published the results of the first optogenetics experiments. Their work [15] reported the ability to control neuronal spiking with a millisecond resolution by expressing a natural occurring membrane localized light-gated ion pump. In further research, Deisseroth explored possibilities of using optogenetics to control brain cells without surgical intervention [16]. With the advancement of optogenetics, its experimental applications have spread to all areas of brain activity research. Optogenetic tools are enabling causal assessment of the roles that different sets of neurons play within neural circuits, and are accordingly being used to reveal how different sets of neurons contribute to emergent computational and behavioral functions of the brain [17].

Currently, optogenetics is actively used to study the neuronal activity of living animals, allowing deep immersion into the brain without destroying its structure. Researchers conduct a variety of optogenetic experiments on mice, including the study of social and feeding behavior [18], False Memory creation in certain parts of the brain [19], and, if possible, activating or suppressing the activity of brain cells with a light flash, while affecting the general behavior of mice [20].

Many researchers have begun to study a possibility of controlling not only the activity but also the behavior of mice using optogenetics. Thus, in [21] the concept of optoception was introduced. The authors called "optoception" a signal internally generated when the brain is perturbed, as it happens with interoception. Using optoception, mice can learn to execute two different sets of instructions based on the laser frequency. The phenomenon of "optoception" can occur on cells of the same type both when activation is turned on and when activation is turned off. Moreover, stimulation of two areas of the brain in one mouse showed that optoception induced by one area of the brain is not necessarily transmitted to a second previously unstimulated area. The development of spatially oriented light technologies significantly improved the spatial and temporal resolution of measurements, which makes the use of optogenetic cellular resolution data possible [22, 23]. These types of studies were conducted both *in vitro* and *in vivo* using various technologies [24, 25].

Light-sheet microscopy (LSM) was developed to allow for fine optical sectioning of thick biological samples without the need for physical sectioning or clearing, which are both time consuming and detrimental to imaging. The functioning principle of LSM is to illuminate the sample while collecting the fluorescent signal at an angle relative to the illuminated plane. Optogenetic manipulation coupled to light-sheet imaging is a powerful tool to monitor living samples [26, 27]. One of the major challenges for optogenetics is that LSM residual objects cause stripe artifacts, which obscure features of interest and, during functional imaging, modulate fluorescence variations related to neuronal activity. In hardware, Bessel beams reduce streaking artifacts and produce high-fidelity quantitative data with 20-fold increasing in accuracy in the detection of activity correlations in functional imaging [28, 29]. LSM is often used in investigation of spontaneous activity, which is observed in complete absence of external stimuli, and therefore a possibility of data cleaning by averaging of repeated trials is not possible [29]. Integrated optogenetic and LSM technologies make it possible to calculate the location of "lost" activity zones in individual slices and identify typical indices of neuronal activity in slices with "cellular" resolution (scan resolution: 0.6 μm per pixel; line scan speed: 0.19 $\mu m \, \mu s^{-1}$ and better) [30].

2.4 Principles of Flocking Method and Its Usage in Detection of Neuronal Ensembles Activities

In this paper, we have extended the application of flocking method to the spatial analysis of optogenetic datasets. Flocking is a common behavior observed in nature, defining the collective behavior of a large number of interacting individuals with a common aim (e.g., going from point A to point B). Nearby members of a flock should move in approximately the same direction and at the same speed. For studying of collective motion or population dynamics in short trajectories is often applied the flocking method, which based on analysis of joint directions and intersections of trajectories with a time lag. Flock methods analyze a behavior of multi-sets of similar elements in research on collective behavior in biology and even in robotics [31–33]. The methods used to calculate a behavior of animals in a flock can also be extended to model neural networks [34]. Individual and collective motion is also investigated for the analysis of cumulative behavior of cells [35–37].

The possibility of organizing parallel calculations by using the processing of activity patterns with spatial reference to individual tiles is shown by Marre et al. [38]; as an extension of this work, the paper [39] shows the possibility of using the CNN model to calculate context dependence for predicting the activity of retinal cells depending on the content of natural images.

In the case of neuronal activity, we are dealing with a set of simultaneously working elements, where behavior of each of the elements depends on both its neighbors and the environment. Doursat suggested that "Neuron flocking" must happen in phase space and across a complex network topology" [40].

The usage of flocking method in tractography helps to reduce noise and improve the accuracy of the analysis. In the paper Aranda et al. [2] have shown that algorithms, based on information about spatial neighborhood such as tractography methods, as well as the flocking paradigm, can improve a calculation of local tracks. Aranda et al. [2] made an assumption for calculations what "the flock members are particles walking in white matter for estimating brain structure and connectivity". The authors applied calculation methods in accordance with Reynolds' rules of flocking behavior [41]. This assumption makes it possible to calculate the behavior of individual sets of elements piece by piece, without using of collective information.

2.5 Processing Model

The application of optogenetic scanning made it possible to identify the main components of neuronal activity and various types of activity changes in the same locations over time.

We assume that the topological properties of distribution of individuals in a moving flock are able to represent information about the environment in the same way as it is realized by a network of neurons. In the methodology used in this paper, we further show that considering a set of elementary components of neuronal activity in the form of a flock improves the extraction of meaningful information.

In calculations to study dynamics of neuronal activity, we used the time delay that occurs when moving from slice to slice during scanning using optogenetic methods. An important advantage of using this method is the identification of locations where a pronounced directionality of neuronal activity trajectories can be observed in a sequence of several adjacent slides, as well as the identification of areas of through intersection of activities.

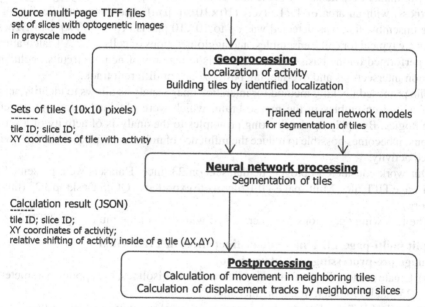

Fig. 2. Flowchart of the neural activity detection process (see text for details)

Flowchart in Fig. 2 shows the process of detecting neuronal activity, which generally consists of the following steps:

– Collecting of raw materials in the form of multi-page TIFF files consisting of slices containing images of the brain, pre-processed by standard tissue cleaning methods.
– Spatial processing using QGIS to identify activity locations. Further, only tiles with the identification of activity locations will be processed. This is much less than the full data from multi-page TIFF sets, and therefore the volume of processed materials is reduced by 10–80 times.

– Post-processing of tiles with identified activities using convolutional neural networks for image segmentation.

The finished results can be uploaded to a JSON file, which contains the numbers of individual slices with the coordinates of activity inside these slices. Further, the results can be used in external applications, both for calculating the movements of active elements, and for constructing tractograms inside 3D multi-page TIFF files.

Methods, source materials, and parameters used in our work for processing multi-page TIFF files with a resolution of 10×10 are described in Sects. 3 and 4.

3   Methods and Applications

3.1   Methods

The registered optogenetic highlighting that we considered is caused by neural ensembles. Illuminated elements are presented in the form of pixels of varying degrees of brightness, with an area of 1×1 pixels (10×10 sq. μm). The characteristic size of a single ensemble that was detected was up to 10×10 pixels (100×100 sq. μm). These ranges are typical for cells, ensembles, and agglomerations of cells [42]. All calculations were performed on the basis of the characteristic features of neural circuits, including common intersections and overlapping buffer zones of different track.

The proposed flocking method for interslice image analysis allows to identify activity of neural ensembles in the mouse brain, which were obtained using optogenetic technologies. By applying the flocking principles to the analysis of activity of a set of neurons, it becomes possible to reduce the influence of noise and replenish the sites with missed activity.

Our work considered optogenetic datasets on 23 mice. Datasets were presented as multi-page TIFF files. Multi-page TIFF files were exported to QGIS Desktop 3.22 (http://qgis.org).

The following operations were performed with each of the multi-page TIFF files:

1. **Split multi-page TIFF files into distinct slices in TIF format**
2. **Image pre-processing and interpolation**
   (a) Create contour lines of intensity (in the form of isolines, the applied parameter is 25 pixels) for each of the distinct slices.
   (b) Apply LF Tools Extend lines plugin to a set of contours in each of the distinct slices and creating extended lines of contours at their start and/or end points, 100 μm in length.
3. **Flocks identifying**
   (a) Create intersection points of extended lines from neighboring slices.
   (b) Search for intersection points of extended lines from neighboring slices that are located at a distance in the range of 0.25–0.5 pixels (Fig. 3).
   (c) Remove all intersection points from the previous item that are present in more than one on the same extended line.
   (d) Search for remaining intersection points from three neighboring slices, which (points) are not more than 0.25 pixels away from each other.

all points are removed if there is more than one
—— point on this line at a distance less than 0.25 or
more than 0.5 pixels from another;

all points are removed if there is more than one point
‑ ‑ ‑ on this line at a distance of 0.25...0.5 pixels from
another;

● layer i-1, with saved point at a distance of 0.25...0.5 from the nearest point;

● layer i-1, with removed nearest points outside distance range 0.25...0.5

● layer i+1, with saved point at a distance of 0.25...0.5 from the nearest point;

● layer i+1, with removed nearest points outside distance range 0.25...0.5

Fig. 3. Data processing to determine whether to save or remove intersection points from the network

(e) Search for all intersection points from the previous item that are more than one on the same extended line.

This operation reveals either a long marginal chain (more than 10 pixels in length) in several neighboring slices or the movement of a large object (10x10 pixels). The spread of intensively of these identified objects occurs over areas of distinct slices.

(f) Remove all intersection points from 3(d) item that are more than one on the same extended line.

(g) Search for remaining intersection points from the previous item.

As a result, small movements (movement within the identified localization, 10x10 pixels) of small objects (3x3 pixels) are revealed.

4. **Plotting of flock trajectories**

(a) Splice of intersection points from 3(g) item (defining the localization of small movement) into a sequence corresponding to the sequence of transitions from slice to slice, if the intersection points from 3(g) item are not more than 10 pixels away from each other.

The parameters used for the calculations were established by selecting and optimizing the number of intersection points connecting lines from two different neighboring slices, taking into account the Nearest neighbor analysis. An extended line is constructed according to the distance between the cells; an extended line is also constructed with length which is conditionally inversely proportional to the distance between intersection points.

3.2 Applications for Spatial Analysis

In our work we processed optogenetic mouse brain images using Open Source Geographic Information System QGIS Desktop 3.22 QGIS applications and special plug-ins (see Table 1) were used for spatial analysis of data both within single slices and between sets of closely spaced slices.

Table 1. Spatial data processing applications.

Plugin	Description
Extracts contour lines https://docs.qgis.org/3.16/en/docs/user_m anual/processing_algs/gdal/rasterextraction. html#gdalcontour	Generate a vector contour from the input raster by joining points with the same parameters. Extracts contour lines from any GDAL-supported elevation raster
Nearest neighbour analysis https://docs.qgis.org/3.16/en/docs/user_m anual/processing_algs/qgis/vectoranalysis. html#qgisnearestneighbouranalysis	Performs nearest neighbour analysis for a point layer. The output presents how data are distributed (clustered, randomly or distributed)
LF Tools https://github.com/LEOXINGU/lftools/wiki/ LF-Tools-for-QGIS	Tools for cartographic production, surveying, digital image processing and spatial analysis (Extended lines)

4 Experiments and Results

4.1 Datasets

Recognition of multipoint activity and spatial analysis of the distribution of neuronal activities according to fluorescence microscopy datasets was performed based on data packages published in an open repository (https://ebrains.eu). As source material, we used fluorescence microscopy datasets:

Set 1 (see Table 2): We used whole-brain datasets [43] from transgenic animals with different interneuron populations (PV, SST and VIP positive cells) which are labeled with fluorescent proteins. These datasets were obtained from 11 mice (male animals, on post-natal day 56). The data was represented in 48 multi-page TIFF files. Each multi-page TIFF included 160 - 288 slices with dorsal or ventral projections of the mouse brain. The data resolution is 10.4x10.4x10 μm.

Set 2 (see Table 2): We used whole-brain datasets [44] obtained using LSM in combination with tissue clearing. These datasets were obtained from 12 mice (male animals, on post-natal day 56). The data was represented in 14 multi-page TIFF files. Each multi-page TIFF file included 800 slices with dorsal or ventral projections of the mice brain. The data resolution is 10x10x10 μm. By processing using CLARITY-TDE method [45, 46] images have been partially cleaned up.

Allen Mouse Common Coordinate Framework [47] served in our work as a frame of data reference to spatial coordinates.

4.2 Results

As a result of our work, localizations of sites (10x10 pixels) with "flocks" were identified based on the intersection points of extended lines in these localizations (Fig. 4).

Table 2. Source material

	Set 1	Set 2
number of mice	11 mice	12 mice
gender and age of animals	male animals, post-natal day 56	male animals, post-natal day 56
parvalbumin-positive interneurons parvalbumin (PV)	4 Animals	5 Animals
somatostatin-positive interneurons somatostatin (SST)	3 Animals	3 Animals
VIP-positive interneurons vasoactive-intestinal peptide (VIP)	4 Animals	4 Animals
number of multi-page TIFF files (several files per mouse)	48 multi-page TIFF files	14 multi-page TIFF files
tissue clearing method	CLARITY/TDE	CLARITY/TDE
resolution	10.4x10.4x10 μm	10x10x10 μm
number of slices in one multi-page TIFF file	about 288 slices	800 slices
size of one slice	about 1200x1500 pixels	1140x1500 pixels
bits per pixel	8BPP	8BPP

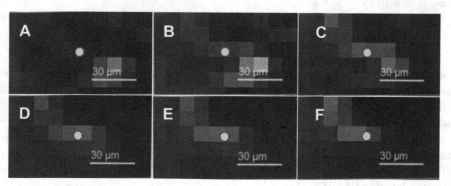

Fig. 4. Slice-by-slice activity near the identified localization (marked by yellow dots). Scale bar: 30 μm.

The values obtained as a result of our work:

– mean number of localization sites in neighboring slices, averaged over all multi-page TIFF files, is 124;

– the percentage of sites with identified "small movement" (movement within the identified localization, 10x10 pixels) relative to the total number of identified localizations, averaged over all multi-page TIFF files, is 73.4%.

The identified sites were further used as segmentation labeling for training U-Net and DeepLabV3Plus neural networks. Gray-level segmentation is performed based on local gray-level variations caused by changes in illumination intensity in a neighboring set of pixels. The training was conducted using test samples from multi-page TIFF files of set 1 and from set 2.

After training used our segmentation labeling, U-Net showed next results in Precision, Recall, and F1-score (F1-score = 2*Precision*Recall/(Precision + Recall)): 81.4%, 76.2%, and 78.7%, respectively; and DeepLabV3Plu showed next results in Precision, Recall, and F1-score: 76.4%, 79.4%, and 77.9% respectively.

5 Conclusion

In our work, we applied the flocking method to analysis of spontaneous brain activity. We selected different cell groups and determined areas occupied by ensembles of cell groups in mouse brain. When performing computational experiments, we analyzed the interslice propagation of neuronal activity for sets of mouse brain images.

In summary, the contributions of this work are as follows:

– We performed a spatial analysis of mouse brain optogenetic images using the flocking method.
– We have shown that using the flocking method, it is possible to detect more accurately both areas and tracks of neuronal activity, identifying the connectivity of extended areas of activity
– We were able to identify localizations of sites with small movements of group activity (stably localized flickering of activity with small movements).

The application of the methodology described in this work improves the processing time of the original images (by parallelizing the processing of individual tiles). The results of image processing can be used both to determine the dynamic processes related to the change in activity over time, and to build 3D activity tracks between slices of multi-page TIFF files.

In the future, the flocking method can be used not only in processing of optogenetic images but also in the analysis of other tracks, including the analysis of data obtained by diffusion-weighted magnetic resonance imaging (DW MRI) + High Angular Resolution Diffusion Imaging (HARDI).

References

1. Patriarchi, T., et al.: Ultrafast neuronal imaging of dopamine dynamics with designed genetically encoded sensors. Science **360** (2018)
2. Aranda, R., Rivera, M., Ramirez-Manzanares, A.: A flocking based method for brain tractography. Med. Image Anal. **18**, 515–530 (2014)

3. Klapoetke, N.C., et al.: Independent optical excitation of distinct neural populations. Nat. Methods **11**, 338–346 (2014)
4. Papp, E.A., Leergaard, T.B., Csucs, G., Bjaalie, J.G.: Brain-wide mapping of axonal connections: workflow for automated detection and spatial analysis of labeling in microscopic sections. Front. Neuroinform. **10**, 11 (2016)
5. Zhan, L., et al.: Comparison of nine tractography algorithms for detecting abnormal structural brain networks in Alzheimer's disease. Front. Aging Neurosci. **7**, 48 (2015)
6. Athey, T., et al.: BrainLine: An Open Pipeline for Connectivity Analysis of Heterogeneous Whole-Brain Fluorescence Volumes (2023). https://doi.org/10.1101/2023.02.28.530429
7. Côté, M.-A., et al.: Tractometer: towards validation of tractography pipelines. Med. Image Anal. **17**, 844–857 (2013)
8. Maier-Hein, L., et al.: Why rankings of biomedical image analysis competitions should be interpreted with care. Nat. Commun. **9**, 5217 (2018)
9. Benedetti, P., Femminella, M., Reali, G.: Mixed-sized biomedical image segmentation based on U-Net architectures. Appl. Sci. **13**, 329 (2022)
10. Falk, T., et al.: U-Net: deep learning for cell counting, detection, and morphometry. Nat. Methods **16**, 67–70 (2019)
11. Furtado, P.: Testing segmentation popular loss and variations in three multiclass medical imaging problems. J. Imaging **7**, 16 (2021)
12. Wang, Y., Wang, C., Wu, H., Chen, P.: An improved Deeplabv3+ semantic segmentation algorithm with multiple loss constraints. PLoS ONE **17**, e0261582 (2022)
13. Ronneberger, O.: Invited talk: U-Net convolutional networks for biomedical image segmentation. In: Maier-Hein, G., et al. (eds.) Bildverarbeitung für die Medizin 2017, p. 3. Springer, Heidelberg (2017). https://doi.org/10.1007/978-3-662-54345-0_3
14. Lee, J., et al.: A pixel-level coarse-to-fine image segmentation labelling algorithm. Sci. Rep. **12**, 8672 (2022)
15. Boyden, E.S., Zhang, F., Bamberg, E., Nagel, G., Deisseroth, K.: Millisecond-timescale, genetically targeted optical control of neural activity. Nat. Neurosci. **8**, 1263–1268 (2005)
16. Deisseroth, K.: Controlling the brain with light. Sci. Am. **303**, 48–55 (2010)
17. Boyden, E.S.: A history of optogenetics: the development of tools for controlling brain circuits with light. F1000 Biol. Rep. **3**, 11 (2011)
18. Jennings, J.H., et al.: Interacting neural ensembles in orbitofrontal cortex for social and feeding behaviour. Nature **565**, 645–649 (2019)
19. Ramirez, S., et al.: Creating a false memory in the hippocampus. Science **341**, 387–391 (2013)
20. Yang, Y., et al.: Wireless multilateral devices for optogenetic studies of individual and social behaviors. Nat. Neurosci. **24**, 1035–1045 (2021)
21. Luis-Islas, J., Luna, M., Floran, B., Gutierrez, R.: Optoception: perception of optogenetic brain perturbations. eNeuro **9** (2022)
22. Ronzitti, E., et al.: Recent advances in patterned photostimulation for optogenetics. J. Opt. **19**, 113001 (2017)
23. Picot, A., et al.: Temperature rise under two-photon optogenetic brain stimulation. Cell Rep. **24**, 1243-1253.e5 (2018)
24. Chen, I.-W., et al.: In vivo sub-millisecond two-photon optogenetics with temporally focused patterned light. J. Neurosci. **39**, 1785–1818 (2019)
25. Shemesh, O.A., et al.: Temporally precise single-cell-resolution optogenetics. Nat. Neurosci. **20**, 1796–1806 (2017)
26. Huisken, J., Stainier, D.Y.R.: Selective plane illumination microscopy techniques in developmental biology. Development **136**, 1963–1975 (2009)
27. Maddalena, L., Pozzi, P., Ceffa, N.G., van der Hoeven, B., Carroll, E.C.: Optogenetics and light-sheet microscopy. Neuromethods **191**, 231–261 (2023)

28. Müllenbroich, M.C., et al.: High-fidelity imaging in brain-wide structural studies using light-sheet microscopy. eNeuro **5** (2018)

29. Müllenbroich, M.C., et al.: Bessel beam illumination reduces random and systematic errors in quantitative functional studies using light-sheet microscopy. Front. Cell. Neurosci. **12**, 1–12 (2018)

30. Prakash, R., et al.: Two-photon optogenetic toolbox for fast inhibition, excitation and bistable modulation. Nat. Methods **9**, 1171–1179 (2012)

31. Ban, Z., Hu, J., Lennox, B., Arvin, F.: Self-organised collision-free flocking mechanism in heterogeneous robot swarms. Mob. Netw. Appl. **26**, 2461–2471 (2021)

32. Vicsek, T., Zafeiris, A.: Collective motion. Phys. Rep. **517**, 71–140 (2012)

33. Papadopoulou, M., et al.: Dynamics of collective motion across time and species. Philos. Trans. R. Soc. B Biol. Sci. **378** (2023)

34. Battersby, S.: News feature: the cells that flock together. PNAS USA **112**, 7883–7885 (2015)

35. Ascione, F., et al.: Collective rotational motion of freely expanding T84 epithelial cell colonies. J. R. Soc. Interface **20** (2023)

36. Ren, H., Walker, B.L., Cang, Z., Nie, Q.: Identifying multicellular spatiotemporal organization of cells with SpaceFlow. Nat. Commun. **13**, 4076 (2022)

37. Tang, W.-C., et al.: Optogenetic manipulation of cell migration with high spatiotemporal resolution using lattice lightsheet microscopy. Commun. Biol. **5**, 879 (2022)

38. Marre, O., et al.: Mapping a complete neural population in the retina. J. Neurosci. **32**, 14859–14873 (2012)

39. Goldin, M.A., et al.: Context-dependent selectivity to natural images in the retina. Nat. Commun. **13**, 5556 (2022)

40. Doursat, R.: Bridging the mind-brain gap by morphogenetic 'neuron flocking': the dynamic self-organization of neural activity into mental shapes. AAAI Fall Symposium. Technical report, FS-13-02, pp. 16–21 (2013)

41. Reynolds, C.W.: Flocks, herds and schools: a distributed behavioral model. ACM SIGGRAPH Comput. Graph. **21**, 25–34 (1987)

42. Bonsi, P., et al.: RGS9–2 rescues dopamine D2 receptor levels and signaling in DYT1 dystonia mouse models. EMBO Mol. Med. **11** (2019)

43. Silvestri, L., et al.: Whole brain images of selected neuronal types (2019). https://doi.org/10.25493/68S1-9R1

44. Silvestri, L., Di Giovanna, A.P., Mazzamuto, G.: Whole-brain images of different neuronal markers (2020). https://doi.org/10.25493/A0XN-XC1

45. Chung, K., et al.: Structural and molecular interrogation of intact biological systems. Nature **497**, 332–337 (2013)

46. Costantini, I., et al.: A versatile clearing agent for multi-modal brain imaging. Sci. Rep. **5**, 9808 (2015)

47. Wang, Q., et al.: The allen mouse brain common coordinate framework: a 3D reference atlas. Cell **181**, 936-953.e20 (2020)

A Machine Learning Approach for Source Code Similarity via Graph-Focused Features

Giacomo Boldini⬡, Alessio Diana⬡, Vincenzo Arceri⬡,
Vincenzo Bonnici(✉)⬡, and Roberto Bagnara⬡

Department of Mathematical, Physical and Computer Sciences, University of Parma,
Parco Area delle Scienze, 53/A, 43124 Parma, Italy
giacomo.boldini@unive.it, alessio.diana@unibo.it,
{vincenzo.arceri,vincenzo.bonnici,roberto.bagnara}@unipr.it

Abstract. Source code similarity aims at recognizing common characteristics between two different codes by means of their components. It plays a significant role in many activities regarding software development and analysis which have the potential of assisting software teams working on large codebases. Existing approaches aim at computing similarity between two codes by suitable representation of them which captures syntactic and semantic properties. However, they lack explainability and generalization for multiple languages comparison. Here, we present a preliminary result that attempts at providing a graph-focused representation of code by means of which clustering and classification of programs is possible while exposing explainability and generalizability characteristics.

Keywords: Code Similarity · Machine Learning · Graph-focused Features · Control-flow Graph

1 Introduction

Source code similarity aims at recognizing common characteristics between two different codes by means of their components. It plays a significant role in many activities regarding software development and analysis, which include plagiarism detection [27], malicious code detection and injection [3,4,6], clone recognition [14], bug identification [12], and code refactoring. In order to obtain precise and reliable code similarity tools, they cannot be based just on the code syntactic structure, but also semantics must be taken into account. For instance, a naive code similarity measure could be given by comparing the syntactic structure of the two compared program fragments (e.g., plain text, keywords, tokens, API calls). Nevertheless, such a trivial similarity technique does not sufficiently capture the semantics of the programs. In particular, two program fragments can have different syntactic structures but they implement the same functionality. On the opposite, they can have similar syntax but they differ in their behaviors.

G. Nicosia et al. (Eds.): LOD 2023, LNCS 14505, pp. 53–67, 2024.
https://doi.org/10.1007/978-3-031-53969-5_5

Thus, more sophisticated representations of the fragments composing a program, and measures based on such representations, are required.

Due to the wide range of features that can be taken into account in this context, it is possible to fine-tune the information embedded in such representations. Based on [24], code similarity is classically split into four main types of increasing complexity, such that each type subsumes the previous one: *Type I* represents a complete syntactic similarity, modulo blank characters, comments, and indentation; *Type II* represents a syntactic similarity, modulo identifiers renaming, literals, and types; *Type III* represents copied fragments with further modifications in statements; *Type IV* represents a functional similarity, i.e., semantic similarity.

At different levels, each type of technique aims at detecting if a program fragment is a *code clone* of another one. From here on, we refer to code clones as portions of code that share some level of similarity with other code segments. The type of clone is defined based on the level of similarity between the fragments.

Tools for identifying code clones are categorized based on the method used to represent source code and the technique for comparing them. The most common traditional categories include those based on text, token, tree, graph, metrics, and hybrid methods [7,24,25]. Recently, there has been a growing interest in applying Machine Learning (ML) approaches, specifically Deep Learning (DL) techniques, to identify code clones [15]. These methods are particularly useful for detecting *Type III* and *Type IV* clones, which are the most challenging to identify. They employ different neural network architectures (e.g., DNN, GNN, RvNN), source code representations (e.g., Abstract Syntax Tree, Control-flow Graph, Data Flow Graph, Program Dependence Graph, or combination of them) and embedding techniques, such as word2vec [18], code2vec [2], graph2vec [20], and RAE [13].

In this paper, we introduce novel techniques for tackling the code similarity problem. Our approach relies on both Control-flow Graph (CFG) [8] and the LLVM Intermediate Representation (LLVM-IR) concepts for representing source code. Furthermore, we employ both supervised and unsupervised Machine Learning methodologies to conduct code similarity analysis.

It is worth noting that the proposed code representation does not focus solely on *Type III* or *Type IV* similarities. Instead, it captures both syntactic and semantic features of the code, allowing for the detection of various types of code clones, exhibiting similarities in terms of design and functionality. By combining these different approaches, the proposed methods could identify a broader range of code clones than traditional techniques that focus solely on either syntax or semantics. In particular, CFG encodes the semantics flow of a program of interest, i.e., how control flows through each code element. Hence, it does not *completely* capture the program semantics (e.g., it does not capture how values flow within the program). Nevertheless, such a representation does not completely fit for *Type IV* similarity, since complete semantic equivalence is required by its definition. Still, it subsumes *Type III* similarity, since it CFG fully encodes the syntax, besides describing the control flow. In addition, for each source language, catching *Type III* similarity is affected by how the language-specific constructs are mapped into LLVM-IR instructions.

Our proposed approach focuses on pursuing an explainable approach to code similarity analysis. Specifically, features used for code similarity can be remapped to fragments of original source code. This provides a more transparent and interpretable analysis: for example, it could be used during the study of the feature importance of the models to refer directly to portions of source code. Furthermore, this can enable a deeper understanding of the similarities and differences between code fragments.

The rest of the manuscript is organized as follows. In Sect. 2 we present the three main phases of the proposed approach to extract graph-based features from C/C++ source code. In addition, both supervised and unsupervised Machine Learning methodologies used are explained here. In Sect. 3 we show some of the obtained results using both clustering and classification methods. In Sect. 4 we discuss all the pros and cons of using this methodology. Section 5 concludes the paper.

2   Methods

In this section, we describe the methods adopted in our solution for facing the code similarity problem. This study focuses on the code similarity of C/C++ translation units.[1] Nevertheless, as we will discuss in Sect. 4, the proposed solution can be also adapted to other programming languages. Before going into details of our contribution, in the following, we describe the toolchain of our solution, from a high-level point of view.

The overall architecture is depicted in Fig. 1. The process consists of three main steps. The first step corresponds to a preprocessing phase that prepares the input source code for further analysis. In particular, this phase translates the program to a lower-level language, namely LLVM-IR [26], an intermediate representation used by the LLVM compiler to represent the source code during all the compilation phases. LLVM-IR is usually employed by compilers to describe and store all the information retrieved from the source code in order to perform a more precise translation into the target language.

The output of this phase is a set of CFGs derived from the LLVM-IR representation of the program of interest, each corresponding to a function/method of the input program. They are the inputs of the second step, which manipulates and enriches the basic CFG structure in order to obtain the so-called Augmented Control-flow Graphs (A-CFGs). This phase also manages the construction of the

---

[1] In C/C++, executable programs are obtained by linking together the code coming from a complete set of *translation units*. A *translation unit* is the portion of a program a compiler operates upon, and is constituted by a main file (typically with a .c/.C/.cxx/.cpp extension) along with all header files (typically with a .h/.H/.hpp extension) that the main file includes, directly or indirectly. A prerequisite of our approach is that the translation units to be analyzed are complete, so that the CLANG compiler can process them without errors.

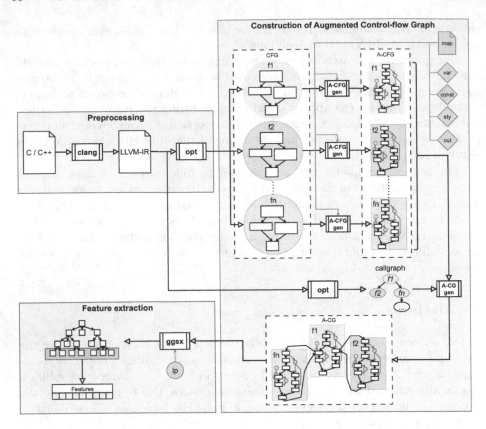

Fig. 1. Features extraction method split in its three main phases: preprocessing, construction of the A-CFGs/A-CG and feature extraction.

Augmented Call Graph (A-CG), namely a single graph made of A-CFGs, where also calling relationships are tracked. From this graph structure, in the last phase, we rely on GRAPHGREPSX [5] for the graph features extraction. Source code is represented by the sets of features previously extracted, used by clustering and classification methods to reveal similarities between programs.

2.1   Preprocessing

The preprocessing stage involves two steps, both implemented by tools coming from the LLVM toolchain.

The first step involves the compilation of a C/C++ translation unit to its corresponding LLVM-IR representation. In order to do this, we rely on the CLANG compiler. During the compilation process, CLANG is instructed to compile to LLVM-IR language, instead of the default object code. Moreover, plain names (the ones appearing in the original source code) are preserved during this compilation phase. Additionally, CLANG is run with all optimizations switched off, so as to ensure that the output code preserves the structure and functionalities of the source code. These steps are crucial for enhancing the CFG and extracting features in the subsequent phases, making it easier to analyze and extract information for our purposes.

From the LLVM-IR code, the second phase manages the construction of CFGs. A CFGs, in LLVM flavour, is a directed graph $G = (N, E)$ where nodes N are basic blocks, and edges $E \subseteq N \times N$ connecting basic blocks correspond to jumps in the control flow between the two linked basic blocks. Each basic block starts with a label (giving the basic block a symbol table entry), contains a list of instructions to be executed in sequence, and ends with a terminator instruction (such as a branch or function return) [26]. In order to generate CFGs, we rely on opt, which is a LLVM tool useful to perform optimization and analysis of LLVM-IR code. Finally, we rely on opt also to create the call graph [8] (CG), a graph that tracks calling relationships between functions in the program.

2.2 Construction of the Augmented Control-Flow Graphs

In this phase, a new graph data structure, called Augmented Control-flow Graph (A-CFG) is built to enhance the information of a traditional CFG.

A A-CFG resembles a single-instruction CFG, where each basic block of the A-CFG corresponds to a single instruction. In order to obtain this, each CFG is analyzed individually. Each basic block of the original CFG is exploded, in the corresponding A-CFG, into a sequence of single LLVM-IR instructions, connected by sequential edges and preserving the original control flow. We call these nodes *instruction nodes*, and we call the edges connecting instructions nodes *flow edges*.

After this initial manipulation, further information is added to a A-CFG. In particular,[2] for each instruction node, variable updates (i.e., assignments) and reads, and used constants are retrieved, and for each variable and constant, a new node is added to the A-CFG. We refer to these nodes as *variable nodes*, which are also labeled with the variable type, and *constant nodes*, labeled with the constant value, respectively. Then, for each instruction node updating a variable x, an edge from the instruction node to the corresponding variable node x is created. Similarly, if the instruction node reads a variable x, an edge from the variable node x to the instruction node is added. The direction of the edge reflects whether the variable is written or read. Similarly, if an instruction node reads a constant, an edge connecting it to the corresponding constant node is added. We refer to these three types of edges as *data edges*.

Instruction nodes labels are assigned through a surjective map $\mathsf{map} : O \to L$, where O denotes the set of LLVM-IR op-codes,[3] and L is an arbitrary set of labels. The *standard map* maps each specific LLVM-IR op-code (e.g., add, store)

---

[2] This phase is managed by using a custom LLVM-IR parser. The parser is generated using ANTLR [21] starting from the LLVM-IR 7.0.0 grammar.

[3] In LLVM-IR, op-code refers to the instruction code (or name) that specifies the operation to be performed by the instruction.

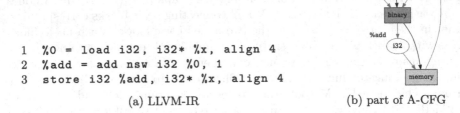

```
1   %0 = load i32, i32* %x, align 4
2   %add = add nsw i32 %0, 1
3   store i32 %add, i32* %x, align 4
```

(a) LLVM-IR   (b) part of A-CFG

Fig. 2. LLVM-IR code and portion of A-CFG for C++ instruction x = x + 1;.

to one of the nine categories defined in the LLVM-IR language manual [26]: *terminator, binary, bitwise, vector, aggregate, memory, conversion, other* and *intrinsic*. For instance, the add op-code at line 2 in Fig. 2a, corresponds to the green node in the A-CFG reported in Fig. 2b, where the standard map has mapped the op-code add to the label *binary*. Instead, the load and store op-codes at lines 1 and 3, respectively, are mapped in the orange nodes, where the standard map has mapped them to memory.

However, in general, a custom map function can be built acting on the set of labels L. Restricting or expanding L, starting from the set of the LLVMIR instruction categories discussed above, one can fine-tune the level of detail regarding the information related to instruction nodes. This allows to meet the specific needs of the feature generation process, where nodes with the same label are indistinguishable.

Figure 2 shows how a part of a single basic block, reported in Fig. 2a, is transformed into its corresponding A-CFG (Fig. 2b) using the standard map discussed above. In particular, black edges represent flow edges, blue edges represent data edges concerning variables, and red edges represent data edges concerning constants.

It is also possible to specify whether to generate variable and constant elements, use simplified types, and remove unnecessary terminator nodes, setting the Boolean parameters var, const, sty, and cut, respectively. Such behavior is provided by specific parameters that can be activated/deactivated by the user.

The A-CFG structure has the granularity of a single function in a program, without tracking calling relationships between different A-CFGs. In order to represent an entire program, starting from the so-built A-CFGs, the idea is to create an Augmented Call Graph (A-CG), namely a new graph consisting of the union of the A-CFGs, linked together by following the function calls within it.

Specifically, for each instruction node n calling a function f, the following edges are added: one edge from n to the entry point instruction node of the

A-CFG of the function f (it is unique) and one for each exit instruction node (i.e., an instruction node terminating f) to n. This process is carried out by searching for calling instruction nodes and analyzing the original CG of the program, generated in the first phase.

2.3 Feature Extraction

The third and last phase consists of extracting all the features that will characterize the analyzed file during the execution of the Machine Learning models. The input of this phase is an augmented graph, both A-CFG and A-CG, created in the previous phase.

For a given graph G, the proposed method uses paths of G of bounded length as features. The indexing phase of GRAPHGREPSX [5] is used to build an index in a prefix-tree format for a graph database. The index is constructed by performing a depth-first search for each node n_j of the graph. During this phase, all paths of length lp or less are extracted, and each path is represented by the labels of its nodes. More formally, each path $(v_1, v_2, \ldots, v_{lp})$ is mapped into a sequence of labels $(l_1, l_2, \ldots, l_{lp})$. All the subpaths $(v_i, \ldots, v_j)$ for $1 \leq i \leq j \leq lp$ of a path $(v_1, v_2, \ldots, v_{lp})$ will be included in the global index, too. The number of times a path appears in the graph is also recorded.

In our proposal, a prefix-tree is built for a A-CG G. Then, only the paths of maximal length (i.e., equal to lp) are considered as features. Each feature is identified by a path p and its value corresponds to the number of occurrences of p in G. As a set of features describing the graph, we extract all the paths leading to the leaves at depth lp of the prefix tree.

2.4 Machine Learning

Let $\mathbb{F}^n$ be a set of features, with $\mathbb{F}^i$ being the i-th feature. Let $\mathbb{O} = \{o_1, o_2, \ldots, o_m\}$ be a set of vectors (or objects) such that $o_i \in \mathbb{F}^n$.

Let $d(o_i, o_j)$ be a real number representing the distance measure between objects o_i and o_j. Usually, distance function d is symmetric, $d(o_i, o_j) = d(o_j, o_i)$, positive separable, $d(o_i, o_j) = 0 \Leftrightarrow o_i = o_j$, and provides triangular inequality, $d(o_i, o_j) \leq d(o_i, o_k) + d(o_k, o_j)$. A clustering method is an unsupervised Machine Learning model which uses the distance function d between objects to organize data into groups, such that there is high similarity (low distance) within members in each group and low similarity across the groups. Each group of data represents a cluster. There are several approaches to clustering, one of which is agglomerative hierarchical clustering. This method works by iteratively merging the closest pair of data points or clusters until all data points belong to a single cluster. The distance between two clusters is computed using the distance function d between data points and the agglomeration method. The resulting cluster hierarchy, or dendrogram, can be cut at different levels to obtain different sets of clusters.

Let $\mathbb{C} = \{c_1, c_2, \ldots, c_t\}$ be a set of classes such that an surjective function $c : \mathbb{O} \mapsto \mathbb{C}$ is defined. Classification is a supervised Machine Learning model used to assign a class to a new object $o \notin \mathbb{O}$ by taking into example $\mathbb{O}$ and the relation between the objects in $\mathbb{O}$ and their assigned classes, the method used to do this is defined by the model.

Let split $\mathbb{O}$ into two subsets $\mathbb{O}^T$ and $\mathbb{O}^V$ such that $\mathbb{O}^T \bigcap \mathbb{O}^V = \emptyset$, which are respectively called the training and the verification set. The training set is used to train a model, while the verification set is used for assessing the performance of the model in correctly assigning the classes to the objects in $\mathbb{O}^V$ without knowing their original/real class. It is essential that the balance between the classes of the original data set must be preserved in the training and verification sets.

Given a Machine Learning model trained on $\mathbb{O}^T$ for all the classes $\mathbb{C}$ and given an object $o \in \mathbb{O}^V$, let $\tilde{c}(o)$ be the class that the model assigns to o and $c(o)$ be the actual class for the object o. We distinguish the object in $\mathbb{O}^V$ into four sets, depending on the accordance between their real class i and the class that is assigned by the model, that are: true positives (TP_i), true negative (TN_i), false positives (FP_i) and false negatives (FN_i). They are defined as: $TP_i = ||\{o \in \mathbb{O}^V \mid c(o) = \tilde{c}(o) = i\}||, TN_i = ||\{o \in \mathbb{O}^V \mid c(o) \neq i \wedge \tilde{c}(o) \neq i\}||, FP_i = ||\{o \in \mathbb{O}^V \mid c(o) \neq i \wedge \tilde{c}(o) = i\}||, FN_i = ||\{o \in \mathbb{O}^V \mid c(o) = i \wedge \tilde{c}(o) \neq i\}||$. The accuracy of the model is defined as $(\sum_{i=1}^{t} ||TP_i||)/(|\mathbb{O}^V|)$. Precision and recall are defined for each class i as $|TP_i|/(|TP_i| + |FP_i|)$ and $|TP_i|/(|TP_i| + |FN_i|)$, respectively. The F1-score for each class i is defined as $(2|TP_i|)/(2|TP_i| + |FP_i| + |FN_i|)$. Because the problem is defined for multiple classes, we can compute a single value for these metrics (precision, recall, F1-score) by averaging over such classes. In particular, let $\mathbb{O}_i = \{o \in \mathbb{O}^V | c(o) = c_i\}$ a subset of verification set and M_i a performance metric (one of precision, recall, F1-score) computed on class c_i. The weighted average version of M uses class cardinality as weights and it is defined as $M = \frac{1}{|\mathbb{O}^V|} \sum_{c_i \in \mathbb{C}} |\mathbb{O}_i| M_i$.

A common approach for exploiting clustering results in a classification task is to assign a class label to each cluster based on the majority class of its constituent data points. Specifically, for each cluster, the class that has the highest number of objects within the cluster is selected as the assigned class label. If ground-truth labels are available, external evaluation measures can be used to evaluate clustering. One commonly used metric is the Rand Index (RI) [11]. It measures the agreement between two clusterings by comparing every pair of data points and counting the number of pairs that are grouped together or separately in both the predicted and ground-truth clusterings. It is defined as the ratio of the sum of agreements (pairs assigned to the same cluster in both clusterings) and disagreements (pairs assigned to different clusters in both clusterings) over the total number of pairs.

This work employs the SciPy library [28] for implementing clustering models, while the Scikit-learn [22] library is used for developing classification models and computing evaluation metrics.

3   Results

We experimentally evaluate our proposal by using it as a basis for clustering and classification models.

We consider as data set a subset of C++ programs, included in the CODENET project [23]. The CODENET project provides a large collection of source files in several programming languages, such as C, C++, Java and Python, and for each of them there are extensive metadata and tools for accessing the dataset to select custom information. The samples come from online systems that allow users to submit a solution to a variety of programming problems, in the form of competitions or courses, ranging from elementary exercises to problems requiring the use of advanced algorithms. Each consists of a single file in which the test cases and printouts of the required results are included.

CODENET includes several benchmark datasets, created specifically to train models and conduct code classification and similarity experiments. These datasets were obtained by filtering the original dataset in order to remove identical problems and nearly duplicate code samples, with the goal of obtaining better training and more accurate metrics [1]. In this work, we used a subset of C++1000 data set (available at https://developer.ibm.com/exchanges/data/all/project-codenet/) consisting of 1000 programming problems, each of them containing 500 C++ programs (submissions).

With respect to the existing taxonomy, the proposed approach can be considered to be between type *III* and type *IV*. Thus, a direct comparison with state-of-the-art approaches is not suitable.

3.1   Clustering

In these experiments, 100 random submissions were considered for each of 10 randomly chosen problems of CODENET dataset. For each (submission) program, the corresponding feature vector is generated using the following generation method's parameters: lp = 3, const = True, var = True, sty = True, cut = True. Also, four atomic categories for store, load, phi, and call instructions are added to the standard map function. Each clustering experiment is designed as follows: N random problems (from 10 available) are selected, whose corresponding submission's feature vectors are taken and merged into a single dataset; outliers are first removed from it using Isolation Forest (IF) [17] and then features are selected by using the Extra Tree Classifier (ETC) [9] supervised method (the ground-truth label are the problem to which each submission belongs); the distance matrix between objects is computed using the Boolean Jaccard index and then a hierarchical clustering algorithm from the SCIPY library tries to find the clusters. For each experiment, four agglomeration methods were tried: *single*, *complete*, *average* and *ward* [19]. 10 experiments were conducted for each combination of N and for each agglomeration method.

Table 1 reports mean and standard deviation of Rand Index (RI) for all the clustering experiments, grouped by the number of considered problems N and by the agglomeration method used in the clustering algorithm.

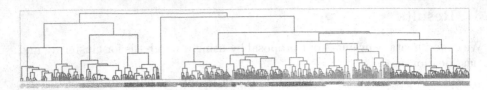

Fig. 3. Dendrogram of an experiment which considers $N = 6$ problems (about 600 submissions) and uses *ward* agglomeration method.

First of all, it shows that there is not actually a link between the variation in the problems considered (N) and the RI metric: in some agglomeration methods, there is a direct growth of both of them; in other methods, there is an inverse growth of them. Second, *ward* agglomeration generally performs better than all the other methods, despite N.

In addition, a dendrogram obtained from an experiment involving $N = 6$ problems (thus about 600 submissions) and using a *ward* agglomeration method is shown in Fig. 3: each leaf in the tree represents a program; the color of the leaf represents the problem to which it belongs, while the color of the subtrees represent the clusters found by performing a cut to obtain at most N clusters.

It can be seen that the partitioning partially respects *Type IV* code similarity: some clusters are pure with regard to the problem they belong to, while others contain programs belonging to different problems. However, in the latter case, almost-pure subtrees are visible, suggesting that programs belonging to the same problem were joined in the same cluster in the early stages of the algorithm.

3.2 Classification

For the classification experiments, the A-CFG generation parameters for executing one experiment are fixed, which are lp $= 3$, const $=$ True, var $=$ False, sty

Table 1. Mean and standard deviation of RI measure regarding the clustering experiments. Results are grouped by the number of different problems solved by the input programs taken into account, N, and by agglomeration method.

N	Single	Complete	Average	Ward
2	0.50 ± 0.00	0.50 ± 0.00	0.50 ± 0.00	0.89 ± 0.08
3	0.39 ± 0.02	0.39 ± 0.07	0.43 ± 0.12	0.68 ± 0.07
4	0.32 ± 0.02	0.53 ± 0.16	0.37 ± 0.09	0.71 ± 0.04
5	0.29 ± 0.03	0.63 ± 0.08	0.48 ± 0.14	0.76 ± 0.04
6	0.25 ± 0.02	0.66 ± 0.08	0.48 ± 0.11	0.81 ± 0.02
7	0.24 ± 0.02	0.73 ± 0.05	0.65 ± 0.07	0.81 ± 0.03
8	0.23 ± 0.01	0.79 ± 0.02	0.67 ± 0.07	0.83 ± 0.01
9	0.22 ± 0.02	0.81 ± 0.02	0.72 ± 0.05	0.85 ± 0.02
10	0.21 ± 0.01	0.83 ± 0.01	0.71 ± 0.04	0.87 ± 0.01

Table 2. Results of the classification experiments by varying the type of employed classifier.

	Accuracy	F1-score	Precision	Recall
KN	0.76 ± 0.03	0.76 ± 0.03	0.79 ± 0.03	0.76 ± 0.03
SVC (C = 1)	0.80 ± 0.06	0.81 ± 0.06	0.83 ± 0.04	0.80 ± 0.06
NuSVC	0.77 ± 0.05	0.78 ± 0.05	0.81 ± 0.03	0.77 ± 0.05
DT	0.74 ± 0.05	0.74 ± 0.05	0.75 ± 0.05	0.74 ± 0.05
RF	0.86 ± 0.03	0.86 ± 0.03	0.87 ± 0.03	0.86 ± 0.03
MLP	0.84 ± 0.01	0.84 ± 0.01	0.85 ± 0.01	0.84 ± 0.01
GNB	0.59 ± 0.07	0.58 ± 0.07	0.69 ± 0.03	0.59 ± 0.07

= False, cut = True and standard map function. As done in clustering experiments, here 10 random problems are used from the CODENET dataset which are the classes during the classification. Every problem has around 100 elements. Table 2 shows mean and standard deviation of the accuracy, F1-score, precision and recall for all the classification methods tried: K-Neighbors (KN), Support Vector Classifier (SVC), Non linear Support Vector Classifier (NuSVC), Decision Tree (DT), Random Forest (RF), MultiLayer Perceptron (MLP), Gaussian Naive Bayes (GNB). All the metrics are computed by averaging the results of a 5-fold cross-validation process. F1-score, precision, and recall are computed as an average between the results of the single classes, weighted using the class cardinality. Almost all the methods obtain a relatively high score, but the best ones are Random Forest and MLP, with a value of accuracy higher than 0.84.

A new test was carried out by doing thousands of experiments with varying generation parameters. Figure 4 shows all methods' mean accuracy trend of all the experiments in regard to the values of the generation parameter analyzed (var, const, cut, sty), one in every plot. In particular, the growth of the value of lp is represented on the horizontal axis and the mean accuracy is represented on the vertical axis. The variables const, sty, and cut have almost the same mean accuracy for both the true and the false values of the parameter. On the other hand, the input parameter var presents a significant change in the mean accuracy. Not taking into account the elements related to variables generally leads to better results than using them. This can be explained by the fact that the classification methods focus on more important characteristics since the variables are omitted in the graph generation.

The time needed for the classification in relation to the value of the generation parameter lp is shown in Fig. 5. Due to the 5-fold cross-validation process, each time corresponds to the sum of five training and five validation processes. Every line in the figure corresponds to the mean for every method for every parameter combination. The lighter color stripe represents 95% of the values around the mean value. The figure shows how MLP is the method with the highest time, below it, there is RF, and below them, we can find all the other methods. However, it can be seen that, for all the methods analyzed, the time increases when the value of lp increases.

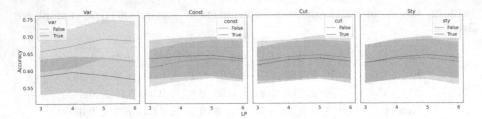

Fig. 4. Variation of the mean accuracy over different values of lp in respect to the change of the parameter values.

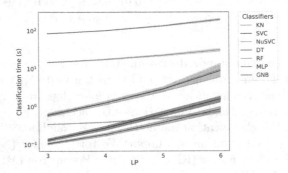

Fig. 5. Trend of log classification times for every classification method.

4   Discussion

The LLVM-IR code represents the actual starting point for the proposed features extraction method. For this reason, two considerations can be done. First, this methodology is easily extended to all those programs written in a language L for which an $L \rightarrow LLVM - IR$ compiler exists. This allows, secondly, a potential comparison between source codes written in different languages. To address this scenario, a more in-depth study on how language constructs are mapped to LLVM-IR is necessary.

The feature extraction phase proposed replaces classic embedding techniques used to describe the code graph representation in a lot of ML approaches seen before. In this case, a variable-length vector is created. It aims to lose as little information on the graph as possible. This set of features depends both on all the graph generation parameters and the path length lp used in feature extraction. In particular, the path length allows to describe the graph in different levels of detail, being more and more precise when the length of extracted path lp increases. Moreover, the map function allows the abstraction level of feature to be raised or lowered even further. This is because the labeling given by the function is reflected in the graph' instruction nodes, and thus in the extracted features. In this study, we analyzed small source code, however, it could be useful to use higher values of lp for analyzing bigger programs. To enable this level of detail, extended versions of GRAPHGREPSX [10,16] will be needed for overcoming current running time and space limitations.

Differently from what happens in modern ML approaches to code similarity, our approach makes a step toward explainability. This is because the features created, during the extraction phase, can be remapped to a set of lines in the code fragment with a certain degree of approximation. By definition, each feature represents a number of identical paths within A-CFG/A-CG graphs, which mainly contain a succession of LLVM-IR instructions. For this reason, using compiler debugging information, it is possible to trace the set of lines of code that generated these successions of LLVM-IR instructions. This explainability feature gives the possibility to visualize which parts of the analyzed code fragments are more responsible for a high value of similarity between them.

5 Conclusions

Software complexity keeps growing, and the functions assigned to software are increasingly critical, in terms of safety and/or security. In addition, there is a general shortage of programmers, which are also characterized by a high turnover rate. While keeping track of all code in typical software projects is difficult in general, it is particularly difficult for developers joining the team at a later stage, and there are not enough senior developers to effectively mentor junior developers. As a result software projects frequently run late and/or enter production without a sufficient level of quality and maturity.

The techniques proposed in this paper for the identification of code similarities have the potential of assisting software teams working on larger codebases in a number of ways: identifying unwanted clones or code illegally copied in violation of open-source licenses; identifying vulnerabilities via similarity with code available in public vulnerability databases; assisting developers in performing tedious tasks (applying a learned recipe multiple times in multiple similar contexts); increasing developers' productivity by identifying regions of code that are amenable to the same treatment (reduction of context switches); capturing to some extent the knowledge of expert developers and project veterans and make it available to newcomers.

The proposed approach is based on a graph representation of the source code on top of which graph-focuses features are extracted for indexing it. In particular, atomic LLVM-IR instructions are ensembled in order to represent the control flow of the indexed program plus other suitable information aimed at better capturing the semantics of the code. Inspired by previous approaches in graph indexing techniques, our methodology extracts paths of fixed length from the formed graph and uses them as features of the indexed program. Because such paths correspond to specific portions of the input code, such features are used to characterize the indexed program as a whole, but they are also capable of identifying small portions of it. This aspect brings explainability to the overall machine learning approach that is applied for clustering programs and for classifying them by computing code similarity via such paths. An evaluation of the proposed approach to an already existing data set of programs developed for solving different problems shows that it is possible to compute clusters of the programs that reflect the actual grouping of them with a feasible approximation. The evaluation also included supervised machine learning for classifying the programs according to the problem they are aimed at solving, showing promising results obtained by statistical evaluation of the performance. The proposed approach is not directly comparable to existing methodologies because it focuses on finding a type of source code similarity that is not currently recognized by any of the existing tools. However, the main advantages of it are the abovementioned explainability and the fact that it works on LLVM-IR instructions. This restricts the similarity calculation to only those source codes that are free of compilation errors. but at the same time, it allows the approach to be potentially applied to any programming language that can be compiled to LLVM-IR. Thus, as a future work, we plan to evaluate its performance in clustering and classifying other programming languages, rather than C/C++, and in exploring a methodology for comparing programs developed in two different languages.

Acknowledgement. This project has been partially founded by the University of Parma (Italy), project number MUR_DM737_B_MAFI_BONNICI. V. Bonnici is partially supported by INdAM-GNCS, project number CUP_E55F22000270001, and by the CINI InfoLife laboratory.

References

1. Allamanis, M.: The adverse effects of code duplication in machine learning models of code. In: Proceedings of the 2019 ACM SIGPLAN International Symposium on New Ideas, New Paradigms, and Reflections on Programming and Software, pp. 143–153 (2019). https://doi.org/10.1145/3359591.3359735
2. Alon, U., et al.: Code2vec: learning distributed representations of code. Proc. ACM Program. Lang. **3**(POPL) (2019). https://doi.org/10.1145/3290353
3. Arceri, V., Mastroeni, I.: Analyzing dynamic code: a sound abstract interpreter for Evil eval. ACM Trans. Priv. Secur. **24**(2), 10:1–10:38 (2021). https://doi.org/10.1145/3426470
4. Arceri, V., Olliaro, M., Cortesi, A., Mastroeni, I.: Completeness of abstract domains for string analysis of javascript programs. In: Hierons, R.M., Mosbah, M. (eds.) ICTAC 2019. LNCS, vol. 11884, pp. 255–272. Springer, Cham (2019). https://doi.org/10.1007/978-3-030-32505-3_15
5. Bonnici, V., et al.: Enhancing graph database indexing by suffix tree structure. In: Dijkstra, T.M.H., Tsivtsivadze, E., Marchiori, E., Heskes, T. (eds.) PRIB 2010. LNCS, pp. 195–203. Springer, Heidelberg (2010). https://doi.org/10.1007/978-3-642-16001-1_17
6. Dalla Preda, M., et al.: Abstract symbolic automata: Mixed syntactic/semantic similarity analysis of executables. In: Proceedings of the 42nd Annual ACM SIGPLAN-SIGACT Symposium on Principles of Programming Languages, pp. 329–341 (2015). https://doi.org/10.1145/2676726.2676986
7. Dhavleesh, R., et al.: Software clone detection: a systematic review. Inf. Softw. Technol. **55**(7), 1165–1199 (2013). https://doi.org/10.1016/j.infsof.2013.01.008

8. Flemming, N., et al.: Principles of Program Analysis. Springer, Heidelberg (2015). https://doi.org/10.1007/978-3-662-03811-6
9. Geurts, P., et al.: Extremely randomized trees. Mach. Learn. **63**(1), 3–42 (2006). https://doi.org/10.1007/s10994-006-6226-1
10. Giugno, R., et al.: Grapes: a software for parallel searching on biological graphs targeting multi-core architectures. PLoS ONE **8**(10), e76911 (2013)
11. Hubert, L.J., Arabie, P.: Comparing partitions. J. Classif. **2**, 193–218 (1985)
12. Jannik, P., et al.: Leveraging semantic signatures for bug search in binary programs. In: Proceedings of the 30th Annual Computer Security Applications Conference, pp. 406–415 (2014). https://doi.org/10.1145/2664243.2664269
13. Jie, Z., et al.: Fast code clone detection based on weighted recursive autoencoders. IEEE Access **7**, 125062–125078 (2019). https://doi.org/10.1109/ACCESS.2019.2938825
14. Krinke, J., Ragkhitwetsagul, C.: Code similarity in clone detection. In: Inoue, K., Roy, C.K. (eds.) Code Clone Analysis, pp. 135–150. Springer, Singapore (2021). https://doi.org/10.1007/978-981-16-1927-4_10
15. Lei, M., et al.: Deep learning application on code clone detection: a review of current knowledge. J. Syst. Softw. **184**, 111141 (2022). https://doi.org/10.1016/j.jss.2021.111141
16. Licheri, N., et al.: GRAPES-DD: exploiting decision diagrams for index-driven search in biological graph databases. BMC Bioinform. **22**, 1–24 (2021)
17. Liu, F.T., et al.: Isolation forest. In: 2008 Eighth IEEE International Conference on Data Mining, pp. 413–422 (2008). https://doi.org/10.1109/ICDM.2008.17
18. Mikolov, T., et al.: Efficient estimation of word representations in vector space. In: 1st International Conference on Learning Representations, ICLR 2013, Scottsdale, Arizona, USA, 2–4 May 2013, Workshop Track Proceedings (2013)
19. Müllner, D.: Modern hierarchical, agglomerative clustering algorithms (2011)
20. Narayanan, A., et al.: graph2vec: learning distributed representations of graphs. CoRR abs/1707.05005 (2017)
21. Parr, T.J., Quong, R.W.: ANTLR: a predicated-LL(k) parser generator. Softw. Pract. Exp. **25**(7), 789–810 (1995). https://doi.org/10.1002/spe.4380250705
22. Pedregosa, F., et al.: Scikit-learn: machine learning in Python. J. Mach. Learn. Res. **12**, 2825–2830 (2011)
23. Puri, R., et al.: Project codenet: a large-scale AI for code dataset for learning a diversity of coding tasks. arXiv preprint arXiv:2105.12655 1035 (2021)
24. Roy, C.K., Cordy, J.R.: A survey on software clone detection research. Queen's Sch. Comput. TR **541**(115), 64–68 (2007)
25. Saini, N., et al.: Code clones: detection and management. Procedia Comput. Sci. **132**, 718–727 (2018). https://doi.org/10.1016/j.procs.2018.05.080. International Conference on Computational Intelligence and Data Science
26. The LLVM Development Team: LLVM Language Reference Manual (Version 7.0.0) (2018)
27. Đurić, Z., Gašević, D.: A source code similarity system for plagiarism detection. Comput. J. **56**(1), 70–86 (2013). https://doi.org/10.1093/comjnl/bxs018
28. Virtanen, P., et al.: SciPy 1.0: fundamental algorithms for scientific computing in python. Nat. Methods **17**, 261–272 (2020). https://doi.org/10.1038/s41592-019-0686-2

Knowledge Distillation with Segment Anything (SAM) Model for Planetary Geological Mapping

Sahib Julka(✉) and Michael Granitzer

Chair of Data Science, University of Passau, 94036 Passau, Germany
{sahib.julka,michael.granitzer}@uni-passau.de

Abstract. Planetary science research involves analysing vast amounts of remote sensing data, which are often costly and time-consuming to annotate and process. One of the essential tasks in this field is geological mapping, which requires identifying and outlining regions of interest in planetary images, including geological features and landforms. However, manually labelling these images is a complex and challenging task that requires significant domain expertise and effort. To expedite this endeavour, we propose the use of knowledge distillation using the recently introduced cutting-edge Segment Anything (SAM) model. We demonstrate the effectiveness of this prompt-based foundation model for rapid annotation and quick adaptability to a prime use case of mapping planetary skylights. Our work reveals that with a small set of annotations obtained with the right prompts from the model and subsequently training a specialised domain decoder, we can achieve performance comparable to state of the art on this task. Key results indicate that the use of knowledge distillation can significantly reduce the effort required by domain experts for manual annotation and improve the efficiency of image segmentation tasks. This approach has the potential to accelerate extra-terrestrial discovery by automatically detecting and segmenting Martian landforms.

Keywords: Segment Anything Model (SAM) · Semantic Segmentation · Knowledge Distillation · Geological Mapping

1 Introduction

We have recently witnessed a paradigm shift in AI with the advent of *foundation models* utilising astronomical amounts of data. The fields of natural language processing and multi-modal learning have been revolutionised with the emergence of ChatGPT and the like [19,22]. The very first foundation models such as CLIP [23], ALIGN [13], and DALLE [24], have focused on pre-training approaches but are not suited to image segmentation. However, recently, Segment Anything (SAM) [18] was released, which is a large vision transformer ViT-based [6] model trained on the large visual corpus (SA-1B) containing more than 11 million images and one billion masks. SAM is designed to generate a valid segmentation result for any prompt. However, SAM is trained on general world case scenarios with popular structures. Recent studies have revealed that SAM can fail on typical medical image segmentation tasks [5,9] and other challenging scenarios [4,11,12,25]. Since SAM's training set mainly contains natural image datasets, it may not be directly transferable to niche tasks on data such as magnetic resonance (MRI), or HiRISE imaging[1], amongst other specialised data formats. Nonetheless, SAM is still a powerful tool that has a powerful image encoder and its prompt functionality can significantly boost the efficiency of manual annotation. In the planetary science domain, where vast amounts of remote sensing data are gathered, annotation is an intensive task. An approach that reduces the effort on the domain experts' end is highly desired. In these scenarios, active learning [15,17] and knowledge distillation [7] via training a specialised model with relatively fewer samples can be highly valuable.

1.1 Segment Anything Model

SAM utilises a vision transformer-based [10] approach to extract image features and prompt encoders to incorporate user interactions for segmentation tasks. The extracted image features and prompt embeddings are then processed by a mask decoder to generate segmentation results and confidence scores. There are four[2] types of prompts supported by SAM, namely *point*, *text*, *box*, and *mask* prompts.

For the points prompt, SAM encodes each point with Fourier positional encoding and two learnable tokens that specify foreground and background. The bounding box prompt is encoded by using the point encoding of its top-left and bottom-right corners. SAM employs a pre-trained text encoder in CLIP for encoding the free-form text prompt. The mask prompt has the same spatial resolution as the input image and is encoded by convolution feature maps.

Finally, SAM's mask decoder consists of two transformer layers with a dynamic mask prediction head and an Intersection-over-Union (IoU) score regression head. The mask prediction head generates three downscaled masks, corresponding to the whole object, part, and subpart of the object. SAM supports three main segmentation modes: fully automatic, bounding box, and point mode.

1.2 Landform Detection on Mars Using HiRISE Images

Mapping planetary landforms plays a crucial role in various tasks such as surveying, environmental monitoring, resource management, and planning. On Earth, for example, the presence of water triggers several geological and geomorphological processes [1]. Conversely, on Mars, researchers have found correlations between the presence of certain landforms such as pits, sinkholes, and landslides and the possible presence of water [2,3]. However, identifying, classifying,

---

[1] "High-Resolution Imaging Science Experiment" is camera aboard the Mars Reconnaissance Orbiter (MRO) spacecraft, which is designed to capture high-resolution images of the Martian surface and provide detailed information about the planet's geology and atmosphere.

[2] Text prompt is currently not released.

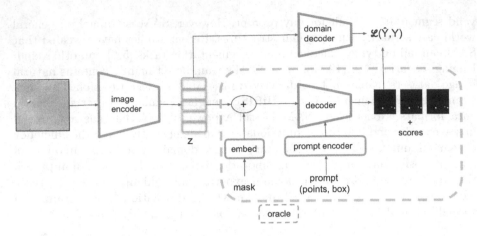

Fig. 1. Overview of our deployed approach by extending SAM. It consists of SAMs image encoder that learns an embedding of the image, and a specialised decoding unit to learn the domain-specific semantics. SAMs prompt encoder and mask decoder, represented within the orange bounding box are utilised only for annotating incrementally the $\Delta(\mathbb{N})$ training samples. While training the domain decoder, the image encoder is frozen so as not to update its weights. (Color figure online)

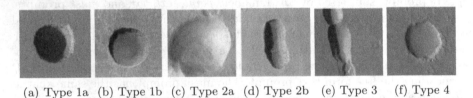

(a) Type 1a   (b) Type 1b   (c) Type 2a   (d) Type 2b   (e) Type 3   (f) Type 4

Fig. 2. Principal types of pits and skylights found on Mars terrain: (a) Skylight with possible cave entrance (Type 1a). (b) Pit with possible relation to a cave entrance (Type 1b). (c) "Bowl" pit with a possible connection to lava tubes (Type 2a). (d) Pit with uncertain connection to lava tubes or dikes (Type 2b). (e) Coalescent pits (Type 3). (f) Pit with a possible connection to lava tubes (Type 4) [20].

and drawing regions of interest manually is a complex and time-consuming process [20]– one that would greatly benefit from automation.

In this regard, the identification and segmentation of various Martian landforms have gained increasing attention in recent years [14,16,20,21]. Figure 2 shows an overview of some of the pits and skylights that can be identified on the Martian terrain. In this study, we focus only on these landforms, utilising a dataset prepared exclusively for it (cf. Sect. 2.1). Automatic detection and segmentation of these landforms have the potential to accelerate the identification of potential landing sites for future missions, study the geological history of Mars, and contribute to a better understanding of the planet's potential habitability. Therefore, this endeavour is of significant importance in planetary science.

In this study, we propose a pipeline that lets the domain experts swiftly annotate a small set of images using the prompt functionality of the SAM model. With this annotated dataset, we train a specialised decoder, leveraging the latent space of SAM. By harnessing the representations obtained from SAM's powerful encoder, we show that even a very small number of annotated samples can achieve performance comparable to state-of-the-art models trained with 100 times the annotated samples.

2 Method

2.1 Dataset

The data used in this work are images acquired by image sensors operating in the visible (VIS) and Near InfraRed (NIR) spectra on board probes orbiting Mars. This data set is composed of images by HiRISE instrument and downloaded both as Reduced Data Record (RDR) and Experiment Data Record (EDR) format from public space archives such as PDS Geosciences Node Orbital Data Explorer (ODE)[3]. With this, Nodjoumi *et al.* [20] released a processed dataset with 486 samples. This dataset is split into 405 images for training, 25 for validation and the rest for testing. In their work, they train a Mask-RCNN using all images annotated manually. In order to explore the applicability of knowledge distillation, we incrementally select train samples for annotation and subsequently train the domain decoder with these. This, in effect, is analogous to learning correct prompts for the task, with the least amount of annotated samples.

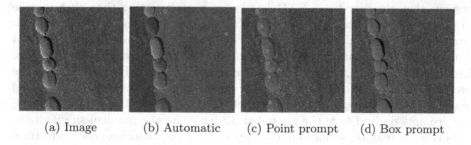

   (a) Image   (b) Automatic   (c) Point prompt   (d) Box prompt

Fig. 3. An overview of generation of segmentation masks with the three different prompt settings in SAM. The box prompt delineates the land mass from the adjacent shadow in comparison to the point prompt.

---

[3] https://ode.rsl.wustl.edu/.

2.2 Prompt-Mode Selection for Annotation

We conducted an evaluation of the SAM model using the three different prompt settings: (a) In the automatic prompt setting, SAM generates single-point input prompts in a grid pattern across the input image and selects high-quality masks using non-maximal suppression. All parameters were set to their default values. In the case of multiple masks being obtained, we selected the mask with the highest returned IoU score. (b) In the point prompt setting, we used the centre of the ground truth regions of interest as the point prompts for SAM. (c) In the box prompt setting, we computed the bounding box for SAM around the ground-truth mask. Figure 3 illustrates the mask generation on an exemplary sample for the three modes. Clearly, the automatic prompt simply segments all regions in a semantic agnostic way. Point and box prompts generate high-quality segmentation masks, with an average image level IoU above 90 %. Although in our case, point and box prompt performed relatively comparably on simpler cases, we empirically found box prompt to be most reliable in occluded and shadowy scenes and thus chose that to be used for final annotations. In practice, the expert would need only a few seconds to draw boxes around all relevant regions of interest on a sample.

2.3 Domain Decoder

Why Not Directly Fine-Tune the SAM Decoder? A recent work [8] from the medical domain corroborates our observation that the model underperforms significantly in comparison to state-of-the-art without training and with just the use of prompts. So fine-tuning the model would be necessary. However, we also observe that the decoder in SAM has learnt patterns other than that specific to the task and is prone to detecting other regions not relevant to our task. In our case, we observed the fine-tuned model[4] to give spurious results. Figure 4 illustrates an exemplary fail-case of fine-tuning the SAM decoder with the labels. Even when fine-tuned, the SAM decoder remains optimal only when prompts are provided, making it challenging to use without human-in-the-loop or additional information from the ground truth. All of the recently developed works [8,9,11] use prompts derived from the ground truth labels for the problem-specific task. This is not a realistic scenario in our application. We, therefore, choose to train a separate decoder to learn the problem-specific semantics.

We employ a lightweight decoder (cf. Fig. 1) comprised of only three upsampling layers using deconvolutions that maps the bottleneck z to an image of the desired size, in our case $(3 \times 1024 \times 1024)$. The bottleneck is obtained by passing the image through SAMs encoder. During training, only the weights of the decoder are updated. We use a sigmoid activation to map the logits in the range $[0, 1]$ for binary segmentation. In this manner, we train the decoder with incremental sets of SAM-annotated images. The incremental function $\Delta(\mathbb{N})$ is used in

---

[4] The SAM decoder is fine-tuned via training with a set of 25 annotated images for 100 epochs.

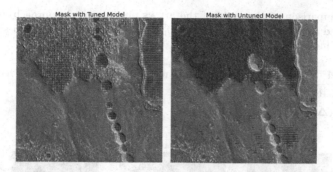

Fig. 4. Landforms of interest are harder to detect without prompts while using the SAM decoder. While the untuned model will segment all surrounding regions, the fine-tuned model still struggles with ignoring the regions of non-interest.

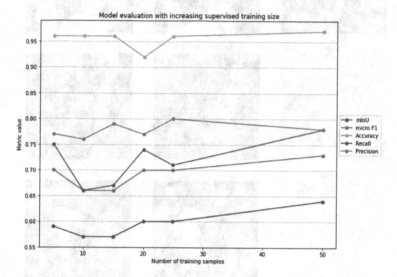

Fig. 5. Development of the evaluation metrics with increasing sizes $\Delta(N)$ of annotated training samples. Increasing training size beyond a handful of samples yields trivial overall improvement.

step sizes with $N \in \{5, 10, 15, 20, 25, 50\}$. All models are trained for a total of 100 training epochs, without additional control. We compare the performance using mean Intersection over Union (mIoU), micro F1, accuracy, and micro-precision and recall. Micro metrics are chosen to better represent the performance under data imbalance. Figure 5 shows the evolution of the metrics. We observe that the performance improvement with additional training samples after a handful is non-significant, with any differences being representative of stochasticity in evaluation rather than true information gain. By observing the metrics above and with qualitative evaluation it can be inferred that depending on the complexity of the domain-specific task, a very small number of annotations can suffice for a representative performance (cf. Fig. 6).

Table 1. Comparison of the state of the art vs our proposed approach trained only with 5 labelled samples. The authors in [20] train their model with 405 samples and report macro metrics.

model	macro F1	accuracy	macro precision	macro recall
Mask-RCNN [20]	0.811	0.774	**0.952**	0.706
ours ($\Delta(5)$)	**0.86**	**0.96**	0.89	**0.93**

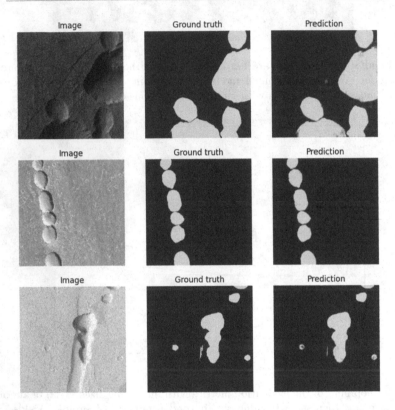

Fig. 6. Example predictions on the test set. The domain decoder identifies all regions of interest reasonably well.

Further, we compare the performance of this approach with $\Delta(5)$ against the Mask-RCNN model proposed in existing literature (cf. Table 1) for the same task, which serves as the benchmark for our comparison. This model is trained with the full training size of 405 manually annotated images. The authors [20] in this work only reported macro metrics and noted that about 1000 positive labels were required for satisfactory performance. We clearly see that knowledge distillation through SAM by utilising relatively minuscule labels surpasses the benchmark on most reported metrics. In spite of the precision being slightly lower, the recall is substantially higher. It is to be noted that in tasks like these, recall should be given a higher importance to precision, since missing a region of interest is more critical than falsely identifying one.

3 Conclusion

In this work, we extended the SAM framework and applied it to the segmentation of landforms like pits and skylights on the surface of Mars using HiRISE images. We observed that SAM has a high accuracy in separating various semantic regions, however, it cannot be directly applied to domain-specific tasks due to a lack of problem-specific bias. To this end, we developed and applied a domain-specific decoder that takes the image embedding generated by SAMs image encoder and learns the problem-specific semantics with substantially fewer labels. By training the domain decoder with only 5 labelled images sampled randomly, we demonstrated an equivalent if not superior performance to the existing Mask-RCNN method for the same task that was trained with over 400 labelled images.

We also explored the applicability of SAMs decoder for annotation using the various out-of-box prompts. We observed that the fully automatic mode is prone to marking irrelevant regions, and further can also miss some regions of interest if it doesn't know where to look. The point-based mode can be ambiguous at times. In contrast, the bounding box-based mode can clearly specify the ROI and obtain reasonable segmentation results without multiple trials and errors. We can therefore conclude that the bounding box-based segmentation mode can be a useful setting for rapid annotation by the domain expert.

In conclusion, our study reveals that SAM can effectively be exploited to accelerate domain-specific segmentation tasks. This work presents the first attempt to adapt SAM to geological mapping by fine-tuning through knowledge distillation. As part of future work, it might be worthwhile to investigate how the process of annotation can be automated, further lowering the load of human-in-the-loop. We hope this work will motivate more studies to build segmentation foundation models in the planetary science domain.

Acknowledgements. The authors acknowledge support from *Europlanet 2024 RI* that has received funding from the European Union's *Horizon 2020* research and innovation programme under grant agreement No. 871149.

References

1. Allemand, P., Delacourt, C., Gasperini, D., Kasperski, J., Pothérat, P.: Thirty years of evolution of the sedrun landslide (swisserland) from multitemporal orthorectified aerial images, differential digital terrain models and field data. Int. J. Remote Sens. Appl **1**, 30–36 (2011)
2. Baker, V.R.: Water and the martian landscape. Nature **412**(6843), 228–236 (2001)
3. Baker, V.R.: Geomorphological evidence for water on mars. Elements **2**(3), 139–143 (2006)

4. Chen, J., Bai, X.: Learning to "segment anything" in thermal infrared images through knowledge distillation with a large scale dataset satir. arXiv preprint arXiv:2304.07969 (2023)
5. Deng, R., et al.: Segment anything model (sam) for digital pathology: assess zero-shot segmentation on whole slide imaging. arXiv preprint arXiv:2304.04155 (2023)
6. Dosovitskiy, A., et al.: An image is worth 16×16 words: transformers for image recognition at scale. arXiv preprint arXiv:2010.11929 (2020)
7. Gou, J., Yu, B., Maybank, S.J., Tao, D.: Knowledge distillation: a survey. Int. J. Comput. Vision **129**, 1789–1819 (2021)
8. He, S., Bao, R., Li, J., Grant, P.E., Ou, Y.: Accuracy of segment-anything model (sam) in medical image segmentation tasks. arXiv preprint arXiv:2304.09324 (2023)
9. Hu, C., Li, X.: When sam meets medical images: an investigation of segment anything model (sam) on multi-phase liver tumor segmentation. arXiv preprint arXiv:2304.08506 (2023)
10. Jaderberg, M., Simonyan, K., Zisserman, A., et al.: Spatial transformer networks. Adv. Neural Inf. Process. Syst. **28** (2015)
11. Ji, G.P., Fan, D.P., Xu, P., Cheng, M.M., Zhou, B., Van Gool, L.: Sam struggles in concealed scenes-empirical study on "segment anything". arXiv preprint arXiv:2304.06022 (2023)
12. Ji, W., Li, J., Bi, Q., Li, W., Cheng, L.: Segment anything is not always perfect: an investigation of sam on different real-world applications. arXiv preprint arXiv:2304.05750 (2023)
13. Jia, C., et al.: Scaling up visual and vision-language representation learning with noisy text supervision. In: International Conference on Machine Learning, pp. 4904–4916. PMLR (2021)
14. Jiang, S., Lian, Z., Yung, K.L., Ip, W., Gao, M.: Automated detection of multitype landforms on mars using a light-weight deep learning-based detector. IEEE Trans. Aerosp. Electron. Syst. **58**(6), 5015–5029 (2022)
15. Julka, S.: An active learning approach for automatic detection of bow shock and magnetopause crossing signatures in mercury's magnetosphere using messenger magnetometer observations. In: Proceedings of the 2nd Machine Learning in Heliophysics, p. 8 (2022)
16. Julka, S., Granitzer, M., De Toffoli, B., Penasa, L., Pozzobon, R., Amerstorfer, U.: Generative adversarial networks for automatic detection of mounds in digital terrain models (mars arabia terra). In: EGU General Assembly Conference Abstracts, pp. EGU21-9188 (2021)
17. Julka, S., Kirschstein, N., Granitzer, M., Lavrukhin, A., Amerstorfer, U.: Deep active learning for detection of mercury's bow shock and magnetopause crossings. In: Amini, M.R., Canu, S., Fischer, A., Guns, T., Kralj Novak, P., Tsoumakas, G. (eds.) Machine Learning and Knowledge Discovery in Databases: European Conference, ECML PKDD 2022, Grenoble, France, 19–23 September 2022, Proceedings, Part IV. pp. 452–467. Springer, Heidelberg (2023). https://doi.org/10.1007/978-3-031-26412-2_28
18. Kirillov, A., et al.: Segment anything. arXiv preprint arXiv:2304.02643 (2023)
19. Lund, B.D., Wang, T.: Chatting about chatgpt: how may ai and gpt impact academia and libraries? Library Hi Tech News (2023)
20. Nodjoumi, G., Pozzobon, R., Sauro, F., Rossi, A.P.: Deeplandforms: a deep learning computer vision toolset applied to a prime use case for mapping planetary skylights. Earth Space Sci. **10**(1), e2022EA002278 (2023)

21. Palafox, L.F., Hamilton, C.W., Scheidt, S.P., Alvarez, A.M.: Automated detection of geological landforms on mars using convolutional neural networks. Comput. Geosci. **101**, 48–56 (2017)
22. Qin, C., Zhang, A., Zhang, Z., Chen, J., Yasunaga, M., Yang, D.: Is chatgpt a general-purpose natural language processing task solver? arXiv preprint arXiv:2302.06476 (2023)
23. Radford, A., et al.: Learning transferable visual models from natural language supervision. In: International Conference on Machine Learning, pp. 8748–8763. PMLR (2021)
24. Ramesh, A., et al.: Zero-shot text-to-image generation. In: International Conference on Machine Learning, pp. 8821–8831. PMLR (2021)
25. Tang, L., Xiao, H., Li, B.: Can sam segment anything? when sam meets camouflaged object detection. arXiv preprint arXiv:2304.04709 (2023)

ContainerGym: A Real-World Reinforcement Learning Benchmark for Resource Allocation

Abhijeet Pendyala, Justin Dettmer, Tobias Glasmachers, and Asma Atamna[✉]

Ruhr-University Bochum, Bochum, Germany
{abhijeet.pendyala,justin.dettmer,tobias.glasmachers,
asma.atamna}@ini.rub.de

Abstract. We present ContainerGym, a benchmark for reinforcement learning inspired by a real-world industrial resource allocation task. The proposed benchmark encodes a range of challenges commonly encountered in real-world sequential decision making problems, such as uncertainty. It can be configured to instantiate problems of varying degrees of difficulty, e.g., in terms of variable dimensionality. Our benchmark differs from other reinforcement learning benchmarks, including the ones aiming to encode real-world difficulties, in that it is directly derived from a real-world industrial problem, which underwent minimal simplification and streamlining. It is sufficiently versatile to evaluate reinforcement learning algorithms on any real-world problem that fits our resource allocation framework. We provide results of standard baseline methods. Going beyond the usual training reward curves, our results and the statistical tools used to interpret them allow to highlight interesting limitations of well-known deep reinforcement learning algorithms, namely PPO, TRPO and DQN.

Keywords: Deep reinforcement learning · Real-world benchmark · Resource allocation

1 Introduction

Supervised learning has long made its way into many industries, but industrial applications of (deep) reinforcement learning (RL) are significantly rare. This may be for many reasons, like the focus on impactful RL success stories in the area of games, a lower degree of technology readiness, and a lack of industrial RL benchmark problems.

A RL agent learns by taking sequential actions in its environment, observing the state of the environment, and receiving rewards [15]. Reinforcement learning aims to fulfill the enticing promise of training a smart agent that solves a complex task through trial-and-error interactions with the environment, without specifying how the goal will be achieved. Great strides have been made in this direction, also in the real world, with notable applications in domains like robotics [6], autonomous driving [6,10], and control problems such as optimizing the power efficiency of data centers [8], control of nuclear fusion reactors [4], and optimizing gas turbines [3].

Yet, the accelerated progress in these areas has been fueled by making agents play in virtual gaming environments such as Atari 2600 games [9], the game of GO [14], and complex video games like Starcraft II [17]. These games provided sufficiently challenging environments to quickly test new algorithms and ideas, and gave rise to a suite of RL benchmark environments. The use of such environments to benchmark agents for industrial deployment comes with certain drawbacks. For instance, the environments either may not be challenging enough for the state-of-the-art algorithms (low dimensionality of state and action spaces) or require significant computational resources to solve. Industrial problems deviate widely from games in many further properties. Primarily, exploration in real-world systems often has strong safety constraints, and constitutes a balancing act between maximizing reward with good actions which are often sparse, and minimizing potentially severe consequences from bad actions. This is in contrast to training on a gaming environment, where the impact of a single action is often smaller, and the repercussions of poor decisions accumulate slowly over time. In addition, underlying dynamics of gaming environments—several of which are near-deterministic—may not reflect the stochasticity of a real industrial system. Finally, the available environments may have a tedious setup procedure with restrictive licensing and dependencies on closed-source binaries.

To address these issues, we present ContainerGym, an open-source real-world benchmark environment for RL algorithms. It is adapted from a digital twin of a high throughput processing industry. Our concrete use-case comes from a waste sorting application. Our benchmark focuses on two phenomena of general interest: First, a stochastic model for a resource-filling process, where certain material is being accumulated in multiple storage containers. Second, a model for a resource transforming system, which takes in the material from these containers, and transforms it for further post-processing downstream. The processing units are a scarce resource, since they are large and expensive, and due to limited options of conveyor belt layout, only a small number can be used per plant. ContainerGym is not intended to be a perfect replica of a real system but serves the same hardness and complexity. The search for an optimal sequence of actions in ContainerGym is akin to solving a dynamic resource allocation problem for the resource-transforming system, while also learning an optimal control behavior of the resource-filling process. In addition, the complexity of the environment is customizable. This allows testing the limitations of any given learning algorithm. This work aims to enable RL practitioners to quickly and reliably test their learning agents on an environment encoding real-world dynamics.

The paper is arranged as follows. Section 2 discusses the relevant literature and motivates our contribution. Section 3 gives a brief introduction to reinforcement learning preliminaries. In Sect. 4, we present the real-world industrial control task that inspired ContainerGym and discuss the challenges it presents. In

Sect. 5, we formulate the real-world problem as a RL one and discuss the design choices that lead to our digital twin. We briefly present ContainerGym's implementation in Sect. 6. We present and discuss benchmark experiments of baseline methods in Sect. 7, and close with our conclusions in Sect. 8.

2   Related Work

The majority of the existing open-source environments on which novel reinforcement learning algorithms could be tuned can be broadly divided into the following categories: toy control, robotics (MuJoCo) [16], video games (Atari) [9], and autonomous driving. The underlying dynamics of these environments are artificial and may not truly reflect real-world dynamic conditions like high dimensional states and action spaces, and stochastic dynamics. To the best of our knowledge, there exist very few such open-source benchmarks for industrial applications. To accelerate the deployment of agents in the industry, there is a need for a suite of RL benchmarks inspired by real-world industrial control problems, thereby making our benchmark environment, ContainerGym, a valuable addition to it.

The classic control environments like mountain car, pendulum, or toy physics control environments based on Box2D are stochastic only in terms of their initial state. They have low dimensional state and action spaces, and they are considered easy to solve with standard methods. Also, the 50 commonly used Atari games in the Arcade Learning Environment [1], where nonlinear control policies need to be learned, are routinely solved to a super-human level by well-established algorithms. This environment, although posing high dimensionality, is deterministic. The real world is not deterministic and there is a need to tune algorithms that can cope with stochasticity. Although techniques like sticky actions or skipping a random number of initial frames have been developed to add artificial randomness, this randomness may still be very structured and not challenging enough. On the other hand, video game simulators like Starcraft II [17] offer high-dimensional image observations, partial observability, and (slight) stochasticity. However, playing around with DeepRL agents on such environments requires substantial computational resources. It might very well be overkill to tune reinforcement learning agents in these environments when the real goal is to excel in industrial applications.

Advancements in the RL world, in games like Go and Chess, were achieved by exploiting the rules of these games and embedding them into a stochastic planner. In real-world environments, this is seldom achievable, as these systems are highly complex to model in their entirety. Such systems are modeled as partially observable Markov decision processes and present a tough challenge for learning agents that can explore only through interactions. Lastly, some of the more sophisticated RL environments available, e.g., advanced physics simulators like MuJoCo [16], offer licenses with restrictive terms of use. Also, environments like Starcraft II require access to a closed-source binary. Open source licensing in an environment is highly desirable for RL practitioners as it enables them to debug the code, extend the functionality and test new research ideas.

Other related works, amongst the very few available open-source reinforcement learning environments for industrial problems, are Real-world RL suite [5] and Industrial benchmark (IB) [7] environments. Real-world RL-suite is not derived from a real-world scenario, rather the existing toy problems are perturbed to mimic the conditions in a real-world problem. The IB comes close in spirit to our work, although it lacks the customizability of ContainerGym and our expanded tools for agent behavior explainability. Additionally, the (continuous) action and state spaces are relatively low dimensional and of fixed sizes.

3 Reinforcement Learning Preliminaries

RL problems are typically studied as discrete-time Markov decision processes (MDPs), where a MDP can be defined as a tuple $\langle \mathcal{S}, \mathcal{A}, p, r, \gamma \rangle$. At timestep t, the agent is in a state $s_t \in \mathcal{S}$ and takes an action $a_t \in \mathcal{A}$. It arrives in a new state s_{t+1} with probability $p(s_{t+1} \mid s_t, a_t)$ and receives a reward $r(s_t, a_t, s_{t+1})$. The state transitions of a MDP satisfy the Markov property $p(s_{t+1} \mid s_t, a_t, \ldots, s_0, a_0) = p(s_{t+1} \mid s_t, a_t)$. That is, the new state s_{t+1} only depends on the current state s_t and action a_t. The goal of the RL agent interacting with the MDP is to find a policy $\pi : \mathcal{S} \to \mathcal{A}$ that maximizes the expected (discounted) cumulative reward. This optimization problem is defined formally as follows:

$$\arg\max_{\pi} \mathbb{E}_{\tau \sim \pi} \left[\sum_{t \geq 0} \gamma^t r(s_t, a_t, s_{t+1}) \right], \tag{1}$$

where $\tau = (s_0, a_0, r_0, s_1, \ldots)$ is a trajectory generated by following π and $\gamma \in (0, 1]$ is a discount factor. A trajectory τ ends either when a maximum timestep count T—also called episode[1] length—is reached, or when a terminal state is reached (early termination).

4 Container Management Environment

In this section, we describe the real-world industrial control task that inspired our RL benchmark. It originates from the final stage of a waste sorting process.

The environment consists of a solid material-transforming facility that hosts a set of *containers* and a significantly smaller set of *processing units* (PUs). Containers are continuously filled with material, where the material flow rate is a container-dependent stochastic process. They must be emptied regularly so that their content can be transformed by the PUs. When a container is emptied, its content is transported on a conveyor belt to a free PU that transforms it into *products*. It is not possible to extract material from a container without emptying it completely. The number of produced products depends on the volume of the material being processed. Ideally, this volume should be an integer multiple of the

---

[1] We use the terms "episode" and "rollout" interchangeably in this paper.

product's size. Otherwise, the surplus volume that cannot be transformed into a product is redirected to the corresponding container again via an energetically costly process that we do not consider in this work. Each container has at least one optimal emptying volume: a global optimum and possibly other, smaller, local optima that are all multiples of container-specific product size. Generally speaking, larger volumes are better, since PUs are more efficient with producing many products in a series.

The worst-case scenario is a container overflow. In the waste sorting application inspiring our benchmark, it incurs a high recovery cost including human intervention to stop and restart the facility. This situation is undesirable and should be actively avoided. Therefore, letting containers come close to their capacity limit is rather risky.

The quality of an emptying decision is a compromise between the number of potential products and the costs resulting from actuating a PU and handling surplus volume. Therefore, the closer an emptying volume is to an optimum, the better. If a container is emptied too far away from any optimal volume, the costs outweigh the benefit of producing products.

This setup can be framed more broadly as a resource allocation problem, where one item of a scarce resource, namely the PUs, needs to be allocated whenever an emptying decision is made. If no PU is available, the container cannot be emptied and will continue to fill up.

There are multiple aspects that make this problem challenging:

- The rate at which the material arrives at the containers is stochastic. Indeed, although the volumes follow a globally linear trend—as discussed in details in Sect. 5, the measurements can be very noisy. The influx of material is variable, and there is added noise from the sensor readings inside the containers. This makes applying standard planning approaches difficult.
- The scarcity of the PUs implies that always waiting for containers to fill up to their ideal emptying volume is risky: if no PU is available at that time, then we risk an overflow. This challenge becomes more prominent when the number of containers—in particular, the ratio between the number of containers and the number of PUs—increases. Therefore, an optimal policy needs to take fill states and fill rates of all containers into account, and possibly empty some containers early.
- Emptying decisions can be taken at any time, but in a close-to-optimal policy, they are rather infrequent. Also, the rate at which containers should be emptied varies between containers. Therefore, the distributions of actions are highly asymmetric, with important actions (and corresponding rewards) being rare.

5   Reinforcement Learning Problem Formulation

We formulate the container management problem presented in Sect. 4 as a MDP that can be addressed by RL approaches. The resulting MDP is an accurate representation of the original problem, as only mild simplifications are made.

Specifically, we model one category of PUs instead of two, we do not include inactivity periods of the facility in our environment, and we neglect the durations of processes of minor relevance. All parameters described in the rest of this section are estimated from real-world data (system identification). Overall, the MDP reflects the challenges properly without complicating the benchmark (and the code base) with irrelevant details.

5.1 State and Action Spaces

State Space. The state s_t of the system at any given time t is defined by the volumes $v_{i,t}$ of material contained in each container C_i, and a timer $p_{j,t}$ for each PU P_j indicating in how many seconds the PU will be ready to use. A value of zero means that the PU is available, while a positive value means that it is currently busy. Therefore, $s_t = (\{v_{i,t}\}_{i=1}^n, \{p_{j,t}\}_{j=1}^m)$, where n and m are the number of containers and PUs, respectively. Valid initial states include non-negative volumes not greater than the maximum container capacity $v_{\max}$, and non-negative timer values.

Action Space. Possible actions at a given time t are either (i) not to empty any container, i.e. to do nothing, or (ii) to empty a container C_i and transform its content using one of the PUs. The action of doing nothing is encoded with 0, whereas the action of emptying a container C_i and transforming its content is encoded with the container's index i. Therefore, $a_t \in \{0, 1, \ldots, n\}$, where n is the number of containers. Since we consider identical PUs, specifying which PU is used in the action encoding is not necessary as an emptying action will result in the same state for all the PUs, given they were at the same previous state.

5.2 Environment Dynamics

In this section, we define the dynamics of the volume of material in the containers, the PU model, as well as the state update.

Volume Dynamics. The volume of material in each container increases following an irregular trend, growing linearly on average, captured by a random walk model with drift. That is, given the current volume $v_{i,t}$ contained in C_i, the volume at time $t+1$ is given by the function f_i defined as

$$f_i(v_{i,t}) = \max(0, \alpha_i + v_{i,t} + \epsilon_{i,t}), \tag{2}$$

where $\alpha_i > 0$ is the slope of the linear upward trend followed by the volume for C_i and the noise $\epsilon_{i,t}$ is sampled from a normal distribution $\mathcal{N}(0, \sigma_i^2)$ with mean 0 and variance σ_i^2. The max operator forces the volume to non-negative values.

When a container C_i is emptied, its volume drops to 0 at the next timestep. Although the volume drops progressively to 0 in the real facility, empirical evidence provided by our data shows that emptying durations are within the range of the time-step lengths considered in this paper (60 and 120 s).

Processing Unit Dynamics. The time (in seconds) needed by a PU to transform a volume v of material in a container C_i is linear in the number of products $\lfloor v/b_i \rfloor$ that can be produced from v. It is given by the function g_i defined in Eq. (3), where $b_i > 0$ is the product size, $\beta_i > 0$ is the time it takes to actuate the PU before products can be produced, and $\lambda_i > 0$ is the time per product. Note that all the parameters indexed with i are container-dependent.

$$g_i(v) = \beta_i + \lambda_i \lfloor v/b_i \rfloor. \tag{3}$$

A PU can only produce one type of product at a time. Therefore, it can be used for emptying a container only when it is free. Therefore, if all PUs are busy, the container trying to use one is not emptied and continues to fill up.

State Update. We distinguish between the following cases to define the new state $s_{t+1} = (\{v_{i,t+1}\}_{i=1}^n, \{p_{j,t+1}\}_{j=1}^m)$ given the current state s_t and action a_t.

$a_t = 0$. This corresponds to the action of doing nothing. The material volumes inside the containers increase while the timers indicating the availability of the PUs are decreased according to:

$$v_{i,t+1} = f_i(v_{i,t}), \qquad\qquad i \in \{1, \ldots, n\}, \tag{4}$$
$$p_{j,t+1} = \max(0, p_{j,t} - \delta), \qquad\qquad j \in \{1, \ldots, m\}, \tag{5}$$

where f_i is the random walk model defined in Eq. (2) and δ is the length of a timestep in seconds.

$a_t \neq 0$. This corresponds to an emptying action. If no PU is available, that is, $p_{j,t} > 0 \forall j = 1, \ldots, m$, the updates are identical to the one defined in Eqs. (4) and (5). If at least one PU P_k is available, the new state variables are defined as follows:

$$v_{a_t,t+1} = 0, \tag{6}$$
$$v_{i,t+1} = f_i(v_{i,t}), \qquad\qquad i \in \{1, \ldots, n\}\backslash\{a_t\}, \tag{7}$$
$$p_{k,t+1} = g_{a_t}(v_{a_t,t}), \tag{8}$$
$$p_{j,t+1} = \max(0, p_{j,t} - \delta), \qquad\qquad j \in \{1, \ldots, m\}\backslash\{k\}, \tag{9}$$

where the value of the action a_t is the index of the container C_{a_t} to empty, '$\backslash$' denotes the set difference operator, f_i is the random walk model defined in Eq. (2), and g_{a_t} is the processing time defined in Eq. (3).

Although the processes can continue indefinitely, we stop an episode after a maximum length of T timesteps. This is done to make ContainerGym compatible with RL algorithms designed for episodic tasks, and hence to maximize its utility. When a container reaches its maximum volume $v_{\max}$, however, the episode is terminated and a negative reward is returned (see details in Sect. 5.3).

5.3 Reward Function

We use a deterministic reward function $r(s_t, a_t, s_{t+1})$ where higher values correspond to better (s_t, a_t) pairs. The new state s_{t+1} is taken into account to return a large negative reward $r_{\min}$ when a container overflows, i.e. $\exists i \in \{1, \dots, n\}$, $v_{i,t+1} \geq v_{\max}$, before ending the episode. In all other cases, the immediate reward is determined only by the current state s_t and the action a_t.

$a_t = 0$. If the action is to do nothing, we define $r(s_t, 0, s_{t+1}) = 0$.

$a_t \neq 0$. If an emptying action is selected while (i) no PU is available or (ii) the selected container is already empty, i.e. $v_{a_t,t} = 0$, then $r(s_t, a_t, s_{t+1}) = r_{\text{pen}}$, where it holds $r_{\min} < r_{\text{pen}} < 0$ for the penalty. If, on the other hand, at least one PU is available and the volume to process is non-zero ($v_{a_t,t} > 0$), the reward is a finite sum of Gaussian functions centered around optimal emptying volumes $v^*_{a_t,i}, i = 1, \dots, p_{a_t}$, where the height of a peak $0 < h_{a_t,i} \leq 1$ is proportional to the quality of the corresponding optimum. The reward function in this case is defined as follows:

$$r(s_t, a_t, s_{t+1}) = r_{\text{pen}} + \sum_{i=1}^{p_{a_t}} (h_{a_t,i} - r_{\text{pen}}) \exp\left(-\frac{(v_{a_t,t} - v^*_{a_t,i})^2}{2w^2_{a_t,i}}\right), \qquad (10)$$

where p_{a_t} is the number of optima for container C_{a_t} and $w_{a_t,i} > 0$ is the width of the bell around $v^*_{a_t,i}$. The Gaussian reward defined in Eq. (10) takes its values in $]r_{\text{pen}}, 1]$, the maximum value 1 being achieved at the ideal emptying volume $v^*_{a_t,1}$ for which $h_{a_t,1} = 1$. The coefficients $h_{a,i}$ are designed so that processing large volumes at a time is beneficial, hence encoding a conflict between emptying containers early and risking overflow. This tension, together with the limited availability of the PUs, yields a highly non-trivial control task. Figure 1 shows an example of Gaussian rewards when emptying a container with three optimal volumes at different volumes in $[0, 40[$.

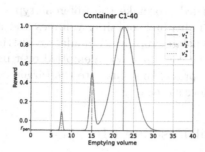

Fig. 1. Rewards received when emptying a container with three optima. The design of the reward function fosters emptying late, hence allowing PUs to produce many products in a row.

6   ContainerGymUsage Guide

In this section, we introduce the OpenAI Gym implementation[2] of our bench-mark environment. We present the customizable parameters of the benchmark, provide an outline for how the Python implementation reflects the theoretical definition in Sect. 5, and show which tools we provide to extend the understanding of an agent's behavior beyond the achieved rewards.

Gym Implementation. Our environment follows the OpenAI Gym [2] frame-work. It implements a step()-function computing a state update. The reset()-function resets time to $t = 0$ and returns the environment to a valid initial state, and the render()-function displays a live diagram of the environment.

We deliver functionality to configure the environment's complexity and difficulty through its parameters such as the number of containers and PUs, as well as the composition of the reward function, the overflow penalty r_{min}, the sub-optimal emptying penalty r_{pen}, the length of a timestep δ, and the length of an episode T. We provide example configurations in our GitHub repository that are close in nature to the real industrial facility.

In the spirit of open and reproducible research, we include scripts and model files for reproducing the experiments presented in the next section.

Action Explainability. While most RL algorithms treat the environment as a black box, facility operators want to "understand" a policy before deploying it for production. To this end, the accumulated reward does not provide sufficient information, since it treats all mistakes uniformly. Practitioners need to understand which types of mistakes a sub-optimal policy makes. For example, a low reward can be obtained by systematically emptying containers too early (local optimum) or by emptying at non-integer multiples of the product size. When a basic emptying strategy for each container is in place, low rewards typically result from too many containers reaching their ideal volume at the same time so that PUs are overloaded. To make the different types of issues of a policy transparent, we plot all individual container volumes over an entire episode. We further provide tools to create empirical cumulative distribution function plots over the volumes at which containers were emptied over multiple episodes. The latter plots in particular provide insights into the target volumes an agent aims to hit, and whether it does so reliably.

7   Performance of Baseline Methods

We illustrate the use of ContainerGym by benchmarking three popular deep RL algorithms: two on-policy methods, namely Proximal Policy Optimization

---

[2] The ContainerGym software is available on the following GitHub repository: https://github.com/Pendu/ContainerGym.

(PPO) [13] and Trust Region Policy Optimization (TRPO) [12], and one off-policy method, namely Deep Q-Network (DQN) [9]. We also compare the performance of these RL approaches against a naive rule-based controller. By doing so, we establish an initial baseline on ContainerGym. We use the Stable Baselines3 implementation of PPO, TRPO, and DQN [11].

7.1   Experimental Setup

The algorithms are trained on 8 ContainerGym instances, detailed in Table 1, where the varied parameters are the number of containers n, the number of PUs m and the timestep length δ.[3] The rationale behind the chosen configurations is to assess (i) how the algorithms scale with the environment dimensionality, (ii) how action frequency affects the trained policies and (iii) whether the optimal policy can be found in the conceptually trivial case $m = n$, where there is always a free PU when needed. This is only a control experiment for testing RL performance. In practice, PUs are a scarce resource ($m \ll n$).

The maximum episode length T is set to 1500 timesteps during training in all experiments, whereas the initial volumes $v_{i,0}$ are uniformly sampled in $[0, 30]$, and the maximum capacity is set to $v_{\max} = 40$ volume units for all containers. The initial PU states are set to free ($p_{j,0} = 0$) and the minimum and penalty rewards are set to $r_{\min} = -1$ and $r_{\text{pen}} = -0.1$ respectively. The algorithms are trained with an equal budget of 2 (resp. 5) million timesteps when $n = 5$ (resp. $n = 11$) and the number of steps used for each policy update for PPO and TRPO is 6144. Default values are used for the remaining hyperparameters, as the aim of our work is to show characteristics of the training algorithms with a reasonable set of parameters. For each algorithm, 15 policies are trained in parallel, each with a different seed $\in \{1, \ldots, 15\}$, and the policy with the best training cumulative reward is returned for each run. To make comparison easier, policies are evaluated on a similar test environment with $\delta = 120$ and $T = 600$.

7.2   Results and Discussion

Table 1 shows the average test cumulative reward, along with its standard deviation, for the best and median policies trained with PPO, TRPO, and DQN on each environment configuration. These statistics are calculated by running each policy 15 times on the corresponding test environment.

PPO achieves the highest cumulative reward on environments with 5 containers. DQN, on the other hand, shows the best performance on higher dimensionality environments (11 containers), with the exception of the last configuration. It has, however, a particularly high variance. Its median performance is also significantly lower than the best one, suggesting less stability than PPO. DQN's particularly high rewards when $n = 11$ could be due to a better exploration of the

---

[3] Increasing the timestep length δ should be done carefully. Otherwise, the problem could become trivial. In our case, we choose δ such that it is smaller than the minimum time it takes a PU to process the volume equivalent to one product.

Table 1. Test cumulative reward, averaged over 15 episodes, and its standard deviation for the best and median policies for PPO, TRPO and DQN. Best and median policies are selected from a sample of 15 policies (seeds) trained on the investigated environment configurations. The highest best performance is highlighted for each configuration.

Config.			PPO		TRPO		DQN	
n	m	δ	best	median	best	median	best	median
5	2	60	**38.55 ± 1.87**	29.94 ± 8.42	37.35 ± 7.13	4.93 ± 4.90	29.50 ± 3.43	1.57 ± 0.93
5	2	120	**51.43 ± 3.00**	49.81 ± 1.56	38.33 ± 8.36	16.86 ± 2.78	42.96 ± 2.80	7.90 ± 4.72
5	5	60	**37.54 ± 1.50**	30.64 ± 2.44	33.62 ± 7.76	7.36 ± 4.40	23.98 ± 13.17	1.96 ± 1.28
5	5	120	**50.42 ± 2.57**	47.23 ± 1.89	47.36 ± 2.07	5.39 ± 4.38	43.26 ± 3.73	8.56 ± 4.12
11	2	60	31.01 ± 8.62	23.25 ± 16.79	26.62 ± 18.94	4.26 ± 3.81	**42.63 ± 21.73**	11.84 ± 10.33
11	2	120	54.30 ± 8.48	49.58 ± 13.54	45.78 ± 18.42	32.08 ± 18.04	**72.19 ± 16.20**	24.07 ± 13.68
11	11	60	27.87 ± 8.89	17.57 ± 13.99	15.66 ± 10.65	4.90 ± 2.68	**28.32 ± 22.42**	8.49 ± 4.92
11	11	120	47.75 ± 10.78	42.37 ± 13.25	**50.55 ± 11.08**	34.29 ± 16.00	29.06 ± 13.10	13.16 ± 8.32

search space. An exception is the last configuration, where DQN's performance is significantly below those of TRPO and PPO. Overall, PPO has smaller standard deviations and a smaller difference between best and median performances, suggesting higher stability than TRPO and DQN.

Our results also show that taking actions at a lower frequency ($\delta = 120$) leads to better policies. Due to space limitations, we focus on analyzing this effect on configurations with $n = 5$ and $m = 2$. Figure 2 displays the volumes, actions, and rewards over one episode for PPO and DQN when the (best) policy is trained with $\delta = 60$ (left column) and with $\delta = 120$.

We observe that with $\delta = 60$, PPO tends to empty containers C1-60 and C1-70 prematurely, which leads to poor rewards. These two containers have the slowest fill rate. Increasing the timestep length, however, alleviates this defect. The opposite effect is observed on C1-60 with DQN, whereas no significant container-specific behavior change is observed for TRPO, as evidenced by the cumulative rewards. To further explain these results, we investigate the empirical cumulative distribution functions (ECDFs) of the emptying volumes per container. Figure 3 reveals that more than 90% of PPO's emptying volumes on C1-60 (resp. C1-70) are approximately within a 3 (resp. 5) volume units distance of the third best optimum when $\delta = 60$. By increasing the timestep length, the emptying volumes become more centered around the global optimum. While no such clear pattern is observed with DQN on containers with slow fill rates, the emptying volumes on C1-60 move away from the global optimum when the timestep length is increased (more than 50% of the volumes are within a 3 volume units distance of the second best optimum). DQN's performance increases when $\delta = 120$. This is explained by the better emptying volumes achieved on C1-20, C1-70, and C1-80.

None of the benchmarked algorithms manage to learn the optimal policy when there are as many PUs as containers, independently of the environment dimensionality and the timestep length. When $m = n$, the optimal policy is known and consists in emptying each container when the corresponding global optimal volume is reached, as there is always at least one free PU. Therefore, achieving the maximum reward at each emptying action is theoretically possible. Figure 4a shows the ECDF of the reward per emptying action over 15 episodes of the best policy for each of PPO, TRPO and DQN when $m = n = 11$. The rewards range from $r_{min} = -1$ to 1, and the ratio of negative rewards is particularly high for PPO. An analysis of the ECDFs of its emptying volumes (not shown) reveals that this is due to containers with slow fill rates being emptied prematurely. TRPO achieves the least amount of negative rewards whereas all DQN's rollouts end prematurely due to a container overflowing. Rewards close to r_{pen} are explained by poor emptying volumes, as well as a bad allocation of the PUs ($r = r_{pen}$). These findings suggest that, when $m = n$, it is not trivial for the tested RL algorithms to break down the problem into smaller, independent problems, where one container is emptied using one PU when the ideal volume is reached.

The limitations of these RL algorithms are further highlighted when compared to a naive rule-based controller that empties the first container whose volume is less than 1 volume unit away from the ideal emptying volume. Figure 4b shows the ECDF of reward per emptying action obtained from 15 rollouts of the rule-based controller on three environment configurations, namely $n = 5$ and $m = 2$, $n = 11$ and $m = 2$, and $n = 11$ and $m = 11$. When compared to PPO in particular ($m = n = 11$), the rule-based controller empties containers less often (17.40% of the actions vs. more than 30% for PPO). Positive rewards are close to optimal (approx. 90% in $[0.75, 1]$), whereas very few negative rewards are observed. These stem from emptying actions taken when no PU is available. These findings suggest that learning to wait for long periods before performing an important action may be challenging for some RL algorithms.

Critically, current baseline algorithms only learn reasonable behavior for containers operated in isolation. In the usual $m \ll n$ case, none of the policies anticipates all PUs being busy when emptying containers at their optimal volume, which they should ideally foresee and prevent proactively by emptying some of the containers earlier. Hence, there is considerable space for future improvements by learning stochastic long-term dependencies. This is apparently a difficult task for state-of-the-art RL algorithms. We anticipate that ContainerGym can contribute to research on next-generation RL methods addressing this challenge.

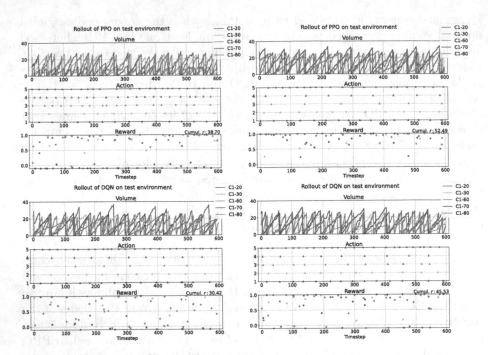

Fig. 2. Rollouts of the best policy trained with PPO (first row) and DQN (second row) on a test environment with $n = 5$ and $m = 2$. Left: training environment with $\delta = 60\,\mathrm{s}$. Right: training environment with $\delta = 120$. Displayed are the volumes, emptying actions and non-zero rewards at each timestep.

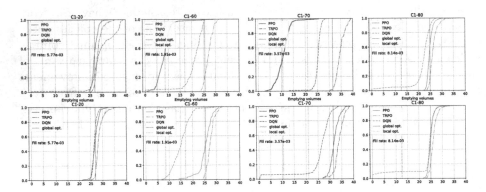

Fig. 3. ECDFs of emptying volumes collected over 15 rollouts of the best policy for PPO, TRPO, and DQN on a test environment with $n = 5$ and $m = 2$. Shown are 4 containers out of 5. Fill rates are indicated in volume units per second. Top: training environment with $\delta = 60$. Bottom: training environment with $\delta = 120$. The derivatives of the curves are the PDFs of emptying volumes. Therefore, a steep incline indicates that the corresponding volume is frequent in the corresponding density.

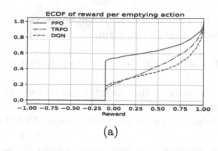
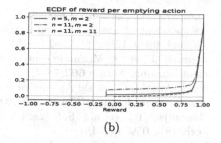

(a)   (b)

Fig. 4. ECDF of the reward obtained for each emptying action over 15 rollouts. (a): Best policies for PPO, TRPO and DQN, trained on an environment with $m = n = 11$ and $\delta = 120$. (b): Rule-based controller on three environment configurations.

8   Conclusion

We have presented ContainerGym, a real-world industrial RL environment. Its dynamics are quite basic, and therefore easily accessible. It is easily scalable in complexity and difficulty.

Its characteristics are quite different from many standard RL benchmark problems. At its core, it is a resource allocation problem. It features stochasticity, learning agents can get trapped in local optima, and in a good policy, the most important actions occur only rarely. Furthermore, implementing optimal behavior requires planning ahead under uncertainty.

The most important property of ContainerGym is to be of direct industrial relevance. At the same time, it is a difficult problem with the potential to trigger novel developments in the future. While surely being less challenging than playing the games of Go or Starcraft II at a super-human level, ContainerGym is still sufficiently hard to make state-of-the-art baseline algorithms perform poorly. We are looking forward to improvements in the future.

To fulfill the common wish of industrial stakeholders for explainable ML solutions, we provide insights into agent behavior and deviations from optimal behavior that go beyond learning curves. While accumulated reward hides the details, our environment provides insights into different types of failures and hence gives guidance for routes to further improvement.

Acknowledgements. This work was funded by the German federal ministry of economic affairs and climate action through the "ecoKI" grant.

References

1. Bellemare, M.G., Naddaf, Y., Veness, J., Bowling, M.: The arcade learning environment: an evaluation platform for general agents. J. Artif. Intell. Res. **47**, 253–279 (2013)
2. Brockman, G., et al.: Openai gym (2016)

3. Compare, M., Bellani, L., Cobelli, E., Zio, E.: Reinforcement learning-based flow management of gas turbine parts under stochastic failures. Int. J. Adv. Manuf. Technol. **99**(9–12), 2981–2992 (2018)
4. Degrave, J., et al.: Magnetic control of tokamak plasmas through deep reinforcement learning. Nature **602**(7897), 414–419 (2022)
5. Dulac-Arnold, G., et al.: An empirical investigation of the challenges of real-world reinforcement learning. CoRR arxiv:2003.11881 (2020)
6. Haarnoja, T., et al.: Soft actor-critic algorithms and applications. CoRR arxiv:1812.05905 (2018)
7. Hein, D., et al.: A benchmark environment motivated by industrial control problems. In: 2017 IEEE Symposium Series on Computational Intelligence (SSCI). IEEE (2017)
8. Lazic, N., et al.: Data center cooling using model-predictive control. In: Proceedings of the Thirty-Second Conference on Neural Information Processing Systems (NeurIPS-2018), Montreal, QC, pp. 3818–3827 (2018)
9. Mnih, V., et al.: Playing atari with deep reinforcement learning. CoRR arxiv:1312.5602 (2013)
10. Osiński, B., et al.: Simulation-based reinforcement learning for real-world autonomous driving. In: 2020 IEEE International Conference on Robotics and Automation (ICRA), pp. 6411–6418 (2020)
11. Raffin, A., Hill, A., Gleave, A., Kanervisto, A., Ernestus, M., Dormann, N.: Stable-baselines3: reliable reinforcement learning implementations. J. Mach. Learn. Res. **22**(268), 1–8 (2021)
12. Schulman, J., Levine, S., Abbeel, P., Jordan, M., Moritz, P.: Trust region policy optimization. In: Bach, F., Blei, D. (eds.) Proceedings of the 32nd International Conference on Machine Learning. Proceedings of Machine Learning Research, vol. 37, pp. 1889–1897. PMLR (2015)
13. Schulman, J., Wolski, F., Dhariwal, P., Radford, A., Klimov, O.: Proximal policy optimization algorithms. CoRR arxiv:1707.06347 (2017)
14. Silver, D., et al.: Mastering the game of go with deep neural networks and tree search. Nature **529**(7587), 484–489 (2016)
15. Sutton, R.S., Barto, A.G.: Reinforcement Learning: An Introduction. MIT press, Cambridge (2018)
16. Todorov, E., Erez, T., Tassa, Y.: Mujoco: a physics engine for model-based control. In: 2012 IEEE/RSJ International Conference on Intelligent Robots and Systems, pp. 5026–5033 (2012)
17. Vinyals, O., et al.: Grandmaster level in StarCraft II using multi-agent reinforcement learning. Nature **575**(7782), 350–354 (2019)

Perceptrons Under Verifiable Random Data Corruption

Jose E. Aguilar Escamilla(✉) and Dimitrios I. Diochnos(✉)

University of Oklahoma, Norman, USA
`jose.efraim.a.e@gmail.com, diochnos@ou.edu`

Abstract. We study perceptrons when datasets are randomly corrupted by noise and subsequently such corrupted examples are discarded from the training process. Overall, perceptrons appear to be remarkably stable; their accuracy drops slightly when large portions of the original datasets have been excluded from training as a response to verifiable random data corruption. Furthermore, we identify a real-world dataset where it appears to be the case that perceptrons require longer time for training, both in the general case, as well as in the framework that we consider. Finally, we explore empirically a bound on the learning rate of Gallant's "pocket" algorithm for learning perceptrons and observe that the bound is tighter for non-linearly separable datasets.

Keywords: Perceptrons · Stability · Verifiable Random Data Corruption · Reduced Sample Size

1 Introduction

The advances in machine learning during the last few years are without a precedent with ChatGPT [4] being perhaps the most remarkable example of the tremendous pace by which these advances are made. Nevertheless, there are concerns about the future of artificial intelligence and machine learning, the extent to which we can control the advances in these disciplines, and ultimately the impact that such advances will have in our daily lives [22].

Supervised machine learning is largely developed around the idea of *probably approximately correct learning* [28], which aims to generate, with high probability, a model that is highly accurate with respect to some ground truth function that is being learnt. For this reason, for the majority of applications, perhaps the most important metric that one cares on optimizing, is that of *accuracy* (or its dual, *error rate*), on a *validation set*. Nevertheless, we are also interested in other aspects of the learnt models, that we typically put under the broader umbrella of *trustworthy machine learning* [29]; the main axes being, *robustness* to noise [21] and adversaries [30], *interpretability* [23], and *fairness* [1], but one may also be interested in other related topics; e.g., *distribution shift* [5], and others.

G. Nicosia et al. (Eds.): LOD 2023, LNCS 14505, pp. 93–103, 2024.
https://doi.org/10.1007/978-3-031-53969-5_8

Hence, within trustworthy machine learning, and especially because of the need of explainable predictions, mechanisms that are inherently interpretable are put further into scrutiny, or attempts are made so that one can substitute components of black-box methods that are difficult to explain with such mechanisms, while still maintaining good predictive accuracy [25]. In this context, a basic, but popular approach, is the use of *linear models* for classification and it is such models that we explore in this work; namely, *perceptrons*.

1.1 Related Work

Different frameworks have been developed in order to study the robustness of classifiers and how one can attack or defend the classifiers in different contexts.

Stability and Reproducibility. Algorithms that do not overfit are *replace-one-stable* [27]; that is, replacing a particular example (x_0, y_0) in the training set, will yield a model that predicts a label on x_0 *similar* to y_0, had the learning algorithm used the original training set which included the example (x_0, y_0); see, e.g., [27]. Other notions of stability are also important and are studied in practice; e.g., how random seeds affect the performance of a learnt model [9]. Moreover, a recent line of work explores *reproducible algorithms* [16], where the idea is that the learning algorithm will generate *exactly the same model* when fed with two different samples drawn from the *same distribution*.

Noise. Beyond stability and reproducibility, data collected can be noisy, and a rich body of literature has been developed in order to understand better what is and what is not possible when datasets are corrupted by noise. A notable noise model is *malicious noise*, where both instances and labels can be corrupted arbitrarily by some adversary. In this model, it can be shown [17] that if we want to allow the possibility of generating a model that has error rate less than ε, then, the original dataset cannot suffer malicious corruption at a rate larger than $\frac{\varepsilon}{1+\varepsilon}$. In other words, even few examples that are carefully crafted and included in the learning process are enough to rule out models with low error rates.

Poisoning Attacks. While noise models typically assume the corruption of the examples in an online manner, *poisoning attacks* [3] typically consider adversaries who have access to the entire training set and decide how to inject or modify examples based on the full information of the dataset. Different models and defense mechanisms can be devised against poisoned data [13].

One approach for poisoning, corresponds to *clean-label* attacks [26], where the adversary may inject malicious examples in the training set, but their labels have to respect the ground truth of the function that is being learnt. In other words, the idea is to change the original distribution of instances to a more malicious one and make learning harder; difficult distributions have also been studied even for perceptrons, where it can be shown that the perceptrons may converge to solutions with significantly different time-complexity rates depending on the distribution [2]. A different line of work studies the *influence* that individual training examples have on the predictions of the learnt model, by being present or absent during learning [18]. Hence, instead of introducing misinformation in the dataset, one tries to understand how important different examples are for the learning process. In other words, how would the model's prediction change if we did not have access to a particular training example? Of course, one can introduce misinformation into the dataset; e.g., [19] considers the problem of designing imperceptible perturbations on training examples that can change the predictions of the learnt model. Overall, poisoning attacks study the brittleness of learning algorithms when training happens in the presence of adversaries.

As a last remark, one can also consider adversaries after training has been completed. In this direction, we find *evasion attacks,* also known as *adversarial examples* [14]. We note, however, that such adversarial settings are outside of the scope of the current work, as we deal with *training-time* attacks.

Missing Data. Concluding our brief literature review, an aspect that is quite orthogonal to the approaches discussed above, is that of dealing with missing data in our datasets. The reasons for the emergence of such datasets can range anywhere from pure chance (e.g., a questionnaire of a study subject is accidentally lost), to strategic information hiding (e.g., university applicants reveal favorable scores and hide potentially unfavorable ones in their applications) [20]. Typically, entries with missing data will either be completely ignored from the dataset and the training process, or some method will be devised in order to fill-in the missing entries based on the information of the rest of the entries in the dataset [12] so that they can subsequently be used for training. Perhaps the main difference with noise and poisoning attacks is the fact that the learner knows that certain entries are completely unknown in the input. Finally, we note that dealing with missing entries may also happen during evasion attacks [7], but we stress again that evasion attacks are outside of the scope of our work.

1.2   Our Setting: Verifiable Random Data Corruption

We explore the robustness of linear models used for classification. In particular, we study the perceptron learning algorithm [24] when the data may, or may not, be linearly separable; we use Gallant's "pocket" algorithm [11] which can deal with non-linearly separable data. Regarding the adversarial setting, we consider situations where the training data has been randomly corrupted by noise (e.g., transmission over an unreliable medium). However, we further assume, as is usually the case in practice, that there are flags (e.g., checksums) that indicate whether or not the information content of individual examples has been modified. In this context, we explore the following question:

> *To what extent are perceptrons tolerant of verifiable random data corruption?*

One natural way of dealing with such a scenario is to simply neglect the verifiably corrupted data altogether from the learning process. Hence, we are interested in understanding the behavior of perceptrons, in the *average case* as well as in a *worst case* sense, under such a simple *sanitization method* that effectively reduces the sample size used for learning. This adversarial setup and the sanitization approach has been considered before in a regression setting [10].

However, there are close connections to other adversarial models. For example, within clean-label attacks, the adversary may duplicate examples that already appear in the dataset. While such an approach may affect dramatically algorithms that rely on statistical properties of the data (e.g., decision trees), using perceptrons is very close to our context. The reason is that the weights of the learnt perceptron are a linear combination of the misclassified examples and hence the same solution can be obtained even when duplicate examples are omitted. Compared to the work of influence functions [18], our work explores the possibility of arbitrarily large amounts of data being removed from the training process, instead of carefully removing few but the most influential ones.

As a summary, we explore a mellow corruption scheme which has the benefits of *(i)* being easier to analyze, and *(ii)* occurring quite naturally in practice.

2   Preliminaries and Background

Notation. We study *binary classification* and we use $\mathcal{Y} = \{-1, +1\}$ to denote the set of *labels*, where -1 (resp. $+1$) corresponds to the *negative class* (resp. *positive class*). We use $\mathcal{X}$ to denote the set of *instances*, corresponding to real vectors; i.e., $\mathcal{X} = \mathbb{R}^n$. Note that throughout the paper we use n to denote the dimension of an instance. For two vectors $\boldsymbol{a}, \boldsymbol{b} \in \mathbb{R}^n$, we denote their inner product $\langle \boldsymbol{a}, \boldsymbol{b} \rangle = \sum_{i=1}^n a_i b_i$. Typically, the learner has access to a collection $\mathcal{T} = ((\boldsymbol{x_1}, y_1), \ldots, (\boldsymbol{x_m}, y_m))$ of *training examples* that are drawn iid from a distribution $\mathcal{D}$ governing $\mathcal{X} \times \mathcal{Y}$. Hence, learning is about selecting one appropriate *model* $h \colon \mathcal{X} \to \mathcal{Y}$ from the *model space* $\mathcal{H}$, based on $\mathcal{T}$. That is, for a learning algorithm $\mathcal{L}$, we have $\mathcal{L}(\mathcal{T}) = h$. We also use the function which returns the sign of its argument; i.e., $\text{sgn}(z) = +1$ if $z > 0$ and $\text{sgn}(z) = -1$ if $z \leq 0$. We consider the *"pocket"* variant of the perceptron, which is suitable also for non-linearly separable data. Furthermore, we need the following definition.

Definition 1 ((α, β)-Stability Against Random Data Corruption). *Let $\alpha, \beta \in [0, 1]$. Let $\mathcal{L}$ be a learning algorithm, let $\mathcal{T}$ be a dataset, and let $\mathcal{L}(\mathcal{T}) = h$. Moreover, let A_h be the accuracy of the learnt model $h \in \mathcal{H}$ on $\mathcal{T}$. We say that $\mathcal{L}$ is (α, β)-stable on $\mathcal{T}$, if when one removes up to an α-fraction of $\mathcal{T}$ chosen uniformly at random, and then use $\mathcal{L}$ to learn a model h' using the remaining examples, then it holds that $A_{h'} \geq A_h - \beta$.*

The fine point is that initially our leaner has access to examples of the form $\mathcal{T}_v = ((x_1, y_1, f_1), \ldots, (x_m, y_m, f_m))$, where $f_i \in \{\boldsymbol{\times}, \boldsymbol{\checkmark}\}$, indicating whether the information content of x_i and y_i has been modified or not. As a consequence, the dataset $\mathcal{T}_v$ is sanitized by simply dropping all the examples of the form $(x_i, y_i, \boldsymbol{\times})$ and learning with the rest. See also Sect. 3.1 for more details.

2.1 The Perceptron Learning Algorithm

A perceptron h maintains a set of weights $w \in \mathbb{R}^n$ and classifies an instance x with label ℓ according to the rule $h(x) = \ell = \text{sgn}(\langle w, x \rangle)$. Upon predicting ℓ, the perceptron updates its weight vector w using the rule $w \leftarrow w + \eta(y - \ell)x$, where η is *the learning rate* and y is the correct label. This rule, when applied on the misclassified examples (weights are updated only when mistakes occur), learns a *halfspace* that correctly classifies all training data assuming linear separability.

2.2 The Pocket Algorithm

The pocket algorithm is more useful as it applies the same update rule as the perceptron, but also stores (in its *pocket*) the best weights that have been discovered during the learning process; thus, allowing us to identify good solutions even for non-linearly separable data. Algorithm 1 has the details, where by following Gallant's notation [11], we have the following variables: π are the current weights of the perceptron; w are the weights of the best perceptron that we have encountered so far; run_π corresponds to the number of consecutive correct classifications using the current weights, when we select examples from T at random; run_π corresponds to the number of consecutive correct classifications that the weights that we have in our pocket (i.e., the best weights that we have encountered so far), when we select examples from T at random; num_ok_π is the true number of examples from T that the current weights π of the perceptron classify correctly; and num_ok_w is the true number of examples from T that the weights w that we have in our pocket classify correctly.

Hence, the "best" pocket weights correspond to to the weights that classified correctly the maximum number of examples from the collection T during training.

Gallant has provided sample complexity bounds for different topologies of perceptron networks. Proposition 1 below is a simplification of a result from [11], adapted to the "single-cell" model (i.e., a single neuron; a perceptron).

Proposition 1 (See [11]). *Let ε be the true error and ε° be the error over T, after learning. Let $S = (\varepsilon - \varepsilon^\circ)/\varepsilon$; i.e., S is the slack given to the algorithm between the true error ε and empirical error ε°. Also, let $L = \|w\|_2 = \sqrt{\sum_{j=0}^{n} w_j^2}$ be the length of the weight vector of the learnt perceptron. Then, with probability at least $1 - \delta$, the learnt perceptron will have true error rate at most ε, if the number of training examples m is larger than*

$$\min\left\{ \begin{array}{l} \frac{8}{S^2\varepsilon} \max\left\{\ln\left(8/\delta\right), \min\{2(n+1), 4(n+1)\log_2(e)\}\ln\left(\frac{16}{S^2\varepsilon}\right)\right\}, \\ \frac{\ln(1/\delta)+(n+1)\ln(2L+1)}{S^2\varepsilon} \min\left\{\frac{1}{2\varepsilon}, 2\right\} \end{array} \right\}.$$

3 Methodology and Resources

3.1 Random Data Corruption and Sanitization

We simulate the process of verifiable random data corruption affecting our dataset, by ignoring random groups of examples in our dataset and using the

---

Algorithm 1: "Pocket" Version of the Perceptron Learning Algorithm

---

Data: Training examples $\mathcal{T}$.
Result: Best weight vector w, in the sense that the induced halfspace classifies
$\mathcal{T}$ with as few misclassifications as possible.

1 $\pi \leftarrow 0$;   /* Initialize to zero all coordinates */
2 $\text{run}_\pi, \text{run}_w, \text{num\_ok}_\pi, \text{num\_ok}_w \leftarrow 0$;
3 Randomly pick a training example (x_i, y_i);
4 **if** π *correctly classifies* (x_i, y_i) **then**
5 $\quad$ $\text{run}_\pi \leftarrow \text{run}_\pi + 1$;
6 $\quad$ **if** $run_\pi > run_w$ **then**
7 $\quad\quad$ Compute num_ok_π by checking every training example;
8 $\quad\quad$ **if** $num\_ok_\pi > num\_ok_w$ **then**
9 $\quad\quad\quad$ $w \leftarrow \pi$;   /* Update the best weight vector found so far */
10 $\quad\quad\quad$ $\text{run}_w \leftarrow \text{run}_\pi$;
11 $\quad\quad\quad$ $\text{num\_ok}_w \leftarrow \text{num\_ok}_\pi$;
12 $\quad\quad\quad$ **if** *all training examples correctly classified* **then**
13 $\quad\quad\quad\quad$ stop (the training examples are separable)
14 **else**
15 $\quad$ $\pi \leftarrow \pi + y_i \cdot x_i$;   /* Form a new vector of perceptron weights */
16 $\quad$ $\text{run}_\pi \leftarrow 0$;

---

rest (where the checksums indicate no tampering) for training. Algorithm 2 summarizes this approach, where the whole evaluation process is repeated R times in order to smooth the results and understand better the expected, as well as the extreme, values of the performance of the learnt model. We test the above method using both synthetic, as well as real-world, datasets, of varying dimensions, in order to better understand the behavior of perceptrons when the available data has been corrupted (verifiably) at random instances.

3.2   Description of Synthetic Datasets

We constructed two synthetic datasets by randomly sampling 3,000 points from a specified dimensional space and labeling them using a pre-selected linear or non-linear model. In particular, for each dataset, we sampled $n \in \{4, 10, 25, 50, 100\}$ dimensional spaces with attribute values ranging between -10 and $+10$, following a uniform distribution in that interval. We then added an extra parameter with a constant value of $+1$ to be used as bias. For the linearly-separable synthetic dataset, we randomly sampled a baseline perceptron's parameters and we used it to produce the label of each point. For the non-linearly separable case, we used an n-degree polynomial of the form $c_0 + c_1 x_1 + c_2 x_2^2 + c_3 x_3^3 + ... + c_n x_n^n$, with randomly-chosen constants between -1 and $+1$. We obtained the labels by using the sign function of the output from the polynomial.

Algorithm 2: Random Data Corruption, Sanitization, and Learning

---

Data: Training examples $\mathcal{T}$, validation examples Γ.

1 **for** R *runs* **do**
2   Split $\mathcal{T}$ into B buckets
3   **for** $b = B$ *down to 1* **do**
4   Select a random permutation of b buckets to form an uncorrupted set of training examples $\mathcal{T}_{\text{clean}}$ and ignore the examples in $\mathcal{T} \setminus \mathcal{T}_{\text{clean}}$ which are assumed to be verifiably corrupted and thus discardedq m
5   Train a perceptron h with the "pocket" algorithm using $\mathcal{T}_{\text{clean}}$ (Algorithm 1)
6   Collect the accuracy of h over the validation examples Γ

3.3 Description of Real-World Datasets

We also performed experiments with five real-world datasets. Two datasets had comparable dimension; 'Iris' is linearly separable while the skin segmentation ('Skin') dataset is not. In order to get a better picture we explored the behavior of three more datasets of larger dimension: 'SPECT', 'Spam', and 'Bank', corre-

Table 1. Summary of real-world datasets used in our experiments.

Dataset	n	Number of Examples	Minority-Class Percentage	Linearly Separable
Iris	4	150	33.3%	yes
Skin	3	245,057	26.0%	no
SPECT	22	267	20.6%	no
Spam	57	4,601	39.4%	no
Bank	95	6,819	3.2%	no

sponding respectively to the SPECT heart dataset, the spambase dataset, and the Taiwanese bankruptcy prediction dataset. Table 1 has details. All datasets are available at the UCI repository.[1]

3.4 Training Set, Validation Set, Buckets, and Smoothing

Apart from the Iris and SPECT datasets which have only 150 and 267 instances respectively, in the other cases (synthetic, or real-world data), we sampled 3,000 data points at random. Such a number is more than sufficient for low dimensions, but is typically too small for traditional values obtained from statistical learning theory for datasets with $n \geq 50$ dimensions, when one wants (excess) error rate $\varepsilon - \varepsilon^\circ \leq 1\%$ with confidence $1 - \delta \geq 99\%$. Nevertheless, this amount seems reasonable given the size, in terms of the number of examples, of the real-world datasets that we explore – and in general, this is a reasonable size for real-world datasets. In every case, we performed an 80-20 split for the creation of the training examples $\mathcal{T}$ and validation examples Γ that are mentioned in Algorithm 2. We then subdivided $\mathcal{T}$ into $B = 100$ buckets of equal size and applied Algorithm 2 for the calculation of the performance (on the validation set Γ) of the models obtained, over $R = 100$ runs, using the pocket algorithm. Each run would process at most $3{,}000|\mathcal{T}|$ training examples.

---

[1] Homepage: https://archive.ics.uci.edu.

Table 2. Worst-case accuracy of perceptrons on synthetic datasets.

(a) Synthetic linearly separable data. Worst-case accuracy shown over 100 runs.

Corruption Level	Data Dimensionality (n)				
	4	10	25	50	100
0%	0.993	0.988	0.980	0.968	0.961
25%	0.990	0.985	0.978	0.968	0.958
50%	0.991	0.983	0.970	0.961	0.936
75%	0.983	0.973	0.961	0.925	0.885
90%	0.958	0.951	0.911	0.850	0.790
95%	0.948	0.886	0.831	0.753	0.681
99%	0.746	0.705	0.623	0.586	0.555

(b) Synthetic non-linearly separable data. Worst-case accuracy shown over 100 runs.

Corruption Level	Data Dimensionality (n)				
	4	10	25	50	100
0%	0.935	0.800	0.676	0.583	0.513
25%	0.935	0.801	0.665	0.591	0.521
50%	0.921	0.770	0.663	0.581	0.536
75%	0.905	0.776	0.643	0.551	0.518
90%	0.891	0.730	0.621	0.558	0.488
95%	0.863	0.706	0.538	0.508	0.478
99%	0.721	0.468	0.491	0.468	0.458

4   Experimental Results and Discussion

Source code is available at github.com/aguilarjose11/Perceptron-Corruption.

4.1   Experimental Results and Discussion on Synthetic Datasets

Table 2 presents the worst-case accuracy attained by perceptrons when learning synthetic datasets of different dimensions, covering both linearly separable as well as non-linearly separable data. Worst-case accuracy drops more than 3.5% only after 50% of the linearly-separable data has been corrupted, or more than 75% of the non-linearly separable data has been corrupted. In other words, by Definition 1, perceptrons are $(0.5, 0.035)$-stable for linearly separable data and $(0.75, 0.035)$-stable for non-linearly separable data, in the worst case. Using a similar table (not shown in the paper) for the average-case accuracy, perceptrons were, *on average*, $(0.5, 0.018)$-stable on linearly separable data and $(0.75, 0.018)$-stable on non-linearly separable data. Another takeaway is that perceptrons are $(0.25, 0.01)$-stable in the worst case, regardless of the synthetic dataset.

4.2   Experimental Results and Discussion on Real-World Datasets

Table 3 shows the average-case and worst-case performance of perceptrons on the datasets that we described in Sect. 3.3. Apart from SPECT, perceptrons are $(0.5, 0.023)$-stable and $(0.75, 0.023)$-stable in the worst case; i.e., perceptrons are similarly stable compared to our observations on the synthetic datasets. However, on SPECT, perceptrons are only $(0.5, 0.068)$-stable in the worst case. Finally, perceptrons, including SPECT, are $(0.25, 0.013)$-stable in the worst case.

Table 3. Mean and worst-case accuracy attained by perceptrons when learning real-world datasets. Values calculated over 100 runs.

(a) Mean accuracy on real data.

Corruption Level	Data Source				
	Iris	Skin	SPECT	Spam	Bank
0%	0.999	0.935	0.753	0.901	0.966
25%	0.998	0.934	0.734	0.898	0.967
50%	0.998	0.933	0.718	0.891	0.966
75%	0.998	0.929	0.709	0.879	0.968
90%	0.986	0.925	0.707	0.856	0.969
95%	0.956	0.922	0.705	0.824	0.971
99%	0.727	0.888	0.715	0.705	0.986

(b) Worst-case accuracy on real data.

Corruption Level	Data Source				
	Iris	Skin	SPECT	Spam	Bank
0%	0.966	0.905	0.643	0.864	0.961
25%	0.966	0.905	0.630	0.864	0.959
50%	0.966	0.910	0.575	0.841	0.959
75%	0.966	0.906	0.602	0.841	0.953
90%	0.666	0.901	0.493	0.766	0.941
95%	0.633	0.878	0.410	0.753	0.931
99%	0.333	0.726	0.205	0.549	0.878

Note that the different classes are represented in an imbalanced way in the datasets; see Table 1 for the ratio of the minority class in every case. Class imbalance can be a nuisance for classification [8,15]. One trivial solution under class imbalance is to predict according to the majority class. Under no corruption, and with the exception of SPECT, perceptrons were able to form a better solution than this trivial approach, whereas for SPECT it was

Table 4. Mean accuracy on real-world data after applying SMOTE on the training set.

Corruption Level	Data Source				
	Iris	Skin	SPECT	Spam	Bank
0%	0.998	0.937	0.703	0.901	0.617
25%	0.999	0.936	0.689	0.899	0.617
50%	0.997	0.935	0.689	0.893	0.616
75%	0.997	0.932	0.654	0.882	0.618
90%	0.996	0.928	0.623	0.858	0.626
95%	0.981	0.925	0.593	0.837	0.637
99%	0.730	0.914	0.534	0.723	0.712

the case that not even such a good solution was found. Nevertheless, in order to better understand how far one can go with imbalanced datasets, we also applied the Synthetic Minority Over-sampling Technique (SMOTE) [6], a popular method for dealing with class imbalance. The experiments using SMOTE had mixed results: SMOTE helped with the accuracy on the Skin dataset, but hurt the accuracy of the models in the other cases. Also, in some cases we identified a tradeoff between less accuracy and higher stability, but this observation was not consistent among all the datasets that we explored. Table 4 has details for the average case when SMOTE was applied in the training dataset.

4.3 Empirical Investigation of Gallant's Bound

In all of our experiments with separable and non-separable data, we compared the empirical accuracy that we obtained in our experiments, with the (*lower bound*) on the *accuracy* that is implied by Proposition 1 when we require failure probability $\delta = 1\%$. Proposition 1 was tighter on non-linearly separable data. Figure 1 gives an example of such a comparison on the non-linearly separable synthetic dataset of dimension $n = 25$.

5   Conclusion

Perceptrons appear to be remarkably stable when learning with reduced sample sizes. In all of our experiments removing up to 25% of the initial training sets would lower the accuracy of the learnt perceptron by not more than 1.3%. Furthermore, SPECT appears to be a harder dataset compared to all the other datasets tested; perceptrons failed to find even trivial solutions such that one can unconditionally predict according to the majority class and achieve better accuracy. This phenomenon complements [2] in the sense that SPECT appears to be a dataset where perceptrons

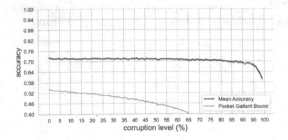

Fig. 1. Mean accuracy of perceptrons learnt on synthetic non-linearly separable datasets ($n = 25$) against the lower bound on the accuracy obtained from Proposition 1 using $\delta = 0.01$.

require an extended amount of time for identifying a good solution. Finally, we observed empirically that Gallant's bound on the pocket algorithm is tighter for non-linearly separable data.

One avenue for future work could be the exploration of penalty mechanisms embedded into the "pocket" algorithm, along the lines of regularization, so that we can obtain even more interpretable solutions for problems of interest, or accelerate learning similar to [10]. Another idea is the investigation of the reproducible weak halfspace learner from [16] and study its stability in this framework.

Acknowledgements. Part of the work was performed at the OU Supercomputing Center for Education & Research (OSCER) at the University of Oklahoma. The work was supported by the second author's startup fund. The first author worked on this topic while he was an undergraduate McNair Sholar.

References

1. Barocas, S., Hardt, M., Narayanan, A.: Fairness and machine learning: limitations and opportunities. fairmlbook.org (2019). http://www.fairmlbook.org
2. Baum, E.: The perceptron algorithm is fast for non-malicious distributions. In: NeurIPS 1989, vol. 2, pp. 676–685. Morgan-Kaufmann (1989)
3. Biggio, B., Nelson, B., Laskov, P.: Poisoning attacks against support vector machines. In: ICML 2012. icml.cc/Omnipress (2012)
4. Brown, T.B., et al.: Language models are few-shot learners. In: NeurIPS 2020, Virtual (2020)
5. Quiñonero Candela, J., Sugiyama, M., Schwaighofer, A., Lawrence, N.D.: Dataset Shift in Machine Learning. The MIT Press, Cambridge (2008)
6. Chawla, N.V., Bowyer, K.W., Hall, L.O., Kegelmeyer, W.P.: SMOTE: synthetic minority over-sampling technique. J. Artif. Intell. Res. **16**, 321–357 (2002)
7. Dekel, O., Shamir, O., Xiao, L.: Learning to classify with missing and corrupted features. Mach. Learn. **81**(2), 149–178 (2010)
8. Diochnos, D.I., Trafalis, T.B.: Learning reliable rules under class imbalance. In: SDM, pp. 28–36. SIAM (2021)

9. Fellicious, C., Weißgerber, T., Granitzer, M.: Effects of random seeds on the accuracy of convolutional neural networks. In: LOD 2020, Revised Selected Papers, Part II. LNCS, vol. 12566, pp. 93–102. Springer, Heidelberg (2020). https://doi.org/10.1007/978-3-030-64580-9_8

10. Flansburg, C., Diochnos, D.I.: Wind prediction under random data corruption (student abstract). In: AAAI 2022, pp. 12945–12946. AAAI Press (2022)

11. Gallant, S.I.: Perceptron-based learning algorithms. IEEE Trans. Neural Netw. **1**(2), 179–191 (1990)

12. García-Laencina, P.J., Sancho-Gómez, J., Figueiras-Vidal, A.R.: Pattern classification with missing data: a review. Neural Comput. Appl. **19**(2), 263–282 (2010)

13. Goldblum, M., et al.: Dataset security for machine learning: data poisoning, backdoor attacks, and defenses. IEEE Trans. Pattern Anal. Mach. Intell. **45**(2), 1563–1580 (2023)

14. Goodfellow, I.J., McDaniel, P.D., Papernot, N.: Making machine learning robust against adversarial inputs. Commun. ACM **61**(7), 56–66 (2018)

15. He, H., Garcia, E.A.: Learning from imbalanced data. IEEE Trans. Knowl. Data Eng. **21**(9), 1263–1284 (2009)

16. Impagliazzo, R., Lei, R., Pitassi, T., Sorrell, J.: Reproducibility in learning. In: STOC 2022, pp. 818–831. ACM (2022)

17. Kearns, M.J., Li, M.: Learning in the presence of malicious errors. SIAM J. Comput. **22**(4), 807–837 (1993)

18. Koh, P.W., Liang, P.: Understanding black-box predictions via influence functions. In: ICML 2017. Proceedings of Machine Learning Research, vol. 70, pp. 1885–1894. PMLR (2017)

19. Koh, P.W., Steinhardt, J., Liang, P.: Stronger data poisoning attacks break data sanitization defenses. Mach. Learn. **111**(1), 1–47 (2022)

20. Krishnaswamy, A.K., Li, H., Rein, D., Zhang, H., Conitzer, V.: Classification with strategically withheld data. In: AAAI 2021, pp. 5514–5522. AAAI Press (2021)

21. Laird, P.D.: Learning from Good and Bad Data, vol. 47. Springer, Heidelberg (2012). https://doi.org/10.1007/978-1-4613-1685-5

22. Marcus, G.: Hoping for the best as AI evolves. Commun. ACM **66**(4), 6–7 (2023). https://doi.org/10.1145/3583078

23. Molnar, C.: Interpretable Machine Learning, 2 edn. Independently Published, Chappaqua (2022). https://christophm.github.io/interpretable-ml-book

24. Rosenblatt, F.: Principles of Neurodynamics. Spartan Books, New York (1962)

25. Rudin, C.: Stop explaining black box machine learning models for high stakes decisions and use interpretable models instead. Nat. Mach. Intell. **1**(5), 206–215 (2019)

26. Shafahi, A., et al.: Poison frogs! targeted clean-label poisoning attacks on neural networks. In: NeurIPS 2018, pp. 6106–6116 (2018)

27. Shalev-Shwartz, S., Ben-David, S.: Understanding Machine Learning - From Theory to Algorithms. Cambridge University Press, Cambridge (2014)

28. Valiant, L.G.: A theory of the learnable. Commun. ACM **27**(11), 1134–1142 (1984)

29. Varshney, K.R.: Trustworthy Machine Learning. Independently Published, Chappaqua (2022)

30. Vorobeychik, Y., Kantarcioglu, M.: Adversarial machine learning. In: Synthesis Lectures on Artificial Intelligence and Machine Learning, # 38. Morgan & Claypool, San Rafael (2018)

Dynamic Soaring in Uncertain Wind Conditions: Polynomial Chaos Expansion Approach

Jiří Novák$^{(\boxtimes)}$ and Peter Chudý

Brno University of Technology, Brno, Czech Republic
`inovak@fit.vutbr.cz`

Abstract. Dynamic soaring refers to a flight technique used primarily by large seabirds to extract energy from the wind shear layers formed above ocean surface. A small Unmanned Aerial Vehicle (UAV) capable of efficient dynamic soaring maneuvers can enable long endurance missions in context of patrol or increased flight range. To realize autonomous energy-saving patterns by a UAV, a real-time trajectory generation for a dynamic soaring maneuver accounting for varying external conditions has to be performed. The design of the flight trajectory is formulated as an Optimal Control Problem (OCP) and solved within direct collocation based optimization. A surrogate model of the optimal traveling cycle capturing wind profile uncertainties is constructed using Polynomial Chaos Expansion (PCE). The unknown wind profile parameters are estimated from observed trajectory by means of a Genetic Algorithm (GA). The PCE surrogate model is subsequently utilized to update the optimal trajectory using the estimated wind profile parameters.

Keywords: Polynomial Chaos Expansion · Surrogate Modeling · Dynamic Soaring · Optimal Control

1 Introduction

The periodic flight technique of a wandering albatross (*Diomedea exulans*) called dynamic soaring caught attention of many biologists and engineers. The potential to stay airborne with little or no propulsion motivates research in guidance and control for dynamic soaring. The capability of computing energy-neutral trajectories as discretized OCP has been demonstrated several times e.g. [1,2]. The stochastic nature of the wind profile parameters has been included in the trajectory generation process in [3] to compute robust dynamic soaring cycles with respect to modeled uncertainties. In comparison, real-time correction technique based on quadratic programming [4] has been proposed to account for wind shear variations in dynamic soaring flight and a control technique such as in [5] was proposed to account for wind variations encountered in flight.

This research considers a surrogate modeling approach for dynamic soaring guidance and wind profile parameter estimation based on PCE and GA. Initially,

G. Nicosia et al. (Eds.): LOD 2023, LNCS 14505, pp. 104–115, 2024.
https://doi.org/10.1007/978-3-031-53969-5_9

the uncertainties associated with the wind profile are modeled and the PCE is evaluated using low number of OCP evaluations. The paper also studies a reverse problem to the PCE surrogate of estimating the wind profile parameters using the observed trajectories.

2   Methodology

2.1   Dynamic Soaring Trajectory Optimization

The system dynamics of the glider adopted from [3] is modeled as three dimensional point mass subject to sigmoid wind profile and wind shear layer. A reference frame $\mathcal{F}_e = (x_e, y_e, z_e, O_e)$ respecting North, East, Down (NED) convention is used throughout this paper. The state of the glider $\mathbf{x} = [x_e \; y_e \; z_e \; V \; \gamma \; \psi]^T$ is defined by its position in NED frame, airspeed, flight path angle relative to airflow and airflow relative heading angle. The control input vector $\mathbf{u} = [C_L \; \phi \; T]^T$ include the lift coefficient, the roll angle and the vehicle thrust which is assumed to be zero in most dynamic soaring studies. The equations of motion of the glider are listed by Eqs. (1)–(6).

$$\dot{x}_e = V \cos \gamma \cos \psi - W_x \tag{1}$$

$$\dot{y}_e = V \cos \gamma \sin \psi - W_y \tag{2}$$

$$\dot{z}_e = -V \sin \gamma \tag{3}$$

$$m\dot{V} = T - D - mg \sin \gamma + m\dot{W}_x \cos \gamma \cos \psi + m\dot{W}_y \cos \gamma \sin \psi \tag{4}$$

$$mV\dot{\gamma} = L \cos \phi - mg \cos \gamma - m\dot{W}_x \sin \gamma \cos \psi - m\dot{W}_y \sin \gamma \sin \psi \tag{5}$$

$$mV\dot{\psi} \cos \gamma = L \sin \phi + m\dot{W}_x \sin \psi - m\dot{W}_y \cos \psi \tag{6}$$

The lift and drag are expressed as $L = \frac{1}{2}\rho V^2 S C_L$ and $D = \frac{1}{2}\rho V^2 S C_D$, where the lift coefficient C_L is assumed to be a control variable and the drag coefficient is computed as $C_D = C_{D0} + kC_L^2$, where C_{D0} is the zero-lift drag coefficient and k is the lift induced factor. Finally, the sigmoid wind profile decomposed to $W_x = \sin \alpha W(z_e)$ and $W_y = \cos \alpha W(z_e)$ direction is described by Eq. (7).

$$W(z_e) = \frac{W_{ref}}{1 + \exp(z_e/\delta)} \tag{7}$$

The quantity W_{ref} represents the wind speed at a reference altitude and δ represents the shear layer thickness. The presence of the wind shear layer at $z_e = 0$, where the wind gradients are strongest are typically utilized by the seabirds maneuvering below 10 m above ocean surface. The wind gradients seen over land are typically not that significant to support energy-neutral flight patterns.

The soaring trajectory is known to have two possible patterns. The traveling pattern (S shape) alternate the turn directions and change the position of the glider in every cycle. In contrast, the non-traveling pattern (O-shape) have constant turn direction and the position at start and end of each cycle match.

The design of the flight trajectory is formulated as an Optimal Control Problem (OCP) and solved within direct collocation based optimization. The OCP is discretized using the Gauss pseudospectral transcription [6] method. Focusing solely on the traveling pattern, the boundary and path constraints defined in (8) were used along with the defined objective function. In this example, the objective was set to minimize thrusting along the trajectory. In all of the above cycles, the objective converges to zero due to sufficient wind gradient and initial speed of the glider [7]. The trajectory is discretized to N collocation points.

$$
\min_{\mathbf{u} \in U} \int_0^{t_f} T(t)\, dt, \quad \mathbf{u} = [C_L(t), \phi(t), T(t)]^T
$$

$$
\text{subj.to} \quad \dot{\mathbf{x}}(t) = \mathbf{f}\left(\mathbf{x}(t), \mathbf{u}(t)\right)
$$

$$
(x_e(0), y_e(0), z_e(0), V(0)) = (x_0, y_0, z_0, V_0)
$$

$$
(x_e(t_f), y_e(t_f)) = (x_f, y_f) \tag{8}
$$

$$
(z_e(t_f), V(t_f)) = (z_e(0), V(0))
$$

$$
V(t), C_L(t), T(t) \geq 0
$$

$$
-\pi < \psi(t) < \pi
$$

$$
-\pi/2 < \gamma(t) < \pi/2
$$

Using the OCP formulation given by (8), traveling dynamic soaring trajectories can be generated for various conditions including conditions which would result in infeasible trajectories without added thrusting. Additionally, the cycle period can be fixed so that $t_f = T_p$. The OCP (8) was discretized using the Gauss pseudospectral transcription method and solved using the IPOPT solver [8].

2.2   Polynomial Chaos Expansion for Optimal Control

The stochastic nature of the environment enabling dynamic soaring motivates the analysis and real-time correction of the maneuver and uncertainty driven guidance solution. The analysis of the underlying OCP with respect to modeled uncertainties would require large number of evaluations without the use of surrogate modeling. The PCE is widely used in many fields and applications such as reliability analysis of structures [9] or Computational Fluid Dynamics (CFD) uncertainty propagation [10]. For the purpose of this paper, the PCE theory will be used to generate a surrogate model of the optimal soaring trajectory with respect to uncertain wind profile parameters. Efficient computation of the optimal traveling cycle using the surrogate model allows fast correction of the cycle. A general scheme for PCE evaluation outlined in Fig. 1 has an evident advantage that the required nodes can be evaluated in parallel.

Assuming a probability space $(\Omega, \mathcal{F}, \mathcal{P})$ with event space Ω, σ-algebra $\mathcal{F}$ and probability measure $\mathcal{P}$, the random input vector can be written as $\mathbf{X}(\omega)$, $\omega \in \Omega$. The random model response $\mathbf{Y}(\omega)$ with unknown probability distribution is a result of transformation $\mathbf{Y} = \mathcal{M}(\mathbf{X})$. A general formulation of the PCE as a weighted sum of orthogonal polynomials [11] is given by Eq. (9), where M is number of random variables, β_α are unknown coefficients and Ψ_α are multivariate basis functions orthonormal with respect to the joint probability density function represented by $\mathbf{X}$.

$$\mathbf{Y} = \mathcal{M}(\mathbf{X}) = \sum_{\alpha \in \mathbb{N}^M} \beta_\alpha \Psi_\alpha(\mathbf{X}) \tag{9}$$

Wiener-Hermite PCE assumes standard normal input variables $X_m \sim \mathcal{N}(0,1)$, which is not satisfied in most applications, therefore probability space $\mathbf{X}$ is transformed by a Nataf transformation in order to be modeled as uncorrelated standard normal space $\boldsymbol{\xi}$. Hermite polynomial basis functions can subsequently be used to model $\mathcal{M}$ with truncated number of terms P. This set is dependent on the polynomial order p used for the expansion, such that the set of basis functions is defined by (10) with cardinality easily computed by a combination formula.

$$\mathcal{A}^{M,p} = \left\{ \alpha \in \mathbb{N}^M : |\alpha| = \sum_{i=1}^{M} \alpha_i \le p \right\} \quad \text{card } \mathcal{A}^{M,p} = \frac{(M+p)!}{M!p!} \tag{10}$$

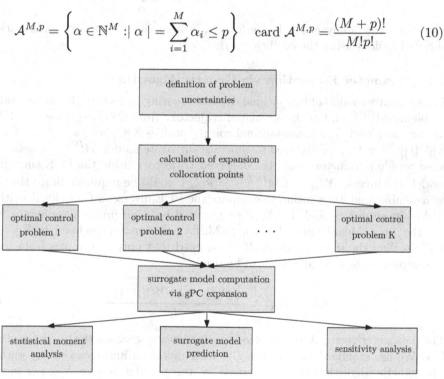

Fig. 1. General approach for PCE based uncertainty analysis for optimal control.

The estimation of PCE coefficients are obtained by minimization of error ϵ using Least-Square Regression (LSR) on the truncated PCE. A data set of evaluations $\mathcal{Y}$ for sample points $\mathcal{X}$ is used to minimize the L_2-norm given by Eq. (11).

$$\hat{\beta} = \arg\min \frac{1}{N} \sum_{i=1}^{N} \left(\beta^T \Psi(\xi^i) - \mathcal{M}(\mathcal{X}^i) \right)^2 \tag{11}$$

The minimization (11) has a solution in the form given by (12), where the data matrix is defined by Eq. (13).

$$\hat{\beta} = (\boldsymbol{\Psi}^T\boldsymbol{\Psi})^{-1}\boldsymbol{\Psi}^T\boldsymbol{y} \tag{12}$$

$$\boldsymbol{\Psi} = \{\Psi_{i,j} = \Psi_j(\xi^i), i = 1, \ldots, M, j = 0, \ldots, P-1\} \tag{13}$$

The term $\boldsymbol{\Psi}^T\boldsymbol{\Psi}$ is decomposed using Singular Value Decomposition (SVD) to avoid ill-conditioned terms in the computation. The resulting polynomial model is structured in a way, such that statistical moments can be easily obtained due to PCE basis orthonormality. The mean and variance can be obtained using Eqs. (14) and (15).

$$\mu = \mathbb{E}\left[\mathbf{Y}^{PCE}\right] = \beta_0 \tag{14}$$

$$\sigma^2 = \mathbb{E}\left[(\mathbf{Y}^{PCE} - \beta_0)^2\right] = \sum_{\alpha \neq 0} \beta_\alpha^2 \tag{15}$$

The sensitivity analysis can be performed in a similar manner by computation of Sobol indices using the coefficients β.

2.3   Parameter Estimation via Genetic Algorithm

The surrogate models of the optimal dynamic soaring trajectory (boundary value problem) $\mathcal{M}^{BVP}$ and of the simulated trajectory (initial value problem) $\mathcal{M}^{IVP}$ can be computed. The mean-optimal control profiles $\mathbb{E}\left[C_L^*[k]\right]$, $k = 1, \ldots N$ and $\mathbb{E}\left[\phi^*[k]\right]$, $k = 1, \ldots N$ are used to generate surrogate model $\mathcal{M}^{IVP}$. To estimate wind profile parameters from the observed trajectory using the PCE surrogate model, the inverse $\boldsymbol{\mathcal{X}}_{est.} = \left(\mathcal{M}^{IVP}\right)^{-1}(\boldsymbol{\mathcal{Y}}_{obs.})$ would be required. Since there is no available analytic solution to compute the PCE inverse, a Genetic Algorithm (GA) was chosen to find the $\boldsymbol{\mathcal{X}}_{est.}$ represented by wind profile parameters. We use the Root Mean Squared Error (RMSE) cost function defined in Eq. (16) while scaling the state variables. It is assumed that only K samples from total N samples of the soaring cycle can be used.

$$G(\boldsymbol{\mathcal{X}}) = \sqrt{\sum_{i=1}^{K} \frac{\|\boldsymbol{\mathcal{Y}}_{obs.}^i - \mathcal{M}^{IVP}(\boldsymbol{\mathcal{X}}^i)\|^2}{K}} \tag{16}$$

The parameter vector $\boldsymbol{\mathcal{X}}$ is converted to a binary sequence for the purposes of the GA. We use standard form of the GA with mutation and crossover operators. The genetic algorithm is parametrized by the population size N_{pop}, crossover probability r_c and mutation probability r_m. The parameter space is bounded as $\boldsymbol{\mathcal{X}}_{min.} \leq \boldsymbol{\mathcal{X}} \leq \boldsymbol{\mathcal{X}}_{max.}$. The estimated set of parameters $\boldsymbol{\mathcal{X}}_{est.}$ is subsequently used by the optimal trajectory surrogate $\mathcal{M}^{BVP}$ to compute the optimal dynamic soaring cycle with respect to updated parameters. Equation (17) is used to update the optimal state and control profiles of the glider.

$$\boldsymbol{\mathcal{Y}}_{est.} = \mathcal{M}^{BVP}(\boldsymbol{\mathcal{X}}_{est.}) \tag{17}$$

The updated control profiles are subsequently applied in simulation and compared to a reference.

3   Experiments and Results

3.1   Uncertainty Propagation

For the reference trajectory, we assume a traveling pattern of the cycle with period $T_p = 8\,\mathrm{s}$. The parameters of a glider are set as $m = 8.5\,\mathrm{kg}$, $C_{D0} = 0.02$, $S = 0.65\,\mathrm{m}^2$ and $k = 0.07$. The reference wind profile parameters are $W_{ref} = 7.8\,\mathrm{m}\cdot\mathrm{s}^{-1}$ and $\delta = 12\,\mathrm{m}$. The initial position of the glider is set to $x_0 = 0\,\mathrm{m}$, $y_0 = 0\,\mathrm{m}$, $z_0 = 0\,\mathrm{m}$. The required final position after performing a cycle are $x_f = 40\,\mathrm{m}$, $y_f = 0\,\mathrm{m}$, $z_0 = 0\,\mathrm{m}$. The Fig. 2 shows a simulation scenario assuming a traveling cycle with the open-loop control input profiles computed by OCP using the reference wind parameters. The effect of wind profile parameter variations can have significant effect on the soaring cycle resulting in mismatch of initial and final conditions.

The performed numerical experiments are summarized in Table 1, where the RMSE values ϵ_p^1 and ϵ_p^3 represent the position mismatch and ϵ_v^1 and ϵ_v^3 represent the velocity mismatch. In order to account for the wind profile uncertainties, the PCE of the soaring cycle is evaluated. The polynomial order $p = 4$ is selected and

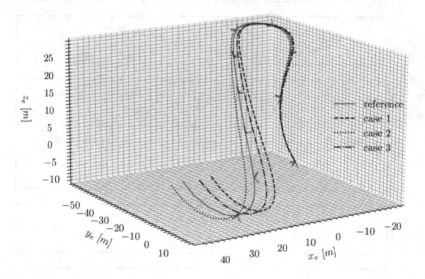

Fig. 2. A single dynamic soaring cycle subject to parametric uncertainties.

Table 1. Overview of dynamic soaring simulations under wind profile variations.

simulation	W_{ref} [m·s^{-1}]	δ [m]	ϵ_v^1 [m·s^{-1}]	ϵ_v^3 [m·s^{-1}]	ϵ_p^1 [m]	ϵ_p^3 [m]
reference	7.80	12.00	–	–	–	–
case 1	8.80	12.00	$3.96 \cdot 10^{-1}$	$3.48 \cdot 10^{-1}$	12.78	17.12
case 2	7.80	13.20	2.71	$1.25 \cdot 10^{-1}$	4.15	5.69
case 3	8.80	13.20	$3.23 \cdot 10^{-1}$	$5.30 \cdot 10^{-1}$	8.30	10.42

the expansion nodes are generated as Gaussian quadrature nodes and weights of order $q = 4$. The wind profile parameters are assumed to be modeled as random normally distributed variables as $W_{ref} \sim \mathcal{N}(\mu_w, \sigma_w^2)$ and $\delta \sim \mathcal{N}(\mu_\delta, \sigma_\delta^2)$. For $\mu_w = 7.8 \, \text{m·s}^{-1}$, $\sigma_w = 0.4 \, \text{m·s}^{-1}$, $\mu_\delta = 12 \, \text{m}$ and $\sigma_\delta = 0.4 \, \text{m}$, the surrogate model was obtained. The Fig. 3 shows the mean and standard deviation of optimal dynamic soaring profiles.

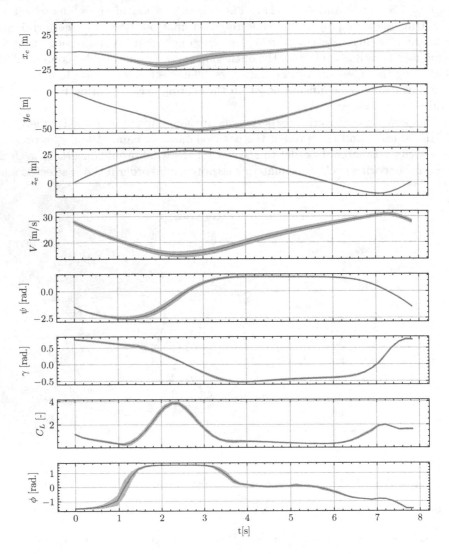

Fig. 3. Mean and standard deviation trajectories computed by PCE.

3.2 Wind Profile Parameter Estimation

For the cases defined in Table 1, the parameter estimation based on $\mathcal{M}^{IVP}$ and genetic algorithm was utilized. The parameter estimation was performed using data from full and half of the cycle while some scenarios assume Gaussian measurement noise $n_m \sim \mathcal{N}(0, \sigma_m^2)$ added to each sample. The population size for the genetic algorithm was set to $N_{pop} = 100$. The two wind profile parameters were converted to $b = 16$ bit array. The crossover probability was set to $r_c = 0.9$ while the mutation probability was set to $r_m = 0.0625$. The number of GA iterations $N_{iter} = 100$ was used. The bounds $\boldsymbol{X}_{min.} = [6 \text{ m} \cdot \text{s}^{-1}, 10 \text{ m}]^T$ and $\boldsymbol{X}_{max.} = [10 \text{ m} \cdot \text{s}^{-1}, 14 \text{ m}]^T$ were set. The Table 2 shows the parameters and results for each experiment assuming case 3 with true wind profile parameters $W_{ref.} = 8.80 \text{ m} \cdot \text{s}^{-1}$ and $\delta = 13.20 \text{ m}$. The GA achieved estimation of the reference speed with lowest error in case 3 while the lowest shear layer thickness error was estimated in case 1. Overall, it can be observed that the added noise has higher impacts the estimation more significantly compared to used trajectory sample size.

Table 2. Parameter estimation scenarios and estimated wind profile values.

estim.	K	σ_m^2	$\hat{W}_{ref}$ [m·s^{-1}]	ΔW_{ref}	$\hat{\delta}$ [m]	$\Delta\delta$	$G(\boldsymbol{X}_{est.})$
case 1	25	$1.00 \cdot 10^{-6}$	8.762	0.038	13.174	0.026	0.624
case 2	25	$2.50 \cdot 10^{-5}$	8.761	0.039	13.122	0.078	2.947
case 3	50	$1.00 \cdot 10^{-6}$	8.824	0.024	13.269	0.069	1.360
case 4	50	$2.50 \cdot 10^{-5}$	8.774	0.026	13.155	0.045	6.305

The convergence of GA in all listed scenarios is shown in Fig. 4. The significance of added noise can be seen comparing case 3 and case 4, where the cost function value reduction is much more significant in case 3 with lower noise.

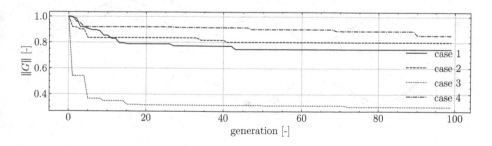

Fig. 4. Progress of GA in all estimation cases.

3.3   Validation of Updated Trajectories

The estimated wind profile values listed in Table 2 have been used to update the optimal state and control trajectories using the surrogate model $\mathcal{M}^{BVP}$. The Table 3 shows the RMSE values of position and velocity after simulating one and three cycles and compared to the reference obtained by IPOPT with the true wind profile values. It can be observed that the position error increases with each cycle compared to the velocity error which can oscillate. The control profiles obtained by the surrogate model with estimated wind profile values were utilized in all simulation cases. The Fig. 5 and Fig. 6 show the evolution of dynamic soaring cycles using the reference optimal control profiles, the control profiles of the initial mean wind profile parameters and the PCE updated trajectories using the estimates from case 4 in Table 3 which closely follow the optimal profiles.

Table 3. Evaluation of the simulations using the updated control profiles.

update	$\hat{W}_{ref}$ [m·s⁻¹]	$\hat{\delta}$ [m]	ϵ_v^1 [m·s⁻¹]	ϵ_v^3 [m·s⁻¹]	ϵ_p^1 [m]	ϵ_p^3 [m]
case 1	8.762	13.174	$2.89 \cdot 10^{-1}$	$1.71 \cdot 10^{-1}$	4.96	7.55
case 2	8.761	13.122	$2.44 \cdot 10^{-1}$	$2.22 \cdot 10^{-1}$	5.12	7.82
case 3	8.824	13.269	$2.86 \cdot 10^{-1}$	$6.48 \cdot 10^{-2}$	3.64	6.92
case 4	8.774	13.155	$2.54 \cdot 10^{-1}$	$1.95 \cdot 10^{-1}$	4.83	7.70

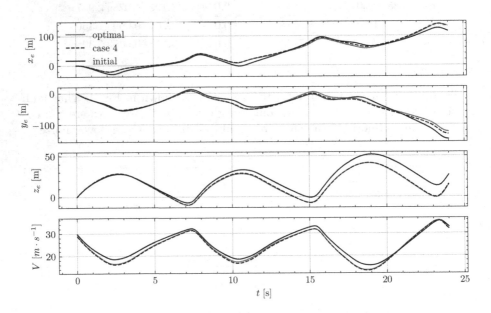

Fig. 5. Position and airspeed in trajectories in simulation of three soaring cycles.

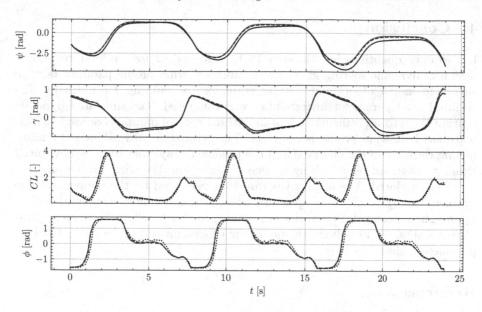

Fig. 6. Attitude and control trajectories in simulation of three soaring cycles.

3.4   Wind Direction Uncertainty

The presented experiment assumed $\alpha = 0$ rad resulting in a wind gradient in y_e axis. In this experiment, both x_e and y_e components of the wind gradient are assumed with direction uncertainty given by $W_\alpha \sim \mathcal{N}(\mu_\alpha, \sigma_\alpha^2)$ and reference wind speed uncertainty $W_{ref} \sim \mathcal{N}(\mu_w, \sigma_w^2)$ where $\mu_\alpha = 0.3$ rad, $\sigma_\alpha = 0.1$ rad and $\mu_w = 8.2$ m $\cdot$ s^{-1}, $\sigma_w = 0.37$ m $\cdot$ s^{-1}. The true values of the wind direction and wind speed for the experiment were $\alpha = 0.2$ rad and $W_{ref} = 8.5$ m $\cdot$ s^{-1}. The same values for the GA parameters were used. As in the first scenario, the estimation using GA is performed in cases with added noise and using different number of trajectory samples. The results are given in Table 4. Even in presence of noise in the simulated trajectory data, the estimation error in wind gradient direction has not exceeded 0.2 deg. This shows a potential to augment additional uncertainties in the model although increasing the number of random variables leads to substantial increase in required number of evaluations to maintain the same accuracy.

Table 4. Parameter estimation of wind direction and reference wind speed.

estim.	K	σ_m^2	$\hat{\alpha}$ [rad]	$\Delta\alpha$	$\hat{W}_{ref}$ [m $\cdot$ s^{-1}]	ΔW_{ref}	$G(\boldsymbol{\mathcal{X}}_{est.})$
case 1	25	$1.00 \cdot 10^{-6}$	0.202	0.002	8.510	0.010	0.626
case 2	25	$2.50 \cdot 10^{-5}$	0.203	0.003	8.414	0.086	3.044
case 3	50	$1.00 \cdot 10^{-6}$	0.199	0.001	8.502	0.002	1.369
case 4	50	$2.50 \cdot 10^{-5}$	0.197	0.003	8.486	0.014	6.662

4   Conclusion

Uncertainty quantification through PCE was studied for optimal trajectory design of dynamic soaring glider given unknown wind profile parameters. The surrogate model of optimal soaring trajectory accounting for normally distributed wind parameter uncertainties was computed. The simulated dynamic soaring trajectories utilizing the mean optimal control inputs were used to estimate the true wind profile parameters. The estimation was achieved by minimizing the error of the observed and the predicted trajectory obtained by initial value problem surrogate. Using a genetic algorithm, the wind profile parameters were updated. The acquired parameters were used by the boundary value problem surrogate model to generate new state and control trajectories. The approximated control trajectories have shown to closely follow the reference trajectories obtained by solving the nonlinear programming problem with the true parameter values.

References

1. Liu, D., Hou, Z., Guo., Z., Yang, X., Gao, X.: Optimal patterns of dynamic soaring with a small unmanned aerial vehicle. Proc. Inst. Mech. Engineers J. Aeros. Eng. **231**, 13593–1608 (2016). https://doi.org/10.1177/0954410016656875
2. Zhao, Y.: Optimal patterns of glider dynamic soaring. Optimal Control Appl. Methods **25**(1), 67–89 (2004)
3. Flanzer, T., Bower, G., Kroo, I.: Robust trajectory optimization for dynamic soaring. In: AIAA Guidance, Navigation, and Control Conference, pp. 1–22 (2012). https://doi.org/10.2514/6.2012-4603
4. Hong, H., Zheng, H., Holzapfel, F., Tang, S.: Dynamic soaring in unspecified wind shear: a real-time quadratic-programming approach. In: 27th Mediterranean Conference on Control and Automation (MED), Akko, Israel, pp. 600–605 (2019). https://doi.org/10.1109/MED.2019.8798573
5. Perez, R.E., Arnal, J., Jansen, W.P.: Neuro-evolutionary control for optimal dynamic soaring. In: AIAA Scitech 2020 Forum (2020). https://doi.org/10.2514/6.2020-1946
6. Benson, D.A., Huntington, G.T., Thorvaldsen, T.P., Tom, P., Rao, A.V.: Direct trajectory optimization and costate estimation via an orthogonal collocation method. J. Guid. Control. Dyn. **29**(6), 1435–1440 (2006). https://doi.org/10.2514/1.20478
7. Sachs, G., Grüter, B.: Optimization of thrust-augmented dynamic soaring. J. Optim. Appl. **192**(1), 960–978 (2022). https://doi.org/10.1007/s10957-021-01999-5
8. Wächter, A., Biegler, L.: On the implementation of an interior-point filter line-search algorithm for large-scale nonlinear programming. Math. Program. **106**, 25–57 (2006). https://doi.org/10.1007/s10107-004-0559-y
9. Novák, L., Novák, D.: Surrogate modelling in the stochastic analysis of concrete girders failing in shear. In: Proceedings of the Fib Symposium 2019: Concrete - Innovations in Materials, Design and Structures, pp. 1741–1747 (2019)

10. Knio, O., Le Maître, O.: Uncertainty propagation in CFD using polynomial chaos decomposition. Fluid Dyn. Res. **38**(1), 616–640 (2006). https://doi.org/10.1016/j.fluiddyn.2005.12.003

11. Novák, L., Novák, D.: Polynomial chaos expansion for surrogate modelling: theory and software. Beton- und Stahlbetonbau **113**(1), 27–32 (2018). https://doi.org/10.1002/best.201800048

Solving Continuous Optimization Problems with a New Hyperheuristic Framework

Nándor Bándi[✉] and Noémi Gaskó

Faculty of Mathematics and Computer Science, Babeş-Bolyai University,
Cluj, Romania
{nandor.bandi,noemi.gasko}@ubbcluj.ro

Abstract. Continuous optimization is a central task in computer science. Hyperheuristics prove to be an effective mechanism for intelligent operator selection and generation for optimization problems. In this paper we propose a two level hyperheuristic framework for continuous optimization problems. The base level is used to optimize the problem with operator sequences that are modeled by a nested Markov chain, while the hyper level searches the operator sequence and parameter space with simulated annealing. The experimental results show that the proposed approach matches the performance of another state-of-the-art hyperheuristic using significantly less operators and computational time. The model outperforms the simple metaheuristic operator approach and the random hyperheuristic search strategy.

Keywords: continuous optimization · hyperheuristic · genetic algorithm · differential evolution · simulated annealing

1 Introduction

Optimization is a key task in computer science, which can be generally formulated as the selection of the best element from a set of available elements, with respect to some criteria. As two main classes of optimization we can distinguish between discrete and continuous optimization, both of them having several application possibilities. Due to the importance of the optimization process, several algorithms were proposed, from exact methods to approximate methods, taking into account the nature of the optimization problem.

Some of the recent studies concern on the choice of the right algorithm for a certain optimization problem. A further idea is to provide an automatic way of selecting or generating heuristics, which is called hyperheuristic optimization (see [17] for a review). The hyperheuristic approach can be seen as the optimization of the optimization process, or in other words 'heuristics to choose a heuristic' (as described in [4]). The majority of existing hyperheuristic algorithms are designed for discrete optimization problems, such as the vehicle routing problem

G. Nicosia et al. (Eds.): LOD 2023, LNCS 14505, pp. 116–130, 2024.
https://doi.org/10.1007/978-3-031-53969-5_10

[13,16], traveling salesman problem [10], bin packing problem [18], etc. Hyperheuristics in the continuous domain have gained focus only recently, although several continuous real-world optimization problems exist, for example engineering problems [6], computational biology problems [21], economical problems [14], where hyperheuristics frameworks can improve significantly the results.

The main goal of the article is to propose a new hyperheuristic framework for continuous optimization problems, a two-level framework, in which the base level models operator sequences using a nested Markov chain and the hyper level uses Simulated Annealing algorithm to find the optimal operator sequences and parameters.

The remainder of the paper is organized as follows. Section two presents the related work, Section three describes the proposed framework, and Section four details the numerical experiments. The last Section contains conclusions and further research directions.

2 Related Work

Several hyperheuristic classifications are proposed in the literature. [1] proposes two main classes of hyperheuristics: selection-based hyperheuristics, where the best performing heuristics are chosen from an existing list, and generation-based hyperheuristics, where new algorithms are designed from existing components. [3] presents four selection classifications (meta-heuristic, random, greedy and reinforcement learning based).

Next we present some existing hyperheuristics where genetic algorithms or differential evolution algorithms are used.

The hypDE framework [20] is based on the Differential Evolution (DE) algorithm, and it is used for constrained optimization problems. [15] proposes a DE solution for mixed-integer non-linear programming problems. In [11] a differential evolution based hyperheuristic (DEHH) is proposed to solve the flexible job-shop scheduling problem with fuzzy processing time. In this algorithm the DE operator is used on the high level to coordinate the low level heuristics.

[8] proposes hyper-TGA, a tabu assisted genetic algorithm for the personnel scheduling problem. In [9] the ALChyper-GA is introduced and used for a course scheduling problem. [22] proposes a hybrid algorithm to solve the job shop scheduling algorithm, where a genetic algorithm is used on the higher level. [23] also proposes a hyperheuristic using a genetic algorithm on the higher level to solve a frequency assignment problem. There are several other hyperheuristic frameworks proposed by the literature such as HyFlex, Hyperion, EvoHyp but these are also designed for solving combinatorial problems [7,19]. The direction of hyperheuristics for continuous problems is less researched which motivates further work in this direction. HyperSPAM [2] is a framework that adaptively alternates between two single point algorithms after initialization with CMA-ES, [5] propose a hyperheuristic framework (CustomHyS) for solving continuous problems by tailoring a heuristic collection to the problem instance.

3   Proposed Hyperheuristic Framework

Our hyperheuristic framework aims to improve on the existing solutions by providing a solution that is able to automatically adapt the operator parameters without relying on a collection of parameters and that is able to match the current state-of-the-art with less operators used. The proposed hyperheuristic framework (SA-PERTURB) is based on two levels. The base level is used to optimize the problem and contains the domain specific problem definition, a set of heuristic operators and a Markov chain defining the sequence of operators that are applied. The hyper level is domain independent and is used to guide and improve the base level by optimizing the applied operator sequence and the operator parameters for the problem. Next we give the formal definition of the proposed framework.

General Formulation. The proposed framework can be defined as the pair (H, B), where H is the hyper level and B is the base level of the system. The base level $B = (\mathcal{C}, f, M, \beta_c)$ contains a set $(\mathcal{C})$ of operator categories, each category (c) having a set of operators, and the objective function definition $f : \mathbb{R}^n \to \mathbb{R}$. Operators are classified into different categories according to the type of exploration or selection that that they provide. The base level is modeling transitions between categories and operators within the categories with the nested Markov chain $M = (M_c, M_{\mathcal{C}})$ where $M_{\mathcal{C}} = (\pi_{\mathcal{C}}, P_{\mathcal{C}})$ and $M_c = \{\pi_c, P_c, \forall c \in \mathcal{C}\}$. The base level optimizes the problem by applying the operators in a sequence parameterized by π_c, P_c, $\pi_{\mathcal{C}}$ and $P_{\mathcal{C}}$ where π_c, P_c are the initial distribution and Markov chain transition probability matrix for the category $c \in \mathcal{C}$, and $\pi_{\mathcal{C}}$ and $P_{\mathcal{C}}$ are the initial category distribution and the transition matrix of the categories. In addition to the parameterization of the operator sequences, the base level is parameterized by the set of operator parameters for each category (β_c).

The hyper level $H = (P \times \beta, S)$ can be defined by the base level configuration search space $P \times \beta$ and the search algorithm S. The search space can be defined as the product of the Markov chain

$$P = \pi_{\mathcal{C}} \times P_{\mathcal{C}} \times \pi_{c_1} \times P_{c_1} \times \pi_{c_2} \times P_{c_2} \cdots \times \pi_{c_k} \times P_{c_k}, \forall c_i \in \mathcal{C}$$

and operator parameter space

$$\beta = \beta_{c_1} \times \beta_{c_2} \cdots \times \beta_{c_k}, \forall c_i \in \mathcal{C}$$

The advantage of this formulation of a hyperheuristic system is that the hyper level is able to simultaneously search in the operator parameter space, and the operator sequence space. Furthermore the upper level is able to easily select for the best performing operator categories which improves the adaptability of the framework. SA-PERTURB uses simulated annealing with linear multiplicative cooling schedule as the search procedure in the hyper level. It makes use of two operator categories and four operators. Genetic Algorithm (GA) operator, Differential Evolution (DE), Grey Wolf Optimizer (GWO) are in the Perturb

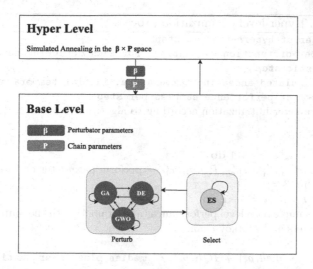

Fig. 1. The structure of the SA-PERTURB architecture. The base level is used to optimize the problem. This level has three operators in the Perturb category: a genetic algorithm (GA), a Differential Evolution (DE), and Grey Wolf Optimizer (GWO), an elitist selector (ES) is used in the select category. The hyper level adapts their parameters and sequence to the problem using Simulated Annealing (SA).

Algorithm 1. Hyperheuristic initialization process

```
    𝒞 - set of operator categories
    β_c - operator parameters of category c
    l_θ, u_θ - lower and upper bounds for operator parameter θ
    π_c, P_c - initial distribution and operator transition probability
matrix for category c
    π_C, P_C - initial distribution and category transition probability
matrix
    for all  c ∈ 𝒞 do
        for all  θ ∈ β_c do
            θ ← l_θ + U(0, 1) · (u_θ − l_θ) // set every parameter randomly within its
            bounds
        end for
        P_c ← U(T) // set a random transition matrix for category c
        π_c ← U(π) // set a random initial distribution for category c
    end for
    P_C ← U(T) // set a random category transition matrix
    π_C ← U(π) // set a random category initial distribution
```

---

category, and Elitist Selection (ES) is in the Select category. The structure of SA-PERTURB is depicted in Fig. 1.

The general formulation of the initialization and optimization process of the proposed hyperheuristic are detailed in Algorithms 1–5.

Algorithm 2. Hyper level optimization process

---

n_s - number of hyperheuristic steps
n_f - number of function evaluations for each offspring per
hyperheuristic step
t, t_0, α - simulated annealing temperature, initial temperature and α
n_p - number of performance samples per step
perform parameter initialization according to algorithm 1
$best_f \leftarrow \infty$
$t \leftarrow t_0$
for $s = 0; s < n_s; s = s + 1$ **do**
$\quad \beta'_{c_1} \ldots \beta'_{c_k}, \pi'_{c_1}, P'_{c_1} \ldots \pi'_{c_k}, P'_{c_k}, \pi'_C, P'_C \leftarrow$ perform parameter mutation according
$\quad$ to algorithm 3
$\quad$ **for** $i = 0; i < n_p; i = i + 1$ **do**
$\quad\quad p_{s,i} \leftarrow$ sample base level performance (algorithm 4) with the mutated operator
$\quad\quad$ parameters and Markov chain
$\quad$ **end for**
$\quad medIQR \leftarrow med(p_s) + IQR(p_s)$ // median plus interquantile range as
$\quad$ performance measure
$\quad$ **if** $medIQR < best_f \vee U(0,1) < e^{\frac{best_f - medIQR}{t}}$ **then**
$\quad\quad best_f \leftarrow medIQR$
$\quad\quad \beta_{c_1} \ldots \beta_{c_k}, \pi_{c_1}, P_{c_1} \ldots \pi_{c_k}, P_{c_k}, \pi_C, P_C \leftarrow \beta'_{c_1} \ldots \beta'_{c_k}, \pi'_{c_1}, P'_{c_1} \ldots \pi'_{c_k}, P'_{c_k}, \pi'_C, P'_C$
$\quad$ **end if**
$\quad t \leftarrow \frac{t_0}{1 + \alpha \cdot s}$
end for

Algorithm 3. Base level parameter mutation process

---

for all $c \in C$ **do**
$\quad \beta'_c \leftarrow \beta_c$
$\quad$ **for all** $\theta' \in \beta'_c$ **do**
$\quad\quad \theta' \leftarrow \theta' + \epsilon, \epsilon \sim \mathcal{N}(0, \frac{(u_{\theta'} - l_{\theta'})}{3})$ // mutate operator parameters
$\quad\quad$ **if** $\theta' < l_{\theta'}$ **then**
$\quad\quad\quad \theta' \leftarrow l_{\theta'}$
$\quad\quad$ **end if**
$\quad\quad$ **if** $\theta' > u_{\theta'}$ **then**
$\quad\quad\quad \theta' \leftarrow u_{\theta'}$
$\quad\quad$ **end if**
$\quad$ **end for**
$\quad$ // mutate the initial distribution and transition matrix of c
$\quad$ keeping the simplex restrictions
$\quad \pi'_c \leftarrow \pi_c + \epsilon$
$\quad P'_c \leftarrow P_c + \epsilon$
end for
// mutate the initial distribution and transition matrix of C keeping
the simplex restrictions
$\pi'_C \leftarrow \pi_C + \epsilon$
$P'_C \leftarrow P_C + \epsilon$

Algorithm 4. Base level optimization process

l_x, u_x - bounds of the optimization problem
X - population of solutions
X' - next generation of solutions
op_c - current operator state in category c
cat - current operator category state
e_f - current number of function evaluations
e_{op} - function evaluation cost of operator op
for all $x \in X$ **do**
    $x \leftarrow U(l_x, u_x)$ // random uniform initialization of offspring
end for
$X' \leftarrow X$
for all $c \in C$ **do**
    $op_c \leftarrow op, op \sim \pi_c$ // initialize the c operator state according to π_c
end for
$cat \leftarrow c, c \sim \pi_C$ // initialize the category state according to π_C
$e_f \leftarrow 0$
while $e_f < n_f$ **do**
    $X, X' \leftarrow op_{cat}(\beta_{cat}, X, X')$ // apply current operator in the current category (op_{cat})
    $e_f \leftarrow e_f + e_{op_{cat}}$
    $cat \leftarrow c, c \sim P_{C,cat}$ // set next operator category state going from cat
    $op_{cat} \leftarrow op, op \sim P_{cat,op_{cat}}$ // set next operator state in category cat going from op_{cat}
end while
return $best(X, X')$ // return the cost of the best offspring

---

Algorithm 5. GA scheme for arithmetic and one point crossover

m - size of population X
for $j = 0, j < m, j = j + 1$ **do**
    $x'_j \leftarrow x_j$ // x'-offspring, x-parent
    if $U(0,1) < GA_{CR}$ **then**
        select best parents from two random GA_{PR} sized parent pools $x_{i_1}, x_{i_2}, i_1 \neq i_2 \neq j$
        for $k = 0, k < n, k = k + 1$ **do**
            if $k < GA_{CP} * n$ **then**
                $x'_j[k] \leftarrow \alpha \cdot x_{i_1}[k] + (1 - \alpha) \cdot x_{i_2}[k]$
            else
                $x'_j[k] \leftarrow \alpha \cdot x_{i_2}[k] + (1 - \alpha) \cdot x_{i_1}[k]$
            end if
            if $U(0,1) < GA_{MR}$ **then**
                $x'_j[k] \leftarrow x_j[k] + \mathcal{N}(0, GA_{MS})$
            end if
        end for
    end if
end for

3.1 Algorithms in Detail

Algorithm 1 details the initalization process of the model. All operator and Markov chain parameters are randomly initialized within their bounds.

Algorithm 2 details the optimization procedure that the hyper level performs while searching for the optimal operator and markov chain parameters. The process starts with the random configuration that is reached via Algorithm 1 and optimizes it via Simulated annealing with linear multiplicative cooling for a predefined number of hyperheuristic steps. At each step the previous configuration is modified according to Algorithm 3 and a statistically significant number of base level performance samples are taken. A function evaluation limit is imposed on the base level for each performance sample. The median and interquantile range of the final objective function samples are compared to select for the best configuration.

Algorithm 3 presents the mutation process that the Simulated Annealing procedure performs at each step. All operator parameters are mutated according to a Gaussian noise factor scaled to the parameter bounds. Furthermore the Markov chain parameters are also mutated with a similar noise factor, preserving their simplex properties.

Algorithm 4 details the optimization process that the base level performs. The procedure starts with a random population of solutions and the operator and Markov chain parameter configuration given by the hyper level. At each optimization step the operator category, and the next operator within that category is selected according to the two levels of the Markov chain. This operator is then applied to each solution in the population. Each operator is given a reference to the current (X) and next population (X') which allows for both selector and exploratory operators to be applied.

Algorithm 5 presents the hybrid crossover used by the GA operator. This operator is parameterized in such a way, that it is able to express both arithmetic and one point crossover. Having the arithmetic crossover rate $\alpha = 0$ results in one point crossover and having the one point crossover point $GA_{CP} = 0$ results in arithmetic crossover.

3.2 Operators

The operators in the Perturb category were chosen with different exploratory functions in mind. The Genetic Algorithm allows the propagation of well performing partial solutions, The Grey Wolf Optimizer [12] is able to express both exploration and exploitation and Differential Evolution which is able to explore the continuous search space well.

The GA operator is parameterized in a way that enables it to express both standard arithmetic and one point crossover (as depicted in Algorithm 5). This operator exposes the arithmetic crossover constant $GA_\alpha \in [0, 1]$, the one point crossover rate $GA_{CR} \in [0, 1]$ and point $GA_{CP} \in [0, 1]$; parent pool ratio $GA_{PR} \in [0.2, 1]$; and mutation rate $GA_{MR} \in [0, 0.1]$ and size $GA_{MS} \in [0, 100]$ to the hyper level.

Table 1. Benchmark functions and their properties

Function	Details	Properties
Qing	$f(x) = \sum\limits_{i=1}^{d}(x^2 - 1)^2,$ $f(x^*) = 0$ $x^* = (\pm\sqrt{i}, ..., \pm\sqrt{i})$ $x_i \in [-500, 500]$	non-convex, separable, multimodal
Rastrigin	$f(x) = 10d + \sum\limits_{i=1}^{d}[x_i^2 - 10cos(2\pi x_i)]$ $f(x^*) = 0$ $x^* = (0, ..., 0)$ $x_i \in [-5.12, 5.12]$	non-convex, separable, multimodal
Rosenbrock	$f(x) = \sum\limits_{i=1}^{d-1}[100(x_{i+1} - x_i^2)^2 + (x_i - 1)^2]$ $f(x^*) = 0$ $x^* = (1, ..., 1)$ $x_i \in [-2.048, 2.048]$	non-convex, non-separable, multimodal
Schweffel 2.23	$f(x) = \sum\limits_{i=1}^{d}x_i^{10}$ $f(x^*) = 0$ $x^* = (0, ..., 0)$ $x_i \in [-10, 10]$	convex, separable, unimodal
Styblinski-Tang	$f(x) = \frac{1}{2}\sum\limits_{i=1}^{d}x_i^4 - 16x_i^2 + 5x_i$ $f(x^*) = -39.16599d$ $x_i^* = (-2.903534, ..., -2.903534)$ $x_i \in [-5, 5]$	non-convex, separable, multimodal
Trid	$f(x) = \sum\limits_{i=1}^{d}(x_i - 1)^2 + \sum\limits_{i=2}^{d}(x_i x_{i-1})$ $f(x^*) = \frac{-d(d+4)(d-1)}{6}$ $x_i^* = i(d + 1 - i)$ $x_i \in [-d^2, d^2]$	convex, non-separable, unimodal

The DE operator has two parameters: crossover rate $DE_{CR} \in [0.9, 1]$ and the scaling factor $DE_F \in [0.4, 0.7]$.

The GWO operator is parameterized by the attack rate $GWO_a \in [0, 2]$. The base level is parameterized by these operator parameters:

$$GA_\alpha, GA_{CR}, GA_{CP}, GA_{PR}, GA_{MR}, GA_{MS}, DE_{CR}, DE_F, GWO_a \in \beta_{perturb}$$

The elitist selector operator (ES) selects the fittest offspring into the next generation and exposes no parameters. The Simulated Annealing uses a high initial temperature of 10000 and cooling ratio of $\alpha = 50$ in order to achieve a better exploration of the search space.

4   Numerical Experiments

Various numerical experiments were performed in order to assess the performance of the proposed method. The method is compared to a state of the art hyperheuristic for continuous problems (CustomHyS), to a state of the art metaheuristic (the coral reef optimizer) and to two random search approaches. The experimental results highlight the importance of intelligently searching for the right parameters and operators for the given optimization problem.

4.1   Benchmarks

The experiments were performed on six well-known single objective continuous benchmarks functions: Qing, Rastrigin, Rosenbrock, Schwefel 2.23, Styblinski-Tang and Trid. These were chosen in order to assess the performance of the methods in various cost landscape shapes. To evaluate performance in diverse scenarios, the functions vary in terms of their dimensionality, convexity, separability, and multimodality. These properties are summarized in Table 1. For every test case a population size of 30 was used, with each offspring limited to 100 function evaluations in each hyperheuristic step. The total number of hyperheuristic steps was limited to 100 with each step having a performance sample size of 30.

4.2   Comparison with Other Approaches

SA-PERTURB was compared with CustomHyS in order to assess whether it can compete with a hyperheuristic that uses significantly more operators. CustomHyS uses twelve search operators on the base level, Simulated Annealing on the hyper level and searches for an operator sequence of fixed length. The suggested heuristic collection and parameters were used, the maximum temperature was set to 200, the cooling rate to 0.05, the sequence cardinality to 3, and the stagnation rate to 0.3. Our approach was also compared to the coral reef optimizer (CRO) which simulates the growth of coral reef in order to assess whether it can outperform a state-of-the-art metaheuristic. The coral reef rate of occupation was set to 0.4, the broadcast/existing rate (F_b) was set to 0.9, the duplication rate (F_a) to 0.1, the depredation rate (F_d) to 0.1, the maximum depredation probability (P_d) to 0.5, the probability of the mutation process (GCR) to 0.1, the mutation process factors $gamma_{min}$ to 0.02, $gamma_{max}$ to 0.2, and the number of attempts of a larvar to set in reef (n_{trials}) to 5.

SA-PERTURB was also compared with two random approaches. The RANDOM-DE, RANDOM-GA are variants which set the operator parameters in a random uniform fashion within the parameter bounds during the hyperheuristic steps.

4.3   Experimental Results

Tables 2 and 3 show the test results highlighting the best solutions after the hyperheuristic optimization procedure. Figures 2 and 3 depict the evolution of

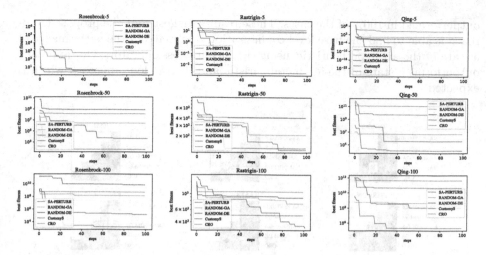

Fig. 2. Performance evolution comparison of the methods. The minimal median plus interquantile range of the performances at each step is shown.

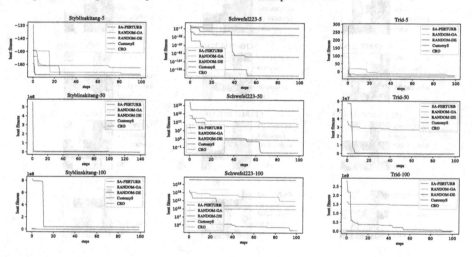

Fig. 3. SA-PERTURB outperforms the simple metaheuristic and random hyperheuristics, and matches the performance of CustomHyS.

the hyperheuristic search and the resulting performance improvement for the problems. The results show that our approach matches the performance of CustomHyS, both approaches found the best solution in 55.6% of the test cases. The CRO approach was outperformed by our method in 88% of the cases. Our approach outperforms RANDOM-DE in 88% of cases and RANDOM-GA is outperformed in 100% of the cases. These results suggest that the proposed app-roach is able to find the base level configuration that matches the performance of a state-of-the-art hyperheuristic, using significantly less operators. Our approach outperforms a state-of-the-art metaheuristic and the random approach which is expected.

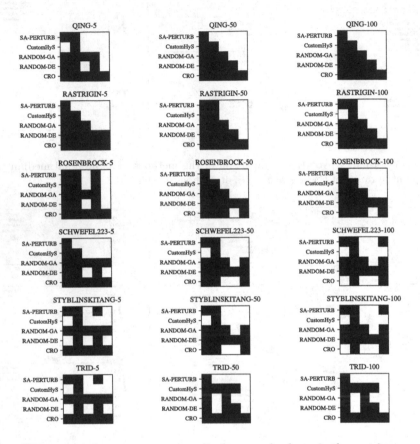

Fig. 4. Wilcoxon rank-sum test matrices. Each row and column in the matrix represents the five methods: SA-PERTURB, CustomHys, Random-GA, Random-DE, and the coral reef optimizer (CRO). White cells depicts statistical difference between results obtained by different methods.

Table 2. The results of the comparison with the random approaches. The average plus standard deviation of the costs is shown. The best solutions according to the Wilcoxon test are highlighted in bold.

problem	dimension	SA-PERTURB	RANDOM-GA	RANDOM-DE
Qing	5	**4.90e−04 ± 8.25e−04**	1.42e+00 ± 8.58e−01	9.39ev02 ± 1.27e−01
	50	**9.05e+03 ± 2.57e+03**	3.11e+06 ± 1.95e+06	3.71e+09 ± 1.05e+09
	100	**1.87e+05 ± 7.00e+04**	1.37e+08 ± 4.51e+07	6.05e+10 ± 9.47e+10
Rastrigin	5	**1.46e−03 ± 2.66e−03**	4.79e+00 ± 2.52e+00	6.06e+00 ± 2.44e+00
	50	**1.01e+02 ± 1.47e+01**	1.81e+02 ± 2.35e+01	3.38e+02 ± 6.29e+01
	100	**4.14e+02 ± 5.49e+01**	6.43e+02 ± 5.61e+01	7.86e+02 ± 8.21e+01
Rosenbrock	5	1.96e+00 ± 1.15e+00	5.20e+02 ± 1.33e+03	**2.91e+00 ± 3.99e+00**
	50	**5.00e+01 ± 2.49e+00**	1.97e+06 ± 8.48e+05	5.72e+08 ± 2.48e+08
	100	**1.91e+03 ± 4.10e+03**	1.14e+08 ± 3.68e+07	1.17e+10 ± 2.74e+10
Schwefel223	5	**3.18e−136 ± 1.71e−135**	3.74e−05 ± 1.25e−04	9.14e−22 ± 2.18e−21
	50	**8.63e+05 ± 3.84e+06**	1.54e+10 ± 3.60e+10	3.20e+16 ± 4.59e+16
	100	**2.84e+11 ± 4.57e+11**	7.98e+12 ± 1.09e+13	4.86e+20 ± 1.43e+20
Styblinski-Tang	5	−1.96e+02 ± 2.59e−02	−1.95e+02 ± 1.23e+00	**−1.95e+02 ± 3.52e+00**
	50	**−1.41e+03 ± 6.65e+01**	7.38e+03 ± 7.28e+03	2.90e+06 ± 1.26e+06
	100	**−1.91e+03 ± 2.37e+02**	2.07e+05 ± 9.09e+04	5.81e+07 ± 1.27e+08
Trid	5	−3.00e+01 ± 6.32e−03	−2.55e+01 ± 3.50e+00	**−3.00e+01 ± 9.05e−06**
	50	**−6.57e+01 ± 4.02e+01**	4.65e+02 ± 5.16e+02	4.30e+03 ± 1.10e+03
	100	**2.58e+01 ± 4.78e+01**	3.30e+03 ± 1.25e+03	2.45e+04 ± 4.39e+03

Table 3. The results of the comparison with CustomHyS and CRO. The average plus standard deviation of the costs is shown. The best solutions according to the Wilcoxon test are highlighted in bold.

problem	dimension	SA-PERTURB	CustomHyS	CRO
Qing	5	4.90e−04 ± 8.25e−04	**3.62e−26 ± 8.07e−26**	2.12e+05 ± 5.70e+05
	50	**9.05e+03 ± 2.57e+03**	2.64e+05 ± 2.07e+05	9.17e+10 ± 2.21e+10
	100	**1.87e+05 ± 7.00e+04**	7.89e+07 ± 3.40e+07	3.80e+11 ± 5.69e+10
Rastrigin	5	**1.46e−03 ± 2.66e−03**	1.32e+00 ± 1.17e+00	7.29e+00 ± 3.55e+00
	50	**1.01e+02 ± 1.47e+01**	**1.08e+02 ± 2.17e+01**	3.71e+02 ± 3.95e+01
	100	4.14e+02 ± 5.49e+01	**2.92e+02 ± 3.50e+01**	1.00e+03 ± 6.27e+01
Rosenbrock	5	**1.96e+00 ± 1.15e+00**	**3.68e+00 ± 4.67e+00**	9.66e+02 ± 1.15e+03
	50	**5.00e+01 ± 2.49e+00**	3.35e+03 ± 1.05e+03	1.21e+08 ± 2.97e+07
	100	**1.91e+03 ± 4.10e+03**	1.20e+05 ± 4.05e+04	4.83e+08 ± 7.28e+07
Schwefel223	5	**3.18e−136 ± 1.71e−135**	6.14e−91 ± 1.95e−90	1.85e+00 ± 6.29e+00
	50	8.63e+05 ± 3.84e+06	**3.05e−04 ± 2.22e−04**	8.85e+08 ± 5.62e+08
	100	2.84e+11 ± 4.57e+11	**1.38e+02 ± 1.21e+02**	7.07e+09 ± 2.61e+09
Styblinski-Tang	5	−1.96e+02 ± 2.59e−02	**−1.95e+02 ± 2.54e+00**	−1.91e+02 ± 8.45e+00
	50	−1.41e+03 ± 6.65e+01	**−1.73e+03 ± 4.14e+01**	−1.30e+03 ± 7.53e+01
	100	−1.91e+03 ± 2.37e+02	**−2.93e+03 ± 7.56e+01**	−2.17e+03 ± 1.24e+02
Trid	5	−3.00e+01 ± 6.32e−03	**−3.00e+01 ± 1.37e−14**	−2.43e+01 ± 6.66e+00
	50	**−6.57e+01 ± 4.02e+01**	2.06e+05 ± 6.97e+04	2.40e+07 ± 4.16e+06
	100	**2.58e+01 ± 4.78e+01**	2.22e+07 ± 8.13e+06	1.26e+09 ± 1.65e+08

Figure 4 presents the Wilcoxon rank-sum statistical test matrices for all test cases. The Wilcoxon rank-sum test determines if there is a statistical difference between two samples (in our case the samples are from different algorithms).

The null hypothesis that two samples come from the same population is rejected with a level of significance $\alpha = 0.05$ if the computed p value is smaller than 0.05 (white cells in figures).

Table 4. Elapsed time of experiments for the compared methods measured in seconds.

problem	dimension	SA-PERTURB	CustomHyS	CRO
Qing	5	**1.20e+02**	4.65e+02	8.45e+03
	50	**1.26e+02**	4.15e+02	8.66e+03
	100	**1.28e+02**	2.00e+03	9.01e+03
Rastrigin	5	**1.52e+02**	1.82e+02	8.04e+03
	50	**1.54e+02**	4.31e+02	8.68e+03
	100	**1.61e+02**	7.83e+03	8.82e+03
Rosenbrock	5	**1.61e+02**	3.97e+02	7.62e+02
	50	**1.67e+02**	8.44e+02	8.67e+03
	100	**1.86e+02**	1.93e+03	8.85e+03

4.4 Computational Time

Table 4 depicts the elapsed computational time of the experiments measured in seconds. The results show that our approach is faster than CustomHyS and CRO. Our approach also scales better with increasing dimensionality than the other approaches.

5 Conclusion

The article presents a two level hyperheuristic for continuous optimization problems. The main innovation of the proposed method lies in the framework being capable of automatic operator parameter and operator sequence tuning to the optimization problem using a small number of operators. The results show that our approach matches the performance of CustomHyS which is a state-of-the-art hyperheuristic for continuous problems using significantly less operators and less computational time. Our approach outperforms the coral reef optimization algorithm (CRO) and the random hyperheuristic search approach. As further work, the upper level of the Markov chain can be extended with more operator categories in order to be able to model more complex operator distributions. Another research direction is the use of better optimization procedures on the hyper level.

References

1. Burke, E.K., Hyde, M.R., Kendall, G., Ochoa, G., Özcan, E., Woodward, J.R.: A Classification of Hyper-Heuristic Approaches: Revisited. Springer, Heidelberg (2019)
2. Caraffini, F., Neri, F., Epitropakis, M.: Hyperspam: a study on hyper-heuristic coordination strategies in the continuous domain. Inf. Sci. **477**, 186–202 (2019). https://doi.org/10.1016/j.ins.2018.10.033
3. Chakhlevitch, K., Cowling, P.: Hyperheuristics: recent developments. In: Cotta, C., Sevaux, M., Sörensen, K. (eds.) Adaptive and Multilevel Metaheuristics. SCI, vol. 136, pp. 3–29. Springer, Heidelberg (2008). https://doi.org/10.1007/978-3-540-79438-7_1
4. Cowling, P., Kendall, G., Soubeiga, E.: A hyperheuristic approach to scheduling a sales summit. In: Burke, E., Erben, W. (eds.) PATAT 2000. LNCS, vol. 2079, pp. 176–190. Springer, Heidelberg (2001). https://doi.org/10.1007/3-540-44629-X_11
5. Cruz-Duarte, J.M., Amaya, I., Ortiz-Bayliss, J.C., Conant-Pablos, S.E., Terashima-Marín, H., Shi, Y.: Hyper-heuristics to customise metaheuristics for continuous optimisation. Swarm Evol. Comput. **66**, 100935 (2021)
6. Csébfalvi, A.: A hybrid meta-heuristic method for continuous engineering optimization. Periodica Polytechnica Civil Eng. **53**(2), 93–100 (2009)
7. Drake, J.H., Kheiri, A., Özcan, E., Burke, E.K.: Recent advances in selection hyper-heuristics. Eur. J. Oper. Res. **285**(2), 405–428 (2020)
8. Han, L., Kendall, G.: An investigation of a tabu assisted hyper-heuristic genetic algorithm. In: The 2003 Congress on Evolutionary Computation, CEC 2003, vol. 3, pp. 2230–2237. IEEE (2003)
9. Han, L., Kendall, G., Cowling, P.: An adaptive length chromosome hyper-heuristic genetic algorithm for a trainer scheduling problem. In: Recent Advances in Simulated Evolution and Learning, pp. 506–525. World Scientific (2004)
10. Kendall, G., Li, J.: Competitive travelling salesmen problem: a hyper-heuristic approach. J. Oper. Res. Soc. **64**(2), 208–216 (2013)
11. Lin, J., Luo, D., Li, X., Gao, K., Liu, Y.: Differential evolution based hyper-heuristic for the flexible job-shop scheduling problem with fuzzy processing time. In: Shi, Y., et al. (eds.) SEAL 2017. LNCS, pp. 75–86. Springer, Cham (2017). https://doi.org/10.1007/978-3-319-68759-9_7
12. Mirjalili, S., Mirjalili, S.M., Lewis, A.: Grey wolf optimizer. Adv. Eng. Softw. **69**, 46–61 (2014)
13. Olgun, B., Koç, Ç., Altıparmak, F.: A hyper heuristic for the green vehicle routing problem with simultaneous pickup and delivery. Comput. Ind. Eng. **153**, 107010 (2021)
14. Patriksson, M.: A survey on the continuous nonlinear resource allocation problem. Eur. J. Oper. Res. **185**(1), 1–46 (2008)
15. Peraza-Vázquez, H., Torres-Huerta, A.M., Flores-Vela, A.: Self-adaptive differential evolution hyper-heuristic with applications in process design. Computación y Sistemas **20**(2), 173–193 (2016)
16. Qin, W., Zhuang, Z., Huang, Z., Huang, H.: A novel reinforcement learning-based hyper-heuristic for heterogeneous vehicle routing problem. Comput. Ind. Eng. **156**, 107252 (2021)
17. Ryser-Welch, P., Miller, J.F.: A review of hyper-heuristic frameworks. In: Proceedings of the EVO20 Workshop, AISB, vol. 2014 (2014)

18. Sim, K., Hart, E., Paechter, B.: A lifelong learning hyper-heuristic method for bin packing. Evol. Comput. **23**(1), 37–67 (2015)
19. Sánchez, M., Cruz-Duarte, J.M., Ortíz-Bayliss, J., Ceballos, H., Terashima-Marin, H., Amaya, I.: A systematic review of hyper-heuristics on combinatorial optimization problems. IEEE Access **8**, 128068–128095 (2020). https://doi.org/10.1109/ACCESS.2020.3009318
20. Villela Tinoco, J.C., Coello Coello, C.A.: hypDE: a hyper-heuristic based on differential evolution for solving constrained optimization problems. In: Schütze, O., et al. (eds.) EVOLVE - A Bridge between Probability, Set Oriented Numerics, and Evolutionary Computation II. AISC, vol. 175, pp. 267–282. Springer, Heidelberg (2013). https://doi.org/10.1007/978-3-642-31519-0_17
21. Weber, G.W., Özöğür-Akyüz, S., Kropat, E.: A review on data mining and continuous optimization applications in computational biology and medicine. Birth Defects Res. C Embryo Today **87**(2), 165–181 (2009)
22. Yan, J., Wu, X.: A genetic based hyper-heuristic algorithm for the job shop scheduling problem. In: 2015 7th International Conference on Intelligent Human-Machine Systems and Cybernetics, vol. 1, pp. 161–164. IEEE (2015)
23. Yang, C., Peng, S., Jiang, B., Wang, L., Li, R.: Hyper-heuristic genetic algorithm for solving frequency assignment problem in TD-SCDMA. In: Proceedings of the Companion Publication of the 2014 Annual Conference on Genetic and Evolutionary Computation, pp. 1231–1238 (2014)

Benchmarking Named Entity Recognition Approaches for Extracting Research Infrastructure Information from Text

Georgios Cheirmpos[1]([✉]), Seyed Amin Tabatabaei[1], Evangelos Kanoulas[2], and Georgios Tsatsaronis[1]

[1] Elsevier BV., Amsterdam, The Netherlands
g.cheirmpos@elsevier.com
[2] University of Amsterdam, Amsterdam, The Netherlands

Abstract. Named entity recognition (*NER*) is an important component of many information extraction and linking pipelines. The task is especially challenging in a low-resource scenario, where there is very limited amount of high quality annotated data. In this paper we benchmark machine learning approaches for *NER* that may be very effective in such cases, and compare their performance in a novel application; information extraction of research infrastructure from scientific manuscripts. We explore approaches such as incorporating Contrastive Learning (*CL*), as well as Conditional Random Fields (*CRF*) weights in BERT-based architectures and demonstrate experimentally that such combinations are very efficient in few-shot learning set-ups, verifying similar findings that have been reported in other areas of NLP, as well as Computer Vision. More specifically, we show that the usage of CRF weights in BERT-based architectures achieves noteworthy improvements in the overall *NER* task by approximately 12%, and that in few-shot setups the effectiveness of *CRF* weights is much higher in smaller training sets.

Keywords: Natural Language Processing · Named Entity Recognition · Few-Shot Learning · Contrastive Learning

1 Introduction

Research Infrastructure (*RI*) is an important enabler for successful and impactful research as it allows universities and research centers to conduct cutting-edge research in their respective fields and to foster innovation. Research centers want to track their investments in *RI*, and to compute metrics such as Return of Investment (*RoI*), for example, by means of the scientific output that their *RI* has enabled, the scientific impact, the volume of funding attracted by research grants, and the number of patents awarded. Such a view can enable research centers and large research infrastructure facilities to plan future investments, as well as to assess historical investments in *RI*. At the same time, the ability to

G. Nicosia et al. (Eds.): LOD 2023, LNCS 14505, pp. 131–141, 2024.
https://doi.org/10.1007/978-3-031-53969-5_11

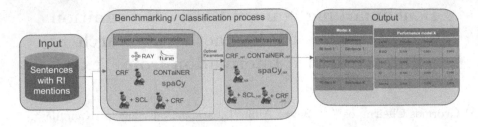

Fig. 1. Overview of the benchmark pipeline. Sentences with *RI* mentions are fed to the hyperparameter optimization process. Along with the optimal parameters, they are then fed to the incremental training step, producing entity predictions and respective scores.

track metrics such as the aforementioned, offers a view on the research collaborations, and the types of interdisciplinary research conducted, since large *RI*s are used by large consortia and big science initiatives and are very important for governance bodies to monitor [1].

A key enabling technology for collecting primary information for the aforementioned metrics is the ability to perform Named Entity Recognition (*NER*) of Research Infrastructure (*RI*), e.g., scientific equipment, that is mentioned in scientific manuscripts and is used in the respective conducted research. Though information extraction has found applications in many domains [14], either as a standalone task or as part of a larger pipeline [10], to the best of our knowledge, this is the first research work that explores *NER* for identifying research equipment used by analyzing a scientific manuscript. We hypothesize that one of the main detractors of building effective *NER* pipelines for this task is the absence of high volume, and high quality training data, and this is precisely the point we address in this paper for this application.

An information extraction pipeline typically consists of text-preprocessing, *NER* and entity linking [31]. In this work we focus only on the *NER* part and our goal is to explore and empirically evaluate whether Contrastive Learning (*CL*) can be effective in low resource environments, i.e., small volume of training data, or even few shot learning; *NER* for *RI* from scientific manuscripts is exactly such an environment. After all, high quality annotations are expensive and time consuming to procure and require deep domain expertise [15].

We provide answers to the following questions: (RQ1) How much Subject Matter Experts (SMEs) agree on what is considered *RI* while annotating such a dataset? (RQ2) How well convolutional approaches do in identifying *RI*s and how does *CL* perform? (RQ3) How robust is *CL* in very small training data set sizes, e.g., few-shot? In order to address these questions, we are introducing a straightforward experimental pipeline that we utilize for the benchmarking of the *NER* approaches that we are comparing, and which we illustrate in Fig. 1. The main components of our information extraction and linking pipeline are a sentence classifier, a *NER* and a linker.

The remaining of the paper is structured as follows; Sect. 2 provides some background that is fundamental for the remaining of the paper; Sect. 3 presents the dataset used, as well as the process followed by the annotators during the annotation; Sect. 4 formalizes experimental setup, including the tested models, hyperparameter tuning and experiments, as well as any pre-processing conducted for the models' inputs; Sect. 5 presents the evaluation of the models' performance; Sect. 6 introduces the few-shot learning simulation benchmark and respective results; finally, we conclude in Sect. 7 and provide pointers to future work.

2   Background

In the following we discuss some related topics that are fundamental for the remaining of the paper. Regarding *NER*, a named entity is a term or phrase that clearly distinguishes one item from a group of other items that have similar attributes [24]. Manufacturer, equipment name and model, research centers and institutes are examples of named entities of interest. An extensive survey on recent advances in *NER* using Deep Learning (*DL*) models is given by Yadav and Bethard in [25]. In [23], the authors tackle a similar entity extraction problem, that of finding funding information from scientific articles using *CRF*, Hidden Markov and Maximum Entropy models, but do not address the issue of having little amounts of training data.

2.1   Conditional Random Fields

Conditional Random Fields (*CRF*) [26] is a popular statistical modeling technique used for *NER* problems and often regarded as a very strong baseline to compare against. In the past, before the introduction of embeddings and representation learning, hand-crafted features such as capitalization, numbers, punctuation, and Part-Of-Speech (*POS*) tags of the words were used as features for the training of *CRF*. Despite the improvements brought by embeddings, feature-based approaches like *CRF* remain relevant in many *NER* applications, as they can still capture important linguistic features of the text.

In the last few years we have also seen the effective combination of *CRF* with Long-Short Term Memory networks (*LSTM*), Bidirectional *LSTM* and Convolutional Neural Networks (*CNN*) for sequence labelling (e.g., [27]). By learning from the sentences' structure, e.g., the expectation of what type of word will follow after seeing the previous one, CRF helps in maximizing predictability performance.

2.2   Few-Shot Learning

Few-shot learning is a type of machine learning that aims to learn from a small amount of labeled data. It involves training models to recognize and classify new objects or concepts based on a few examples, rather than requiring a large labeled dataset. It is particularly useful in situations where the labeling cost of new data points is high, such as the use case we explore in this work.

The extensive survey of Wang [13] lists common examples of few-shot learning tasks such as classification in computer vision and sentiment analysis in NLP. Yang [12] propose a few-shot *NER* system based on nearest neighbor (*NN*) learning using contextual token representation where a *NN* classifier achieves State-Of-the-Art results in few-shot learning tasks. Wang et al. [11] address the formulation of strategies for fine-tuning NLP models for few-shot tasks by aligning the fine-tuning with pre-training objective unleashing more benefits from the pre-trained language models.

2.3   Contrastive Learning

Most of the work using *CL*, for self-supervised, supervised or semi-supervised learning has been conducted in the computer vision, and more recently in NLP. *CL* assists in the model training by *"contrasting"* positive with negative pairs of samples, which clusters the same class samples closer [18]. Gunel et al. [19] propose a way to include a supervised *CL* term in the process of fine-tuning pre-trained language models. They introduce a contrastive objective with a new type of loss function (Supervised Contrastive Loss) which is a weighted sum of the cross entropy loss with the contrastive loss. This assists in solving the problem that persists with the cross entropy loss, which is known to perform poorly on generalization [20,21]. The Supervised Contrastive Loss function is explained below.

$$\mathcal{L} = (1 - \lambda)\mathcal{L}_{CE} + \lambda\mathcal{L}_{SCL} \tag{1}$$

$$\mathcal{L}_{CE} = -\frac{1}{N}\sum_{i=1}^{N}\sum_{c=1}^{C} y_{i,c} \cdot log\,\hat{y}_{i,c} \tag{2}$$

$$\mathcal{L}_{SCL} = \sum_{i=1}^{N} -\frac{1}{N_{y_i} - 1}\sum_{j=1}^{N} \mathbb{1}_{i\neq j}\mathbb{1}_{y_i = y_j} \log \frac{exp(\frac{\Phi(x_i)\cdot\Phi(x_j)}{\tau})}{\sum_{k=1}^{N} \mathbb{1}_{i\neq k}exp(\frac{\Phi(x_i)\cdot\Phi(x_k)}{\tau})} \tag{3}$$

The overall model loss (1) is computed by combining the canonical cross-entropy (*CE*) loss (2) and the Supervised Contrastive Loss (*SCL*) loss (3) with a weighting factor λ that determines their relative contribution.

As we see, in Eq. (2), L_{CE} is the normal cross entropy loss that measures the difference between the predicted probability distribution and the true probability distribution, and aims to minimize this difference during training. On the other hand, Eq. (3) explains the *SCL* (L_{SCL}) which aims to push similar examples closer together and dissimilar examples farther apart in the learned embedding space by minimizing the distance between similar pairs and maximizing the distance between dissimilar pairs. The parameter τ is the scalar temperature parameter which controls the separation of classes while $\Phi(x)$ denotes the BERT

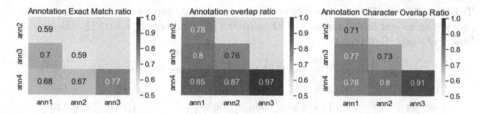

Fig. 2. Cohen's κ Annotation agreement heatmap IAA metrics, on 3 dimensions. Exact annotation match, Annotation overlap and Annotation character overlap. Pairwise evaluation (annX = annotator X)

encoder output before the softmax layer. The contrastive loss calculation requires the output of the last hidden layer of the BERT model. The cross entropy uses the output of the prediction layer of BERT (logits).

One of the most related works that utilize *CL* in a few-shot learning setup, and which we also compare experimentally in this paper, is the work by Sarathi Das et al. (CONTaiNER) [22], who propose a way to counter overfitting in a set of training domains. This leads to a better generalization in unseen target domains. This technique also optimizes inter-token distribution distance for few-shot *NER* and tokens are differentiated based on their Gaussian-distributed embeddings.

3   Dataset

This section outlines the process of constructing the dataset for our *NER* task. A total of 84 scientific articles were randomly selected from the corpus of articles affiliated with our university. These articles were then distributed for annotation in equal-sized batches to pairs of Subject Matter Experts (*SME*). The annotation process was conducted at the sentence level and involved two steps: first, the *SME* identified and flagged the sentences that indicated a reference or use of an *RI*; then, they annotated the *RI* at the token level.

To ensure the quality and consistency of the annotations, each document was annotated by two *SMEs*, and any discrepancies were harmonized by a third annotator. In total, four *SMEs* were involved in the annotation process. The articles were provided in a custom annotation environment, and in plain text format for ease of annotation.

One problem for annotating *RI* is the absence of a standarized way to refer to it in an article. For example, the following mentions can all refer to the same *RI*: 'AccuChek glucometer (Roche Diagnostics, Mannheim, Germany)', 'Accu-Chek glucometer (Roche, Germany)', 'Accu-Chek ŏ Performa Glucometer', 'Accu-Chek ŏ glucometer'. Symbols, punctuation and special characters are part of the *RI* mention and need to be retained during the sentence tokenization process.

$$length(span(RI_1) \cap span(RI_2)) = length(span(RI_1) \cup span(RI_2)) \quad (4)$$

Table 1. Dataset EDA on number of sentences and BIO tag count. Last three columns provide an insight on the number of RI mentions per sentence. B, I, O, EQ = Beginning, Inner, Outside, Equipment

	sentences	RIs	B-EQ	I-EQ	O-EQ	max RIs	mean RIs	std RIs
Train set	231	255	255	1487	4679	5	1.1	1.13
Test set	123	139	139	946	2663	4	1.13	0.423

$$length(span(RI_1) \cap span(RI_2)) > 0 \tag{5}$$

$$\frac{length(span(RI_1) \cap span(RI_2))}{length(span(RI_1) \cup span(RI_2))} \tag{6}$$

To assess the difficulty of the task we measured the level of Inter-Annotator Agreement (IAA) using the Fleiss' κ [16] and Cohen's κ [17] metrics for pairwise and group-wise agreement, respectively. We measured agreement on three dimensions: annotation exact match, annotation overlap and annotation character overlap ratio, presented in Eqs. (4), (5) and (6) respectively, where RI_1, RI_2 represent the span of text annotated by two *SME*. Regarding the Eqs. (4–6): RI_1, RI_2 is the RI annotated by $SME_{1,2}$, $span()$ denotes the offets of the RI in the sentence, $length()$ denotes the size of offset overlap in the cases of union ($\cup$) or intersection ($\cap$).

To answer RQ1 we report the aforementioned metrics in Fig. 2. These plots give us an idea of how difficult the annotation task is. The high overlap score indicates that despite some disagreement on the span of the *RI* in text, *SME* overall refer to the same *RI* and that is also validated by the high ratio score between the pairs. However, in many cases even *SME* do not agree which part of the text should be annotated as an *RI*, though there is some overlap between what they have annotated. For data quality assurance a third subject matter expert addresses the disagreement between each pair of annotators.

The name of an RI contains punctuation and symbols, so these characters should be treated as separate tokens. TreeBankWordTokenizer [29], which uses regular expressions to tokenize text was the tokenizer of choice as it treats most punctuation characters as separate tokens. All sentences are split into tokens using the NLTK TreeBankWordTokenizer implementation [8]. For token labels, the *BIO* format [28] is used. Table 1 shows some statistics regarding the training and test datasets. We follow a 70:30 document split[1].

4   Experimental Setup

This section describes the *NER* models used in the comparison, the hyperparameter optimization process and how the best performing versions of the models were chosen. In total, we compare 6 *NER* models:

---

[1] DOI: 10.17632/ty73wxgtpx.1.

(1) **spaCy** [9] *NER* model with the English-core-web-small pretrained pipeline, on news and blogs type of text. Each instance was trained for 50 epochs, batch of compounding size (4, 32, 1.001) and 0.3 dropout.

(2) **CRF**: each instance was trained for 2250 iterations, L-BFGS algorithm, backtracking linesearch, $3e^{-3}$ c1 and $1.4e^{-1}$ c2.

(3) **CONTaiNER**, using the library of the authors [22]: train batch size of 8, max sequence length 64, learning rate $5e^{-5}$, weight decay $1e^{-3}$, 46 epochs, embedding dimensions 16, temperature 0.6 with Euclidean loss.

(4) **BertForTokenClassification** (BERT_NER) [2,6]: with the BERT-base-uncased pretrained model, 20 epochs, max sequence length of 128, learning rate $1e^{-5}$, train batch size of 4 and max gradient norm 10.

(5) **SCL loss function BertForTokenClassification** (BERT_NER_CL): 22 epochs, max sequence length of 128, train batch size of 8, λ 0.5, learning rate $5e^{-5}$, temperature 1.0 and max gradient norm 18.

(6) **CRF layer BertForTokenClassification** (BERT_NER_CRF): 100 epochs, max sequence length of 512, train batch 16, learning rate $5e^{-6}$, weight decay $1e^{-2}$.

All models were hyper parameter optimized using the Tune framework, Python library for Scalable Hyperparameter Tuning [7], with an extensive space search on a seeded environment for reproducibility. BERT models were downloaded from HuggingFace [2], and for vanilla CRF we used the sklearn-CRFsuite [3]. Experiments were run using a g5.4xlarge EC2 AWS instance, using Python [5] and PyTorch [4].

5   Evaluation

To answer RQ2 we evaluated the models in terms of Precision, Recall, and F1 score on the macro and token level. The evaluation scores are extracted from the average scores of 20 models trained at the 100% (whole dataset) increment run and results are presented on tag and macro level in Table 2. To verify the statistical significance on the macro F1 level we compared all models increment 100% 20 scores using the Student's T-test [30]. The p-values are displayed in Fig. 3. For validation purposes we also used Student's T-test.

Spacy and CRF model already show satisfactory performance with F1 0.82 and 0.85 respectively. While we can say that CRF produces False positives labels for the B-tag, technically it means that its output will need more post-processing than Spacy. As with BERT_NER_CL, which also lacks behind on the B-tag, the False positives are in most cases isolated tokens such as parentheses or equations fragments. This is partially caused by TreeBankWordTokenizer which treats special symbols as entities during the annotation process.

BERT_NER, BERT_NER_CL and CONTaiNER all have similar performance on the macro level with the later achieving the second best Precision score across the models. CONTaiNER appears to be a more precision oriented model while the BERT models are more balanced between Precision and Recall.

Table 2. Average performance of the 20 NER models at 100% increment, and BEST performance at 100% increment

Tag	Spacy			CRF			BERT_NER		
–	Precision	Recall	F1	Precision	Recall	F1	Precision	Recall	F1
O	0.9	0.95	0.93	0.95	0.95	0.95	0.97	0.96	0.97
B-EQ	0.77	0.72	0.75	0.74	0.72	0.73	0.83	0.86	0.84
I-EQ	0.87	0.8	0.82	0.87	0.86	0.87	0.89	0.92	0.9
Macro average	0.85	0.8	0.82	0.85	0.84	0.85	0.89	0.92	0.9
Macro average (BEST)	0.85	0.85	0.85	0.88	0.87	0.87	0.92	0.92	0.92
Tag	CONTaiNER			BERT_NER_CL			BERT_NER_CRF		
–	Precision	Recall	F1	Precision	Recall	F1	Precision	Recall	F1
O	0.96	0.98	0.97	0.97	0.96	0.97	0.98	0.98	0.98
B-EQ	0.85	0.76	0.8	0.83	0.86	0.84	0.90	0.94	0.92
I-EQ	0.94	0.9	0.92	0.9	0.92	0.91	0.91	0.95	0.93
Macro average	0.91	0.88	0.89	0.9	0.91	0.91	**0.93**	**0.96**	**0.94**
Macro average (BEST)	**0.93**	0.90	0.92	**0.93**	0.93	0.93	**0.93**	**0.96**	**0.94**

With the help of Fig. 3 their difference in macro score is significant, but the calculated p-values are close to *alpha*.

BERT_NER_CRF has the slight edge over the previous models, having the best Precision score at 0.93, but with a Recall score of 0.96, it balances out on an F1 score of 0.94 and at least 0.9 score on all token levels. Figure 3 confirms the significance of difference in scores for Spacy, CRF and NER_CRF with all models. On the micro level, selecting the best in performance model on the 100% increment out of the 20 generated, all models performed competitively, with CONTaiNER, BERT_NER_CL, BERT_NER_CRF achieving best Precision score. While these scores represent an *outlier* capability of the models, it can also provide an insight on their stability and reproducibility of experiments. Scores are displayed in Table 2. Overall, convolutional approaches perform well in identifying RIs but their performance is impacted positively when combined with an additional features, such as *CL* or *CRF*.

6   Few-Shot Simulation

The goal of this simulation is to benchmark the robustness of each model in a very data scarce scenario. To answer RQ3 we also focus on the *CL* performance in various increments. The optimal parameters for each model were used for the Few-shot simulation experiment. We sample the training dataset randomly 20 times, at each increment step (of 5%), to train 20 models. Every increment step amounts to increasing the dataset by 11 samples. At the 100% increment (full dataset) the dataset is simply shuffled. Thic process produces 400 trained models for each type (20 increments * 20 data sampling at each increment)

Each model's performance is analyzed on the macro level for Precision, Recall and F1. Figure 4 contains the results on the model performance on each

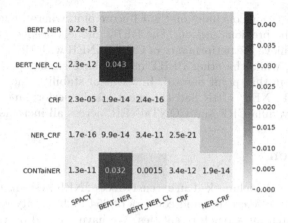

Fig. 3. Student's T-test measuring statistical significance of model's performance difference on the F1 metric (a = 0.05)

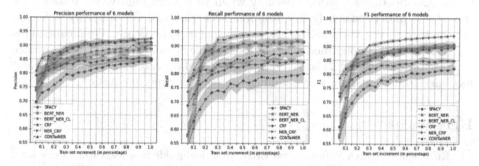

Fig. 4. Performance of models (Precision, Recall, F1) across training on various dataset increments. Scores are calculated as the average of 20 when on an incremental step (points) and 1 step standard deviation (transparent region).

increment. The figure shows dashed lines with points as average scores and 1 step standard deviation is represented by surrounding transparent region.

Regarding Precision, CONTaiNER and BERT_NER_CRF perform better that all the other models already from increment 0.1 having little to no deviation, making them quite stable in their reproducibility. The other models perform adequately and their deviation from the mean tends to get small while the dataset increases showing there is need for enough data to achieve good generalization. On the contrary, high Recall is harder for models to achieve when data is low in number. BERT_NER_CL with CONTaiNER have a strong start with the other BERT models lagging quite behind. From increment 0.3 BERT_NER_CRF manages to generalize well on this few-shot environment with the other two BERT model following. CRF is quite stable on deviation from the mean the larger the dataset gets while Spacy becomes more unstable.

All this information concludes on the F1 score plot, where we notice a similar situation as in the previous section. The BERT models needed at least 15% of the dataset to achieve the performance of CONTaiNER with BERT_NER_CRF performing best, while the other BERT models have similar performance with CONTaiNER from that point onward. In terms of stability and reproducibility of results, BERT_NER_CRF had the most stable performance for each run followed by the vanilla CRF and CONTaiNER, across all increments.

7   Conclusion

In this research, we benchmarked an extensive set of NER systems using a labeled dataset which we have created and consists of 354 high-quality sentences annotated with research infrastructure entities. We have focused on the specific use case of extracting mentions of Research Infrastructure from academic articles. Transformer, Convolutional and statistical model performed satisfactory in a token classification scenario, especially when they work jointly. We tested their robustness under extremely low resource scenarios and the majority achieved satisfactory scores with half of the scores being statistically significant. This suggests that 1) simpler model can be used while achieving similar performance to state-of-the-art, and 2) scarcity of annotated data can still provide sufficient feedback while researching or planning to invest in experimentation.

References

1. European Research & Innovation ERIC practical guidelines legal framework for a European Research Infrastructure Consortium. Publications Office (2015)
2. Wolf, T., et al.: Huggingface's transformers: state-of-the-art natural language processing. arXiv Preprint arXiv:1910.03771 (2019)
3. sklearn - CRFSuite. https://github.com/TeamHG-Memex/sklearn-crfsuite/
4. Paszke, A., Gross, S., et al.: PyTorch: An Imperative Style, High-Performance Deep Learning Library (2019)
5. Van Rossum, G., Drake, F.L.: Python 3 Reference Manual (2009)
6. Devlin, J., Chang, M., Lee, K., Toutanova, K.: BERT: pre-training of deep bidirectional transformers for language understanding. arXiv Preprint arXiv:1810.04805
7. Liaw, R., Liang, E., Nishihara, R., Moritz, P., Gonzalez, J., Stoica, I.T.: A Research Platform for Distributed Model Selection and Training (2018)
8. Loper, E., Bird, S.: NLTK: The Natural Language Toolkit. CoRR cs.CL/0205028 (2002). http://dblp.uni-trier.de/db/journals/corr/corr0205.html#cs-CL-0205028
9. Honnibal, M., Montani, I.: spaCy 2: Natural language understanding with Bloom embeddings, convolutional neural networks and incremental parsing (2017)
10. Lee, H., Peirsman, Y., Chang, A., Chambers, N., Surdeanu, M., Jurafsky, D.: Stanford's multi-pass sieve coreference resolution system at the CoNLL-2011 shared task (2011)
11. Wang, Z., Zhao, K., Wang, Z., Shang, J.: Formulating few-shot fine-tuning towards language model pre-training: a pilot study on named entity recognition. arXiv preprint arXiv:2205.11799 (2022)

12. Yang, Y., Katiyar, A.: Simple and effective few-shot named entity recognition with structured nearest neighbor learning (2020)
13. Wang, Y., Yao, Q., Kwok, J.T., Ni, L.M.: Generalizing from a few examples: a survey on few-shot learning. ACM Comput. Surv. **53**(3), 34, Article no. 63 (2021). https://doi.org/10.1145/3386252
14. Clinical information extraction applications: a literature review. J. Biomed. Inform. **77** (2018)
15. Snow, R., O'connor, B., Jurafsky, D., Ng, A.Y.: Cheap and fast-but is it good? Evaluating non-expert annotations for natural language tasks (2008)
16. Fleiss, J.L.: Measuring nominal scale agreement among many raters. Psychol. Bull. **76** (1971)
17. Banerjee, M., Capozzoli, M., McSweeney, L., Sinha, D.: Beyond kappa: a review of interrater agreement measures. Can. J. Stat. **27** (1999)
18. Hadsell, R., Chopra, S., LeCun, Y.: Dimensionality reduction by learning an invariant mapping. In: 2006 IEEE Computer Society Conference on Computer Vision and Pattern Recognition (CVPR 2006), vol. 2. IEEE (2006)
19. Gunel, B., Du, J., Conneau, A., Stoyanov, V.: Supervised contrastive learning for pre-trained language model fine-tuning (2020)
20. Cao, K., Wei, C., Gaidon, A., Arechiga, N., Ma, T.: Learning imbalanced datasets with label-distribution-aware margin loss. In: Advances in Neural Information Processing Systems, vol. 32 (2019)
21. Liu, W., Wen, Y., Yu, Z., Yang, M.: Large-margin softmax loss for convolutional neural networks. In: ICML, vol. 2 (2016)
22. Das, S.S.S., Katiyar, A., Passonneau, R.J., Zhang, R.: CONTaiNER: few-shot named entity recognition via contrastive learning (2021). https://github.com/psunlpgroup/CONTaiNER
23. Kayal, S., et al.: Tagging funding agencies and grants in scientific articles using sequential learning models. In: BioNLP 2017, pp. 216–221 (2017)
24. Li, J., Sun, A., Han, J., Li, C.: A survey on deep learning for NER. IEEE Trans. Knowl. Data Eng. **34**, 50–70 (2022)
25. Yadav, V., Bethard, S.: A survey on recent advances in named entity recognition from deep learning models (2019)
26. Lafferty, J., McCallum, A., Pereira, F.C.N.: Conditional random fields: probabilistic models for segmenting and labeling sequence data (2001)
27. Ma, X., Hovy, E.: End-to-end sequence labeling via bi-directional LSTM-CNNs-CRF. arXiv preprint arXiv:1603.01354 (2016)
28. Ramshaw, L.A., Marcus, M.P.: Text chunking using transformation-based learning. In: Armstrong, S., Church, K., Isabelle, P., Manzi, S., Tzoukermann, E., Yarowsky, D. (eds.) TLTB. Text, Speech and Language Technology, vol. 11, pp. 157–179. Springer, Dordrecht (1999). https://doi.org/10.1007/978-94-017-2390-9_10
29. Bird, S., Klein, E., Loper, E.: Natural Language Processing with Python: Analyzing Text with the Natural Language Toolkit. O'Reilly Media Inc. (2009)
30. Dror, R.B., et al.: The Hitchhiker's guide to testing statistical significance in natural language processing. In: Proceedings of the 56th Annual Meeting of the ACL (2018)
31. Tabatabaei, S.A., et al.: Annotating research infrastructure in scientific papers: an NLP-driven approach. In: Proceedings of the 61st Annual Meeting of the ACL (Volume 5: Industry Track), pp. 457–463 (2023)

Genetic Programming with Synthetic Data for Interpretable Regression Modelling and Limited Data

Fitria Wulandari Ramlan[(✉)] and James McDermott

School of Computer Science, University of Galway, Galway, Ireland
{f.wulandari1,james.mcdermott}@universityofgalway.ie

Abstract. A trained regression model can be used to create new synthetic training data by drawing from a distribution over independent variables and calling the model to produce a prediction for the dependent variable. We investigate how this idea can be used together with genetic programming (GP) to address two important issues in regression modelling, *interpretability* and *limited data*. In particular, we have two hypotheses. (1) Given a trained and non-interpretable regression model (e.g., a neural network (NN) or random forest (RF)), GP can be used to create an interpretable model while maintaining accuracy by training on synthetic data formed from the existing model's predictions. (2) In the context of limited data, an initial regression model (e.g., NN, RF, or GP) can be trained and then used to create abundant synthetic data for training a second regression model (again, NN, RF, or GP), and this second model can perform better than it would if trained on the original data alone. We carry out experiments on four well-known regression datasets comparing results between an initial model and a model trained on the initial model's outputs; we find some results which are positive for each hypothesis and some which are negative. We also investigate the effect of the limited data size on the final results.

Keywords: Interpretable Regression Model · Limited Data · Synthetic Data · Symbolic Regression · Distillation

1 Introduction

Regression is a supervised machine learning technique for modelling the relationship between explanatory variables and a target variable. While the purpose of most regression methods is to optimise the parameters for a certain model structure, symbolic regression is a type of machine learning which also learns the structure. It focuses on developing a model from the given data and expressing it as a mathematical equation. Symbolic regression allows scientists to learn more about the phenomena underlying the data. Symbolic regression makes the model interpretable and offers insight into the data. Symbolic regression problems are frequently solved using genetic programming (GP).

© The Author(s), under exclusive license to Springer Nature Switzerland AG 2024
G. Nicosia et al. (Eds.): LOD 2023, LNCS 14505, pp. 142–157, 2024.
https://doi.org/10.1007/978-3-031-53969-5_12

GP is an evolutionary-inspired technique for computer program generation [14]. In GP, computer programs are randomly evolved over many generations, and each generation is assessed by measuring its fitness. Fitness is task-specific. When measuring fitness, the most fit is selected for further evolution, and the rest are discarded. The objective is to obtain a better collection of programs with only the most fit. A characteristic of GP is that it is easily interpretable, making it a good candidate for interpretable regression modelling. The programs (equations) are represented in a tree format where the nodes (functions) represent operators, and the leaf nodes (terminals) represent variables and numerical constants. Fitness is the root mean square error (RMSE).

One issue commonly faced by researchers and users of regression models is **interpretability**. A neural network (NN) is seen as a black box model in machine learning because we do not know what the weight values represent without lots of calculations or how the input is transformed as it propagates through the network. Random forest (RF) tends to become very deep, reusing features in sub-trees, and requires a lot of time to read through. Like the NN, RF is also not interpretable.

A second issue commonly faced in regression is **limited data**. Limited data refers to the situation in which the amount of available data is relatively small or insufficient to perform certain tasks effectively. Machine learning algorithms often require large amounts of data to learn patterns and make accurate predictions. When there is only a limited amount of data available, it can be difficult to train algorithms effectively, and the resulting models may not perform as well as they would with more data.

In this paper, we seek to address both of the above issues using an approach based on **synthetic data**. This idea is used in many areas of machine learning. NN are commonly trained using data augmentation. Generative adversarial networks (GANs) are sometimes used to generate synthetic data. When a tabular dataset contains some missing values, sometimes they are filled in by *imputation* (i.e., synthesis). Machine learning explainability methods such as LIME [15] also take advantage of synthetic data, in particular data created with the help of predictions by an existing classification or regression model. Distillation methods [7,9] sometimes work by training a small model (the "student") using data synthesized by a larger model (the "teacher"). Of course, all such methods are limited by the quality of the synthetic data.

We address the issue of interpretability by using a non-interpretable regression model (e.g., an NN or RF) to create synthetic data. This can then be used to help train an interpretable model (e.g., GP). See Fig. 1 (top). Our hypothesis is that the new model maintains accuracy while improving interpretability.

We address the issue of limited data using a similar idea. Before training the model we intend to use as the final model, we first train an initial model and use it to generate synthetic data. This is used to augment the training set for the final model. See Fig. 1 (bottom). Our hypothesis is that the accuracy of the final model is improved, thanks to the extra data, relative to the model we would have achieved without the initial model and its predictions.

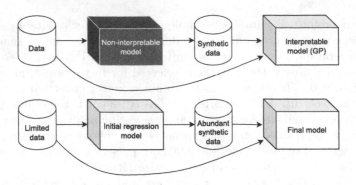

Fig. 1. System summary. We address the issue of interpretability by using a given non-interpretable model to generate synthetic data to be used in training an interpretable model (top). We address the issue of limited data in a similar way, by training an initial model and using it to generate abundant synthetic data to be used in training a final model (bottom).

2   Literature Review

Table 1. Comparison between existing interpretability and limited-data techniques and our technique

Technique	Problem	Original model M1	Explainer model M2	Sampling X for M2	Goal
LIME [15]	Classification	NN, RF	Logistic Regression	Noise near instance	Local Explanation
ELA [11]	Regression	GP	Linear Regression	Dataset neighbours of instance	Local Explanation
GPX [5]	Regression, Classification	NN, RF, SVM	GP	Noise near instance	Local Explanation
Cao et al. [2]	Anomaly detection	KDE	GP	Noise near dataset	Deal with limited data
Ours	Regression	NN, RF	GP	Noise near dataset	Global interpretability, deal with limited data

Interpretability refers to the ability of humans to understand the outputs and the decision-making processes of AI and ML models. With an interpretable ML model, a human may be able to identify causal relationships between the system's inputs and outputs [6,17]. Interpretable models are often preferred over more complex models because they provide a clear understanding of how the model makes predictions, which can be essential for building trust and understanding with stakeholders [15]. The aim is to provide insights into the relationship between the independent and dependent variables in a simple and intuitive way, which our model output will express in a simple mathematical equation generated from GP.

Table 1 shows a comparison between some machine learning studies that developed interpretable models, such as LIME [15], ELA [11], and GPX [5], and our approach. LIME aims to provide an explanation for a single decision, i.e., the prediction $\hat{y}$ for a single point X, as opposed to providing a global explanation of the model's behavior. It works by sampling values near X, and weighing them by distance to X. It then creates an interpretable model, such as a simple linear algorithm, that approximates the behavior of the black-box model in a local region of its complex decision function. By interpreting the linear model we can identify a subset of the features that contribute most to the prediction and improve the explainability.

Explanation by Local Approximation (ELA) [11] is a method used to explain the predictions of GP model at a local level. It starts by using GP to identify the function that describes the relationship between the input features and the output variable in the training set. The identified function is then used to make predictions on the test data. For a specific point of interest in the test data, ELA identifies the nearest neighbors of that point in the training set. These neighbors are used to fit a local linear regression model that explains the prediction of the machine learning model for the point of interest.

Genetic Programming Explainer (GPX) [5] is a method for explaining the predictions of a complex machine learning models, such as NN, by generating symbolic expressions that reflect the local behavior of the model. It achieves this by generating a noise set near the points of interest, and then using GP to evolve a symbolic expression that best fits the behavior of the model in that region.

In conclusion, LIME, ELA, and GPX are techniques used for generating local explanations for machine learning models. They all generate perturbed points or take existing points around the instance of interest to understand the behavior of the model in that particular region. However, the main difference between LIME, ELA, and GPX lies in how they create these samples and create local explanations. LIME generates a dataset of perturbed instances around the instance being explained and then trains a simpler model (e.g., logistic regression) on this dataset to approximate the behavior of the original model. On the other hand, ELA identifies the k-nearest neighbors of the point of interest in the training set and fits a local linear regression model using these neighbors to explain the prediction of the machine learning model for the point of interest. GPX creates a noise set near the points of interest and then uses genetic programming to evolve a symbolic expression that best fits the behavior of the model in that region.

In this study, we use two regression algorithms, random forest (RF) and neural network (NN). RFs create estimates by building an ensemble of decision trees, with each tree trained on a random subset of the data, and averaging the predictions over the estimators. NNs are composed of neurons stacked in lists, with each neuron in a layer connected to every neuron in the subsequent layer. Both algorithms are considered black-box, making predictions involves a large amount of randomness and parameter adjustments that are very difficult to interpret. The methods are also non-deterministic and quite robust. Methods have been developed for interpreting NNs and they can provide valuable insights into how a model arrived at a particular decision. However, such approaches are typically focused on single predictions identified in a small region of the models complex decision function, e.g. LIME, so the confidence level of the interpretation is only guaranteed locally. Our aim is to achieve this at a global level using a method inspired by knowledge distillation.

In machine learning, knowledge distillation is defined as the transferring of knowledge from a larger to a smaller model [7]. The idea is to train the smaller model using the output predictions of the larger after training [9]. The method we propose applies a similar concept, with a slight change, the model sizes are not limited to larger and smaller, they may be equal, and the model types may differ. We input the predictions of a black-box algorithm, such as a NN or RF, as the training data to GP algorithm with the goal of creating an interpretable model that achieves similar or better performance.

In cases where limited data is available, generating synthetic data can help increase the quantity and diversity of data available for training. Cao [2] proposes a method that involves sampling from probability distributions learned using Kernel Density Estimation (KDE) on the original dataset to generate additional data points. GP is then used to create a model of the data by evolving a program that takes input data and outputs an estimate of the density at any given point. We use a similar approach by sampling from one distribution at each data point and creating a new data point with a slight change in some direction.

3 Methods

3.1 Two Goals, a Single Method

In this work, we have two goals, both using synthetic data with GP to address two important issues in regression modelling, *interpretability* and *limited data*.

We hypothesize that:

(1) Given a trained and non-interpretable regression model (e.g., a neural network (NN) or random forest (RF)), GP can be used to create an interpretable model while maintaining accuracy by training on synthetic data formed using the existing model's predictions.

(2) In the context of limited data, an initial regression model (e.g., NN, RF, or GP) can be trained and then used to create abundant synthetic data for training a second regression model (again, NN, RF, or GP), and this second model can perform better than it would if trained on the original data alone.

However, both goals are accomplished by following the same method: starting with a limited training dataset, an initial regression model M1 is trained. Synthetic data is created by drawing from a distribution over independent variables and calling the model to produce predictions for the dependent variable. A new model M2 is then trained on the combination of the original data and the synthetic data. The idea has the same limitation as all synthetic data approaches. If we get bad results from the first model, the second model will be bad. However, if we get good results from the first model, the second model may be good. Our hypotheses do not require the second model to be better than the first model which would be impossible in principle.

3.2   Generating Synthetic Data

We generate synthetic X and synthetic y separately. We follow a procedure similar to that of Cao et al. [2].

We first generate synthetic X by randomly choosing one of the X points from the limited data. Then we make a small change in some direction by adding a small random normal vector with a mean of zero and a standard deviation ϵ.

Second, we generate synthetic y by predicting, using a pre-trained model, a y value for the X that has been generated. We then concatenate the synthetic data X and y with the limited data. This procedure differs from that of Cao et al. [2] in that we do not assume a global distribution (such as a normal) of the data X and draw synthetic points from that distribution. Each synthetic point x' is drawn from a local normal centred at one of the true data points x, with variance ϵ. This makes no global distribution assumption other than that the data has been standardised (which justifies using a constant ϵ). We have experimented with different values of ϵ and the value 0.3 is the best for our standardised datasets. Figure 2 shows the resulting synthetic dataset X.

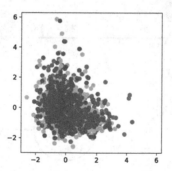

Fig. 2. Distributions of data points with respect to two variables, namely fixed acidity (X_1) and volatile acidity (X_2). Original, limited, and synthetic data points. Original data is in red, limited data in green, and synthetic data in orange. The goal is to use the limited data to generate synthetic data in a distribution similar to that of the original data. Here the standard deviation for each synthetic point is $\epsilon = 0.3$. (Color figure online)

3.3   Multi-objective GP

We use multi-objective GP, which optimises both model error on training data and model complexity. The result of a GP run is thus a Pareto front. We consider two ways of choosing a single equation from the Pareto front: the lowest error, and a method based on a heuristic combining both accuracy and complexity. This heuristic is described in detail by [4]. Below, we will refer to the equation picked by the heuristic as GPp and the lowest error as GPe.

4   Experiments

To test Hypothesis 1, we do the following steps: we train a non-interpretable model (e.g., RF) on the limited dataset, referred to as M1. We then train a GP model, referred to as M2, on the limited data and the synthetic data produced by M1. The final step is observing if the performance of M2 after M1 is improved, maintained, or disimproved, versus M1 alone. See Fig. 3.

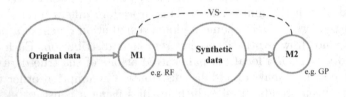

Fig. 3. Hypothesis 1. To test our hypothesis that GP can add interpretability while maintaining accuracy, we compare M2 after M1 versus M1 alone.

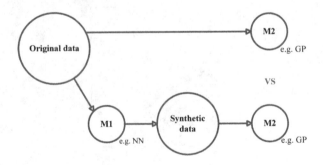

Fig. 4. Hypothesis 2. To test our hypothesis related to dealing with limited data, we compare M2 after M1 versus M2 alone.

To test Hypothesis 2, we do the following steps: we train a non-interpretable model (e.g., NN) on the limited dataset, referred to as M1. We then train a GP model, referred to as M2, on the limited data and the synthetic data produced by M1. The final step is observing if the performance of M2 after M1 is improved, maintained, or disimproved, versus M2 alone. See Fig. 4.

4.1   Experimental Design

Thus, the experiments we require are to train a model from each model class as single models (which we will see as M1, see Fig. 3), then use all of these trained models to generate synthetic data, then train a model from each model class as the final model (M2). The model classes we investigate are NN, RF, and GP. However, because we have two methods of choosing an equation from the GP Pareto front, we have effectively four model classes, NN, RF, GPp, and GPe. We also have four datasets and four dataset subset sizes (described below).

All models replicate each run five times. This is lower than is typical for a one-versus-one experimental design (where thirty or more is recommended), however it is more than sufficient to give clear trends in the context of multiple subset sizes and a four-cross-four experimental design.

For each dataset and each subset size, we thus train all four model classes and obtain four test RMSE values. Then we use each of these models to produce synthetic data, giving four new datasets. We then train each of the four model classes on each of these. This gives 16 test RMSE values for M2 (for each named dataset and each subset size).

4.2   Datasets

The datasets that we use in this work are given in Table 2. These datasets have previously been widely used in the field of machine learning. The Red wine quality dataset was collected from the UCI repository [3], and the other three were collected from the PMLB repository on GitHub [12].

We take 20% of each dataset as the test subset. We standardize each dataset to zero mean and unit variance based on the training set only. We then create several small training sets in each case by randomly selecting subsets of sizes [100, 200, 300, 500] in order to investigate the effect of different training dataset sizes in the context of limited data. This is for experimental purposes only. In a real-world application of our method, we would use the entire limited training data, not a subset, to train an initial model.

Table 2. Benchmark datasets

Dataset	Features	Number of Samples		
		Total	Train	Test
Red wine quality	11	1599	1279	320
feynman-I-9-18 (AF1)	9	100000	80000	20000
feynman-II-36-38 (AF2)	8	100000	80000	20000
529-Pollen	4	3848	3078	770

4.3   Hyperparameters

We take advantage of the well-known scikit-learn [13] library to train NN and RF models. When tuning the NN, we varied the number of neurons in a single layer, the learning rate of the optimizer, and the L2 regularisation term. We tested 10, 50, and 100 neurons for the single and multiple layers and values 1e−3, 1e−4, and 1e−5 for both the learning rate and regularisation value. The models were trained with the Adam optimizer. Each model was trained five times with a different random state for 5000 iterations while varying the hyperparameters. The hyperparameter settings with the highest average R^2 value over the five repetitions were selected. The final model used for experimentation included a single hidden layer with 50 neurons using the tanh activation function and no output activation. The parameters for the optimizer are learning rate = 1e−3, $\beta1$ = 0.9, $\beta2$ = 0.999, and alpha = 1e−3 for L2 regularisation. The results reported were averaged over five seeds ranging from 0–4.

For RF, we tested 3, 10, 100, and 1000 as the number of estimators. The scikit-learn RF package selects the optimal number of features to consider for each split based on the square root of the total number of features. The minimum number of samples required at a leaf node is calculated as 0.18 * N samples, and the maximum depth was set to 3. Each model was also trained five times with a different random state, ranging from 0–4. The final model used for the experiment included 1000 number of trees. The results reported in this work regarding the selected model were also averaged over five seeds ranging between 0–4.

We also take advantage of PySR [4] library to train GP model. PySR uses tournament selection. Mutation in PySR performed by randomly changing a part of the equation. For crossover performed by selecting two equations and exchanging parts of the equations. PySR does multi-objective optimisation: minimizing the error while also minimizing the complexity of the resulting expression. To measure the complexity we count the number of nodes. PySR uses multiple populations and there is migration between the populations.

In order to optimize our GP algorithm, we conducted pilot experiments with various configurations, including different numbers of populations, population sizes, and iterations. In the pilot experiments, we evaluated the performance of 4, 10, and 20 populations, with population sizes of 200, 100, and 40 individuals per population. We adopted a model selection strategy that selects the best individual from all populations to build the final model, which refers to the equation picked by the heuristic as GPp. The binary operators used were ["+", "-", "*", "/"], while the unary operators were ["log" and "sin"] functions. We trained each GP model five times with different random states, ranging from 0 to 4. To select the optimal hyperparameters, we calculated the R^2 value for each configuration and chose the set with the highest average R^2 value over the five repetitions. The final model used for experimentation had 4 populations with a population size of 200 individuals and was trained for 100 iterations.

4.4   Implementation

All models are developed in Python v3.9.7. The Random forest and MLP regressors are implemented using scikit-learn [13] v1.0.2, and the genetic programming was executed using the PySR library [4] v0.11.12. PySR runs with Julia [1] v1.8.5 in the background. Another library we use to support our experiments is numpy [8] v1.21.5, which is used for numerical operations. Pandas [16] v1.4.2, which is used for data manipulation and analysis. Matplotlib [10] v3.5.1, which is used for data visualization. We also use the StandardScaler function from preprocessing of scikit-learn for data normalization. For splitting the data into training and testing sets, we use a model selection from scikit-learn. For evaluating the model performance, RMSE, we use the metrics from the scikit-learn. Finally, we use the time to measure the execution time of certain operations.

5   Results

This section compares the performance of four model classes on four datasets, and four dataset subset sizes. The main concern of this work is the effort of the abovementioned on our hypotheses in Sect. 3.1. However, since the results in four dataset subset sizes are similar, for easier comparison, we first focus on the testing results where the training part comes from a limited data of 500 subsets.

Table 3. Test RMSE values of M1

Model	Wine	AF1	AF2	Pollen
NN	0.966	0.156	0.277	0.467
RF	0.797	0.803	0.827	0.896
GPp	0.810	0.770	0.899	0.610
GPe	0.807	0.371	0.403	0.466

We first present raw results. Table 3 shows the test RMSE results of each single model trained using the limited data, referred to as M1. Table 4 shows the test RMSE results of each model, referred to as M2, which is trained using the limited data and the synthetic data produced by M1. We get the results by train the model (e.g., NN, RF, and GP) on the original datasets (M1), we then generate synthetic data from M1, and we train a second regression model (again, NN, RF, and GP) on the combine original data and synthetic data.

5.1   Interpretable Model

To confirm our hypothesis 1 (interpretability), we compare the M2 after M1 (Fig. 4) versus M1 alone (Table 3). The results are presented in the Fig. 5 by subtracting the M1 alone from the M2 after M1.

For example, we train NN on the original Wine data, we then generate synthetic data from NN, and we train GP on the combined original data and synthetic data to produce an interpretable model. Then we compare GP (M2 after M1) with NN (M1 alone) by subtracting. This gives the result -0.199 in the top right of Fig. 5. This is negative (shown in green) which shows GP has improved accuracy while adding interpretability.

Table 4. Test RMSE of M2 after M1

Wine		M2			
		NN	RF	GPp	GPe
M1	NN	0.894	0.805	0.893	0.767
	RF	0.767	0.816	0.807	0.752
	GPp	0.765	0.812	0.810	0.759
	GPe	0.782	0.811	0.833	0.804

AF1		M2			
		NN	RF	GPp	GPe
M1	NN	0.161	0.824	0.435	0.371
	RF	0.670	0.904	0.734	0.708
	GPp	0.652	0.879	0.770	0.681
	GPe	0.344	0.851	0.371	0.371

AF2		M2			
		NN	RF	GPp	GPe
M1	NN	0.294	0.874	0.771	0.439
	RF	0.719	0.916	0.771	0.741
	GPp	0.784	0.923	0.899	0.799
	GPe	0.386	0.947	0.403	0.403

Pollen		M2			
		NN	RF	GPp	GPe
M1	NN	0.467	0.906	0.610	0.462
	RF	0.804	0.906	0.860	0.802
	GPp	0.576	0.905	0.610	0.574
	GPe	0.465	0.909	0.690	0.465

Looking at all the results, we see mostly red for NN (M1) to GP (M2 after M1). This shows GP has lost accuracy. However, for RF (M1) to GP (M2 after M1) we see either white or green on all four datasets. It shows that GP has succeeded in maintaining or improving accuracy while adding interpretability.

The other results in this figure (e.g. NN to RF) are less interesting because they are not about improving interpretability. However, we note that training an NN with data synthesized by some other model often gives an improvement in performance compared to that other model.

As we mentioned earlier that we have two methods of choosing an equation from the GP Pareto front: GPp and GPe. We presents all the list output of GP model on NN data of M1 in Table 5 from dataset 500 subsets of Wine dataset. This table shows the end of the PySR run. The output is the Pareto front, a set of solutions representing the trade-off between the different objectives. This represents the set of solutions that are the best compromises between accuracy and complexity. The Pareto front can be used to choose the best solution according to the specific requirements of the problem, e.g., a solution that is less error but more complex that referred to our GPe.

5.2   Limited Data

To confirm our hypothesis 2 (limited data), we compare M2 after M1 (Fig. 4) versus M2 alone (Table 3). The results are presented in the Fig. 6 by subtracting the M2 alone from the M2 after M1.

Wine		M2			
		NN	RF	GPp	GPe
M1	NN	-0.072	-0.161	-0.073	-0.199
	RF	-0.029	0.019	0.010	-0.044
	GPp	-0.045	0.003	0.000	-0.051
	GPe	-0.024	0.004	0.026	-0.002

AF1		M2			
		NN	RF	GPp	GPe
M1	NN	0.005	0.669	0.279	0.215
	RF	-0.134	0.101	-0.070	-0.095
	GPp	-0.118	0.109	0.000	-0.089
	GPe	-0.027	0.480	0.000	0.000

AF2		M2			
		NN	RF	GPp	GPe
M1	NN	0.017	0.597	0.494	0.161
	RF	-0.109	0.089	-0.056	-0.087
	GPp	-0.115	0.024	0.000	-0.100
	GPe	-0.017	0.544	0.000	0.000

Pollen		M2			
		NN	RF	GPp	GPe
M1	NN	0.000	0.439	0.143	-0.005
	RF	-0.093	0.010	-0.036	-0.095
	GPp	-0.034	0.295	0.000	-0.036
	GPe	-0.002	0.443	0.223	-0.002

Fig. 5. Adding interpretability while maintaining accuracy. M1 is the initial model and M2 the final model. We report Test RMSE of M2 minus Test RMSE of M1. Subset size 500. Top to bottom: Wine, AF1, AF2, Pollen. The orange rectangle indicates we only focus on non-interpretable models NN and RF and making interpretable models using GP. Negative numbers (green) indicate that RMSE has reduced, i.e. the new model M2 has better accuracy than the original, while zero (white) indicates accuracy is maintained and positive (red) indicates it has disimproved. Where the new model M2 is GP, the new model is interpretable. (Color figure online)

For example, we train NN on the original Wine data, we then generate synthetic data from NN, and we train GP on the combined original data and synthetic data. However, in this, we compare this GP after NN (M2 after M1) with GP (M2 alone) by subtracting. E.g., if we look at Wine data, we see the original result GPe is 0.807 and the result of GPe after NN is 0.767. By subtracting this gives the result −0.040 in the top right of Fig. 6. This value is negative (shown in light green) which shows that using an NN to generate synthetic data as a preliminary step improves the performance of GP.

In the case of limited data, we focus on performance only, not on interpretability. So we are interested in improvements in accuracy, and maintaining accuracy is not sufficient. Figure 6 shows that all models (M1) improve the performance of NN and GPe (M2 after M1) on Wine data. Also, all models (M1) improve the performance of GPp (M2 after M1) on AF1 and AF2 data. However, in all other cases, using an extra model (M1) to generate abundant synthetic data does not improve performance. For each dataset, the diagonal line in Fig. 5 represents the same values as in Fig. 6. This makes sense, because in each case we are carrying

Wine		M2			
		NN	RF	GPp	GPe
M1	NN	-0.072	0.008	0.083	-0.040
	RF	-0.199	0.019	-0.003	-0.054
	GPp	-0.201	0.016	0.000	-0.048
	GPe	-0.184	0.014	0.023	-0.002

AF1		M2			
		NN	RF	GPp	GPe
M1	NN	0.005	0.021	-0.335	0.000
	RF	0.514	0.101	-0.036	0.337
	GPp	0.496	0.075	0.000	0.310
	GPe	0.188	0.048	-0.399	0.000

AF2		M2			
		NN	RF	GPp	GPe
M1	NN	0.017	0.047	-0.128	0.036
	RF	0.441	0.089	-0.128	0.338
	GPp	0.507	0.096	0.000	0.396
	GPe	0.109	0.119	-0.496	0.000

Pollen		M2			
		NN	RF	GPp	GPe
M1	NN	0.000	0.010	0.000	-0.005
	RF	0.337	0.010	0.251	0.335
	GPp	0.109	0.009	0.000	0.108
	GPe	-0.002	0.013	0.080	-0.002

Fig. 6. Using an initial model M1 to generate abundant synthetic data. We report Test RMSE of M2 after M1, minus Test RMSE of M2 alone. Subset size 500. Top to bottom: Wine, AF1, AF2, Pollen. Again, negative numbers (green) are better. They indicate cases where using an extra model M1 improves the performance of the final model M2 versus what M2 would have achieved alone. (Color figure online)

out the same calculation (see Sect. 3.1): e.g. for NN, we subtract the RMSE of NN after NN (M2 after M1) from the RMSE of NN (M2 alone) in both figures.

For GPe after GPe (M2 after M1) and GPp after GPp (M2 after M1) are interesting: the change in RMSE is always 0 or almost 0 (−0.002) across both hypotheses. This shows that the equation found by M2 after M1 is the same or almost the same as the equation found by M2 alone. This makes sense, as if a GP equation is the best for the limited data, it will also be the best for the combined data, since the synthetic data is perfectly explained by the GP equation.

Table 5. Multi-objective GP equations results, e.g., from M2 after M1 NN model of 500 subsets wine dataset in the first loop

Complexity	Equations	RMSE Train	RMSE Test
1	x_{10}	0.93	0.98
2	$sin(x_{10})$	0.86	0.89
3 (GPp)	$sin(sin(x_{10}))$	0.81	0.81
5	$((x_9 + x_{10}) * 0.3550342)$	0.76	0.79
6	$((sin(x_{10}) - x_1) * 0.42619416)$	0.75	0.79
7	$((x_{10} - (x_6 - x_9)) * 0.30185044)$	0.75	0.79
8	$((sin(x_{10}) - (x_6 - x_9)) * 0.34612435)$	0.73	0.78
9	$sin(((x_9 + x_0) * 0.28875753) + sin(x_{10}))$	0.73	0.78
10	$((sin(x_{10}) - log\_abs(sin(x_1) + 1.1223447)) * 0.4547724)$	0.72	0.78
11	$sin(((x_9 - (x_6 - x_0)) * 0.32662305) + sin(x10))$	0.72	0.77
12	$((x_9 - log\_abs(x_9 + (x_9/(x_1 + 0.28769046)))) * 0.33636877)$	0.71	0.77
13	$((((x_9 - x_6) - log\_abs(x_9 + x_3)) + sin(x_{10})) * 0.3037547)$	0.71	0.77
14	$((x_9 - log\_abs(((x_9/(x_1 + 0.2879827)) - 0.02217112) + x_9)) * 0.3519976)$	0.71	0.76
15	$(((x_9 - log\_abs((x_9/(x_1 + 0.28458524)) + x_9)) + sin(x_{10})) * 0.3037547)$	0.70	0.76
16	$(((x_9 - log\_abs(((x_9/(x_1 + 0.28458524)) - 0.022169303) + x_9)) + x_{10}) * 0.2869915)$	0.70	0.76
17	$(((x_9 - log\_abs(((x_9/(x_1 + 0.28769046)) - 0.022171093) + x_9)) + sin(x_{10})) * 0.31854948)$	0.69	0.75
19	$(((x_9 - log\_abs((((x_9/(x_1 + 0.283546)) - 0.010319713) /1.2340765) + x_9)) + sin(x_{10})) * 0.30185044)$	0.69	0.75
20 (GPe)	$((x_9 - (log\_abs(((x_9/(x_1 + 0.28397667)) - 0.021826793) + x_9)) + sin(x_6))) + sin(x_{10})) * 0.2853589)$	0.69	0.75

6 Conclusion and Future Work

We have presented a method based on using one trained regression model to generate synthetic data for use in training a second regression model, similar to that introduced by Cao et al. [2].

We experimented with four well-known regression datasets. We have partly confirmed two hypotheses. The first hypothesis relates to the effectiveness of GP in maintaining or improving accuracy while adding interpretability. We train NN and RF on the dataset, we then generate synthetic data from the model, and train GP on the combined original data and synthetic data to produce an interpretable model. The results show that GP has succeeded for RF to GP. However the results for NN were not as promising. It might be because adding GP may not provide significant benefits for NN.

The second hypothesis relates to the ability of all models to maintain or improve accuracy while using limited data. The results of the model comparisons show variability in the accuracy of the models when they are trained with limited data from other models. However, the extent of this effect varies depending on the specific models and datasets involved. Specifically, some models have maintained or improved their accuracy on certain datasets when trained with limited data from other models, while others have lost accuracy. Overall, the results are mixed but do show that there is potential for the combination of GP and other ML models to deal with the issues of interpretability and limited data.

For future work, we aim to investigate the extrapolation behaviour of these models. We will investigate a significant challenge of some regression models, which is their difficulty in extrapolating beyond unseen data. And we will use our method of generating synthetic data to generate new data outside the range of training data, and investigate which combinations of models (M1 and M2) is useful for.

Acknowledgement. This publication has emanated from research conducted with the financial support of Science Foundation Ireland under Grant number 18/CRT/6223.

References

1. Bezanson, J., Edelman, A., Karpinski, S., Shah, V.B.: Julia: a fresh approach to numerical computing. SIAM Rev. **59**(1), 65–98 (2017). https://doi.org/10.1137/141000671
2. Cao, V.L., Nicolau, M., McDermott, J.: One-class classification for anomaly detection with kernel density estimation and genetic programming. In: Heywood, M.I., McDermott, J., Castelli, M., Costa, E., Sim, K. (eds.) EuroGP 2016. LNCS, vol. 9594, pp. 3–18. Springer, Cham (2016). https://doi.org/10.1007/978-3-319-30668-1_1
3. Cortez, P., Cerdeira, A., Almeida, F., Matos, T., Reis, J.: Modeling wine preferences by data mining from physicochemical properties. Decis. Support Syst. **47**(4), 547–553 (2009)
4. Cranmer, M.: Interpretable machine learning for science with PySR and SymbolicRegression.jl. arXiv preprint arXiv:2305.01582 (2023)
5. Ferreira, L.A., Guimarães, F.G., Silva, R.: Applying genetic programming to improve interpretability in machine learning models. In: 2020 IEEE Congress on Evolutionary Computation (CEC), pp. 1–8. IEEE (2020)
6. Gilpin, L.H., Bau, D., Yuan, B.Z., Bajwa, A., Specter, M., Kagal, L.: Explaining explanations: an overview of interpretability of machine learning. In: 2018 IEEE 5th International Conference on Data Science and Advanced Analytics (DSAA), pp. 80–89. IEEE (2018)
7. Gou, J., Yu, B., Maybank, S.J., Tao, D.: Knowledge distillation: a survey. Int. J. Comput. Vision **129**, 1789–1819 (2021)
8. Harris, C.R., et al.: Array programming with NumPy. Nature **585**(7825), 357–362 (2020). https://doi.org/10.1038/s41586-020-2649-2
9. Hinton, G., Vinyals, O., Dean, J.: Distilling the knowledge in a neural network. arXiv preprint arXiv:1503.02531 (2015)
10. Hunter, J.D.: Matplotlib: a 2D graphics environment. Comput. Sci. Eng. **9**(3), 90–95 (2007). https://doi.org/10.1109/MCSE.2007.55
11. Miranda Filho, R., Lacerda, A., Pappa, G.L.: Explaining symbolic regression predictions. In: 2020 IEEE Congress on Evolutionary Computation (CEC), pp. 1–8. IEEE (2020)
12. Olson, R.S., La Cava, W., Orzechowski, P., Urbanowicz, R.J., Moore, J.H.: PMLB: a large benchmark suite for machine learning evaluation and comparison. Bio-Data Min. **10**(36), 1–13 (2017). https://doi.org/10.1186/s13040-017-0154-4

13. Pedregosa, F., et al.: Scikit-learn: machine learning in Python. J. Mach. Learn. Res. **12**, 2825–2830 (2011)
14. Poli, R., Langdon, W.B., McPhee, N.F.: A field guide to genetic programming (2008). Published via http://lulu.com and freely available at http://www.gp-field-guide.org.uk (With contributions by J. R. Koza)
15. Ribeiro, M.T., Singh, S., Guestrin, C.: Why should I trust you? Explaining the predictions of any classifier. In: Proceedings of the 22nd ACM SIGKDD International Conference on Knowledge Discovery and Data Mining, pp. 1135–1144 (2016)
16. The pandas development team: pandas-dev/pandas: Pandas (2020). https://doi.org/10.5281/zenodo.3509134
17. Watson, D.S.: Conceptual challenges for interpretable machine learning. Synthese **200**(2), 65 (2022)

A FastMap-Based Framework for Efficiently Computing Top-K Projected Centrality

Ang Li[1($\boxtimes$)], Peter Stuckey[2,3], Sven Koenig[1], and T. K. Satish Kumar[1]

[1] University of Southern California, Los Angeles, CA 90007, USA
{ali355,skoenig}@usc.edu, tkskwork@gmail.com
[2] Monash University, Wellington Road, Clayton, VIC 3800, Australia
peter.stuckey@monash.edu
[3] OPTIMA ARC Industrial Training and Transformation Centre,
Melbourne, Australia

Abstract. In graph theory and network analysis, various measures of centrality are used to characterize the importance of vertices in a graph. Although different measures of centrality have been invented to suit the nature and requirements of different underlying problem domains, their application is restricted to explicit graphs. In this paper, we first define implicit graphs that involve auxiliary vertices in addition to the pertinent vertices. We then generalize the various measures of centrality on explicit graphs to corresponding measures of projected centrality on implicit graphs. We also propose a unifying framework for approximately, but very efficiently computing the top-K pertinent vertices in implicit graphs for various measures of projected centrality. Our framework is based on FastMap, a graph embedding algorithm that embeds a given undirected graph into a Euclidean space in near-linear time such that the pairwise Euclidean distances between vertices approximate a desired graph-based distance function between them. Using FastMap's ability to facilitate geometric interpretations and analytical procedures in Euclidean space, we show that the top-K vertices for many popularly used measures of centrality—and their generalizations to projected centrality—can be computed very efficiently in our framework.

Keywords: Projected Centrality · FastMap · Graph Embedding

1 Introduction

Graphs are used to represent entities in a domain and important relationships between them: Often, vertices represent the entities and edges represent the relationships. However, graphs can also be defined implicitly by using two kinds of vertices and edges between the vertices: The *pertinent* vertices represent the main entities, i.e., the entities of interest; the *auxiliary* vertices represent the hidden entities; and the edges represent relationships between the vertices. For

© The Author(s), under exclusive license to Springer Nature Switzerland AG 2024
G. Nicosia et al. (Eds.): LOD 2023, LNCS 14505, pp. 158–173, 2024.
https://doi.org/10.1007/978-3-031-53969-5_13

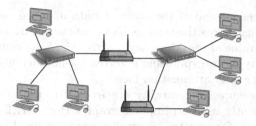

Fig. 1. Shows a communication network with user terminals, routers, and switches; solid lines show direct communication links. Depending on the application, the user terminals may be considered as the pertinent vertices while the routers and the switches may be considered as the auxiliary vertices.

example, in an air transportation domain, the pertinent vertices could represent international airports, the auxiliary vertices could represent domestic airports, and the edges could represent flight connections between the airports. In a social network, the pertinent vertices could represent individuals, the auxiliary vertices could represent communities, and the edges could represent friendships or memberships. In a communication network, as shown in Fig. 1, the pertinent vertices could represent user terminals, the auxiliary vertices could represent routers and switches, and the edges could represent direct communication links.

Explicit and implicit graphs can be used to model transportation networks, social networks, communication networks, and biological networks, among many others. In most of these domains, the ability to identify the "important" pertinent vertices has many applications. For example, the important pertinent vertices in an air transportation network could represent transportation hubs, such as Amsterdam and Los Angeles for Delta Airlines. The important pertinent vertices in a social network could represent highly influential individuals. Similarly, the important pertinent vertices in a communication network could represent admin-users, and the important pertinent vertices in a properly modeled biological network could represent biochemicals critical for cellular operations.

The important pertinent vertices in a graph (network) as well as the task of identifying them depend on the definition of "importance". Such a definition is typically domain-specific. It has been studied in explicit graphs and is referred to as a *measure of centrality*. For example, the *page rank* is a popular measure of centrality used in Internet search engines [20]. In general, there are several other measures of centrality defined on explicit graphs, such as the *degree centrality*, the *closeness centrality* [13], the *harmonic centrality* [2], the *current-flow closeness centrality* [5,22], the *eigenvector centrality* [3], and the *Katz centrality* [16].

The degree centrality of a vertex measures the immediate connectivity of it, i.e., the number of its neighbors. The closeness centrality of a vertex is the reciprocal of the average shortest path distance between that vertex and all other vertices. The harmonic centrality resembles the closeness centrality but reverses the sum and reciprocal operations in its mathematical definition to be able to handle disconnected vertices and infinite distances. The current-flow closeness centrality also resembles the closeness centrality but uses an "effective resistance" between two vertices instead of the shortest path distance between them. The eigenvector centrality scores the vertices based on the eigenvector corresponding to the largest eigenvalue of the adjacency matrix. The Katz centrality generalizes the degree centrality by incorporating a vertex's k-hop neighbors with a weight α^k, where $\alpha \in (0, 1)$ is an attenuation factor.

While many measures of centrality are frequently used on explicit graphs, they are not frequently used on implicit graphs. However, for any measure of centrality, a measure of "projected" centrality can be defined on implicit graphs. The measure of projected centrality is equivalent to the regular measure of centrality applied on a graph that "factors out" the auxiliary vertices from the implicit graph. Auxiliary vertices can be factored out by conceptualizing a clique on the pertinent vertices, in which an edge connecting two pertinent vertices is annotated with a graph-based distance between them that, in turn, is derived from the implicit graph.[1] The graph-based distance can be the shortest path distance or any other domain-specific distance. If there are no auxiliary vertices, the graph-based distance function is expected to be such that the measure of projected centrality reduces to the regular measure of centrality.

For a given measure of projected centrality, the projected centrality of a pertinent vertex is referred to as its *projected centrality value*. Identifying the important pertinent vertices in a network is equivalent to identifying the top-K pertinent vertices with the highest projected centrality values. Graph theoretically, the projected centrality values can be computed for all pertinent vertices of a network in polynomial time, for most measures of projected centrality. However, in practical domains, the real challenge is to achieve scalability to very large networks with millions of vertices and hundreds of millions of edges. Therefore, algorithms with a running time that is quadratic or more in the size of the input are undesirable. In fact, algorithms with any super-linear running times, discounting logarithmic factors, are also largely undesirable. In other words, modulo logarithmic factors, a desired algorithm should have a near-linear running time close to that of merely reading the input.

Although attempts to achieve such near-linear running times exist, they are applicable only for certain measures of centrality on explicit graphs. For example, [6,9] approximate the closeness centrality using sampling-based procedures. [1] maintains and updates a lower bound for each vertex, utilizing the bound to skip the analysis of a vertex when appropriate. It supports fairly efficient approximation algorithms for computing the top-K vertices for the closeness and the harmonic centrality measures. [4] provides a survey on such approximation algorithms. However, algorithms of the aforementioned kind are known only for a few measures of centrality on explicit graphs. Moreover, they do not provide a general framework since they are tied to specific measures of centrality.

In this paper, we generalize the various measures of centrality on explicit graphs to corresponding measures of projected centrality on implicit graphs. Importantly, we also propose a framework for computing the top-K pertinent

---

[1] The clique on the pertinent vertices is a mere conceptualization. Constructing it explicitly may be prohibitively expensive for large graphs since it requires the computation of the graph-based distance between every pair of the pertinent vertices.

vertices approximately, but very efficiently, using a graph embedding algorithm called FastMap [7,18], for various measures of projected centrality. FastMap embeds a given undirected graph into a Euclidean space in near-linear time such that the pairwise Euclidean distances between vertices approximate a desired graph-based distance function between them. In essence, the FastMap framework allows us to conceptualize the various measures of centrality and projected centrality in Euclidean space. In turn, the Euclidean space facilitates a variety of geometric and analytical techniques for efficiently computing the top-K pertinent vertices. The FastMap framework is extremely valuable because it implements this reformulation for different measures of projected centrality in only near-linear time and delegates the combinatorial heavy-lifting to analytical techniques that are better equipped for efficiently absorbing large input sizes.

Computing the top-K pertinent vertices in the FastMap framework for different measures of projected centrality often requires interpreting analytical solutions found in the FastMap embedding back in the original graphical space. We achieve this via nearest-neighbor queries and Locality Sensitive Hashing (LSH) [8]. Through experimental results on a comprehensive set of benchmark and synthetic instances, we show that the FastMap+LSH framework is both efficient and effective for many popular measures of centrality and their generalizations to projected centrality. For our experiments, we also implement generalizations of some competing algorithms on implicit graphs. Overall, our approach demonstrates the benefits of drawing power from analytical techniques via FastMap.

2   Background: FastMap

FastMap [10] was introduced in the Data Mining community for automatically generating Euclidean embeddings of abstract objects. For many real-world objects such as long DNA strings, multi-media datasets like voice excerpts or images, or medical datasets like ECGs or MRIs, there is no geometric space in which they can be naturally visualized. However, there is often a well-defined distance function between every pair of objects in the problem domain. For example, the edit distance[2] between two DNA strings is well defined although an individual DNA string cannot be conceptualized in geometric space.

FastMap embeds a collection of abstract objects in an artificially created Euclidean space to enable geometric interpretations, algebraic manipulations, and downstream Machine Learning algorithms. It gets as input a collection of abstract objects $\mathcal{O}$, where $D(O_i, O_j)$ represents the domain-specific distance between objects $O_i, O_j \in \mathcal{O}$. A Euclidean embedding assigns a κ-dimensional point $p_i \in \mathbb{R}^\kappa$ to each object O_i. A good Euclidean embedding is one in which the Euclidean distance χ_{ij} between any two points p_i and p_j closely approximates $D(O_i, O_j)$. For $p_i = ([p_i]_1, [p_i]_2 \ldots [p_i]_\kappa)$ and $p_j = ([p_j]_1, [p_j]_2 \ldots [p_j]_\kappa)$, $\chi_{ij} = \sqrt{\sum_{r=1}^{\kappa}([p_j]_r - [p_i]_r)^2}$.

---

[2] The edit distance between two strings is the minimum number of insertions, deletions, or substitutions that are needed to transform one to the other.

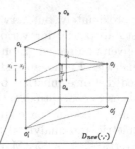

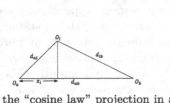

(a) the "cosine law" projection in a triangle

(b) projection onto a hyperplane that is perpendicular to $\overline{O_aO_b}$

Fig. 2. Illustrates how coordinates are computed and recursion is carried out in FastMap, borrowed from [7].

FastMap creates a κ-dimensional Euclidean embedding of the abstract objects in $\mathcal{O}$, for a user-specified value of κ. In the very first iteration, FastMap heuristically identifies the farthest pair of objects O_a and O_b in linear time. Once O_a and O_b are determined, every other object O_i defines a triangle with sides of lengths $d_{ai} = D(O_a, O_i)$, $d_{ab} = D(O_a, O_b)$ and $d_{ib} = D(O_i, O_b)$, as shown in Fig. 2a. The sides of the triangle define its entire geometry, and the projection of O_i onto the line $\overline{O_aO_b}$ is given by

$$x_i = (d_{ai}^2 + d_{ab}^2 - d_{ib}^2)/(2d_{ab}). \tag{1}$$

FastMap sets the first coordinate of p_i, the embedding of O_i, to x_i. In the subsequent $\kappa - 1$ iterations, the same procedure is followed for computing the remaining $\kappa - 1$ coordinates of each object. However, the distance function is adapted for different iterations. For example, for the first iteration, the coordinates of O_a and O_b are 0 and d_{ab}, respectively. Because these coordinates fully explain the true domain-specific distance between these two objects, from the second iteration onward, the rest of p_a and p_b's coordinates should be identical. Intuitively, this means that the second iteration should mimic the first one on a hyperplane that is perpendicular to the line $\overline{O_aO_b}$, as shown in Fig. 2b. Although the hyperplane is never constructed explicitly, its conceptualization implies that the distance function for the second iteration should be changed for all i and j in the following way:

$$D_{new}(O_i', O_j')^2 = D(O_i, O_j)^2 - (x_i - x_j)^2. \tag{2}$$

Here, O_i' and O_j' are the projections of O_i and O_j, respectively, onto this hyperplane, and $D_{new}(\cdot, \cdot)$ is the new distance function.

FastMap can also be used to embed the vertices of a graph in a Euclidean space to preserve the pairwise shortest path distances between them. The idea is to view the vertices of a given graph $G = (V, E)$ as the objects to be embedded. As such, the Data Mining FastMap algorithm cannot be directly used for

Algorithm 1. FASTMAP: A near-linear-time graph embedding algorithm.

Input: $G = (V, E)$, κ, and ϵ
Output: $p_i \in \mathbb{R}^r$ for all $v_i \in V$

1: **for** $r = 1, 2 \ldots \kappa$ **do**
2:     Choose $v_a \in V$ randomly and let $v_b = v_a$.
3:     **for** $t = 1, 2 \ldots C$ (a small constant) **do**
4:         $\{d_{ai} : v_i \in V\} \leftarrow \text{ShortestPathTree}(G, v_a)$.
5:         $v_c \leftarrow \text{argmax}_{v_i} \{d_{ai}^2 - \sum_{j=1}^{r-1}([p_a]_j - [p_i]_j)^2\}$.
6:         **if** $v_c == v_b$ **then**
7:             Break.
8:         **else**
9:             $v_b \leftarrow v_a; v_a \leftarrow v_c$.
10:        **end if**
11:    **end for**
12:    $\{d_{ai} : v_i \in V\} \leftarrow \text{ShortestPathTree}(G, v_a)$.
13:    $\{d_{ib} : v_i \in V\} \leftarrow \text{ShortestPathTree}(G, v_b)$.
14:    $d_{ab}^r \leftarrow d_{ab}^2 - \sum_{j=1}^{r-1}([p_a]_j - [p_b]_j)^2$.
15:    **if** $d_{ab}' < \epsilon$ **then**
16:        $r \leftarrow r - 1$; Break.
17:    **end if**
18:    **for** each $v_i \in V$ **do**
19:        $d_{ai}' \leftarrow d_{ai}^2 - \sum_{j=1}^{r-1}([p_a]_j - [p_i]_j)^2$.
20:        $d_{ib}' \leftarrow d_{ib}^2 - \sum_{j=1}^{r-1}([p_i]_j - [p_b]_j)^2$.
21:        $[p_i]_r \leftarrow (d_{ai}' + d_{ab}' - d_{ib}')/(2\sqrt{d_{ab}'})$.
22:    **end for**
23: **end for**
24: **return** $p_i \in \mathbb{R}^r$ for all $v_i \in V$.

generating an embedding in linear time. This is because it assumes that the distance d_{ij} between any two objects O_i and O_j can be computed in constant time, independent of the number of objects. However, computing the shortest path distance between two vertices depends on the size of the graph.

The issue of having to retain (near-)linear time complexity can be addressed as follows: In each iteration, after we heuristically identify the farthest pair of vertices O_a and O_b, the distances d_{ai} and d_{ib} need to be computed for *all* other vertices O_i. Computing d_{ai} and d_{ib} for any single vertex O_i can no longer be done in constant time but requires $O(|E|+|V|\log|V|)$ time instead [12]. However, since we need to compute these distances for all vertices, computing two shortest path trees rooted at each of the vertices O_a and O_b yields all necessary shortest path distances in one shot. The complexity of doing so is also $O(|E| + |V|\log|V|)$, which is only linear in the size of the graph[3]. The amortized complexity for computing d_{ai} and d_{ib} for any single vertex O_i is therefore near-constant time.

The foregoing observations are used in [18] to build a graph-based version of FastMap that embeds the vertices of a given undirected graph in a Euclidean

---

[3] unless $|E|$ is $O(|V|)$, in which case the complexity is near-linear in the size of the input because of the $\log|V|$ factor.

space in near-linear time. The Euclidean distances approximate the pairwise shortest path distances between vertices. Algorithm 1 presents the pseudocode for this algorithm. Here, κ is user-specified, but a threshold parameter ϵ is introduced to detect large values of κ that have diminishing returns on the accuracy of approximating pairwise shortest path distances.

3   Measures of Projected Centrality

In this section, we generalize measures of centrality to corresponding measures of projected centrality. Consider an implicit graph $G = (V, E)$, where $V^P \subseteq V$ and $V^A \subseteq V$, for $V^P \cup V^A = V$ and $V^P \cap V^A = \emptyset$, are the pertinent vertices and the auxiliary vertices, respectively. We define a graph $G^P = (V^P, E^P)$, where, for any two distinct vertices $v_i^P, v_j^P \in V^P$, the edge $(v_i^P, v_j^P) \in E^P$ is annotated with the weight $\mathcal{D}_G(v_i^P, v_j^P)$. Here, $\mathcal{D}_G(\cdot, \cdot)$ is a distance function defined on pairs of vertices in G. For any measure of centrality $\mathcal{M}$ defined on explicit graphs, an equivalent measure of projected centrality $\mathcal{M}^P$ can be defined on implicit graphs as follows: $\mathcal{M}^P$ on G is equivalent to $\mathcal{M}$ on G^P.

The distance function $\mathcal{D}_G(\cdot, \cdot)$ can be the shortest path distance function or any other domain-specific distance function. If it is a graph-based distance function, computing it would typically require the consideration of the entire graph G, including the auxiliary vertices V^A. For example, computing the shortest path distance between v_i^P and v_j^P in V^P requires us to utilize the entire graph G. Other graph-based distance functions are the probabilistically-amplified shortest path distance (PASPD) function [17] and the effective resistance between two vertices when interpreting the non-negative weights on edges as electrical resistance values.

4   FastMap for Top-K Projected Centrality

In this section, we show how to use the FastMap framework, coupled with LSH, for efficiently computing the top-K pertinent vertices with the highest projected centrality values in a given graph (network), for various measures of projected centrality. This subsumes the task of efficiently computing the top-K vertices in explicit graphs, for various regular measures of centrality. We note that the FastMap framework is applicable as a general paradigm, independent of the measure of projected centrality: The measure of projected centrality that is specific to the problem domain affects only the distance function used in the FastMap embedding and the analytical techniques that work on it. In other words, the FastMap framework allows us to interpret and reason about the various measures of projected centrality by invoking the power of analytical techniques. This is in stark contrast to other approaches that are tailored to a specific measure of centrality or its corresponding measure of projected centrality.

In the FastMap framework, any point of interest computed analytically in the Euclidean embedding may not map to a vertex in the original graph. Therefore, we use LSH [8] to find the point closest to the point of interest that corresponds to a vertex. In fact, LSH answers nearest-neighbor queries very efficiently in near-logarithmic time. It also efficiently finds the top-K nearest neighbors of a query point. The efficiency and effectiveness of FastMap+LSH allow us to rapidly switch between the original graphical space and its geometric interpretation.

We assume that the input is an edge-weighted undirected graph $G = (V, E, w)$, where V is the set of vertices, E is the set of edges, and for any edge $e \in E$, $w(e)$ specifies a non-negative weight associated with it. We also assume that G is connected since several measures of centrality and projected centrality are not very meaningful for disconnected graphs.[4] For simplicity, we further assume that there are no self-loops or multiple edges between any two vertices.

In the rest of this section, we first show how to use the FastMap framework for computing the top-K vertices in explicit graphs, for some popular measures of centrality. We then show how to use the FastMap framework more generally for computing the top-K pertinent vertices in implicit graphs, for the corresponding measures of projected centrality.

4.1   FastMap for Closeness Centrality on Explicit Graphs

Let $d_G(u, v)$ denote the shortest path distance between two distinct vertices $u, v \in V$. The closeness centrality [13] of v is the reciprocal of the average shortest path distance between v and all other vertices. It is defined as follows:

$$C_{clo}(v) = \frac{|V| - 1}{\sum_{u \in V, u \neq v} d_G(u, v)}. \tag{3}$$

Computing the closeness centrality values of all vertices and identifying the top-K vertices with the highest such values require calculating the shortest path distances between all pairs of vertices. All-pair shortest path computations generally require $O(|V||E| + |V|^2 \log |V|)$ time via the Floyd-Warshall algorithm [11].

The FastMap framework allows us to avoid the above complexity and compute the top-K vertices using a geometric interpretation. We know that given N points $q_1, q_2 \ldots q_N$ in Euclidean space $\mathbb{R}^\kappa$, finding the point q that minimizes $\sum_{i=1}^{N}(q - q_i)^2$ is easy. In fact, it is the centroid given by $q = (\sum_{i=1}^{N} q_i)/N$. Therefore, we can use the distance function $\sqrt{d_G(\cdot, \cdot)}$ in Algorithm 1 to embed the square-roots of the shortest path distances between vertices. This is done by returning the square-roots of the shortest path distances found by ShortestPathTree() in lines 4, 12, and 13. Computing the centroid in the resulting embedding minimizes the sum of the shortest path distances to all vertices. This centroid is mapped back to the original graphical space via LSH.

Overall, we use the following steps to find the top-K vertices: (1) Use FastMap with the square-root of the shortest path distance function between

---

[4] For disconnected graphs, we usually consider the measures of centrality and projected centrality on each connected component separately.

vertices to create a Euclidean embedding; (2) Compute the centroid of all points corresponding to vertices in this embedding; and (3) Use LSH to return the top-K nearest neighbors of the centroid.

4.2   FastMap for Harmonic Centrality on Explicit Graphs

The harmonic centrality [2] of a vertex v is the sum of the reciprocals of the shortest path distances between v and all other vertices. It is defined as follows:

$$C_{har}(v) = \sum_{u \in V, u \neq v} \frac{1}{d_G(u, v)}. \tag{4}$$

As in the case of closeness centrality, the time complexity of computing the top-K vertices, based on shortest path algorithms, is $O(|V||E| + |V|^2 \log |V|)$. However, the FastMap framework once again allows us to avoid this complexity and compute the top-K vertices using analytical techniques. Given N points $q_1, q_2 \ldots q_N$ in Euclidean space $\mathbb{R}^\kappa$, finding the point q that maximizes $\sum_{i=1}^{N} \frac{1}{\|q - q_i\|}$ is not easy. However, the Euclidean space enables gradient ascent and the standard ingredients of local search to avoid local maxima and efficiently arrive at good solutions. In fact, the centroid obtained after running Algorithm 1 is a good starting point for the local search.

Overall, we use the following steps to find the top-K vertices: (1) Use Algorithm 1 to create a Euclidean embedding; (2) Compute the centroid of all points corresponding to vertices in this embedding; (3) Perform gradient ascent starting from the centroid to maximize $\sum_{i=1}^{N} \frac{1}{\|q - q_i\|}$; and (4) Use LSH to return the top-K nearest neighbors of the result of the previous step.

4.3   FastMap for Current-Flow Centrality on Explicit Graphs

The current-flow closeness centrality [5, 22] is a variant of the closeness centrality based on "effective resistance", instead of the shortest path distance, between vertices. It is also known as the *information centrality*, under the assumption that information spreads like electrical current. The current-flow closeness centrality of a vertex v is the reciprocal of the average effective resistance between v and all other vertices. It is defined as follows:

$$C_{cfc}(v) = \frac{|V| - 1}{\sum_{u \in V, u \neq v} R_G(u, v)}. \tag{5}$$

The term $R_G(u, v)$ represents the effective resistance between u and v. A precise mathematical definition for it can be found in [5].

Computing the current-flow closeness centrality values of all vertices and identifying the top-K vertices with the highest such values are slightly more expensive than calculating the shortest path distances between all pairs of vertices. The best known time complexity is $O(|V||E| \log |V|)$ [5].

Once again, the FastMap framework allows us to avoid the above complexity and compute the top-K vertices by merely changing the distance function used in Algorithm 1. We use the PASPD function presented in Algorithm 1 of [17]. (We ignore edge-complement graphs by deleting Line 14 of this algorithm.) The PASPD function computes the sum of the shortest path distances between two vertices in a set of graphs G_{set}. G_{set} contains different lineages of graphs, each starting from the given graph. In each lineage, a fraction of probabilistically-chosen edges is progressively dropped to obtain nested subgraphs. The PASPD captures the effective resistance between two vertices for the following two reasons: (a) The larger the $d_G(u, v)$, the larger the PASPD between u and v, as the effective resistance between them should be larger; and (b) The larger the number of paths between u and v in G, the smaller the PASPD between them, as the effective resistance between them should be smaller.

Overall, we use the following steps to find the top-K vertices: (1) Use FastMap with the PASPD function[5] between vertices to create a Euclidean embedding; (2) Compute the centroid of all points corresponding to vertices in this embedding; and (3) Use LSH to return the top-K nearest neighbors of the centroid.

4.4   Generalization to Projected Centrality

We now generalize the FastMap framework to compute the top-K pertinent vertices in implicit graphs for different measures of projected centrality. There are several methods to do this. The first method is to create an explicit graph by factoring out the auxiliary vertices, i.e., the explicit graph $G^P = (V^P, E^P)$ has only the pertinent vertices V^P and is a complete graph on them, where for any two distinct vertices $v_i^P, v_j^P \in V^P$, the edge $(v_i^P, v_j^P) \in E^P$ is annotated with the weight $\mathcal{D}_G(v_i^P, v_j^P)$. This is referred to as the All-Pairs Distance (APD) method. For the closeness and the harmonic centrality measures, $\mathcal{D}_G(\cdot, \cdot)$ is the shortest path distance function. For the current-flow closeness centrality measure, $\mathcal{D}_G(\cdot, \cdot)$ is the PASPD function. The second method also constructs the explicit graph G^P but computes the weight on each edge only approximately using differential heuristics [24]. This is referred to as the Differential Heuristic Distance (DHD) method. The third method is similar to the second, except that it uses the FastMap heuristics [7, 17] instead of the differential heuristics. This is referred to as the FastMap Distance (FMD) method.

The foregoing three methods are inefficient because they construct G^P explicitly by computing the distances between all pairs of pertinent vertices. To avoid this inefficiency, we propose the fourth and the fifth methods. The fourth method is to directly create the FastMap embedding for all vertices of G but apply the analytical techniques only to the points corresponding to the pertinent vertices. This is referred to as the FastMap All-Vertices (FMAV) method. The fifth method is to create the FastMap embedding only for the pertinent vertices of G and apply the analytical techniques to their corresponding points. This is referred to as the FastMap Pertinent-Vertices (FMPV) method.

---

[5] It is also conceivable to use the square-root of the PASPD function.

Table 1. Results for various measures of centrality. Entries show running times in seconds and nDCG values.

Instance	Size ($|V|$, $|E|$)	Closeness			Harmonic			Current-Flow		
		GT	FM	nDCG	GT	FM	nDCG	GT	FM	nDCG
myciel5	(47, 236)	0.01	0.01	0.8810	0.01	0.08	0.8660	0.00	0.06	0.7108
games120	(120, 638)	0.06	0.03	0.9619	0.06	0.21	0.9664	0.02	0.12	0.9276
miles1500	(128, 5198)	0.41	0.09	0.9453	0.42	0.29	0.8888	0.05	0.79	0.9818
queen16_16	(256, 6320)	1.06	0.12	0.9871	1.07	0.49	0.9581	0.11	0.84	0.9381
le450_5d	(450, 9757)	3.13	0.23	0.9560	3.11	0.91	0.9603	0.30	1.59	0.8648
myciel4	(23, 71)	0.00	0.01	0.9327	0.00	0.04	0.8299	0.00	0.02	0.7697
games120	(120, 638)	0.07	0.03	0.8442	0.07	0.21	0.8004	0.02	0.14	0.9032
miles1000	(128, 3216)	0.28	0.07	0.9427	0.28	0.25	0.8510	0.04	0.47	0.7983
queen14_14	(196, 4186)	0.56	0.09	0.8866	0.55	0.38	0.8897	0.06	0.73	0.9188
le450_5c	(450, 9803)	3.32	0.24	0.8843	3.27	0.88	0.9203	0.29	2.09	0.8196
kroA200	(200, 19900)	2.59	0.52	0.9625	2.54	0.50	0.7275	0.14	2.95	0.7589
pr226	(226, 25425)	4.01	0.51	0.9996	4.06	0.63	0.6803	0.18	3.52	0.7978
pr264	(264, 34716)	6.87	0.62	0.9911	6.95	0.84	0.6506	0.30	6.30	0.8440
lin318	(318, 50403)	13.62	1.00	0.9909	13.16	1.19	0.9537	0.43	8.07	0.7243
pcb442	(442, 97461)	39.97	1.91	0.9984	39.25	2.01	0.9757	0.84	17.77	0.7283
orz203d	(244, 442)	0.11	0.06	0.9975	0.11	0.41	0.9943	0.05	0.13	0.8482
den404d	(358, 632)	0.23	0.08	0.9969	0.23	0.58	0.8879	0.10	0.14	0.9471
isound1	(2976, 5763)	18.19	0.63	0.9987	18.55	4.86	0.9815	6.18	1.95	0.9701
lak307d	(4706, 9172)	46.74	1.02	0.9996	48.56	7.66	0.9866	15.82	2.90	0.9845
ht_chantry_n	(7408, 13865)	131.30	1.51	0.9969	134.42	12.29	0.9144	37.92	3.64	0.9189
n0100	(100, 99)	0.01	0.02	0.9171	0.01	0.17	0.8102	0.01	0.05	0.9124
n0500	(500, 499)	0.34	0.10	0.8861	0.33	0.79	0.6478	0.18	0.16	0.9466
n1000	(1000, 999)	1.33	0.19	0.9125	1.38	1.63	0.9477	0.70	0.34	0.7292
n1500	(1500, 1499)	3.00	0.28	0.8856	3.15	2.39	0.6360	1.61	0.41	0.8118
n2000	(2000, 1999)	5.25	0.37	0.9078	5.56	3.21	0.8925	2.71	0.75	0.9516
n0100k4p0.3	(100, 262)	0.02	0.03	0.9523	0.03	0.17	0.9308	0.01	0.08	0.9326
n0500k6p0.3	(500, 1913)	0.83	0.11	0.9340	0.85	0.83	0.8951	0.23	0.51	0.8411
n1000k4p0.6	(1000, 3192)	3.00	0.24	0.9119	3.07	1.68	0.8975	1.13	0.83	0.8349
n4000k6p0.6	(4000, 19121)	86.37	1.09	0.9095	85.18	6.70	0.9160	214.99	8.13	0.8054
n8000k6p0.6	(8000, 38517)	387.25	2.86	0.9240	392.76	14.65	0.9368	474.30	11.08	0.7466

5   Experimental Results

We used six datasets in our experiments: DIMACS, wDIMACS, TSP, movingAI, Tree, and SmallWorld. The DIMACS dataset[6] is a standard benchmark dataset of unweighted graphs. We obtained edge-weighted versions of these graphs, constituting our wDIMACS dataset, by assigning an integer weight chosen uniformly at random from the interval $[1, 10]$ to each edge. We also obtained edge-weighted graphs from the TSP (Traveling Salesman Problem) dataset [21]

---

[6] https://mat.tepper.cmu.edu/COLOR/instances.html.

Table 2. Results for various measures of projected centrality. Entries show running times in seconds and nDCG values.

Instance	Running Time (s)						nDCG				
	APD	DHD	FMD	ADT	FMAV	FMPV	DHD	FMD	ADT	FMAV	FMPV
Closeness											
queen16_16	0.55	0.32	0.12	0.05	0.12	0.10	0.9827	0.9674	0.9839	0.9707	0.9789
le450_5d	1.58	0.98	0.29	0.09	0.27	0.26	0.9576	0.9621	0.9872	0.9577	0.9521
queen14_14	0.29	0.19	0.08	0.03	0.08	0.09	0.9274	0.8996	0.9639	0.9377	0.8898
le450_5c	1.71	1.00	0.32	0.09	0.28	0.23	0.9169	0.9231	0.9700	0.8959	0.8798
lin318	7.01	0.68	0.50	0.45	0.97	1.05	0.9285	1.0000	0.9645	0.9867	1.0000
pcb442	18.03	1.31	0.93	0.84	2.26	1.97	0.9455	1.0000	0.9663	0.9969	0.9950
lak307d	24.78	116.69	25.85	1.98	0.36	0.35	0.8991	0.9928	0.9387	0.9994	0.9960
ht_chantry_n	67.59	266.07	58.18	4.88	0.47	0.45	0.8956	0.9952	0.9522	0.9879	0.9879
n1500	1.71	10.87	2.35	0.19	0.06	0.06	0.7990	0.9485	0.9830	0.7759	0.7758
n2000	2.96	19.45	4.32	0.35	0.08	0.09	0.8187	0.9691	0.9720	0.9284	0.9229
n4000k6p0.6	44.54	78.68	16.96	1.37	0.84	0.68	0.9403	0.9172	0.9582	0.9146	0.9299
n8000k6p0.6	207.24	319.56	67.23	5.39	1.37	1.58	0.9432	0.9376	0.9522	0.9274	0.9296
Harmonic											
queen16_16	0.54	0.31	0.12	0.00	0.40	0.32	0.9730	0.9602	0.9729	0.9466	0.9730
le450_5d	1.59	0.95	0.31	0.00	0.65	0.52	0.9467	0.9499	0.9901	0.9447	0.9578
queen14_14	0.31	0.18	0.08	0.00	0.29	0.22	0.9235	0.8912	0.9903	0.9381	0.8767
le450_5c	1.74	0.98	0.30	0.00	0.65	0.51	0.8905	0.8694	0.9962	0.8485	0.8866
lin318	6.93	0.72	0.52	0.00	1.22	1.06	0.9922	1.0000	0.8974	0.8128	0.8289
pcb442	20.79	1.39	1.04	0.01	2.07	1.85	0.9924	1.0000	0.9859	0.9274	0.9755
lak307d	24.76	105.98	23.11	0.16	4.09	3.94	0.9265	0.9915	0.9908	0.9819	0.9762
ht_chantry_n	65.83	262.83	58.63	0.37	6.34	6.37	0.7549	0.6484	0.9947	0.8909	0.9925
n1500	1.74	10.83	2.41	0.01	1.33	1.22	0.6280	0.7079	0.9858	0.6575	0.7469
n2000	2.95	19.09	4.15	0.01	1.71	1.61	0.6695	0.6647	0.9354	0.9227	0.9252
n4000k6p0.6	44.53	78.08	17.17	0.10	3.74	3.76	0.9252	0.9083	0.9971	0.9163	0.9072
n8000k6p0.6	215.14	314.85	67.36	0.43	7.94	7.39	0.9351	0.9111	0.9999	0.9074	0.9439
Current-Flow											
queen16_16	5.92	0.63	0.72	-	1.21	1.26	0.9638	0.9542	-	0.9690	0.9683
le450_5d	17.50	1.47	1.45	-	2.02	2.05	0.9463	0.9678	-	0.9529	0.9632
queen14_14	3.25	0.38	0.59	-	1.29	1.16	0.9243	0.8916	-	0.8867	0.8946
le450_5c	30.90	2.03	2.58	-	5.55	5.03	0.9275	0.9372	-	0.8832	0.8778
lin318	61.02	3.29	5.22	-	11.69	10.99	0.8995	1.0000	-	0.9993	0.9990
pcb442	180.43	6.96	12.75	-	24.14	23.77	0.9781	1.0000	-	0.9997	0.9997
lak307d	265.38	105.91	24.77	-	3.28	2.71	0.6864	0.6158	-	0.6800	0.6316
ht_chantry_n	499.27	260.40	58.07	-	1.88	3.54	0.4634	0.4568	-	0.4314	0.4997
n1500	5.70	10.62	2.45	-	0.55	0.30	0.6731	0.6820	-	0.6436	0.6487
n2000	10.01	19.05	4.21	-	0.58	0.63	0.6689	0.6847	-	0.6487	0.6468
n4000k6p0.6	448.20	76.73	19.42	-	7.74	7.65	0.9218	0.9130	-	0.9320	0.9126
n8000k6p0.6	2014.92	306.53	73.18	-	16.04	14.40	0.9325	0.9174	-	0.9189	0.9229

and large unweighted graphs from the movingAI dataset [23]. In addition to these benchmark datasets, we synthesized Tree and SmallWorld graphs using the Python library NetworkX [14]. For the trees, we assigned an integer weight chosen uniformly at random from the interval $[1, 10]$ to each edge. We generated the small-world graphs using the Newman-Watts-Strogatz model [19]. For the regular measures of centrality, the graphs in the six datasets were used as such. For the projected measures of centrality, 50% of the vertices in each graph were randomly chosen to be the pertinent vertices. (The choice of the percentage of pertinent vertices need not be 50%. This value is chosen merely for presenting illustrative results.)

We note that the largest graphs chosen in our experiments have about 18500 vertices and 215500 edges.[7] Although FastMap itself runs in near-linear time and scales to much larger graphs, some of the baseline methods used for comparison in Tables 1 and 2 are impeded by such large graphs. Nonetheless, our choice of problem instances and the experimental results on them illustrate the important trends in the effectiveness of our approach.

We used two metrics for evaluation: the normalized Discounted Cumulative Gain (nDCG) and the running time. The nDCG [15] is a standard measure of the effectiveness of a ranking system. Here, it is used to compare the (projected) centrality values of the top-K vertices returned by an algorithm against the (projected) centrality values of the top-K vertices in the ground truth (GT). The nDCG value is in the interval $[0, 1]$, with higher values representing better results, i.e., closer to the GT. We set $K = 10$. All experiments were done on a laptop with a 3.1GHz Quad-Core Intel Core i7 processor and 16GB LPDDR3 memory. We implemented FastMap in Python3 and set $\kappa = 4$.

Table 1 shows the performance of our FastMap (FM) framework against standard baseline algorithms that produce the GT for various measures of centrality. The standard baseline algorithms are available in NetworkX.[8] The rows of the table are divided into six blocks corresponding to the six datasets in the order: DIMACS, wDIMACS, TSP, movingAI, Tree, and SmallWorld. Due to limited space, only five representative instances are shown in each block. For all measures of centrality, we observe that FM produces high-quality solutions and is significantly faster than the standard baseline algorithms on large instances.

Table 2 shows the performances of APD, DHD, FMD, FMAV, and FMPV for various measures of projected centrality. An additional column, called "Adapted" (ADT), is introduced for the closeness and harmonic measures of projected centrality. For the closeness and harmonic measures, ADT refers to our intelligent adaptations of state-of-the-art algorithms, presented in [6] and [1], respectively, to the projected case. The rows of the table are divided into three blocks corresponding to the three measures of projected centrality. Due to limited space, only twelve representative instances are shown in each block: the largest two from each block of Table 1. The nDCG values for DHD, FMD, ADT, FMAV, and FMPV are computed against the GT produced by APD. We observe that all our algorithms produce high-quality solutions for the various measures of projected centrality on most instances. While the success of ADT is attributed to the intelligent adaptations of two separate algorithms, the success of FMAV and FMPV is attributed to the power of appropriate analytical techniques used in the same FastMap framework. The success of DHD and FMD is attributed to their ability to closely approximate the all-pairs distances. We also observe that FMAV and FMPV are significantly more efficient than APD, DHD, and FMD

---

[7] Tables 1 and 2 show only representative instances that may not match these numbers.

[8] https://networkx.org/documentation/stable/reference/algorithms/centrality.html.

since they avoid the construction of explicit graphs on the pertinent vertices. For the same reason, ADT is also efficient when applicable.

For all measures of centrality and projected centrality considered in this paper, Tables 1 and 2 demonstrate that our FastMap approach is viable as a unified framework for leveraging the power of analytical techniques. This is in contrast to the nature of other existing algorithms that are tied to certain measures of centrality and have to be generalized to the projected case separately.

6   Conclusions and Future Work

In this paper, we generalized various measures of centrality on explicit graphs to corresponding measures of projected centrality on implicit graphs. Computing the top-K pertinent vertices with the highest projected centrality values is not always easy for large graphs. To address this challenge, we proposed a unifying framework based on FastMap, exploiting its ability to embed a given undirected graph into a Euclidean space in near-linear time such that the pairwise Euclidean distances between vertices approximate a desired graph-based distance function between them. We designed different distance functions for different measures of projected centrality and invoked various procedures for computing analytical solutions in the resulting FastMap embedding. We also coupled FastMap with LSH to interpret analytical solutions found in the FastMap embedding back in the graphical space. Overall, we experimentally demonstrated that the FastMap+LSH framework is both efficient and effective for many popular measures of centrality and their generalizations to projected centrality.

Unlike other methods, our FastMap framework is not tied to a specific measure of projected centrality. This is because its power stems from its ability to transform a graphical problem into Euclidean space in only near-linear time close to that of merely reading the input. Consequently, it delegates the combinatorics tied to any given measure of projected centrality to various kinds of analytical techniques that are better equipped for efficiently absorbing large input sizes.

In future work, we will apply our FastMap framework to various other measures of projected centrality not discussed in this paper, strengthening the confluence of discrete and analytical algorithms for graphical problems.

Acknowledgments. This work at the University of Southern California is supported by DARPA under grant number HR001120C0157 and by NSF under grant number 2112533. The views, opinions, and/or findings expressed are those of the author(s) and should not be interpreted as representing the official views or policies of the sponsoring organizations, agencies, or the U.S. Government. This research is also partially funded by the Australian Government through the Australian Research Council Industrial Transformation Training Centre in Optimisation Technologies, Integrated Methodologies, and Applications (OPTIMA), Project ID IC200100009.

References

1. Bergamini, E., Borassi, M., Crescenzi, P., Marino, A., Meyerhenke, H.: Computing top-k closeness centrality faster in unweighted graphs. ACM Trans. Knowl. Disc. Data **13**, 1–40 (2019)
2. Boldi, P., Vigna, S.: Axioms for centrality. Internet Math. **10**, 222–262 (2014)
3. Bonacich, P.: Power and centrality: a family of measures. Am. J. Sociol. **92**(5), 1170–1182 (1987)
4. Bonchi, F., De Francisci Morales, G., Riondato, M.: Centrality measures on big graphs: exact, approximated, and distributed algorithms. In: Proceedings of the 25th International Conference Companion on World Wide Web (2016)
5. Brandes, U., Fleischer, D.: Centrality measures based on current flow. In: Diekert, V., Durand, B. (eds.) STACS 2005. LNCS, vol. 3404, pp. 533–544. Springer, Heidelberg (2005). https://doi.org/10.1007/978-3-540-31856-9_44
6. Cohen, E., Delling, D., Pajor, T., Werneck, R.F.: Computing classic closeness centrality, at scale. In: Proceedings of the 2nd ACM Conference on Online Social Networks (2014)
7. Cohen, L., Uras, T., Jahangiri, S., Arunasalam, A., Koenig, S., Kumar, T.K.S.: The FastMap algorithm for shortest path computations. In: Proceedings of the 27th International Joint Conference on Artificial Intelligence (2018)
8. Datar, M., Immorlica, N., Indyk, P., Mirrokni, V.S.: Locality-sensitive hashing scheme based on p-stable distributions. In: Proceedings of the 20th Annual Symposium on Computational Geometry (2004)
9. Eppstein, D., Wang, J.: Fast approximation of centrality. Graph Algorithms Appl. **5**(5), 39 (2006)
10. Faloutsos, C., Lin, K.I.: FastMap: a fast algorithm for indexing, data-mining and visualization of traditional and multimedia datasets. In: Proceedings of the 1995 ACM SIGMOD International Conference on Management of Data (1995)
11. Floyd, R.W.: Algorithm 97: shortest path. Commun. ACM **5**(6), 345 (1962)
12. Fredman, M.L., Tarjan, R.E.: Fibonacci heaps and their uses in improved network optimization algorithms. J. ACM **34**(3), 596–615 (1987)
13. Freeman, L.: Centrality in social networks conceptual clarification. Soc. Netw. **1**, 238–263 (1979)
14. Hagberg, A., Swart, P., S Chult, D.: Exploring network structure, dynamics, and function using NetworkX. Technical report, Los Alamos National Lab, Los Alamos, NM (United States) (2008)
15. Järvelin, K., Kekäläinen, J.: Cumulated gain-based evaluation of IR techniques. ACM Trans. Inf. Syst. **20**(4), 422–446 (2002)
16. Katz, L.: A new status index derived from sociometric analysis. Psychometrika **18**(1), 39–43 (1953)
17. Li, A., Stuckey, P., Koenig, S., Kumar, T.K.S.: A FastMap-based algorithm for block modeling. In: Proceedings of the International Conference on the Integration of Constraint Programming, Artificial Intelligence, and Operations Research (2022)
18. Li, J., Felner, A., Koenig, S., Kumar, T.K.S.: Using FastMap to solve graph problems in a Euclidean space. In: Proceedings of the International Conference on Automated Planning and Scheduling (2019)
19. Newman, M.E., Watts, D.J.: Renormalization group analysis of the small-world network model. Phys. Lett. A **263**(4–6), 341–346 (1999)
20. Page, L., Brin, S., Motwani, R., Winograd, T.: The PageRank citation ranking: Bringing order to the web. Technical report, Stanford InfoLab (1999)

21. Reinelt, G.: TSPLIB-A traveling salesman problem library. ORSA J. Comput. **3**(4), 376–384 (1991)
22. Stephenson, K., Zelen, M.: Rethinking centrality: methods and examples. Soc. Netw. **11**(1), 1–37 (1989)
23. Sturtevant, N.: Benchmarks for grid-based pathfinding. Trans. Comput. Intell. AI Games **4**(2), 144–148 (2012)
24. Sturtevant, N.R., Felner, A., Barrer, M., Schaeffer, J., Burch, N.: Memory-based heuristics for explicit state spaces. In: Proceedings of the 21st International Joint Conference on Artificial Intelligence (2009)

Comparative Analysis of Machine Learning Models for Time-Series Forecasting of *Escherichia Coli* Contamination in Portuguese Shellfish Production Areas

Filipe Ferraz[1,2,3], Diogo Ribeiro[1,2,3], Marta B. Lopes[4,5], Sónia Pedro[6], Susana Vinga[3,7], and Alexandra M. Carvalho[1,2,7(✉)]

[1] Instituto Superior Técnico (IST), Lisbon, Portugal
alexandra.carvalho@tecnico.ulisboa.pt
[2] Instituto de Telecomunicações (IT), Lisbon, Portugal
[3] INESC-ID, Instituto Superior Técnico, Universidade de Lisboa, Lisbon, Portugal
[4] Center for Mathematics and Applications (NOVA Math),
Department of Mathematics, NOVA SST, 2829-516 Caparica, Portugal
[5] UNIDEMI, Department of Mechanical and Industrial Engineering, NOVA SST,
2829-516 Caparica, Portugal
[6] IPMA - Instituto Português do Mar e da Atmosfera, 1495-006 Lisbon, Portugal
[7] Lisbon Unit for Learning and Intelligent Systems, Lisbon, Portugal

Abstract. Shellfish farming and harvesting have experienced a surge in popularity in recent years. However, the presence of fecal bacteria can contaminate shellfish, posing a risk to human health. This can result in the reclassification of shellfish production areas or even prohibit harvesting, leading to significant economic losses. Therefore, it is crucial to establish effective strategies for predicting contamination of shellfish by the bacteria *Escherichia coli* (*E. coli*). In this study, various univariate and multivariate time series forecasting models were investigated to address this problem. These models include autoregressive integrated moving average (ARIMA), vector autoregressive (VAR), and long short-term memory (LSTM) networks. The data used for this study consisted of measurements of both *E. coli* concentrations and meteorological variables, which were obtained from the Portuguese Institute of Sea and Atmosphere (IPMA) for four shellfish production areas. Overall, the ARIMA models performed the best with the lowest root mean squared error (RMSE) compared to the other models tested. The ARIMA models were able to accurately predict the concentrations of *E. coli* one week in advance. Additionally, the models were able to detect the peaks of *E. coli* for all areas, except for one, with recall values ranging from 0.75 to 1. This work represents the initial steps in the search for candidate forecasting models to help the shellfish production sector in anticipating harvesting prohibitions and hence supporting management and regulation decisions.

Keywords: Time Series · Forecasting · Shellfish Contamination · *E.coli*

G. Nicosia et al. (Eds.): LOD 2023, LNCS 14505, pp. 174–188, 2024.
https://doi.org/10.1007/978-3-031-53969-5_14

1   Introduction

Over the past few years, shellfish farming and harvesting in aquaculture have experienced significant growth in both quantity and economic value, primarily due to the continuous rise in demand for fish protein among humans [1]. Shellfish farming can help reduce ocean eutrophication – an excessive richness of nutrients in a body of water, often caused by runoff from land, resulting in dense plant growth and animal death due to lack of oxygen. This is because shellfish are filter-feeding organisms that rely on the plankton present in the water column to grow [2,3]. However, microbiological contamination and the proliferation of toxic phytoplankton can compromise water quality, resulting in the prohibition of harvesting or reclassification of production areas into worse sanitary statuses. This threatens the sustainability of shellfish farming businesses. Furthermore, consuming contaminated shellfish, whether from fecal bacteria or biotoxins, poses a risk to human health. Faecal coliforms, including *Escherichia coli* (*E. coli*), are useful indicators for evaluating the bacterial quality of shellfish harvesting areas [4]. In fact, in the European Union, the sanitary quality of these areas is determined by monitoring *E. coli* levels in the shellfish flesh [5]. Although most contaminants may naturally depurate and dissipate over time while shellfish are still living in the water, it is often not financially feasible to return them to the water once they have been harvested [6].

To ensure public health safety, the Portuguese Institute of Sea and Atmosphere (IPMA) regularly monitors shellfish, classifies the production areas and prohibits their harvest and commercialization if the *E. coli* concentrations, expressed as most probable number (MPN) per 100 g, exceed the safety limits defined in EU Regulations [5] (refer to Table 1).

Although the current strategy is effective in consumer protection, it is reactive, thus responding only after shellfish are harvested, resulting in severe economic losses. Therefore, it is imperative to develop predictive strategies for shellfish contamination.

Table 1. Regulatory limits for *E. coli* concentrations in shellfish [5].

Sanitary Status	*E. coli* Regulatory Limits	Observations
A	80% of the results ≤ 230 (MPN/100 g) and 100% of the results ≤ 700 (MPN/100 g)	Bivalves can be caught and marketed for direct human consumption
B	90% of the results ≤ 4600 (MPN/100 g) and 100% of the results ≤ 46000 (MPN/100 g)	Bivalves may be harvested and destined for purification, transposition or processing into an industrial unit
C	100% of the results ≤ 46000 (MPN/100 g)	Bivalves may be harvested and intended for prolonged transposition or transformation into an industrial unit only
Forbidden	Any result > 46000 (MPN/100 g)	The harvest of bivalves is not permitted

Artificial intelligence can play a crucial role in predicting shellfish contamination. Thanks to its ability to analyze vast and varied datasets, it can identify patterns and trends that might otherwise go unnoticed. Therefore, based on environmental conditions, it can be used to assess the risk of *E. coli* contamination in different regions. This information can help authorities and local producers prioritize resources and take preventive measures to reduce the consequences of contamination.

Several modeling approaches have been proposed to forecast environmental phenomena in aquatic ecosystems, as is the case of predicting chrorophyll a (*chl-a*) concentration, extensively studied for water quality purposes. Chen et al. (2015) [7] developed an ARIMA model to predict the daily *chl-a* concentrations with data from Taihu Lake in China. The ARIMA model outperformed a multivariate linear regression (MVLR) model in terms of predictive accuracy. In addition, Cho et al. (2018) [8] applied an LSTM to predict the concentration of *chl-a* using daily measured water quality data from the Gongju observation station of the Geum River (South Korea). The authors conducted a comparison of their results with previous approaches in predicting *chl-a* concentration one and four days in advance. The model used in this study demonstrated superior performance. In another work, Lee and Lee (2018) [9] developed deep learning models to predict one-week-ahead *chl-a* concentration, a well-established indicator for algal activity, in four major rivers of South Korea. This was accomplished by utilizing water quality and quantity data. The structure of the models consisted of nine input variables, three hidden layers (the first with 32 nodes and the remaining two with 64 nodes), and one output layer, all of which were completely connected. The findings revealed that the LSTM model outperformed the other two models, following the trend line even when variations in *chl-a* were large.

Closely related to the topics of this work, Schmidt et al. (2018) [6] developed generalized linear models (GLMs), averaged GLMs, and generalized additive models (GAMs) for predicting short-term variations in shellfish concentrations of *E. coli* and biotoxin (okadaic acid and its derivates dinophysistoxins and pectenotoxins) in St Austell Bay and Turnaware Bar, United Kingdom. Through a metadata analysis, the following key variables were identified and used to develop the models: rainfall, river flow, solar radiation, sea surface temperature (SST) and wind speed and direction. The best biotoxin model provided 1-week forecasts with an accuracy of 86%, 0% false positive rate, and 0% false discovery rate for predicting the closure of shellfish beds due to biotoxin. Finally, the best *E. coli* models were able to obtain an accuracy of 99% and 98% for St Austell Bay and Turnaware Bar, respectively.

Regarding the continental Portuguese coast, Cruz et al. (2022) [10] developed multiple forecasting methods to predict mussel contamination by *diarrhetic shellfish poisoning* (DSP) toxins. The data used consisted of DSP toxin concentration, toxic phytoplankton cell counts, meteorological variables and remote sensing data such as SST and chl-a concentration. The results showed that the artificial neural network models (ANN) outperformed the VAR and ARIMA models. Surprisingly, the LSTM model, trained only on the biotoxin variable, outperformed the multivariate models by accurately predicting DSP concentration one-week-ahead.

Several studies have highlighted the relationship between *E. coli* concentrations and environmental variables, many of which are utilized in this work. Ciccarelli et al. (2017) confirmed that *E. coli* concentrations in clams harvested from the south coast of Marche Region (Italy) varied in correlation with rainfall events [11]. Temperature, salinity, pH, and solar radiation are also indicated as key factors [12–14].

This study attempts to predict shellfish contamination by *E. coli* in Portugal, which has not been done before. Based on the positive outcomes of previous literature, this work utilizes VAR, ARIMA, and LSTM models to forecast *E. coli* contamination in various shellfish production areas along the continental Portuguese coast.

The work was developed in *Python* and the paper is structured as follows. Section 2 provides a theoretical overview of the concepts necessary to properly understand the conducted study. Section 3 describes the data used for the study and the steps taken to prepare it for the prediction models. The development of the models is reported in Sect. 4. Section 5 details the experiments performed and the obtained results. Finally, the conclusions and suggestions for future work are presented in Sect. 6.

2   Background

Time Series. A time series (TS) is a set of observations, each being recorded at a specific time. A TS can be either discrete or continuous depending on whether the observations are recorded only at specific time intervals or continuously over some period of time [15].

TS can be univariate (UTS) if they contain only values of a single variable over time or multivariate (MTS) when m variables are measured over time [17]. A UTS can be represented as $\{X_t\}_{t\in\{1,\ldots,T\}}$ and an MTS as a family $\{\boldsymbol{X}_t\}_{t\in\{1,\ldots,T\}}$ containing multiple UTS, each denoted by the vector $\boldsymbol{X}_t = (X_{1t}, X_{2t}, \ldots, X_{mt})$, where X_{it} is the i-th component variable at time t [16]. The key feature of an MTS is that its observations are not only dependent on component i, but also on time t [16]. This implies that both the correlation between observations of a single component variable X_{it} at different times and the interdependence between different component series X_{it} and $X_{jt'}$ must be taken into account. Even when t and t' are not the same, these series can be correlated if $i \neq j$ [16].

Autoregressive Integrated Moving Average Model (ARIMA). ARIMA is a UTS model which combines an autoregressive model (AR) with a moving average model (MA). A TS $\{X_t\}_{t\in\{1,\ldots,T\}}$ is said to be an AR(p) if it is a weighted linear sum of the past p values plus a random shock ϵ_t with zero mean and constant variance [17], while a MA(q) corresponds to a TS that is a weighted linear sum of the last q random shocks (with zero mean and constant variance) [17]. By adding together the terms of an AR(p) model and a MA(q)

model, we obtained an autoregressive moving average (ARMA) process of order (p, q), denoted ARMA(p, q) [18], given by:

$$X_t = \phi_1 X_{t-1} + \phi_2 X_{t-2} + \cdots + \phi_p X_{t-p} + \epsilon_t + \theta_1 \epsilon_{t-1} + \cdots + \theta_q \epsilon_{t-q}. \quad (1)$$

In practice, most TS exhibit non-stationarity, making it unsuitable for applying AR, MA, or ARMA models directly. These models are only appropriate for data without trends and seasonality. However, a non-stationary TS can be transformed into a stationary one by differencing adjacent terms. If the original data series is differenced d times before fitting an ARMA(p, q) model, then the model for the original undifferenced series is said to be an ARIMA(p, d, q) model [17].
Vector Autoregressive Model (VAR). The models presented so far are only applicable to UTS. The VAR model extends the AR model to the multivariate context. A MTS $\{X_t\}_{t \in \{1,...,T\}}$ follows a VAR model of order p, VAR(p), if

$$X_t = \phi_0 + \sum_{i=1}^{p} \phi_i X_{t-i} + a_t, \quad (2)$$

where ϕ_0 is a m-dimensional constant vector, ϕ_i is a $m \times m$ matrix for $i > 0$ and a_t is a sequence of independent and identically distributed random vectors with zero mean and a positive definite covariance matrix [19].
Long Short-Term Memory (LSTM). LSTM networks are a type of recurrent neural network (RNN) that can learn long-term dependencies. An LSTM cell consists of a hidden state, which represents the short-term memory component, and an internal cell state, which represents the long-term memory [20–22].

Each cell is equipped with a set of gating units that regulate the flow of information, namely the input, forget, and output gates. The input and forget gates work in tandem to decide the amount of previous information to preserve in the current cell state and the amount of current context to transmit to future time steps [21,22].
Performance Measuring Methods. Evaluating the performance of the aforementioned models is essential. The mean squared error (MSE) is a loss function that calculates the average squared difference between estimated and true values, while the root mean squared error (RMSE) represents its square root, given by

$$RMSE = \sqrt{MSE} = \sqrt{\frac{1}{n} \sum_{i=1}^{n} (y_i - \hat{y}_i)^2}, \quad (3)$$

where n represents the number of samples in the set, and y_i and $\hat{y}_i$ are the observed and estimated values for samples i, respectively.

Table 2. The chosen production areas and their respective most commonly represented species.

Production Area	Code Name	Commercia name/Species
Estuário do Tejo	ETJ1	Japanese carpet shell/*Ruditapes philippinarum*
Estuário do Sado	ESD1	Peppery furrow/*Scrobicularia plana*
Ria de Aveiro	RIAV2	Cockle/*Cerastoderma edule*
Litoral Lagos - Albufeira	L7c2	Mussel/*Mytilus spp*

Table 3. The variables from both datasets.

Variable	Description	Unit
mean_temp	Mean air temperature	$°C$
max_temp	Maximum air temperature	$°C$
min_temp	Minimum air temperature	$°C$
mean_wind_intensity	Mean wind intensity	ms^{-1}
mean_wind_dir	Mean wind direction	$°$
wind_dir	Wind direction (N, NE, ...)	-
rainfall	Rainfall	mm
E. coli	*E. coli* concentrations	$MPN/100\,g$

3   Data Preprocessing

Data Description. This study considered two datasets provided by IPMA, both containing time series. One dataset pertains to *E. coli* concentrations, spanning from January 2014 to December 2020, while the other pertains to meteorological variables, both for each of the continental Portuguese shellfish production areas. The meteorological data consists of daily measurements from several meteorological stations along the Portuguese coast, spanning from January 2015 to December 2020. The data used in this study is private; however, the source code can be made available upon request via email.

Data Selection. Upon acquiring the data, it was observed that certain production areas and species had more measurements available than others. As a result, an analysis was conducted to determine which areas and species were most suitable for further analysis based on the amount of available data. The distribution of samples over the years for the most commonly represented species in each production area enabled the selection of the most viable ones, which are listed in Table 2.

Table 3 provides a detailed description of the variables used by the models for forecasting. In this study, missing values were imputed using linear interpolation of the available data.

Datasets Integration. To ensure an even distribution of data, required for time series analysis, only one *E. coli* measurement per week was considered for each production area. In cases where multiple *E. coli* concentrations were recorded in a week, only the highest count was retained. Additionally, weeks without any recorded measurements were marked as missing values in the *E. coli* series. For each production area, the online meteorological station closest to each sampling site was selected, and its variables were merged according to the date of *E. coli* series, resulting in a dataset containing all the variables listed in Table 3.

In cases where *E. coli* measurements were recorded but some meteorological variables were missing, the average of the previous and following three days was used. If no value was present within this range, the corresponding point in the time series was marked as a missing value. In weeks without recorded *E. coli* measurements, the mean of the most prevalent sampling site in the area was assigned to each meteorological variable on that week. For the categorical variables *mean_wind_dir* and *wind_dir* the mode was used instead of the mean.

4 Model Development

In this section, we describe the ARIMA, VAR, and LSTM models that were obtained and utilized for the forecast.

ARIMA. For the ARIMA models, two versions were created. In the first model, the original values of the TS were used (ARIMA1); in the second one, a logarithmic transformation was applied (ARIMA2). By applying the *log* operator, negative values are avoided, as the exponential operator is then used to convert the predicted values back to the original scale.

The data was split into training and test sets. The function *auto_arima* from the package *pmdarima.arima* was used to determine the best parameters (p, d, q) for the model. This function was applied to the training set and conducted the Augmented Dickey-Fuller (ADF) test to determine the order of differencing d. Then, different models were fitted for combinations of p and q, both ranging from 0 to 5. The *auto_arima* function determined the best model by finding the parameters combination that gave the lowest Akaike Information Criterion (AIC) value.

VAR. We started by conducting the Granger's Causality Test to determine which variables provided predictive value for *E. coli* prediction; this was done by utilizing the function *grangercausalitytests* from the *statsmodels.tsa.stattools* package to each meteorological variable. The following variables were found to be relevant: for the ETJ1 production area, the *mean_temp*, *max_temp*, and *min_temp* variables; for the L7c2 production area, the *mean_temp*, *min_temp*, *wind_dir*, and *rainfall* variables. For the ESD1 and RIAV2 areas, no features showed predictive value for the *E. coli* variable. The data was then split into training and test sets. Since the VAR model requires the time series to be stationary, the ADF test was conducted on each series of the training set to verify that, and in case of being non-stationary, differencing was applied. Two VAR models were built for each production area: VAR1, trained with all variables; and VAR2, trained only with those considered important by the Granger's Causality

Table 4. Selection criteria values obtained for the VAR model in the ESD1 production area with p ranging from 0 to 9. The lowest value for each selection criterion is represented in bold.

p	0	1	2	3	4	5	6	7	8	9
AIC	27.40	**25.57**	25.92	26.17	26.48	26.45	26.77	26.96	27.05	26.91
BIC	27.56	**27.04**	28.70	30.26	31.87	33.15	34.78	36.37	37.59	38.85
HQIC	27.47	**26.17**	27.05	28.67	28.67	29.17	30.02	30.84	31.28	31.76

Test. These VAR models were created with the function *VAR*, from the package *statsmodels.tsa.api*, applied to the training set. The function *select_ order* was used to select the right order p of each model. This choice was made based on AIC, Bayesian Information Criterion (BIC) and Hannan-Quinn Information Criterion (HQIC) presented in Table 4, which refers to the training set of the ESD1 area between 2015 and 2017.

LSTM. The LSTM models were built using *Keras* in *Python*, and both univariate and multivariate models were constructed. These models differ in their internal LSTM parameters, as well as in the input variables and respective processing. Two univariate models were built, named LSTM1-U and LSTM2-U, utilizing only the *E.coli* variable; in addition, two multivariate models, named LSTM1-M and LSTM2-M, were built. While LSTM1-M employed all available variables, the LSTM2-M model used the same variables as the VAR2 model.

For these networks, the data were divided into two independent folds. The first was divided into a training and validation set, with the objective of training and adjusting the best hyperparameters, respectively; and the second was the test set. Inside each set, a sliding window with a length of 5 (approximately one month of past values to predict the next one) was applied. It should be noted that the folds were imputed independently, and then normalization was performed by *minmaxScaler* so that variables with different scales could contribute equally to the model fitting.

The *KerasTuner* function was utilized to determine the optimal combination of hyperparameters. The model was provided with the option of having 1 or 2 layers, with each layer containing $1, 101, 201, \ldots, 1001$ neurons, and either ReLU or Tanh as the activation function. Additionally, the function selected between Adam, SGD, or RMSprop as optimizers, with a learning rate of 0.01, 0.001, or 0.0001. Due to the regressive nature of the problem, the MSE was chosen as the loss function. Table 5 provides the optimal hyperparameters for the ETJ1 production area predictions between 2018 and 2020; note that each production area obtained a different set of hyperparameters.

5   Results and Discussion

In this section, we present and discuss the results of the one-week-ahead forecast of *E. coli* in the species listed in Table 2, within their respective production areas. Each model was trained with data up to a certain year and tested on

Table 5. Optimal hyperparameter configuration used for testing from 2018 to 2020 in the ETJ1 production area.

Hyper parameters	Neurons per Hidden Layer	Activation Function	Learning Rate	Optimizer	Loss
LSTM1-U	701, 101	Tanh	0.001	Adam	MSE
LSTM2-U	101, 701	Tanh	0.001	Adam	MSE
LSTM1-M	301, 101	Tanh	0.001	Adam	MSE
LSTM2-M	501, 801	ReLU	0.001	RMSprop	MSE

Table 6. RMSE values obtained from the one-week-ahead forecast for different test sets in the ETJ1 production area.

Models	2016–2020	2017–2020	2018–2020	2019–2020	2020
ARIMA1	7815.16	9261.11	**7874.29**	**8159.57**	**6584.05**
ARIMA2	7815.16	8992.02	8318.10	8900.62	7658.22
VAR1	7418.83	8244.86	8247.41	8378.79	6654.62
VAR2	**7316.17**	8447.36	8380.50	8286.52	6880.19
LSTM1-U	8189.48	**8207.88**	9617.57	8814.77	7081.10
LSTM2-U	-	10624.93	7961.84	10330.80	8799.10
LSTM1-M	-	10485.92	9660.671	10735.39	9505.59
LSTM2-M	-	9154.24	10271.60	10958.67	8234.36

the remaining data. Specifically, each model was trained with data prior to 2016 and tested on data between 2016 and 2020. The same process was repeated for training with data prior to 2017 and testing on data between 2017 and 2020, and so on. Table 6 displays the RMSE values obtained in the ETJ1 production area.

It should be noted that the LSTM2-U model was validated using a separate validation set, consisting of the two years preceding the first year in the test set. Therefore, in the last column of Table 6, where the test set consists of data from the year 2020, the years 2018 and 2019 were used as the validation set, and previous years were used as the training set. The remaining LSTM models were validated using the same principle, but only for one year of data. Additionally, the *E. coli* data used by the multivariate models (VAR1, VAR2, LSTM1-M, and LSTM2-M) only spanned from the year 2015 onwards, due to the limited availability of meteorological variables data from that year until 2020, as previously mentioned. Consequently, the first column of the multivariate LSTM models is empty, as there was insufficient data to construct a training and validation set.

Recall that Table 1 provides four possible categories for the predicted values of *E. coli* concentrations. Therefore, the problem at hand can be formulated as a multiclass classification problem instead of a regression problem. In this study, we adopted a restrictive scenario by not considering the permitted tolerances, prioritizing public health. This means that for class A, 100% of the results had to be less than or equal to 230 (MPN/100 g) to receive that label, and for class B, all results had to fall between 4600 and 46000 (MPN/100 g). Figure 1 displays the resulting confusion matrices for all models tested on the data from the ETJ1 production area between 2018 and 2020.

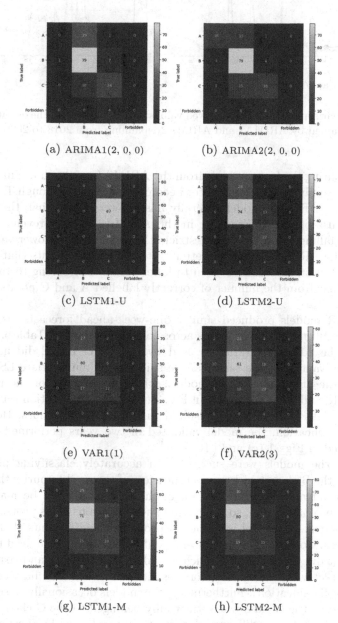

(a) ARIMA1(2, 0, 0)   (b) ARIMA2(2, 0, 0)

(c) LSTM1-U   (d) LSTM2-U

(e) VAR1(1)   (f) VAR2(3)

(g) LSTM1-M   (h) LSTM2-M

Fig. 1. Confusion matrices of the one-week-ahead predictions obtained by the models for the test set between 2018 and 2020 in the ETJ1 production area. The classes come from Table 1.

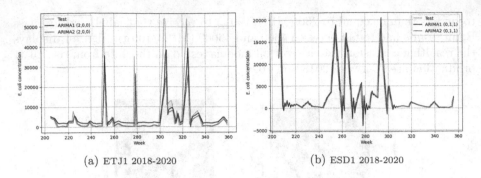

(a) ETJ1 2018-2020   (b) ESD1 2018-2020

Fig. 2. Comparison between the observed values for the *E. coli* and the one-week-ahead forecasts obtained by ARIMA1 and ARIMA2, spanning from 2018 to 2020.

By comparing the forecast plots from the ARIMA models, shown in Fig. 2, the usefulness of applying the *log* operator becomes apparent. Although Table 6 indicates that the ARIMA1 model generally has a lower RMSE than the ARIMA2 model, the use of the *log* operator in the ARIMA2 model prevents predicted values from falling below zero and restricts the predictions to lower values. Consequently, the ARIMA2 model is better suited to classify points that belong to the A class but struggles more often to label points that belong to the C class. This is evident from the number of correctly labelled A and C points in Figs. 1 (a) and (b).

The VAR models produced similar one-week-ahead forecasts, as shown in Fig. 3 (a), resulting in similar errors across all experiments in Table 6. This suggests that the additional variables used by the VAR1 model did not provide significant advantages. The same can be said for the multivariate LSTM models. As for the univariate LSTM models, whose predictions for the ETJ1 area between 2018 and 2020 are shown in Fig. 3 (b), using a validation set with only one year of data instead of two resulted in better RMSE values. However, in terms of classification, the model validated in two years performed better, as demonstrated in Figs. 1 (c) and (d).

Overall, the models were successful in accurately classifying points that belonged to the B class, which warns producers of the need to purify the bivalves one week in advance. During this stage, the shellfish cannot be marketed for direct human consumption, so identifying such situations helps protect the population. However, no model was able to properly classify points that belonged to the A class. Consequently, producers would be alerted to the need to purify a product that did not require it, which would be inconvenient and costly for the production sector(the choice of a more restrictive scenario may have contributed to this poor classification). Furthermore, the models occasionally misclassified a point as part of the B class when it actually belonged to the C class. The main implication of these two different classifications resides in the shorter commercial circuit of B class bivalves, which can go to purification besides transposition

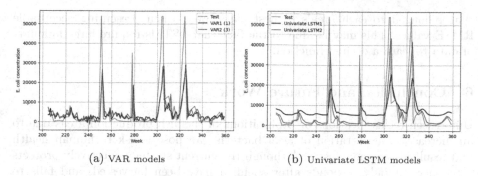

(a) VAR models   (b) Univariate LSTM models

Fig. 3. Comparison between the observed values for the *E. coli* and the one-week-ahead forecasts obtained by the VAR and univariate LSTM models, spanning from 2018 to 2020 in the ETJ1 production area.

Table 7. RMSE and TPR obtained from all the models for a test set between 2017–2020 in ESD1, RIAV2 and L7c2.

	ESD1		RIAV2		L7c2	
Models	RMSE	TPR	RMSE	TPR	RMSE	TPR
ARIMA1	1641.38	**0.84**	1305.48	0.62	95.75	1
ARIMA2	1800.02	0.76	1278.20	0.62	142.18	1
VAR1	**1532.99**	0.76	**1160.94**	**0.75**	116.34	0.875
VAR2	-	-	-	-	128.39	0.875
LSTM1-U	1622.79	0.76	1892.11	0	**63.10**	0.25
LSTM2-U	1571.56	0.8	1762.10	0.375	75.92	0.125
LSTM1-M	3837.23	0	2861.73	0.125	96.46	0
LSTM2-M	-	-	-	-	124.98	0

or industrial transformation. So this classification mistake is not critical for the safety of consumers since the bivalves would not be destined for direct human consumption.

It should also be noted that no model was able to accurately label the 'Forbidden' class (see Table 1) for the ETJ1 production area. This area presented the most and highest peaks and was the only one with *E. coli* concentrations in the 'Forbidden' class. Table 7 presents the RMSE values and true positive rates (TPR) for the test set spanning from 2017 to 2020 in the remaining areas. The TPR measures the percentage of actual positives which are correctly identified, being used when overlooked classes (false negatives) are more costly than false alarms (false positives). The TPR in Table 7 corresponds to the highest class in each area: C, C and B for ESD1, RIAV2 and L7c2, respectively.

The best models for the ESD1 and RIAV2 areas achieved a reasonable TPR, while in the L7c2 area, they were able to identify all of its peaks. However, the LSTM models, particularly the multivariate ones, struggled to label the points that belonged to each area's highest class, even when presenting acceptable RMSE values. This may be due to the fact that LSTMs require large amounts of data to learn a desired function.

6   Conclusion and Future Work

Unforeseen reclassifications or prohibitions of shellfish production areas due to an increased concentration of fecal bacteria can pose a risk to human health and result in economic losses. Although the current strategy effectively protects consumers, it only responds after shellfish have been harvested, and fails to prevent the monetary losses of local producers, a challenging problem that has not been previously addressed along the Portuguese coast. In light of this, we constructed several machine learning forecasting models, both univariate and multivariate, to provide predictions of faecal bacteria contamination and address this issue. The conducted experiments revealed that a simple ARIMA model yielded good forecasting results, outperforming the other models in most cases, achieving a TPR of 0.84 in a troublesome area and 1 in an area without high peaks.

Future work could involve conducting additional experiments, such as using more robust imputation methods to address the missing data problem. Other machine learning models, such as feed-forward neural networks (FFNN) and convolutional neural networks (CNN), could also be employed to tackle this forecasting problem. However, like LSTMs, these models require a large amount of data to achieve good results. To address this issue, data from multiple production areas could be merged. In terms of forecast analysis, in addition to the one-week-ahead forecast, two- and three-week-ahead forecasts could be applied. Knowing in advance which weeks the production zones will be reclassified or prohibited enables local producers and businesses to take necessary precautions at an earlier stage.

Acknowledgements. This work was supported by national funds through Fundação para a Ciência e a Tecnologia (FCT) through projects UIDB/00297/2020 and UIDP/00297/2020 (NOVA Math), UIDB/00667/2020 and UIDP/00667/2020 (UNIDEMI), UIDB/50008/2020 (IT), UIDB/50021/2020 (INESC-ID), and also the project MATISSE (DSAIPA/DS/0026/2019), and CEECINST/00042/2021, PTDC/CCI-BIO/4180/2020, and PTDC/CTM-REF/2679/2020. This project has received funding from the European Union's Horizon 2020 research and innovation programme under grant agreement No 951970 (OLISSIPO project).

References

1. Mateus, M., et al.: Early warning systems for shellfish safety: the pivotal role of computational science. In: Rodrigues, J.M.F., et al. (eds.) ICCS 2019. LNCS, vol. 11539, pp. 361–375. Springer, Cham (2019). https://doi.org/10.1007/978-3-030-22747-0_28
2. Matarazzo Suplicy, F.: A review of the multiple benefits of mussel farming. Rev. Aquac. **12**(1), 204–223 (2020)
3. Hallegraeff, G., Anderson, D., Cembella, A., Enevoldsen, H.: Manual on Harmful Marine Microalgae, 2nd edn. UNESCO (2004)
4. Mok, J.S., Shim, K.B., Kwon, J.Y., Kim, P.H.: Bacterial quality evaluation on the shellfish-producing area along the south coast of Korea and suitability for the consumption of shellfish products therein. Fisheries Aquatic Sci. **21**(36), (2018)
5. European Union: Commission Implementing Regulation (EU) 2019/ 627 - of 15 March 2019 - Laying down Uniform Practical Arrangements for the Performance of Official Controls on Products of Animal Origin Intended for Human Consumption in Accordance with Regulation (EU) 2017. Offic. J. Eur. Union, **131**, 51–100, (2019)
6. Schmidt, W., et al.: A generic approach for the development of short-term predictions of Escherichia coli and biotoxins in shellfish. In: Aquaculture Environment Interactions, vol. 10, pp. 173–185 (2018)
7. Chen, Q., Guan, T., Yun, L., Li, R., Recknagel, F.: Online forecasting chlorophyll a concentrations by an auto-regressive integrated moving average model: feasibilities and potentials. In: Harmful Algae, Elsevier B. V., vol. 43, pp. 58–65 (2015)
8. Cho, H., Choi, U.-J., Park, H.: Deep learning application to time-series prediction of daily chlorophyll-a concentration. In: WIT Transactions on Ecology and the Environment, vol. 215, pp. 157–163. https://doi.org/10.2495/EID180141
9. Lee, S., Lee, D.: Improved prediction of harmful algal blooms in four Major South Korea's rivers using deep learning models. Int. J. Environ. Res. Public Health **15** (2018)
10. Cruz, R.C., Costa, P.R., Krippahl, L., Lopes, M.B.: Forecasting biotoxin contamination in mussels across production areas of the Portuguese coast with artificial neural networks. Knowl. Based Syst. **257** (2022)
11. Ciccarelli, C., et al.: Assessment of relationship between rainfall and Escherichia coli in clams (Chamelea gallina) using the Bayes Factor. Italian J. Food Saf. **6**(6826) (2017)
12. Jang, J., Hur, H.G., Sadowsky, M.J., Byappanahalli, M.N., Yan, T., Ishii, S.: Environmental Escherichia coli: ecology and public health implications-a review. J. Appl. Microbiol. **123**(3), 570–581 (2017)
13. Anacleto, P., Pedro, S., Nunes, M.L., Rosa, R., Marques, A.: Microbiological composition of native and exotic clams from Tagus estuary: effect of season and environmental parameters. Mar. Pollut. Bull. **74**(1), 116–124 (2013)
14. Campos, C.J.A., Kershaw, S.R., Lee, R.J.: Environmental influences on faecal indicator organisms in coastal waters and their accumulation in bivalve shellfish. Estuaries Coasts **36**, 834–853 (2013)
15. Brockwell, P.J., Davis, R.A.: Introduction to Time Series and Forecasting, 2nd edn. Springer, Berlin (2002)
16. Wei, W.W.S.: Multivariate Time Series Analysis and Applications, 1st edn. Wiley, Hoboken (2019)
17. Chatfield, C.: Time-Series Forecasting. CHAPMAN & HALL/CRC (2001)

18. Cowpertwait, P.S.P., Metcalfe, A.V.: Introductory Time Series with R. Springer, Berlin (2009)
19. Tsay, R.S.: Multivariate Time Series Analysis: With R and Financial Applications, 1st edn. Willey, Hopboken (2014)
20. Hochreiter, S., Schmidhuber, J.: Long short-term memory. Neural Comput. **9**, 1735–80 (1997)
21. Goodfellow, I., Bengio, Y., Courville, A.: Deep Learning. MIT Press, Cambridge (2016)
22. Hewamalage, H., Bergmeir, C., Bandara, K.: Recurrent neural networks for time series forecasting: current status and future directions. Int. J. Forecast. **37**(1), 388–427 (2021)

The Price of Data Processing: Impact on Predictive Model Performance

Gail Gilboa Freedman(✉) (iD)

Reichman University, Herzliya, Israel
gail.gilboa@runi.ac.il
https://www.runi.ac.il/en/faculty/ggilboa

Abstract. The study investigates how data processing affects the performance of a predictive model trained on processed (vs. unprocessed) data. By unprocessed data, we refer to data that includes information on the influence of each mixture of transmissions from the input neurons. By processed data, we refer to data where the transmission from each input neuron is treated independently, and its co-activity behaviors with other neurons are processed into a single signal. It is intuitive that predicting the output may be less accurate when the input data undergo such processing. However, it is not immediately clear what factors determine the degree of accuracy loss.

Employing the simplest structure for the described data processing, namely two input neurons and one output neuron, we built predictive models that can forecast the activity of the output neuron based on the states of the input neurons, using a synthetic processed (vs. unprocessed) data set of historical system states. It has been discovered that a significant decrease in accuracy occurs when the conditional output firing probabilities are high for opposite inputs. The prediction of significant effects of data processing was demonstrated by utilizing a real data set of breast cancer. The study emphasizes the significance of evaluating the degree of impact caused by data processing, with possible uses for predictive models and applications, such as explainable AI.

Keywords: machine learning · data processing · predictive model

1 Introduction

Fire-Together-Wire-Together (FTWT) is a well-established doctrine [Hebb (1949)] used routinely as a basis for adjusting connection weights in neural networks [Kempter (1999)]. A simplistic layer that only takes into account the relationships between individual inputs and the output introduces a bias by oversimplifying the actual information processing pattern. This approach fails to consider other factors, such as co-firing frequencies, and the influence of co-firing neurons on the output. As an example, consider early detection of breast cancer recurrence, where machine learning classification algorithms are used to predict the event while the patient is still asymptomatic. Predictor variables

G. Nicosia et al. (Eds.): LOD 2023, LNCS 14505, pp. 189–196, 2024.
https://doi.org/10.1007/978-3-031-53969-5_15

such as menopause status and radiation therapy exert a joint influence on the outcome [Zhang et al. (2011)], and simply studying the individual influence of each can result in a loss of information.

The current study aims to explore these nuances and their impact on information processing in computational neural networks. The study is inspired by the literature on neural networks [Yao (1993)] and graph entropy measures, which uses information theory to determine the structural information content of graphs [Dehmer and Mowshowitz (2011)]. The study explores the use of information measures to determine the efficiency of adjustments between connections in multilayer perceptrons [Murtagh (1991)]. Specifically, it examines the impact of data processing on the performance of a predictive model trained on the data, focusing on the simplest system: two input neurons affecting the activity of an output neuron (Fig. 2).

Two options for training the model are evaluated: training on unprocessed-data, where each mixture of transmissions from the two input neurons is treated as a single signal (see the hidden layer in Fig. 1); and training on processed-data, where co-activity behaviors are processed and the activity of each input neuron is treated as a signal (Fig. 2). It is easy to understand that predicting the output may be less accurate when the input data undergoes processing and co-activity patterns are not taken into consideration in the learning process. However, it is not immediately clear what factors determine the degree of accuracy loss. This is exactly what this study aims to identify. For this purpose, I used computational simulation to generate multiple examples of data sets and used each set to compare the performance of the two predictive models. The comparison results are organized into a descriptive model to identify patterns of association between data set characteristics and the impact of data processing. The descriptive model was built by a machine learning algorithm [Quinlan (1986)] which is a decision tree (for a review see [Safavian and Landgrebe (1991)]). Decision trees are widely used for the purpose of predicting a label (a class) from some input features, but also used as descriptive tools [Delen et al. (2013)], as in the present case.

The simulation results demonstrate that the impact of data processing on the accuracy of output prediction is strongly linked to the behavior of the system in its two states when the input values are opposite, meaning one input neuron fires while the other does not, and vice versa. For instance, data processing has a high influence on accuracy when these states both yield similar low conditional probabilities and a lesser influence when these states yield different conditional probabilities, indicating one is low and one is high. This result can be explained by the fact that when the system behaves similarly in these states, it cannot be attributed to the activity of a single neuron, but rather to the fact that the activities of the two neurons differ from one another. This information is lost during data processing, resulting in a higher loss of accuracy of output prediction. These insights were then tested using real breast cancer data. The results showed that data processing can indeed have a significant impact on the performance of predictive model, and that the extent of this impact is influenced by the conditional probabilities of the outputs. the findings in this paper have the potential to be beneficial for the field of developing explainable AI applications.

2   Methods

The study focuses on the simplest structure with co-activity behaviors – a three-neuron system, comprising two input neurons and one output neuron. This system allows researchers to build a predictive model that can forecast the activity of the output neuron based on the states of the input neurons, using a data set of historical system states.

The three neurons system has eight possible *system states* (a state is a combination of the three neurons). We define a *setup* of this system as the distribution over its states. The setup is useful for generating a data set with the same tendency for each state to occur.

The objective is to examine two training processes that rely on this data set. The examination focuses on the resulting predictive models, the difference in their performance, and the impact of the setup properties on the extent of this difference.

The Analysis consists of three steps.

First, in step 1, a computational simulation is used to generate multiple (10000) setups by selecting eight probabilities from a uniform distribution and normalizing them to sum to 1. This results in randomly generated distributions, each of which is used to create a data set of multiple (1000) system states. Each of these data sets is represented by a matrix of eight counters (Table 1) showing the number of times each combination of states for the three neurons occurs in the data set.

Step 2 involves constructing two predictive models for each data set and comparing their accuracy metrics. The first model is trained on unprocessed data and learns the relationship between the combined activity of the input neurons and the likelihood of the output neuron firing. The second model is trained on processed data and learns the independent relationship between the activity of each input neuron and the likelihood of the output neuron firing. Table 2 presents the matrix of four counters representing the relationship between the activity of a single input neuron and that of the output neuron.

To compare the accuracy metrics for the two models, we examine the two input-output connections, making the common assumptions that (a) the strengths of each input-output connection sum to 1, and (b) the relative strength of each is determined by its relative potential to improve the model's efficiency. For simplicity, we presume that the strengths of both connections are positive, and that when an input neuron fires, the likelihood of the output neuron firing increases; however, this assumption is not required, as the results would be equivalent if the state names of the input neurons were simply relabeled.

To quantify the strengths of the connections, we define the input-output connection w as the ratio of the number of times input neuron 1 and the output neuron fire together to the number of times they fire either together or independently. This definition is grounded in twofold reasoning. On the one hand, it is based on the assumption that greater concurrence between the activity of neuron 1 and the output neuron signifies a stronger connection between these two neurons. At the same time, it is informed by the development of computational neural networks, where the weight w is optimized by reducing the loss function [Ding et al. (2013)], which is the cross-entropy [Namdari and Zhaojun (2019)].

Finally, in Step 3, we calculate the performance measure, namely loss of accuracy, for each data set by taking the difference between the accuracy of the two models described in Step 2, and dividing it by the accuracy of the first model.

We then categorize the data sets into three groups based on their loss of accuracy values: those below the 20th percentile are labeled as LOW, those above the 80th percentile as $HIGH$, and all others as $MEDIUM$.

We are interested in four properties of the data set and how they are related to the states of the system in the simulation results ($LOW, MEDIUM, HIGH$). These properties are the conditional probabilities that the output neuron fires, given that:

- P_{00}: Neither input neuron fires.
- P_{01}: The first input neuron does not fire, and the second input neuron fires.
- P_{10}: The first input neuron fires, and the second input neuron does not fire.
- P_{11}: Both input neurons fire.

To gauge the impact of these parameters, I employ a machine learning algorithm that generates a decision tree in which P_{00}, P_{01}, P_{10} and P_{11} are used to classify the states of the system as ($LOW, MEDIUM, HIGH$).

3   Results

The influence of the data set parameters ($P_{00}, P_{01}, P_{10}, P_{11}$) on the labels ($LOW, MEDIUM, HIGH$) generates aggregated patterns organized in the descriptive decision tree in Fig. 3.

The tree maps a set of rules in a hierarchical structure that specifies the properties of its nodes. Specifically, for each leaf (a node with no arrow issuing from it), the tree specifies the following properties: the number of samples sorted into this leaf; their distribution in terms of how many samples fall into each class (LOW, MEDIUM and HIGH), and the class prediction assigned to this leaf.

For each internal node (i.e., all nodes that are not leaves), the graph specifies the aforementioned properties, and also the splitting criterion. The labels serve as indicators for the data sets, reflecting the impact of data processing on the effectiveness of predictive models that utilize the data set. Through analysis of each decision tree, we glean insights into the relationship between the setup properties and various aspects of performance.

The main insights characterize classes of setups that exhibit HIGH and LOW loss of accuracy, and can be summarized as follows:

- The branch that leads to the fourth rightmost identifies a high tendency (775 of 1046 samples) to show HIGH loss of accuracy given the following conditional probabilities that the output neuron fires: low ($P_{00} \leq 0.42$) when neither input neuron fires; not low ($P_{01} > 0.37$) when only the second input neuron fires; and not high ($P_{10} \leq 0.49$) when only the first input neuron fires.
- The branch that leads to the leftmost leaf identifies a high tendency (612 of 774 samples) tendency to show LOW loss of accuracy given the following conditional probabilities that the output neuron fires: low ($P_{01} \leq 0.37$) when only the second input neuron fires; and low ($P_{11} < 0.39$) when both input neurons fire.
- The branch that leads to the third leftmost leaf identifies a high tendency (179 of 299 samples) tendency to show LOW loss of accuracy given the following conditional probabilities that the output neuron fires: low ($P_{01} \leq 0.37$) when only the second input neuron fires; not low ($P_{11} > 0.39$) when both input neurons fire; and not low ($P_{00} \geq 0.716$) when both neurons don't fire.

4   Figures, Tables and Schemes

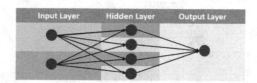

Fig. 1. A multilayer perceptron with a hidden layer.

Table 1. The eight counters in the matrix indicate the system states in the data set that fall into the eight different categories. Each matrix entry at position (i, j) is a counter, denoted as C_{ij}, representing the number of instances in the data set where the two input neurons take the values shown in row i and the output neuron takes the value in column j.

	0	1
00	C_{000}	C_{001}
01	C_{010}	C_{011}
10	C_{100}	C_{101}
11	C_{110}	C_{111}

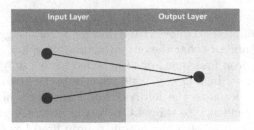

Fig. 2. A simple perceptron, with no hidden layer.

Table 2. The eight counters in the matrix indicate the system states in the data set that fall into the eight different categories. Each matrix entry at position (i, j) is a counter, denoted as C_{ij}, representing the number of instances in the data set where the two input neurons take the values shown in row i and the output neuron takes the value in column j.

	0	1
0	C_{00}	C_{00}
1	C_{10}	C_{11}

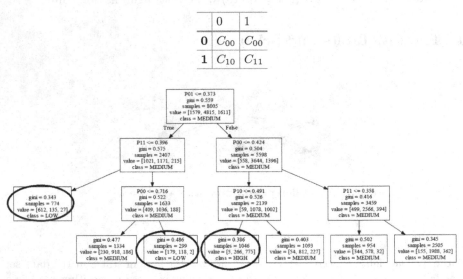

Fig. 3. For each leaf (a node with no arrow issuing from it), the tree specifies the following properties: the number of samples sorted into this leaf; their distribution in terms of how many samples fall into each label; and the class prediction assigned to this leaf. For each internal node (i.e., not a leaf), the graph specifies the aforementioned properties, and also the splitting criterion.

5   Illustration: Predicting Breast Cancer Recurrence

To illustrate the implications of the findings, let us revisit the breast cancer scenario mentioned earlier, and use the approach outlined here to analyze real-world data.

The data used comprise 286 real patient records obtained from the Breast Cancer Data Set at the Institute of Oncology in Ljubljana, accessible through the

UCI ML Repository (https://archive.ics.uci.edu/ml/datasets/Breast+Cancer). We are interested in building a predictive model for the Class variable, with the possible values "no-recurrence-events" and "recurrence-events". The data were first organized into subsets based on two binary predictors: Menopause (indicating that the patient was pre-menopausal at the time of diagnosis) and Irradiation (indicating that the patient had undergone radiation therapy, a treatment that uses high-energy x-rays to eliminate cancer cells). Both predictors align with the assumption that when an input neuron fires, the likelihood of the output neuron firing increases.

We calculate the properties of the data set ($P_{00} = 0.180, P_{01} = 0.624, P_{10} = 0.296, P_{11} = 0.395$), and use them to classify the data set based on the decision tree depicted in Fig. 3. The data set meets the criteria for tending to exhibit HIGH loss of accuracy, with the following characteristics: $P_{00} < 0.4, P_{01} > 0.37$, and $P_{10} < 0.5$. Indeed, a positive loss is observed, as the predictive model based on the conditional probabilities of the raw data displays an accuracy level of 0.724, while the model using the alternative approach (processed data) has lower accuracy (0.7).

6   Discussion

The aim of this study was to explore the impact of data processing on the performance of predictive models, focusing on a simple system of two input neurons and one output neuron. The study acknowledges the simplicity of the scenario with only two input neurons and recognizes the need for further exploration in more complex scenarios. To address this, future studies could investigate the impact of data processing on predictive model performance using a wider range of neural network architectures and diverse data sets. Additionally, evaluating the experiments on multiple data sets would provide a more comprehensive understanding of the generality and robustness of the findings. The results of the study demonstrate that the degree of accuracy loss caused by data processing is strongly linked to the behavior of the system in its two states when the input values are opposite, suggesting applications for explainable AI.

Data Availibility Statement. Publicly archived Python code and data sets analyzed or generated in this study can be accessed here

References

Delen, D., Kuzey, C., Uyar, A.: Measuring firm performance using financial ratios: a decision tree approach. Expert Syst. Appl. **40**(10), 3970–3983 (2013)

Ding, S., Li, H., Chunyang, S., Junzhao, Yu., Jin, F.: Evolutionary artificial neural networks: a review. Artif. Intell. Rev. **39**(3), 251–260 (2013)

Dehmer, M., Mowshowitz, A.: A history of graph entropy measures. Inf. Sci. **181**(1), 57–78 (2011)

Hebb, D.O.: The organization of behavior: a neuropsychological theory. Inf, Sci (1949)

Kempter, R., Wulfram, G., Van Hemmen, J.L.: Hebbian learning and spiking neurons. Phys. Rev. E **59**(4), 4498 (1999)

Murtagh, F.: Multilayer perceptrons for classification and regression. Neurocomputing **2**(5–6), 183–197 (1991)

Namdari, A., Zhaojun, L.: A review of entropy measures for uncertainty quantification of stochastic processes. Adv. Mech. Eng. 11(6) (2019)

Quinlan, J.R.: Induction of decision trees. Mach. Learn. **1**(1), 81–106 (1986)

Safavian, S.R., David, L.: A survey of decision tree classifier methodology. IEEE Trans. Syst. Man Cybern. **21**(3), 660–674 (1991)

Yao, X.: A review of evolutionary artificial neural networks. Int. J. Intell. Syst. **1**(1), 539–567 (1993)

Zhang, W., Becciolini, A., Biggeri, A., Pacini, P., Muirhead, C.R.: Second malignancies in breast cancer patients following radiotherapy: a study in Florence. Italy. Breast Cancer Res. **13**(2), 1–9 (2011)

Reward Shaping for Job Shop Scheduling

Alexander Nasuta[✉][iD], Marco Kemmerling[iD], Daniel Lütticke[iD],
and Robert H. Schmitt[iD]

Institute of Information Management in Mechanical Engineering (WZL-MQ/IMA),
RWTH Aachen University, Aachen, Germany
alexander.nasuta@ima.rwth-aachen.de

Abstract. Effective production scheduling is an integral part of the success of many industrial enterprises. In particular, the job shop problem
(JSP) is highly relevant for flexible production scheduling in the modern era. Recently, numerous approaches for the JSP using reinforcement
learning (RL) have been formulated. Different approaches employ different reward functions, but the individual effects of these reward functions
on the achieved solution quality have received insufficient attention in the
literature. We examine various reward functions using a novel flexible RL
environment for the JSP based on the disjunctive graph approach. Our
experiments show that a formulation of the reward function based on
machine utilization is most appropriate for minimizing the makespan of
a JSP among the investigated reward functions.

Keywords: Production Scheduling · Job Shop Scheduling ·
Disjunctive Graph · Reinforcement Learning · Reward Shaping

1 Introduction

Production planning is an everyday challenge in many companies. Allocating
resources and tasks is often a complex decision. Depending on the application
and industry, there are various complex formulations for the resulting optimization problem. For example, planning problems in the service sector have different
constraints on a schedule than in manufacturing industries. Efficient scheduling
is essential for producing customized products and small quantities in the manufacturing industry. The formulation as a job shop problem (JSP) offers one of
the most flexible forms of description in the context of workshop production and
is therefore particularly well suited for this case [3]. Each product is considered
as a job in a JSP. A job consists of a number of tasks that must be completed on
a certain machine in a particular order. The JSP features a high computational

This work has been supported by the FAIRWork project (www.fairwork-project.eu)
and has been funded within the European Commission's Horizon Europe Programme
under contract number 101049499. This paper expresses the opinions of the authors
and not necessarily those of the European Commission. The European Commission is
not liable for any use that may be made of the information contained in this paper.

G. Nicosia et al. (Eds.): LOD 2023, LNCS 14505, pp. 197–211, 2024.
https://doi.org/10.1007/978-3-031-53969-5_16

complexity since the problem is NP-hard. Small JSP instances can be solved with exact solution methods, such as integer programming [9]. Due to the NP-hardness of the problem, finding optimal solutions using exact methods within a viable amount of time becomes infeasible as the problem size increases. Thus, heuristics are often utilized to solve larger JSP instances. Heuristics typically consist of simplistic rules and generally cannot provide formal guarantees about the quality of the solution [16]. Rarely does a set of fixed rules yield intelligent behavior. Reinforcement learning (RL) has the potential to develop solution strategies that are more sophisticated than simple heuristics by repeatedly solving JSP instances. Once a Policy is trained by RL, it can generate solutions for a JSP quickly, similar to a heuristic. Furthermore, RL in production scheduling has the potential to incorporate reliability aspects into the scheduling process [10]. However, one crucial challenge in applying RL to a specific problem is finding a suitable reward function. In job shop scheduling, the makespan, the time needed to process all tasks, is often chosen as an optimization metric. A statement about the makespan of a schedule is generally only possible for a whole schedule and not a partial one. The natural formulation of a makespan-oriented reward function is therefore sparse, i.e. only produces non-zero rewards once a problem has been solved to completion by the agent. While sparse reward functions can be used in RL, dense rewards are known to be more conducive to training success [5]. Thus, a dense reward function is desirable. Designing the reward function to encourage RL agents to solve a RL problem more efficiently, incorporating domain knowledge, is also known as reward shaping [7]. Waubert de Puiseau et al. [10] conduct a survey about RL approaches in production scheduling, while focusing on reliability aspects. Samsonov et al. [12] outline numerous JSP approaches with different optimization goals, like tardiness minimization, machine utilization and make span minimization amongst others. Samsonov et al. [12] contrast different approaches regarding state representation, the action space design, the RL algorithm, and the specific use case. This paper compares reward functions of several state-of-the-art approaches to conclude which type of shaping is well suited for the JSP with makespan minimization.

2   Related Work

Sparse and dense reward functions for the JSP have been proposed. Samsonov et al. [13] introduced a sparse reward function. Dense reward functions have been proposed by Zhang et al. [18] and Tassel et al. [17]. Each approach embeds its reward function in a specific setting. Observation space, action space, and the modeling approach vary significantly among the particular methods. It is difficult to state to what degree the reward function contributed to the method's success and to what extent other modeling factors played a role. Our work compares the introduced reward functions with fixed action and observation space.

Samsonov et al. [13] implemented a discrete time simulation for the JSP. Individual tasks are processed on simulated machines. The agent is queried for input when decisions about assigning tasks to machines have to be made. The proposed action space maps available jobs to a predefined set of fixed-sized normalized processing times. This way, the size of the action space is independent of the problem size. The designed reward function is sparse. A reward is given only at the last step of an episode, and the reward for all intermediate steps is zero. The return is qualitatively inversely proportional to the makespan, resulting in improvements near the optimal solution being rewarded especially high.

Tassel et al. [17], similar to Samsonov et al. [13], proposes a time-discrete formalization of the JSP. However, the state representation, the action space, and the reward function differ. At each time step, a decision is made on assigning available jobs (or tasks respectively) to available machines. The reward function is based on the machine utilisation. A reward is determined based on the area a task takes up in the Gantt chart and the size of any idle times introduced on machines. Tassel et al. do not directly optimize the makespan, but the scheduled area. Since the sum of the processing times of all tasks of an instance is always the same, the makespan depends substantially on the waiting times of individual tasks. Waiting times are expressed in the Gantt chart as free areas or holes. Therefore, a reduction of the holes or an increase of the scheduled area leads to lower makespans. An advantage of this reward design is that the scheduled area can be determined in each step, resulting in a dense reward function.

Zhang et al. [18] leverages the formulation of the JSP as a disjunctive graph rather than a discrete time simulation. Moreover, Zhang et al. utilize disjunctive graphs with a graph neural network (GNN) to transform the state into a fixed size. The policy incorporates information from the nodes transformed by the GNN and the whole embedded graph at each scheduling step. The reward function is based on the increase in the critical path during the scheduling process. Through the GNN, the agent can solve instances of different sizes. Zhang et al. trained the agent with many automatically generated JSP instances and investigated to what extent the agent was able to learn a generalized solution procedure for JSP instances.

3   Methodology

There are several ways to formalize the JSP as a RL problem regarding the reward function, observation space, and action space. This paper introduces a novel flexible RL environment that allows specifying the reward function, observation space, and action space to enable the analysis of different design decisions for RL. The environment was used to evaluate the reward functions outlined in Sect. 2, amongst a newly introduced one, and assess how they affected learning behavior and aims to address the question how the reward function can be shaped efficiently for the JSP. In the following sections, the JSP is first formally introduced. Afterwards, the essential aspects of the environment, particularly the reward functions, are described.

3.1   Job-Shop Problem Formalization

The JSP consists of a set of tasks $T = \{T_1, T_2, \ldots, T_N\}$ that needs processing on a set of machines $\mathcal{M} = \{M_i\}_{i=1}^m$. A job is a sequence of tasks, corresponding to the production of a specific product, while a task corresponds to a production step. A JSP considers a set of jobs $\mathcal{J} = \{J_j\}_{j=1}^n$. The number of tasks is determined by the number of jobs n and the number of machines: $N = n \cdot m$. A solution of the JSP, a feasible schedule, assigns a start time $\hat{s}_\alpha$ to each task T_α such that none of the constraints above is violated. Our work utilizes a disjunctive graph approach for the JSP. A disjunctive graph G consists of a set of nodes $\mathcal{V}$, a set of directed edges $\mathcal{A}$, and a set of undirected edges $\mathcal{E}$: $G = (\mathcal{V}, \mathcal{A}, \mathcal{E})$. The set of nodes $\mathcal{V}$ is composed of the set of tasks $\mathcal{T}$ and two fictitious nodes, the source T_0 and the sink $T_\star$: $\mathcal{V} = \mathcal{T} \cup \{T_0, T_\star\}$. $\mathcal{A}$ is the set of directed edges representing precedence relations. These edges are called conjunctive edges. Initially $\mathcal{A}$ only contains precedence relations introduced by jobs. $(T_\beta, T_\alpha) \in \mathcal{A}$ means that T_α is the immediate predecessor of T_β. The edges of a directed graph can be weighted by the duration of a task, as shown in Błażewicz et al. [4]. For each possible pair of tasks, processed on the same machine, an undirected edge is inserted into the graph. These edges are called disjunctive edges. In Fig. 1, these are represented by dashed lines. All disjunctive edges associated with a specific machine M_i constitute the set $\mathcal{E}_i$ and all disjunctive edges form the set $\mathcal{E}$. In Fig. 1, $\mathcal{E}$ corresponds to all dashed edges, whereas $\mathcal{E}_i$ is the set of all dashed edges of a specific color. A disjunctive edge $(T_\beta, T_\alpha) \in \mathcal{E}_i$ encapsulates that it is not yet decided whether task T_β will be processed before T_α on machine i or the other way around. Therefore all undirected edges must be transformed into directed ones to obtain a valid schedule. For each disjunctive edge in $\mathcal{E}$, one has to decide in which direction it points and whether this edge shall be present in the final schedule. The orientation must be such that the graph remains cycle-free. Cycles in a graph lead to infeasible schedules as illustrated in Fig. 1d. When the graph is fully scheduled, i.e., when all disjunctive edges are turned into conjunctive ones, the makespan can be determined by finding the longest path in the graph, the critical path, which by construction is always a path from the source node T_0 to the sink node $T_\star$. The length of the critical path is equal to the makespan. In Fig. 1c, the critical path is highlighted in red. The discrete optimization problem for minimizing the makespan can be stated as follows. Minimize $s_\star$ subject to [3]:

$$\hat{s}_\beta - \hat{s}_\alpha \geq p_\alpha \qquad\qquad \forall\, (T_\beta, T_\alpha) \in \mathcal{A} \tag{1}$$

$$\hat{s}_\alpha \geq 0 \qquad\qquad \forall\, T_\alpha \in \mathcal{T} \tag{2}$$

$$\hat{s}_\beta - \hat{s}_\alpha \geq p_\alpha \vee \hat{s}_\alpha - \hat{s}_\beta \geq p_\beta \qquad\qquad \forall\, \{T_\beta, T_\alpha\} \in \mathcal{E}_i, \forall\, M_i \in \mathcal{M} \tag{3}$$

3.2 State Space

RL demands that the state representations satisfy the Markov property. Using the disjunctive graph approach, it is trivial to adhere to this requirement since the graph contains all the information. Disjunctive edges can be realized mainly in two ways. One could represent a disjunctive edge as two directed edges. This representation is called explicit disjunctive edges in the following. However, the disjunctive edges $\mathcal{E}_i$ belonging to a particular machine can also be acquired by

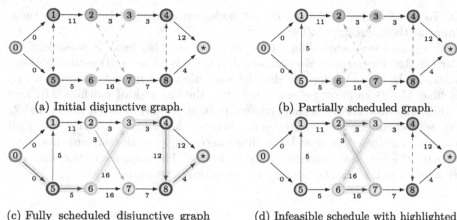

(a) Initial disjunctive graph.

(b) Partially scheduled graph.

(c) Fully scheduled disjunctive graph with highlighted critical path.

(d) Infeasible schedule with highlighted cycle.

Fig. 1. Disjunctive graph scheduling.

storing in each node to which machine it belongs to. This can be represented graphically by coloring the nodes. This representation is called implicit disjunctive edges in the following. Figure 2 illustrates both formulations.

Any data structure capable of constructing Fig. 2 is a valid representation of the state and satisfies the Markov property. The environment provides several ways to capture the state of the graph utilizing implicit disjunctive edges. All options are based on the adjacency matrix, that is extended with information concerning the machine and the processing duration of a task.

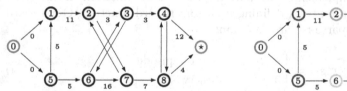

(a) Disjunctive graph using explicit disjunctive edges.

(b) Disjunctive graph using implicit disjunctive edges.

Fig. 2. Possible disjunctive graph representations.

3.3   Action Space

For the action space, two modes are provided: one based on tasks and one based on jobs. For the tasks case, the action space consists of the set of all tasks $\mathcal{T}$, in the job case the action space is the set of all jobs $\mathcal{J}$. Internally, the environment is based on tasks and maps a job to the next unscheduled task within the specified job. To avoid cycles in the graph, the nodes are scheduled from left to right, similar to the approach of Zhang et al.

In certain cases, scheduling a task between two other tasks is possible. Consider Fig. 3 as an example. For machine M_1, there is a long waiting time between T_1 and T_5. In the next step, T_7 shall be scheduled, which must be processed on machine M_1. T_7 can now be appended after the last task of machine M_1, task T_5 (Fig. 3b). However, T_7 is short enough to be scheduled between the T_1 and T_5 (Fig. 3c). Scheduling a task between two others is known as a *left shift*. For a left shift, the machine edges must be modified accordingly. In the implemented environment, the left shift can be enabled if desired. The implementation executes left shifts only if a task fits strictly between two other tasks.

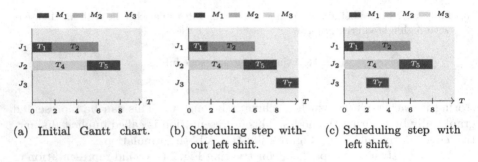

(a) Initial Gantt chart.   (b) Scheduling step without left shift.   (c) Scheduling step with left shift.

Fig. 3. Left shifts in Gantt charts.

3.4 Reward Functions

There are four preimplemented options for the reward function in the environment and the possibility of defining a custom reward function. Our reward function is a trivial sparse reward function:

$$r(s_t) = \begin{cases} -\dfrac{C}{C_{LB}} & \text{end of episode} \\ 0 & \text{otherwise} \end{cases} \qquad (4)$$

Here C_{LB} denotes a lower bound of the makespan for a specific instance. In this form, the reward is always proportional to the instance's makespan and does not introduce initial learning difficulties. If $C_{LB} = C_{opt}$ the reward r will approach -1 as the agent finds better and better solutions. Another option is the reward function by Samsonov et al.:

$$r(s_t) = \begin{cases} 1000 \, \dfrac{\gamma^{C_{opt}}}{\gamma^C} & \text{end of episode} \\ 0 & \text{otherwise} \end{cases} \qquad (5)$$

Here C_{opt} denotes the optimal makespan, C denotes the makespan of the solution and $\gamma \in (1, \infty)$ is parameter of the reward function. The resulting reward

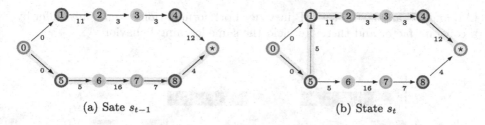

(a) Sate s_{t-1} (b) State s_t

Fig. 4. Illustration of Zhang et al.'s reward function using implicit disjunctive edges.

difference of two near-optimal solutions is overproportionally high. With Samsonov et al.'s reward function the sign does not have to be reversed, since the division transforms the minimization into a maximization. The third option is Zhang et al.'s reward function:

$$r(s_t) = H(s_{t-1}) - H(s_t) \tag{6}$$
$$H(s_t) = \max_{\alpha}\{C_{LB}(T_\alpha, s_t)\} \tag{7}$$

In contrast to the previous reward functions it is not sparse. Visually speaking it corresponds to the increment in the critical path multiplied by -1. Figure 4 illustrates this concept. When scheduling task T_1 respectively, the length of the critical path in the graph increases from $H(s_{t-1}) = 32$ to $H(s_t) = 34$ and the reward for that timestep results in $r(s_t) = -2$. The increments sign is flipped so that an increment in the critical path is penalized. More formally, the function H can be treated as a quality measure for a specific state. Unlike above, $C_{LB}(T, s)$ denotes a function here. It maps a specific task T and the state s of the disjunctive graph to the estimated time of completion of task T. The last option is based on the machine utilisation and inspired by Tassel et al.'s reward function:

$$r(s_t) = \frac{\sum\limits_{\substack{\alpha \\ \exists \hat{s}_\alpha}} p_\alpha}{|\mathcal{M}| \max\limits_{\substack{\alpha \\ \exists \hat{s}_\alpha}} \hat{s}_\alpha + p_\alpha} \tag{8}$$

Visually speaking, it is the scheduled area of the Gantt chart divided by the total area spanned by the Gantt chart. Figure 5 illustrates this approach. The numerator of the fraction sums up the duration of all scheduled tasks, which corresponds to the scheduled area, whereas the denominator is the largest finishing time multiplied by the number of machines which constitutes the total area. This reward function is not exactly the same as Tassel et al.'s reward function, although it adapts the concept in the disjunctive graph representation. In principle the Gantt chart can be constructed with jobs or machine on the y-axis. Both are valid representation of a schedule. Equation 8 is constructed with respect to a Gantt chart with machines (Fig. 5a). One could reformulate Eq. 8 with respect to jobs. In that case $|\mathcal{M}|$ would have to be replaced by $|\mathcal{J}|$. Since

$|\mathcal{M}|$ and $|\mathcal{J}|$ are fixed for a JSP instance both formulations would only differ by a constant factor and therefore yield the same learning behavior.

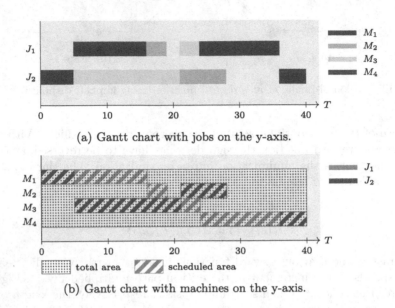

(a) Gantt chart with jobs on the y-axis.

(b) Gantt chart with machines on the y-axis.

Fig. 5. Visualisation of the graph-tassel reward function.

4   Experimental Setup

For comparing the reward functions, instances with different sizes taken from the literature were used. Namely, the instances ft06 [6], ft10 [6], orb04 [1], ta01 [15], ta02 [15] are considered. The Proximal Policy Optimization (PPO) algorithm by Schulman et al. [14] with action masking has been applied to learn the policy. All experiments were set up with Stable Baselines3 [11] and tracked using Weights and Biases [2]. In addition to our implementation, the JSSEnv by Tassel et al. [17] and OR-Tools [8] were considered as a reference. For all combinations of the presented reward functions and left shift functionality turned on or off, a hyperparameter tuning was performed on a concrete instance for each instance size considered. Namely, hyperparameter tuning was performed for the instances ft06, orb04, and ta01. The setup only considers one specific instance at a time. In general, finding a parameter set that leads to good results on various instances would be more appropriate. However, with the comparison of different reward functions, numerous runs have to be performed per construction. Therefore, only individual instances are considered here. For each JSP instance, 100 runs were created with random hyperparameters, 50 with enabled left shifts and 50 without. The optimality gap, which is proportional to the makespan but allows comparability among instances, was chosen as an optimization metric and is defined as follows:

$$\text{optimality gap} = \frac{C}{C_{opt}} - 1 \tag{9}$$

Here C_{opt} denotes the optimal makespan, C denotes the makespan of the solution. Additionally, the number of left shifts and the reward is tracked. For evaluation, we use the left shift percentage:

$$\text{left shifts} = \frac{\#t_{ls}}{t} \tag{10}$$

t denotes the timestep, and $\#t_{ls}$ is the number of timesteps that resulted in the environment performing a left shift. To investigate their long term behavior, the PPO parameterizations of the runs with the smallest optimality gap were used to perform a run with 2500000 timesteps. Furthermore, for the 10×10 and 15×15 sizes, the tuned PPO parameterizations were applied to other instances, namely to the ft10 and ta02 instances. All obtained data is accessible on Weights and Biases[1]. The code of the environment[2] and the experimental setup[3] can be accessed on GitHub.

5   Results and Discussion

The optimality gaps after 2500000 timesteps training are listed in Table 1. Instances on which no hyperparameter tuning took place are marked with an asterisk ($\star$). The column *optimality gap left shift* lists the numerical value of the optimality gap according to Eq. 9 for the solution, the schedule of the instance, and indicates that left shifts according to Sect. 3.3 were performed. Analogously *optimality gap no left shift* indicates that left shifts were not performed to obtain the solution schedule. It is inconclusive from the table data alone which reward function is most appropriate, since for every instance size a different reward function resulted in the lowest optimality gap. It is evident, nevertheless, that the left shift functionality has a major impact on the optimality gap and yields improved solutions. This is not surprising since a left shift, as constructed here, never leads to a degradation of the critical path within a timestep.

   Examining training data allows further insights into the effectiveness of the individual reward functions. Figures 6, 7, 8, 9 and 10 show the timestep behaviour of the optimality gap, the percentage of timesteps with left shifts, and the reward. In the legend left shift is abbreviated by ls and indicates that the left shifts according to Sect. 3.3 were performed. For the graphs of the optimality gaps, a running average smoothing with a window size of 20 data points was performed to show the general trend of the curve more clearly. The plots were created by sampling 750 data points. The reward graph was also smoothed with a running

---

[1] https://wandb.ai/querry/reward-functions-comparison.

[2] https://github.com/Alexander-Nasuta/graph-jsp-env.

[3] https://github.com/Alexander-Nasuta/Reward-Shaping-for-Job-Shop-Scheduling.

Table 1. Optimality gaps after 2500000 timesteps.

instance	size	environment	reward function	optimality gap no left shift	optimality gap left shift
ft06	6 × 6	graph-jsp-env	graph-tassel	0.10910	0.12730
			zhang	0.09091	0.03636
			samsonov	0.25450	0.07273
			our	0.10910	**0.0**
		JSSEnv	tassel	0.0	–
		OR-Tools	–	0.0	–
orb04	10 × 10	graph-jsp-env	graph-tassel	0.3264	0.1652
			zhang	0.6846	0.3701
			samsonov	0.1582	**0.11240**
			our	0.60500	0.2119
		JSSEnv	tassel	0.07463	–
		OR-Tools	–	0.0	–
ft10*	10 × 10	graph-jsp-env	graph-tassel	0.36450	0.23330
			zhang	0.81510	0.34620
			samsonov	0.23010	**0.10220**
			our	0.66990	0.30750
		JSSEnv	tassel	0.04194	–
		OR-Tools	–	0.0	–
ta01	15 × 15	graph-jsp-env	**graph-tassel**	0.98940	**0.15920**
			zhang	3.58900	0.39480
			samsonov	1.25000	0.44920
			our	1.23900	0.35820
		JSSEnv	tassel	0.2059	–
		OR-Tools	–	0.0	–
ta02*	15 × 15	graph-jsp-env	**graph-tassel**	1.0920	**0.13120**
			zhang	2.57700	0.45660
			samsonov	1.25100	0.22990
			our	1.14100	0.28140
		JSSEnv	tassel	0.09164	–
		OR-Tools	–	0.0	–

average with a size of 200 datapoints for dense reward functions. For sparse reward functions only the non zero data points were considered and plotted with a running average of 20 data points. Some PPO parameterizations do not exhibit appropriate learning behavior. Examples are the Samsonov reward function with left shifts in Fig. 6 and the Zhang reward function without left shifts in Fig. 9. The latter evolved to optimality gaps higher than the initial optimality gap instead of

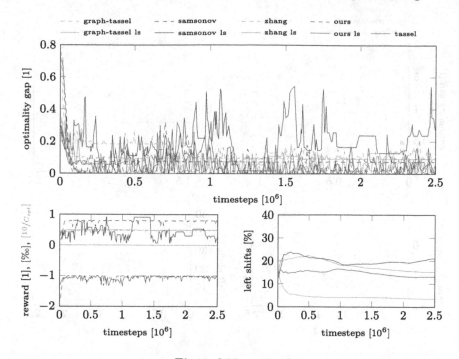

Fig. 6. ft06 metrics.

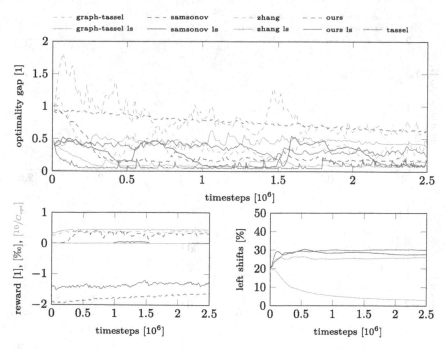

Fig. 7. orb04 metrics.

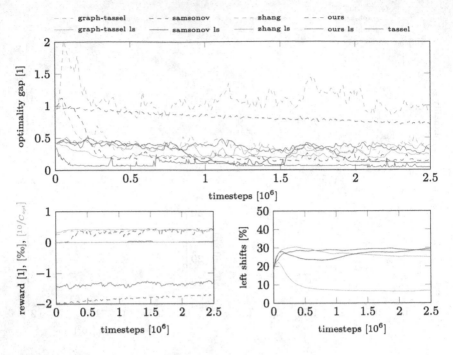

Fig. 8. ft10 metrics.

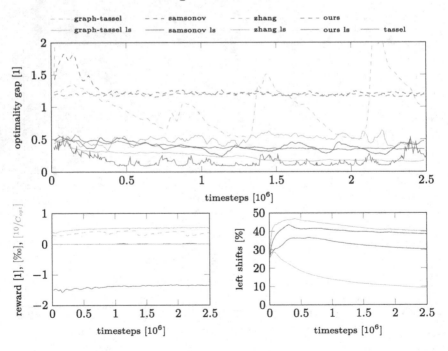

Fig. 9. ta01 metrics.

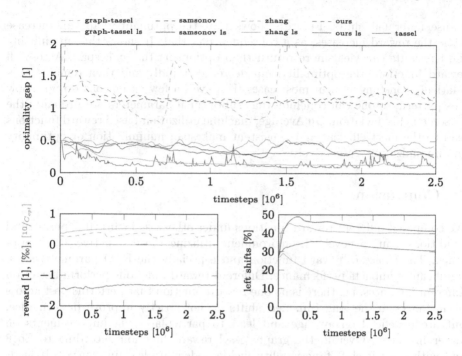

Fig. 10. ta02 metrics.

minimizing the optimality gap. It should be noted that this parameterization was the result of hyperparameter tuning, so there are at least 49 parameterization that perform worse. Therefore, it can be concluded that the PPO algorithm in this setup is very sensitive to the input parameters. The reward functions of Samsonov, Zhang, as well as our reward functions use the left shift functionality extensively. In Figs. 6, 7, 8, 9 and 10 it is unclear whether the percentage of left shifts changes systematically over time. This makes sense since both the Samsonov and our reward function are sparse. The return of an episode always remains the same for a schedule regardless of how many left shifts were performed in the episode. With the Zhang reward function, a left shift leads to a reward of 0 instead of a negative reward. If the discount factor is equal to 1, which is the case for all performed runs here, left shifts do not change the return of an episode with the same resulting schedule. One can therefore call the return of a schedule in Samsonov, Zhang, and our reward function indifferent to left shifts. There is no incentive to increase or decrease the number of left shifts in the scheduling process. It should be noted, however, that left shifts still help to lead to schedules with lower or equal makespan. The adapted Tassel reward function (graph-tassel) interacts differently with the left shift functionality. Graphically speaking, left shifts in the present form densify a Gantt chart without increasing the makespan, thus increasing machine utilization. Episodes which result in the same schedule extract more return without left shifts, so there is an incentive to use fewer left shifts. Figure 6 shows that the optimality gap can increase while the reward increases, as a consequence of that. It shows that minimizing the left shifts can yield more return than optimizing the makespan. Across all reward functions the optimality gap decreases rapidly and then remains at a relatively constant level in most cases. However, a few exceptions do not show this pattern, so one can conclude that suboptimal parameters are probably the reason for this deviation. On Average, machine utilization-based reward functions lead to the most efficient and consistent makespan minimization or optimality gap minimization, respectively.

6 Conclusion

We compared different reward functions under otherwise identical experimental conditions using a novel RL environment utilizing the disjunctive graph approach. The choice of reward function and especially the PPO parametrisation significantly impacts performance. Different reward functions perform better on different instances, i.e. there is no one reward function that performs best across all instances in the long run. Left shifts do not strictly improve performance, but are beneficial on average and lead to particularly big improvements on larger instances. Overall, the graph-tassel reward function according to Eq. 8 and with the left shift functionality yielded adequate learning curves. It leads to initially near optimal solutions on the investigated instances. However, the graph-tassel reward function minimizes left shifts after a certain time which leads to more return but also may increase makespan. This issue could be addressed by considering only the area of a machine or a job instead of the whole Gantt chart. This approach is used in the JSSEnv reward function with respect to the area of a single job with a timestep (cf. Fig. 5a). During the experiments, the Tassel reward function combined with the JSSEnv provided consistently strong results. Therefore, it can be assumed that the unchanged formulation of the Tassel reward function leads to even better results than the one using Eq. 8. However, this remains to be confirmed experimentally. A reward function based on the machine utilization appears to be a promising approach for makespan minimization. However, it must be carefully designed to ensure that maximizing the machine utilization strictly corresponds to minimizing the makespan.

References

1. Applegate, D., Cook, W.: A computational study of the job-shop scheduling problem. ORSA J. Comput. **3**(2), 149–156 (1991)
2. Biewald, L.: Experiment tracking with weights and biases (2020)
3. Błażewicz, J., Ecker, K.H., Pesch, E., Schmidt, G., Sterna, M., Weglarz, J.: Handbook on Scheduling: From Theory to Practice. Springer, Heidelberg (2019). https://doi.org/10.1007/978-3-319-99849-7
4. Błażewicz, J., Pesch, E., Sterna, M.: The disjunctive graph machine representation of the job shop scheduling problem. Eur. J. Oper. Res. **127**(2), 317–331 (2000)

5. Burda, Y., Edwards, H., Pathak, D., Storkey, A., Darrell, T., Efros, A.A.: Large-scale study of curiosity-driven learning. In: International Conference on Learning Representations (2018)
6. Fisher, H.: Probabilistic learning combinations of local job-shop scheduling rules. Industr. Sched. 225–251 (1963)
7. Grzes, M.: Reward shaping in episodic reinforcement learning (2017)
8. Perron, L., Furnon, V.: OR-tools (2022)
9. Pinedo, M.: Planning and Scheduling in Manufacturing and Services. Springer, Heidelberg (2005). https://doi.org/10.1007/b139030
10. de Puiseau, C.W., Meyes, R., Meisen, T.: On reliability of reinforcement learning based production scheduling systems: a comparative survey. J. Intell. Manuf. **33**(4), 911–927 (2022)
11. Raffin, A., Hill, A., Gleave, A., Kanervisto, A., Ernestus, M., Dormann, N.: Stable-baselines3: reliable reinforcement learning implementations. J. Mach. Learn. Res. **22**(268), 1–8 (2021)
12. Samsonov, V., Hicham, K.B., Meisen, T.: Reinforcement learning in manufacturing control: baselines, challenges and ways forward. Eng. Appl. Artif. Intell. **112**, 104868 (2022)
13. Samsonov, V., et al.: Manufacturing control in job shop environments with reinforcement learning. In: ICAART (2), pp. 589–597 (2021)
14. Schulman, J., Wolski, F., Dhariwal, P., Radford, A., Klimov, O.: Proximal policy optimization algorithms. arXiv preprint arXiv:1707.06347 (2017)
15. Taillard, E.: Benchmarks for basic scheduling problems. Eur. J. Oper. Res. **64**(2), 278–285 (1993)
16. Talbi, E.G.: Metaheuristics: From Design to Implementation. Wiley, Hoboken (2009)
17. Tassel, P.P.A., Gebser, M., Schekotihin, K.: A reinforcement learning environment for job-shop scheduling. In: 2021 PRL Workshop-Bridging the Gap Between AI Planning and Reinforcement Learning (2021)
18. Zhang, C., Song, W., Cao, Z., Zhang, J., Tan, P.S., Chi, X.: Learning to dispatch for job shop scheduling via deep reinforcement learning. Adv. Neural. Inf. Process. Syst. **33**, 1621–1632 (2020)

A 3D Terrain Generator: Enhancing Robotics Simulations with GANs

Silvia Arellano[1], Beatriz Otero[1](✉)(iD), Tomasz Piotr Kucner[2](iD),
and Ramon Canal[1](iD)

[1] Universitat Politècnica de Catalunya, Barcelona, Spain
silvia.arellano@estudiantat.upc.edu, {beatriz.otero,ramon.canal}@upc.edu
[2] Aalto University, Espoo, Finland
tomasz.kucner@aalto.fi

Abstract. Simulation is essential in robotics to evaluate models and techniques in a controlled setting before conducting experiments on tangible agents. However, developing simulation environments can be a challenging and time-consuming task. To address this issue, a proposed solution involves building a functional pipeline that generates 3D realistic terrains using Generative Adversarial Networks (GANs). By using GANs to create terrain, the pipeline can quickly and efficiently generate detailed surfaces, saving researchers time and effort in developing simulation environments for their experiments. The proposed model utilizes a Deep Convolutional Generative Adversarial Network (DCGAN) to generate heightmaps, which are trained on a custom database consisting of real heightmaps. Furthermore, an Enhanced Super-Resolution Generative Adversarial Network (ESRGAN) is used to improve the resolution of the resulting heightmaps, enhancing their visual quality and realism. To generate a texture according to the topography of the heightmap, chroma keying is used with previously selected textures. The heightmap and texture are then rendered and integrated, resulting in a realistic 3D terrain. Together, these techniques enable the model to generate high-quality, realistic 3D terrains for use in robotic simulators, allowing for more accurate and effective evaluations of robotics models and techniques.

Keywords: GAN · Terrain rendering · 3D image generation · Robotics simulators · Sim2Real

1 Introduction

Simulation plays a crucial role in robotics by providing a controlled setting for testing algorithms and prototypes before conducting experiments on physical agents. These tools allow researchers to test robots in unusual, inaccessible, or even impossible scenarios, improving safety and performance. However, developing 3D virtual environments is a demanding and time-consuming task. These environments have become increasingly complex and detailed over the years, making the process even more challenging. To enhance the user's experience and

G. Nicosia et al. (Eds.): LOD 2023, LNCS 14505, pp. 212–226, 2024.
https://doi.org/10.1007/978-3-031-53969-5_17

generate more realistic environments, researchers have been exploring procedural modeling as an ongoing line of research in recent years [19]

Procedural modeling is a well-established technique in the video game industry, used in popular games like Minecraft [12] to generate diverse, never-ending worlds. Some commercial applications related to animation and video games provide users with the ability to procedurally generate landscapes and scenes for games and driving simulations. However, there is a lack of applications for automatic scenario generation prepared to be applied in robotics or robotics simulators. The requirements for 3D virtual environments in robotics are often different from those in video games or other industries. In robotics, the environments must be physically accurate and realistic to ensure that the results obtained in simulation can be extrapolated to real-world scenarios. This level of fine-tuning requires a high degree of control over the environment and the models used, which is often not possible with traditional terrain modeling techniques.

This paper proposes a complete pipeline to generate a realistic terrain, including both mesh and texture. Besides, the output of this pipeline aims to make it easy for the user to apply various surface parameters in each zone of the mesh based on the terrain type. These parameters, such as friction, contact forces, and other critical factors in robotics, are essential for achieving accurate simulations.

The pipeline is divided into four primary components, detailed in Fig. 1. The first component involves heightmap generation, which is critical for obtaining a realistic mesh. This will be further explained in Sect. 3. Enhancing the quality of the heightmap is essential for rendering a terrain precisely. For that reason, a super resolution algorithm is applied, increasing four times the original resolution of the image. This procedure is explained in Sect. 4. Once the heightmap is generated and improved, the third component of the pipeline focuses on generating a texture that meets the user's requirements and is consistent with the generated map's topography. The user is expected to specify how many textures should be combined and how the combination should be done. After that, the system generates the texture. This procedure will be further amplified in Sect. 5. Finally, the output from the previous steps is integrated by rendering the heightmap to obtain a mesh, which is then combined with the texture. The resulting mesh is then exported and can be used in any preferred robotic simulator. The integration step is further explained in Sect. 6.

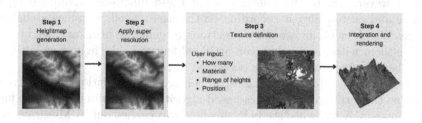

Fig. 1. Workflow pipeline of the 3D terrain generator, showing the various stages involved in generating a terrain from scratch.

The main contributions of this work are the following:

- Creation of an open-source heightmap dataset by extracting data directly from various online sources, as there was no comprehensive publicly available dataset at the time.
- Selection of an appropriate combination of neural networks to achieve diverse and high quality resources.
- A comparative analysis between the proposed pipeline and other existing techniques for generating 3D terrains for robotics simulations.
- Development of a software capable of generating a wide range of combinations of shapes and terrains to provide robots with diverse training scenarios.

The remainder of this paper is organized as follows. Section 2 provides a review of related work on procedural terrain generation. Section 3 and 4 illustrate the generation procedure of both meshes and textures, explaining the techniques and the Neural Network used in the process. Then, Sect. 5 presents the integration process, and Sect. 6 describes the potential uses and applications of this software in robotics. We draw the main conclusions in Sect. 7.

2   Background

2.1   Current Terrain Generation Strategies

The aim of procedural modeling is, instead of designing content by hand, to design a procedure that creates content automatically [19]. Some of its advantages include the reduction of cost and time needed for scene generation, a high level of detail in the output without human intervention and low memory requirements. However, the main drawbacks are the difficulty in manipulating and controlling the generation process and the significant amount of computational resources required [7,19].

Our proposal focuses on procedural modeling of outdoor terrain, which can be approached using three strategies: noise functions, erosion simulation, and texture synthesis. All of these strategies make use of height maps, which are a representation of a terrain's surface as a grid of elevation values.

The first way to generate new height maps is using bandwidth-limited noise functions like Perlin [16], which can capture the nature of mountainous structures. However, due to the randomness of these functions, manipulating the outcome and customizing it according to the user's needs becomes a challenging task. Gasch et al. [6] use noise functions to generate arbitrary heights, and combine them with previously given artificial features, like figures or letters, in a way that the final result can be perceived as real.

Another strategy for generating terrains is erosion simulation, which involves modifying the terrain using physics-based algorithms that aim to resemble natural phenomena. A renowned study in this field was carried out by Musgrave et al. [13], where they computed the surface erosion caused by water, by considering the quantity of water that would accumulate at each vertex of a fractal height field and the impact of thermal weathering.

The last procedure considered is texture synthesis, which aims to generate new surfaces by studying and finding patterns in existing terrains. In this way, it can capture their realism, so that they obtain compelling results. Panagiotou et al. [15] used a combination of GAN and Conditional GAN (cGAN) to create images that resemble satellite images and Digital Elevation Models that can match the output of the first GAN.

Our approach can be included in the last type of procedure, as we use a Deep Convolutional Generative Adversarial Network (DCGAN), a type of generative model that has shown great success in generating high-quality images, to create heightmaps of terrain that will be rendered as 3D meshes. We combine this method with an Enhanced Super-Resolution Generative Adversarial Network (ESRGAN) to improve the quality of the generated samples, and a technique that is based on chroma keying to assign different textures to the terrain depending on its height.

2.2 Related Existing Applications

In recent years, we can observe a rapid acceleration in research on data-driven methods across multiple fields. However, they are heavily dependent on the amount of available data. As mentioned before, there is a lack of applications for automatic scenario generation in the context of robotics or robotics simulators. Nevertheless, there are similar approaches in the field of video games and animation. Some remarkable commercial applications in this fields are Procedural Worlds [2], which provides Unity [24] users with the ability to procedurally generate landscapes and scenes for games and driving simulation, and iClone 8 [17], a 3D animation software with a special focus on nature terrain generation.

There are also several design programs available, such as Terragen [21], and E-on Vue [20], that allow users to create computer-generated environments. Nevertheless, these programs require expertise in 3D design tools and lack the automatic generation feature that our tool aims to provide.

3 Heightmap Generation

The first step in the pipeline aims to create a terrain mesh. To create varied surfaces, this approach takes advantage of heightmaps, which are grayscale images that represent the elevation of a terrain and can be transformed into 3D surfaces once they are rendered. Although these images are in grayscale, we chose to represent them in the RGB color space for our dataset. Our objective for this part of the pipeline is to generate diverse and realistic heightmaps while avoiding those that resemble flat planes with little height variation, as they may not be of great interest. For this purpose, we have studied two different methods: unconditional image generation with diffusion models and Generative Adversarial Networks (GANs).

On the one hand, Denoising Diffusion Probabilistic Models (DDPM), also known as diffusion models, were introduced by Ho et al. [10]. A DDPM is a generative model used to produce samples that match a given dataset within a finite time. The diffusion model consists of two processes: a forward diffusion process that gradually adds Gaussian noise during several timesteps to the input data until the signal has become completely noisy, and a reverse denoising process that learns to generate data by removing the noise added in the forward process.

The diffusion process uses small amounts of Gaussian noise, which allows for a simple neural network parameterization. Diffusion models are known for their efficiency in training and their ability to capture the full diversity of the data distribution. However, generating new data samples with a diffusion model requires the model to be forward-passed through the diffusion process several times, which can be computationally expensive and result in a slower process compared to GANs. In contrast, GANs only require one pass through the generator network to generate new data samples, making them faster and less computationally expensive for this particular task. However, it is important to remark that the computational efficiency of both models can vary depending on factors such as dataset size, model complexity, and implementation details.

On the other hand, GANs, first proposed by Goodfellow et al. in [8], are a type of neural network architecture that include two models: a generator and a discriminator. The generator creates fake data that is meant to be similar to the real data, while the discriminator tries to differentiate between the real and fake data. The objective of the generator is to create synthetic images so that the discriminator perceives them as real. In this study, we utilized Deep Convolutional GANs (DCGANs) to generate heightmaps. The model used in this study was adapted from the code provided by Tang's DCGAN256 repository on GitHub [22]. As one of our goals is to make this tool practical and accessible to a wide range of users, we have chosen to prioritize the use of GANs over DDPM for our task. By doing so, we aim to reduce the computational resources required and increase the speed of the generation process, which will make the software more efficient and agile.

3.1   Architecture of the GAN

The GAN architecture utilizes a generator that receives a random noise vector of size 4096 as an input and transforms it into a fake heightmap with a size of $256 \times 256 \times 3$. It is composed of eight transposed convolutional layers, as detailed in Fig. 2. The discriminator in this study takes in images of size $256 \times 256 \times 3$ and processes them through six convolutional layers, as shown in Fig. 3. The output of the discriminator is a sigmoid activation function that indicates the probability of the input image being real or fake.

3.2   Obtention of the Training Data

To ensure accurate results during neural network training, it is crucial to use a sufficient number of training samples that are diverse enough to avoid overfitting. However, as there were no open-source heightmap datasets available, we had

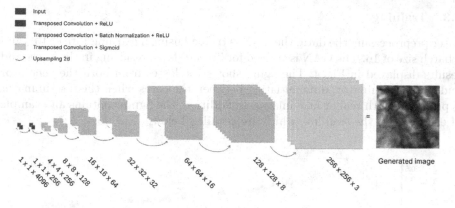

Fig. 2. Structure of the DCGAN generator

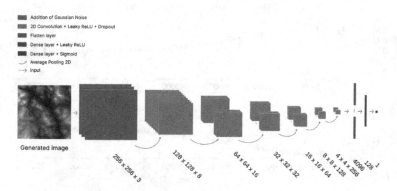

Fig. 3. Structure of the DCGAN discriminator

to create our own. Our dataset consists of 1000 real heightmap images from various terrains worldwide. We obtained the images entirely from the web *Cities: Skylines online heightmap generator* [1] using web scraping techniques. Each heightmap represents an area of 17.28 km^2 and has a resolution of 1081×1081 px.

To enhance efficiency and facilitate experimentation, we preprocess the samples before training. The images are initially represented with 16 bits per pixel, but we represent them with 8 bits in RGB format to achieve a reduced file size and ensure compatibility with a broader range of tools and software. Furthermore, to improve computer speed and efficiency, we reduce the resolution of the samples to 256×256 px. To avoid a biased dataset with too many plain terrains, which may be of less interest to users, we filter out images based on their standard deviation. To determine the threshold for plain terrains, we compute the standard deviation of a noise image and keep those images that have a standard deviation of 35% or more of that value. This way, we ensure the presence of plain terrains while maintaining diversity.

3.3   Training

After preprocessing the data, the GAN is trained using a learning rate of 10^{-4} and a batch size of 16. The GAN is trained for 75,000 steps, resulting in the losses and results displayed in Fig. 4. The figure shows the losses from both the generator and the discriminator, differentiating between the cases when the discriminator is presented with real or fake images. In addition, the figure contains an example of each part of the training, which are detailed below.

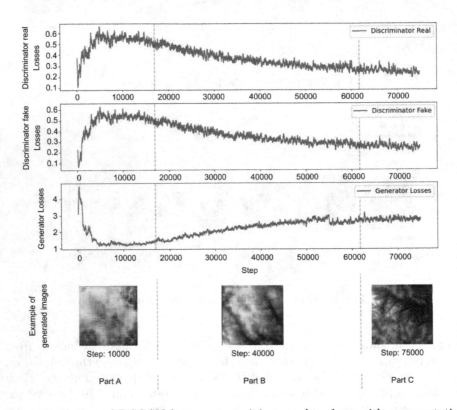

Fig. 4. Evolution of DCGAN losses over training epochs, along with representative images from each stage. From top to bottom, the figure displays the losses of the discriminator when given real data, the losses of the discriminator when given fake data, and the losses of the generator.

The graph on Fig. 4 illustrates that the discriminator and the generator play a min-max game, where if one of them decreases its loss, the other one will increase it. To explain this further, we can divide the training into three parts, separated by dashed lines on the plot. It's worth noting that monitoring the loss values is primarily useful for tracking the GAN's progress over the steps and ensuring that the training is not diverging, rather than measuring the improvement in the quality of the generated images.

In the first part of the training (Part A), corresponding to the early steps of the training, the discriminator starts guessing randomly (since it hasn't received enough training), causing an increase in its loss and a small value in the generator's loss. In the second part (Part B), the rate of change in the losses becomes smoother, and the generated images start to become more realistic and clear, although there are still some areas with blurry patterns. Finally, in the latter part of the training (Part C), the losses become stable, and the generated images are realistic and visually diverse, which satisfies the requirements for our application.

3.4   Evaluation

We aimed to evaluate the quality of the generated images using metrics commonly used in the image generation field, such as Inception Score (IS) and Fréchet Inception Distance (FID). However, due to the nature of our images, neither of these metrics may be the most appropriate for our evaluation.

The Inception Score (IS), defined by Salimans et al. [18] is an algorithm used to assess the quality of images created by a generative image model such as a generative adversarial network (GAN). It uses an Inception v3 Network pre-trained on ImageNet and calculates a statistic of the network's outputs when applied to generated images. As stated by Barratt et al. [3], in order to obtain a good IS, the images should be sharp. Besides, it is expected that the evaluated images belong to a great variety of classes in ImageNet. However, our images don't belong to any of the ImageNet classes, and our dataset isn't organized in classes, like the would IS expect. Therefore, the IS may not be the most appropriate metric to use for evaluating image quality.

The FID, defined by Heusel et al. [9] is also used to assess the quality of images generated by a GAN. The FID quantifies the similarities between two image datasets by comparing their distributions, specifically by calculating the distance between feature vectors calculated for real and generated images. A lower FID indicates that the generated images are more similar to the real ones and hence indicates better image quality and a better model.

We compared a dataset of 1000 real heightmaps with a dataset of 1000 generated heightmaps and obtained an FID score of 274.54, which suggests that the two datasets are not very similar. However, it's important to note that the images in our datasets have a wide range of standard deviations, which can have an impact on the FID score. This is because FID measures the similarity between two sets of images based on their feature statistics, and the feature statistics used in FID computation are sensitive to variations in pixel intensities. Therefore, this would mean that in our case the FID score may not accurately reflect the visual similarity between the real and generated images.

4   Super Resolution

To render the result with the appropriate resolution, we studied two different methods of increasing the resolution of the heightmap, including ESRGAN proposed by Wang et al. [25] and bicubic interpolation. We evaluated the methods considering metrics such as the Peak Signal-to-Noise Ratio (PSNR) and the Structural Similarity Index (SSIM), defined below.

$$\textbf{PSNR} = 20\log_{10}\left(\frac{MAX_I}{\sqrt{MSE}}\right) \quad (1) \qquad \textbf{MSE} = \frac{1}{n}\sum_{i=1}^{n}(x_i - \hat{x}_i)^2 \quad (2)$$

$$\textbf{SSIM}(\textbf{x}, \textbf{y}) = \frac{2\mu_x\mu_y + C_1}{\mu_x^2 + \mu_y^2 + C_1} + \frac{2\sigma_{xy} + C_2}{\sigma_x^2 + \sigma_y^2 + C_2} \quad (3)$$

PSNR measures the ratio of the maximum possible power of a signal to the power of corrupting noise that affects the fidelity of its representation. In Eq. (1), the maximum pixel value of the evaluated image is denoted as MAX_I, with the minimum level assumed to be 0. The noise level is modelled as square root of Mean Squared Error (MSE).

The Mean Squared Error (MSE) is given by Eq. 2, where x_i represents the value of a pixel in the original image and $\hat{x}_i$ represents the corresponding pixel value in the processed image. A higher PSNR value indicates a lower level of noise in the reconstructed image, which in turn indicates a higher image quality.

The SSIM compares the luminance, contrast, and structure of the two images by taking into account their mean values (μ_x and μ_y) and variances (σ_x^2 and σ_y^2), as well as their covariance (σ_{xy}). It consists of two terms: the first term computes the similarity of the luminance and contrast between the two images, while the second term computes the similarity of their structure. The constants C_1 and C_2 are used to avoid division by zero errors and are typically small positive values. A higher SSIM value indicates a higher similarity between the two images.

We evaluated the algorithms using a real heightmap database consisting of 1000 1024 × 1024 pixel heightmaps that were previously downsized to 256 pixels. These downsized images represent examples of images that could potentially have been generated by the GAN model. We intentionally downsized the images to evaluate the ability of the enhancing techniques and the quality of the results when working with lower resolution images. Both algorithms were employed to enhance these downsized images and restore them to their original size.

The evaluation results indicate that, in 54% of cases, bicubic interpolation outperformed ESRGAN with an average difference of 0.5 dB. Even though most images generated by ESRGAN are acceptable, it occasionally produced small unexpected artifacts, such as regions with colored pixels, as it can be seen in the example of Fig. 5(a), which resulted in slight lumps or peaks in the final mesh and a decrease of up to 9 dB in the PSNR, like in Fig. 5(b). Only a 1.2% of the images had a significant impact. The difference in SSIM values between the two techniques was found to be on the order of 10^{-3}. The time taken to produce results was also considered, with bicubic interpolation taking 2 min and ESRGAN taking less than a second.

Despite the artifacts produced by ESRGAN in a subset of images, we decided to use it to enhance the resolution of the generated heightmaps since most of the images were acceptable, and ESRGAN was significantly faster. However, further investigation and refinement may be necessary to address the artifacts in the subset of images where a significant impact was observed. The code used followed the original paper. It was obtained from the TensorFlow Hub and was initially developed by Dey [4].

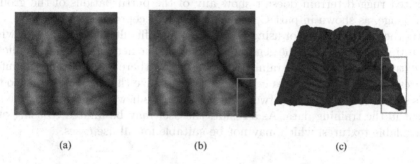

(a)   (b)   (c)

Fig. 5. Example of a real heightmap (a), the output after enhancing it with ESRGAN (b) and the rendered result (c). The green bounding boxes indicate the perceivable artifacts generated by the ESRGAN. (Color figure online)

5   Texture Generation

As depicted in Fig. 1, after the heightmap is generated and its resolution is improved, the next step is to generate its texture. The ability to customize textures based on height and position is a key feature of this application, as it allows users to create terrains that suit their specific needs. Two methods have been studied to achieve a realistic result that fits the topography of the surface, which are presented below.

5.1   Pix2Pix

The first approach used is Pix2Pix, a GAN presented by Isola et al. [11] that is used in image-to-image translation, which involves generating an output image that is a modification of a given input image. In this case, image-to-image translation refers to generating a texture that fits the topography of the environment, having the heightmap as an input of the network. This can lead to realistic results, as it is expected to make good associations between heights and material.

To implement this technique, we utilized the Pix2Pix network provided by TensorFlow [23]. The network requires a database consisting of images that contain both the heightmap and its associated real texture. Following the same approach as in Sect. 3.2, we created the necessary database using web-scraping techniques. However, such dataset is not diverse nor balanced enough. There are more textures for the low height areas than for the high height areas. This leads to a bias that favours materials from flat areas over those in rugged areas, as shown in Figs. 6(a) and 6(b), which show a plain and a rugged texture generated by the Pix2Pix network, respectively.

The generated plain texture has similar features to the ground truth image, such as the brownish perturbations in parts A and B of Fig. 6. However, the generated rugged terrain doesn't show any of the perturbations of the ground truth image, as shown in part C, making it an unacceptable result.

Another disadvantage of using Pix2Pix is the difficulty users may face when attempting to personalize the generated textures. The network can only replicate the textures present in the training database, and there is a limited amount of textures available. This makes it difficult to customize the map according to the user's specific needs, as the network cannot generate new textures that were not present in the training data. As a result, the user may be limited to using only the available textures, which may not be suitable for all use cases.

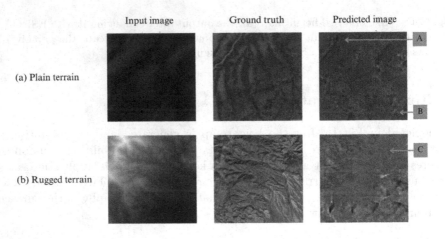

Fig. 6. Comparison between the ground truth textures of a plain and a rugged terrain with the predicted textures produced by the pix2pix network, given the corresponding heightmap as input.

5.2   Chroma Keying

The second approach studied was the application of chroma keying. This technique involves using color information to select and isolate pixels that fall within a range of colors previously defined, known as the "key colors". In our case, the application of chroma keying involved dividing the heightmap into different ranges of grey levels and assigning a different material to each range. To allow for user customization, the user selected the range of greys in which each material should be placed. To create varied combinations of textures, we have chosen eight materials and found an image from the Internet for each one of them. The selected materials were rock, grass, snow, clay and moss, grass with rock, rock with snow and grass with sand. Then, the material images were resized to 1024×1024 so that they could cover the entire heightmap in case needed. Additionally to this height division, a feature was added to include position as a parameter to further customize the texture of the heightmap. However, the application of this technique resulted in evident contours between neighboring textures, creating an unnatural look, as shown in Fig. 7(b). To address this issue, we applied post-processing to the image. Firstly, we added a Gaussian blur filter from the OpenCV library [14] to a copy of the texture image. The key parameters, σ_x and σ_y, were set to 2 to achieve moderate smoothing. Next, a binary mask was created using the detected contours from the original image, which was then dilated with a kernel of (5, 5) pixels to select the surrounding areas of the contours. Then, we replaced the pixels in the original image that were selected by the mask with their corresponding pixels in the blurry image. These procedures resulted in an image with smoothed contours. However, the axis changes still appeared unnaturally straight. To remedy this, we added a swirl effect in both axes using 16 swirls uniformly distributed. The strength of the swirl was set to 2, and the radius to 100 pixels to avoid over-prominence. The resulting image is shown in Fig. 7(c).

(a)   (b)   (c)

Fig. 7. Example of the different phases involved in texture generation: (a) the original heightmap used as input for generating the texture, (b) the texture after applying chroma keying, and (c) the texture after post-processing.

6 Integration and Rendering

To integrate the texture to the mesh, we used the Blender API [5]. However, instead of integrating the texture as a whole, it is integrated material by material.

Table 1. Combination of values that the user needs to input to generate the textures given the corresponding heightmaps depicted in Fig. 8.

Figure	Height Value	Position (px)	Texture
7(a)	(0, 55)	(0, 1023), (0, 1023)	clay
	(55, 105)	(0, 1023), (0, 1023)	clay and moss
	(105, 155)	(0, 1023), (0, 1023)	rock
	(135, 255)	(511, 1023), (0, 511)	snow
	(155, 255)	(0, 1023), (0, 1023)	rock with snow
6 and 7(b)	(0, 75)	(0, 511), (0, 511)	rock
	(0, 75)	(511, 1023), (0, 511)	rock with sand
	(0, 75)	(0, 511), (511, 1023)	clay and moss
	(75, 150)	(0, 511), (0, 1023)	grass and rock
	(75, 150)	(0, 1023), (0, 511)	grass
	(150, 255)	(0, 255), (0, 255)	snow
	(150, 255)	(255, 511), (0,255)	rock with snow
	(150, 255)	(255, 511), (0, 255)	rock with snow
	(150, 255)	(255, 511), (255, 511)	snow
7(c)	(0, 35)	(0, 1023), (0, 1023)	clay
	(35, 105)	(0, 1023), (0, 1023)	grass and rock
	(35, 105)	(511, 1023), (511, 1023)	clay and moss
	(105, 165)	(0, 1023), (0, 1023)	rock with snow
	(165, 255)	(0, 1023), (0, 1023)	snow

In this way, the map can be exported in pieces to the chosen simulator and assign different properties to each type of material. The resulting files are exported with the extension .obj and .mtl. Some results can be seen in Fig. 8.

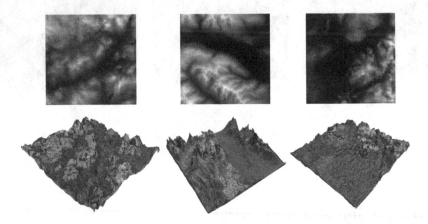

Fig. 8. Final rendering of three different terrains, generated by integrating the corresponding heightmaps shown in the top row of this Figure with the textures defined in Table 1.

7   Conclusions

Simulation environments are commonly utilized in robotics, but the process of creating realistic 3D terrains for robotic simulation purposes can be challenging and laborious. In this paper, we introduce a system that automatically generates 3D terrains from scratch, generating both heightmaps and textures. The system is composed of four main parts: heightmap generation, heightmap resolution enhancement, texture generation, and integration. After comparing Diffusion Models and DCGANs, we used DCGANs for heightmap generation, as it is computationally less expensive and faster, making it accessible to a wider range of users. For heightmap resolution improvement, we employed ESRGAN, which outperformed bicubic interpolation in execution time and most of the outputs were adequate. For texture generation, we studied two methods: Pix2Pix and chroma keying. Although Pix2Pix was limited in terms of available textures and biased towards flat areas, chroma keying allowed for greater customization by dividing the heightmap into different grey ranges, differentiating between positions, and assigning a different material chosen by the user to each range. To achieve a more natural look, we also used post-processing to smooth the contours and add swirl effects. Finally, we integrated and rendered both the mesh and texture, resulting in .obj and .mtl files that together, represent the generated terrain in the chosen robotics simulator engine. In this way, the user can customize the parameters of each material separately.

As future work, we suggest exploring new techniques for customizing the terrain, so the result can be more similar to the user's idea. Additionally, enhancing the databases to provide the neural networks with more and better quality data could result in further improvements. Besides, addressing the artifacts observed in the ESRGAN's outputs may require further investigation and refinement.

Acknowledgment. This work is partially supported by the Spanish Ministry of Science and Innovation under contract PID2021-124463OB-IOO, by the Generalitat de Catalunya under grant 2021-SGR-00326. Finally, the research leading to these results also has received funding from the European Union's Horizon 2020 research and innovation programme under the HORIZON-EU VITAMIN-V (101093062) project.

References

1. Cities: Skylines online heightmap generator. https://heightmap.skydark.pl/. Accessed 5 Apr 2023
2. Procedural worlds. https://www.procedural-worlds.com/. Accessed 13 Mar 2023
3. Barratt, S., Sharma, R.: A note on the inception score (2018)
4. Dey, A.: Image enhancing using ESRGAN (2019). https://github.com/tensorflow/hub/blob/master/examples/colab/image_enhancing.ipynb. Accessed 31 Mar 2023
5. Foundation, B.: Blender. https://www.blender.org/. Accessed 19 Mar 2023
6. Gasch, C., Chover, M., Remolar, I., Rebollo, C.: Procedural modelling of terrains with constraints. Multimed. Tools Appl. **79**, 31125–31146 (2020)

7. González-Medina, D., Rodríguez-Ruiz, L., García-Varea, I.: Procedural city generation for robotic simulation. In: Robot 2015: Second Iberian Robotics Conference. AISC, vol. 418, pp. 707–719. Springer, Cham (2016). https://doi.org/10.1007/978-3-319-27149-1_55

8. Goodfellow, I.J., et al.: Generative adversarial networks (2014)

9. Heusel, M., Ramsauer, H., Unterthiner, T., Nessler, B., Hochreiter, S.: GANs trained by a two time-scale update rule converge to a local nash equilibrium (2018)

10. Ho, J., Jain, A., Abbeel, P.: Denoising diffusion probabilistic models (2020)

11. Isola, P., Zhu, J.Y., Zhou, T., Efros, A.A.: Image-to-image translation with conditional adversarial networks. In: 2017 IEEE Conference on Computer Vision and Pattern Recognition (CVPR), pp. 5967–5976 (2017). https://doi.org/10.1109/CVPR.2017.632

12. Mojang Studios: Minecraft. https://www.minecraft.net/en-us/. Accessed 11 Mar 2023

13. Musgrave, F.K., Kolb, C.E., Mace, R.S.: The synthesis and rendering of eroded fractal terrains. ACM Siggraph Comput. Graph. **23**(3), 41–50 (1989)

14. OpenCV: OpenCV library. https://opencv.org/. Accessed 6 Apr 2023

15. Panagiotou, E., Charou, E.: Procedural 3D terrain generation using generative adversarial networks. arXiv preprint arXiv:2010.06411 (2020)

16. Perlin, K.: An image synthesizer. ACM Siggraph Comput. Graph. **19**(3), 287–296 (1985)

17. Reallusion: iclone8. https://www.reallusion.com/iClone/. Accessed 19 Mar 2023

18. Salimans, T., Goodfellow, I., Zaremba, W., Cheung, V., Radford, A., Chen, X.: Improved techniques for training GANs (2016)

19. Smelik, R.M., Tutenel, T., Bidarra, R., Benes, B.: A survey on procedural modelling for virtual worlds. In: Computer Graphics Forum, vol. 33, pp. 31–50. Wiley Online Library (2014)

20. E-on Software: Vue - overview. https://info.e-onsoftware.com/vue/overview. Accessed 19 Mar 2023

21. Software, P.: Terrage. https://planetside.co.uk/. Accessed 19 Mar 2023

22. Tang, G.: DCGAN256. https://github.com/t0nberryking/DCGAN256. Accessed 27 Mar 2023

23. TensorFlow: pix2pix: image-to-image translation with a conditional GAN. https://github.com/tensorflow/docs/blob/master/site/en/tutorials/generative/pix2pix.ipynb. Accessed 3 Apr 2023

24. Unity Technologies: Unity real-time development platform. https://unity.com/. Accessed 27 Mar 2023

25. Wang, X., et al.: ESRGAN: enhanced super-resolution generative adversarial networks. In: Proceedings of the European Conference on Computer Vision (ECCV) Workshops (2018)

Hybrid Model for Impact Analysis
of Climate Change on Droughts
in Indian Region

Ameya Gujar[1], Tanu Gupta[2] , and Sudip Roy[1,2(✉)]

[1] Dept. of Computer Sc. and Engg., IIT Roorkee, Roorkee, India
[2] CoEDMM, IIT Roorkee, Roorkee, India
sudip.roy@cs.iitr.ac.in

Abstract. Droughts are prolonged periods of dry weather that have
become more frequent and severe due to climate change and global warming. It can have devastating impacts on agriculture, water resources,
and ecosystems. Hence, a framework for the prediction of droughts
is necessary for mitigating its impact, as it enables authorities to
prepare and respond effectively. This paper presents a hybrid model
comprised of the Convolutional Neural Network and Gated Recurrent
Units (called CNN-GRU) to predict the Standardized Precipitation-
Evapotranspiration Index (SPEI), which is used to measure drought
intensity. We use India Meteorological Department (IMD) rainfall and
temperature data of Maharasthra state of India during the years 1960–
2021 as the historical dataset. Whereas, for the future projections, we use
the Coupled Model Intercomparison Project Phase 6 (CMIP6) dataset of
the same region during the years 2015-2100 for different Shared Socioeconomic Pathways (SSP) scenarios. Both these datasets include the daily
precipitation, minimum temperature and maximum temperature values.
The proposed model is trained and validated using IMD dataset and the
final evaluation of its ability to predict the future droughts is conducted
on the CMIP6 dataset. We confirm that it outperforms in terms of
mean squared error, mean absolute error, and root mean squared error
over both IMD and CMIP6 datasets based on the comparative study
with the existing deep learning models.

Keywords: Climate change · Drought · Deep learning · CMIP6 ·
SPEI · IMD

1 Introduction

The intersection of climate change and disasters has been a growing concern
in recent years. Climate change refers to the long-term change in the average
weather patterns of the local, regional, and global climates of earth. It is caused
primarily by the emission of greenhouse gases, particularly carbon dioxide and
methane [25]. This has resulted in more rapid and intense natural disasters,
which are serious problems occurring over a short or long period of time. It

G. Nicosia et al. (Eds.): LOD 2023, LNCS 14505, pp. 227–242, 2024.
https://doi.org/10.1007/978-3-031-53969-5_18

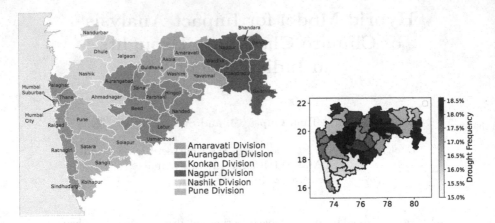

Fig. 1. (a) Political map of Maharashtra with all districts. (b) District-wise frequency distributions of drought occurrences in Maharashtra over the last 60 years [23].

causes widespread human, material, economic, or environmental losses. Climate-induced disasters like floods, droughts, cyclones, etc. have become increasingly common in recent years. Climate change is causing more frequent and severe droughts in many regions worldwide [10]. Rising temperatures increase evaporation and dry out soils, leading to water scarcity and crop failures. As the temperature of earth rises, precipitation patterns are changing. Some areas are experiencing more intense rainfall, while others are experiencing longer periods of drought. Droughts are a major environmental hazard that can cause significant social, economic, and environmental losses [30]. With climate change becoming an increasingly pressing issue, the prediction of droughts is crucial for effective water resource management and disaster risk reduction. In India, droughts are recurrent phenomena and can have severe consequences on agriculture, food security, and the livelihoods of millions of people [1].

Over many years, various indices have been developed to predict droughts, including the Standardized Precipitation-Evapotranspiration Index (SPEI) [29]. The SPEI is a multi-scalar drought index that takes into account both precipitation and potential evapotranspiration, providing a comprehensive measure of water balance. In this study, we focus on predicting droughts in an Indian state namely Maharashtra using SPEI based on the data obtained from the India Meteorological Department (IMD). The topography of Maharashtra is varied, encompassing coastal zones and mountainous terrain, presenting an intriguing and complex landscape for investigating droughts. The diverse topography of the region generates distinct microclimates and ecosystems that are influenced in distinct ways by droughts. A political map of Maharashtra with all the districts is shown in Fig. 1(a) and the district-wise frequency distribution of drought occurrences over the past 60 years is shown in Fig. 1(b) [23]. As shown in Fig. 1(b), Maharashtra experiences a range of climatic conditions, and its central region is particularly vulnerable to recurrent droughts and water scarcity challenges.

To assess the probability of future drought occurrences in this region, we use the data from the Coupled Model Inter-comparison Project Phase 6 (CMIP6) [13]. Specifically, we use the bias-corrected Beijing Climate Center Climate System Model Version 2 of Medium Resolution (BCC-CSM2-MR) model to simulate the future climate scenarios under four Shared Socioeconomic Pathways (SSPs). These scenarios represent different socio-economic and environmental conditions that can influence future climate change and drought occurrences. This paper aims to assess the proposed framework to predict the SPEI, given the past data, and to predict future droughts in different future SSP scenarios. The main contributions of this paper are as follows:

o A hybrid model is proposed that is comprised of a CNN and GRU for drought prediction in the Maharashtra state.
o The capability of the proposed framework is compared with different time series deep learning models, namely Long Short Term Memory (LSTM), Gated Recurrent Units (GRU), Recurrent Neural Network (RNN), CNN-LSTM, CNN-RNN.
o The proposed model is used to predict the future droughts in Maharashtra using CMIP6 dataset.

The remainder of the paper is organised as follows. Section 2 discusses review of recent related work and Sect. 3 provides the background of the work. Section 4 presents the motivation of this work and problem statement. Section 5 provides dataset details, and the architecture of the proposed methodology is presented in Sect. 6. Simulation results with the comparative study are discussed in Sect. 7 and finally, Sect. 8 concludes the paper.

2   Literature Survey

Droughts are monitored and assessed using a variety of indices, most of which are based on precipitation and temperature. There are many drought indices, including (but not limited to) Standardized Precipitation Index (SPI) [18], Palmer Drought Severity Index (PDSI) [5] and SPEI [29]. SPI only requires precipitation data, while PDSI and SPEI require both precipitation and temperature data to calculate them. Nair et al. [22] have addressed the topic of drought assessment in the Vidarbha region of Maharashtra. They evaluated the spatial and temporal pattern of drought risk in that region highlighting their frequency, duration, and intensity.

In literature, many research articles investigated the use of regression or data-driven techniques such as Artificial Neural Network (ANN) and Adaptive Neuro-Fuzzy Inference System (ANFIS) to predict drought indices [4,7,20]. ANFIS has been tested for its ability to predict SPI-based drought at different time scales and has been compared with feed-forward neural networks [6]. Deo et al. have found that M5-tree and multivariate adaptive regression outperformed Least-Squares Support-Vector Machines (LSSVM) in predicting SPEI [12]. Convolutional Neural Networks (CNN) has been used in predicting drought using

SPEI [3, 11] and using satellite images [8, 26]. Pei et al. have used Gated Recurrent Units (GRU) in predicting droughts using Meteorological Drought Index (MCI) drought index [19]. Recently, a novel hybrid model called Convolutional Long Short Term Memory (CNN-LSTM) was proposed for predicting SPEI [11]. Their research uses the feature extraction capability of Convolutional Neural Network (CNN) and the sequence modeling ability of Long Short Term Memory (LSTM). They have compared Genetic Programming (GP), ANN, CNN, LSTM, and CNN-LSTM models for predicting droughts and come to the conclusion that CNN-LSTM performed the best in predicting the SPEI for two districts in Ankara province, Turkey.

In summary various drought indices have been developed to monitor and assess droughts, among which SPI, PDSI, and SPEI are frequently utilized. More recently, researchers explored the potential of using CNN-GRU, a hybrid model that combines the feature extraction capability of CNNs with the sequence modeling ability of GRUs, for drought prediction. CNN-GRU model has been used successfully in predicting soil moisture [31] and wind power [32].

3 Background

This section discusses the indices used to predict drought based on the SPEI.

3.1 Drought Indices

Drought indices are used to measure the extent and severity of drought, which is an important aspect of managing water resources and mitigating the effects of drought on agriculture, ecosystems, and society. SPI [18] is a drought index that considers precipitation only and is based on converting precipitation data to probabilities using various probability distributions. PDSI [5] is another drought index that calculates relative dryness using temperature and precipitation data. Still, it lacks the multi-timescale properties of the indices like SPI or SPEI, making it challenging to associate with particular water resources like runoff, snowpack, reservoir storage, etc. SPEI [29] is an extension of the SPI that takes into account both precipitation (P) and Potential EvapoTranspiration (PET) and can be calculated across different ranges, allowing for the analysis of short-term weather effects and long-term effects like the average intensity of droughts.

3.2 Calculation of SPEI

SPEI is a climatic indicator representing the water balance by considering both P and PET. The probability distribution of the differences between P and PET is modeled by SPEI using a log-logistic distribution. To calculate SPEI, PET is estimated using either by the Thornthwaite equation [28] or by the Hargreaves equation [14]. These equations use factors such as temperature, humidity, and extraterrestrial radiation to estimate the amount of water that could potentially be evaporated from the land surface. Thornthwaite equation requires only monthly average temperature data, whereas PET calculation based on the Hargreaves equation is more reliable than the Thornthwaite equation [22] as it uses both $Tmin$ (minimum mean temperature) and $Tmax$ (maximum mean temperature) data as the inputs. Hargreaves equation for calculating PET is as follows:

$$PET = 0.0023 \times R_a \left[\frac{T_{\max} + T_{\min}}{2} + 17.8 \right] \sqrt{(T_{\max} + T_{\min})} \tag{1}$$

where R_a is the extra-terrestrial radiation and T_{max} and T_{min} are the maximum and minimum temperatures.

The standardized deficit (D) between P and PET time series can be obtained using Eq. 2.

$$D = P - PET \tag{2}$$

Then the cumulative probability function of D is obtained by fitting a log-logistic probability function, which is given by Eq. 3, where W is defined by Eq. 4. Here, p is the probability of exceeding a given D and the sign of the resultant $SPEI$ is reversed for $p > 0.5$.

$$SPEI = W - \frac{2.515517 + 0.802853\,W + 0.010328\,W^2}{1 + 1.432788\,W + 0.189269\,W^2 + 0.001308\,W^3} \tag{3}$$

$$W = \begin{cases} \sqrt{-2\ln p} & \text{for } p \leqslant 0.5 \\ \sqrt{-2\ln(1-p)} & \text{for } p > 0.5 \end{cases} \tag{4}$$

4 Motivation and Problem Statement

In this section, we have discussed the motivation of this work on predicting droughts in an Indian state and the problem statement.

4.1 Motivation

Maharashtra is a state which faces a wide set of climates, and Central Maharashtra is known for its frequent droughts and water scarcity issues. Maharashtra has a diverse topography ranging from coastal regions to hilly areas, which makes it a unique and challenging area for studying droughts. The diverse topography of the region creates micro-climates and diverse ecosystems that are affected differently by droughts. Drought is a regular problem that keeps occurring and has a major effect on people living there, mostly farmers who depend on agriculture for their livelihood. Maharashtra experiences frequent droughts, with more than 25 droughts recorded in the last 45 years [1]. These droughts severely impact agricultural productivity, water resources, and the livelihood of the people. This motivates us to work on predicting droughts in any district of Maharashtra with any future scenario based on the past data.

4.2 Problem Statement

In order to study the impact of climate change on drought duration and intensity in the Maharashtra state of India, need train the machine learning models on the past 60 years of data and see how accurately we can predict the SPEI values under different future CMIP6 SSP scenarios during the period from 2021 to 2100.

5 Description of Datasets

In this section, we discuss about the study area, and the IMD and CMIP6 datasets used in this work.

5.1 Study Area

Maharashtra is the third largest state of India and is located on the western side between the latitudes 16° N and 22° N and longitudes 72° E and 81° E. Maharashtra is known for its frequent droughts and water scarcity issues. Drought is a recurring phenomenon in this state and has become a major concern for the farmers and residents, who heavily rely on agriculture for their livelihood. The state is divided into 35 districts and six subdivisions: Konkan, Pune, Nashik, Aurangabad, Amravati, and Nagpur. Out of these, Konkan comprises the coastal area and is on the windward side of the ghats, therefore, receiving heavy rain, while the other divisions are on the leeward side of the ghats. Pune and Nashik divisions receive moderate rainfall, and the region is known for its hilly terrain. The districts in the Aurangabad division receive less rainfall than other divisions and are known for their arid and semi-arid regions, making them prone to droughts. Amravati and Nagpur divisions are located on Maharashtra's eastern side and receive moderate rainfall. The Aurangabad division has faced severe droughts in recent years, leading to crop failures, water scarcity, and distress migration. Other regions such as Amravati, Nagpur, Pune, and Nashik have also faced droughts in the past, although not as severe as the Aurangabad division. The Konkan division, on the other hand, is less prone to droughts but faces flooding and landslides during the monsoon season. Therefore, varied micro-climates in different regions make Maharashtra a suitable study area for drought prediction.

5.2 IMD Dataset

In order to predict the droughts in Maharashtra state, in this study we utilized a daily gridded historical dataset provided by IMD, Pune [23]. The data had a spatial resolution of 0.25° × 0.25°. The data for temperature is recorded in degrees Celcius (°C), and for precipitation P is in millimeters (mm). IMD datasets can be used to study long-term trends in precipitation and temperature, monitor drought and flood conditions, predict crop yields, and develop early warning systems for extreme weather events. The IMD provides detailed documentation

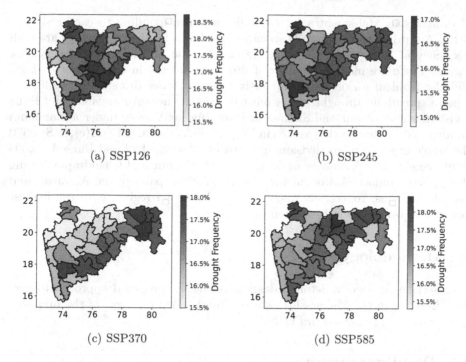

Fig. 2. District-wise drought frequency distributions in Maharashtra state during 2015-2100 for scenario (a) SSP126, (b) SSP245, (c) SSP370, and (d) SSP585.

and metadata for its datasets, which can be helpful in understanding the data quality and limitations. In this work, we have extracted the data of precipitation, *Tmin*, and *Tmax* for each district of Maharashtra from the year 1960 to the year 2021.

5.3   CMIP6 Dataset

In this study, the World Climate Research Programme's (WCRP) Coupled Model Intercomparison Project Phase 6 (CMIP6) [13] is used to create a dataset for future drought prediction. CMIP6 has simulations from the latest state-of-the-art climate models. In this work, we use bias-corrected climate projections from CMIP6 for South Asia [21], which would give better simulations. There are 13 models for every country, each with five scenarios, namely SSP126, SSP245, SSP370, SSP585, and historical. Each scenario has daily precipitation P, *Tmin*, and *Tmax* from 2015 to 2100. Out of these 13 models, we choose BCC-CSM2-MR model as it is one of the best-performing models for India, validated by Konda et al. [16] having normalized root mean square error less than 0.7 and Taylor skill score greater than 0.75.

From the BCC-CSM2-MR CMIP6 model, we use the four SSP scenarios, namely SSP126, SSP245, SSP370, and SSP585. Figure 2 shows the frequency of droughts in Maharashtra for each district corresponding to each SSP scenario. According to the SSP126 scenario (as shown in Fig. 2(a)), there are high frequency of droughts in the Aurangabad division and in the central Nashik region. There are medium number of droughts months in the Nagpur division and the Konkan region has comparatively much fewer droughts. SSP245 sees a more spread-out drought across most districts, where Aurangabad and Pune divisions with Solapur and Satara (in Pune division) being under a maximum number of drought months (as seen in Fig. 2(b)). As shown in Fig. 2(c), in SSP370 the whole of the Nagpur division and parts of Aurnagabad and Pune divisions would receive a high number of droughts with Nagpur and Chandrapur having the highest number of drought months. In SSP585, parts of the Amravati and Nashik divisions, collectively known as the Vidarbha region, have the highest number of drought months as seen in Fig. 2(d).

6   Methodology

This section presents a detailed description of the proposed approach to predict droughts in the Maharashtra state. The detailed overview of the proposed methodology is presented in Fig. 3.

6.1   Data Preprocessing

Data preprocessing has been divided into the following steps.

1. **Data extraction:** We extracted the values of precipitation P, *Tmin*, and *Tmax* for each district of Maharashtra from the IMD dataset. Also, the values of *P*, *Tmin*, and *Tmax* corresponding to each SSP scenario are extracted for each district of Maharashtra from the CMIP6 dataset.
2. **Converting daily data to monthly:** As the SPEI works on monthly aggregated data only, we converted the daily values of the precipitation, *Tmin*, and *Tmax* into monthly *P*, *Tmin*, and *Tmax* values. For this, we sum up the daily

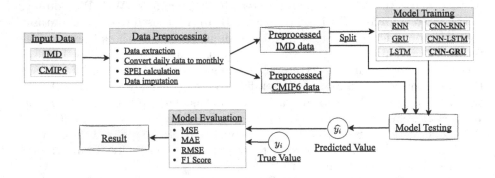

Fig. 3. Overview of the proposed methodology.

P values and average out *Tmin*, and *Tmax* values for each month. Then three mentioned parameters were consolidated into a single file per district for the IMD dataset and a single file per district per SSP scenario for the CMIP6 datasets.

3. **Calculation of SPEI:** After converting the daily values to monthly, we have calculated the SPEI values as discussed in Sect. 3 for all districts and all SSP scenarios.

4. **Data imputation:** When computing the SPEI for historical IMD dataset and CMIP6 dataset, we encountered an issue where the first five months of SPEI calculations were not possible. This is due to the fact that SPEI is calculated using a 6-month time window and as a result, the first five months did not have enough data to perform the calculation. To address this issue, we used a missing value imputation method called K-nearest neighbor (KNN) imputation to fill in the missing values for these first five months.

Following the preprocessing of the IMD dataset, we split the data into training and testing datasets using a 7:3 ratio. The purpose of this is to train and validate the models on the training dataset before evaluating their performance on the testing dataset. Once the model is trained and validated using the IMD dataset and the final evaluation of its ability to predict future drought scenarios is conducted on the CMIP6 dataset. This dataset serves as an external validation dataset, allowing us to test the generalization ability of the model on an unseen dataset.

6.2   Proposed Model

In order to predict the drought in Maharashtra state, we propose a model which is a combination of two popular deep learning algorithms: Convolutional Neural Networks (CNN) [17] and Gated Recurrent Units (GRU) [9].

The CNN layers help extract deep features from the input data. One-dimensional convolutional layer (CNN-1D) layers are commonly used for feature extraction in deep learning models that process sequential data. These CNN-ID layers capture local patterns or features in the input sequence and produce a more compact representation of the input that can be processed more efficiently by the subsequent layers. The GRU layers are then used to learn higher-level features from the CNN-ID output. The GRU layers are recurrent neural network layers that can capture temporal dependencies and patterns in the input sequence. By stacking three GRU layers on top of each other with dropout regularization, the network can learn more complex and abstract features from the CNN-ID output. Table 1 represents the network structure of the proposed model. The Adam optimization algorithm [15] is chosen as the optimizer. The model was run with 100 epochs and a batch size of 32.

Table 1. Details of the structure of designed CNN-GRU model.

Layer (type)	#neurons	output-shape	#param
CNN-ID	8	(None, 6,8)	48
Batch Normalization		(None, 6,8)	32
CNN-ID	8	(None, 1,8)	392
Batch Normalization		(None, 1,8)	32
GRU with 0.1 dropout	9	(None,1,9)	513
GRU with 0.1 dropout	9	(None,1,9)	540
GRU with 0.1 dropout	9	(None,1,9)	540
Dense	8	(None,1,8)	80
Dense	1	(None,1,1)	9

6.3 Evaluation Mertics

To evaluate the performance of the proposed model, we use four commonly used evaluation metrics: mean squared error (MSE), mean absolute error (MAE), root mean squared error ($RMSE$) and F1 score. These metrics are chosen based on their common use in machine learning evaluations and their ability to capture different aspects of model performance. MSE measures the average squared difference between the predicted values and the actual values and defined by Eq. 5, where y_i and $\hat{y}_i$ is the actual and the predicted value, respectively, and n is the number of observations.

$$MSE = \frac{1}{n} \sum_{i=1}^{n} (y_i - \hat{y}_i)^2 \tag{5}$$

On the other hand, MAE measures the average absolute difference between the predicted values and the actual values. It is defined by Eq. 6.

$$MAE = \frac{\sum_{i=1}^{n} |y_i - \hat{y}_i|}{n} \tag{6}$$

As defined by Eq. 7, $RMSE$ takes the square root of the average squared difference between the predicted and actual values.

$$RMSE = \sqrt{\sum_{i=1}^{n} \frac{(y_i - \hat{y}_i)^2}{n}} \tag{7}$$

Finally, using Eq. 8 the $F1$ score values are computed to show the performance of the proposed framework, where TP = True Positives, TN = True Negatives, FP = False Positives, and FN = False Negatives.

$$Precision = \frac{TP}{TP + FP}$$

$$Recall = \frac{TP}{TP + FN} \tag{8}$$

$$F1\ score = 2 \times \frac{Precision \times Recall}{Precision + Recall}$$

7 Experimental Setup and Results

This section provides a detailed description of the experimental setup and the results of the proposed model.

7.1 Experimental Setup

For all sets of experiments, we use Google Colab [2] with TensorFlow version 2.11.0 and the SPEI calculation has been done on R Studio [27] using the SPEI module.

7.2 Experimental Results

In order to assess the performance of the proposed model, we conducts a comparative study with five existing models. As described in Sect. 6, the model is trained on the preprocessed IMD training dataset and evaluated on both the IMD testing dataset and the preprocessed CMIP6 dataset.

Comparative Results on IMD Testing Dataset: In order to evaluate the models, we calculate the evaluation metrics for all districts for each model and then average the results for all 35 districts. Table 2 shows the average training and testing results. It can be observed from Table 2 that the proposed model produces comparatively better results than existing models. The values of the *MSE*, *MAE* and *RMSE* of the proposed model are 0.434, 0.483 and 0.652, respectively.

Table 2. Average training and testing results on IMD dataset.

Model	Training			Testing		
	MSE	*MAE*	*RMSE*	*MSE*	*MAE*	*RMSE*
LSTM	0.284	0.366	0.531	0.465	0.507	0.675
GRU	0.256	0.348	0.505	0.469	0.501	0.669
RNN	0.298	0.383	0.544	0.435	0.479	0.653
CNN-LSTM	0.352	0.437	0.592	0.441	0.488	0.657
CNN-RNN	0.376	0.457	0.613	0.437	0.485	0.655
Proposed	0.360	0.444	0.599	0.434	0.483	0.652

Table 3. Average testing results for CMIP6 dataset.

Model	SSP126			SSP245			SSP370			SSP585		
	MSE	MAE	$RMSE$	MSE	MAE	$RMSE$	MSE	MAE	$RMSE$	MSE	MAE	$RMSE$
LSTM	0.430	0.492	0.654	0.412	0.484	0.638	0.473	0.525	0.686	0.459	0.504	0.675
GRU	0.422	0.488	0.649	0.406	0.480	0.634	0.464	0.519	0.680	0.453	0.500	0.671
RNN	0.407	0.476	0.637	0.385	0.466	0.618	0.445	0.507	0.666	0.436	0.487	0.658
CNN-LSTM	0.401	0.475	0.633	0.383	0.466	0.616	0.452	0.512	0.671	0.432	0.489	0.656
CNN-RNN	0.401	0.475	0.632	0.384	0.468	0.617	0.449	0.510	0.668	0.433	0.489	0.656
Proposed	0.399	0.472	0.630	0.380	0.464	0.614	0.447	0.509	0.667	0.429	0.486	0.653

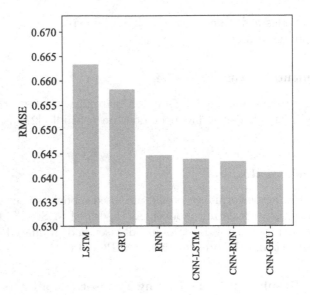

Fig. 4. Average $RMSE$ value of each model for CMIP6 dataset.

Comparative Results on CMIP6 Dataset: Table 3 shows the average testing results for each SSP scenarios of CMIP6 dataset. The result analysis for each scenario is as follows.

- **SSP126:** It is an SSP-based RCP scenario with low radiative forcing by the end of the century. For this scenario, the values of the MSE, MAE and $RMSE$ are 0.399, 0.472 and 0.630, respectively, corresponding to the proposed model.
- **SSP245:** It is an SSP-based RCP scenario with medium radiative forcing by the end of the century. For this scenario, the proposed model obtains values of 0.380, 0.464 and 0.614 for MSE, MAE and $RMSE$, respectively.
- **SSP370:** SSP370 is a Gap-filling baseline scenario with a medium to high radiative forcing by the end of the century. The proposed model has a MSE, MAE and $RMSE$ of 0.447, 0.509 and 0.667, respectively. For this scenario, the RNN model is able to make slightly better predictions for the Sangli district

with a difference in RMSE of 0.03, which led to slightly better results, *MSE*, *MAE* and *RMSE* of 0.445, 0.507 and 0.666, respectively.

o **SSP585:** SSP585 is an SSP-based RCP scenario with high radiative forcing by the end of the century. The proposed model has a *MSE*, *MAE*, and *RMSE* of 0.429, 0.486, and 0.653, respectively.

The results confirm that the proposed model exhibits the least prediction error, giving an average*MSE*, *MAE*, and *RMSE* as 0.414, 0.4827 and 0.6411, respectively, for all four prediction models as shown in Fig. 4. In conclusion, the proposed model exhibits optimal performance in the SSP245 scenario with medium radiative forcing. However, it demonstrates less accurate predictions for droughts in scenarios with high radiative forcing, while showing relatively improved performance in situations characterized by low radiative forcing.

Table 4. Drought classification based on SPEI values.

Level	Drought Category	SPEI Values
0	Non-drought	$-1.0 <$ Index
1	Moderate drought	$-1.5 <$ Index < -1.0
2	Severe drought	$-2.0 <$ Index < -1.5
3	Extreme drought	Index < -2.0

Table 5. Confusion matrix for the proposed model on IMD dataset.

		Actual			
		ND	MD	SD	ED
Predicted	ND	5272	294	1	0
	MD	504	650	2	0
	SD	4	67	193	0
	ED	2	4	4	3

Table 6. Confusion matrix for the proposed model on CMIP6 dataset.

SSP126		Actual				SSP245		Actual			
		ND	MD	SD	ED			ND	MD	SD	ED
Predicted	ND	27855	1533	17	0	Predicted	ND	28320	1436	16	0
	MD	2668	2641	35	0		MD	2370	2342	33	0
	SD	46	118	641	1		SD	50	197	747	0
	ED	0	2	8	30		ED	0	5	14	65
SSP370		Actual				SSP585		Actual			
		ND	MD	SD	ED			ND	MD	SD	ED
Predicted	ND	28020	1606	12	0	Predicted	ND	27902	1561	21	0
	MD	2572	2349	32	0		MD	2750	2447	33	0
	SD	49	216	654	0		SD	36	169	617	0
	ED	0	5	15	65		ED	0	6	8	45

Table 7. F1 score value for the proposed model on CMIP6 dataset.

Scenarios	F1 score			
	ND	MD	SD	ED
SSP126	0.929	0.548	0.851	0.845
SSP245	0.936	0.538	0.828	0.873
SSP370	0.930	0.515	0.801	0.867
SSP585	0.927	0.520	0.827	0.866

Classification of Droughts: Using SPEI, drought can be classified [24] as no drought (ND), moderate (MD), severe (SD), and extreme drought (ED) as shown in Table 4. Table 5 shows the confusion matrix for the proposed model for the IMD dataset. The value of the F1 scores for ND, MD, SD, and ED classes for the IMD dataset are 0.929, 0.617, 0.828, and 0.429, respectively. Further, the confusion matrix and corresponding F1 score for the different CMIP6 scenarios are shown in Table 6 and Table 7, respectively. The results confirm that the proposed model performs exceptionally well during ND, SD, and ED drought scenarios but exhibits relatively weaker performance during MD.

8 Conclusions

This paper aims to use a historical dataset of Maharashtra collected from IMD and train a model to predict future droughts across each district for four probable future scenarios. To figure out a solution to this problem, taking into consideration that the data is in a time-series format, we took into consideration algorithms like LSTM, GRU, and RNN. The proposed approach combining the convolutional neural network with the gated recurrent units confirms better accuracy in drought forecasting.

References

1. Drought Management Plan, November 2017, Ministry of Agriculture. https://agricoop.nic.in/. Accessed 19 Apr 2023
2. Google Colaboratory. https://colab.research.google.com/notebooks/intro.ipynb. Accessed 19 Apr 2023
3. Adikari, K.E., Shrestha, S., Ratnayake, D.T., Budhathoki, A., Mohanasundaram, S., Dailey, M.N.: Evaluation of artificial intelligence models for flood and drought forecasting in arid and tropical regions. Environ. Model. Softw. **144**, 105136 (2021)
4. Ali, Z., et al.: Forecasting drought using multilayer perceptron artificial neural network model. Adv. Meteorol. **2017** (2017)
5. Alley, W.M.: The Palmer drought severity index: limitations and assumptions. J. Appl. Meteorol. Climatol. **23**(7), 1100–1109 (1984)
6. Bacanli, U.G., Firat, M., Dikbas, F.: Adaptive neuro-fuzzy inference system for drought forecasting. Stoch. Env. Res. Risk Assess. **23**, 1143–1154 (2009)

7. Barua, S., Ng, A., Perera, B.: Artificial neural network-based drought forecasting using a nonlinear aggregated drought index. J. Hydrol. Eng. **17**(12), 1408–1413 (2012)

8. Chaudhari, S., Sardar, V., Rahul, D., Chandan, M., Shivakale, M.S., Harini, K.: Performance analysis of CNN, Alexnet and VGGNet models for drought prediction using satellite images. In: Proceedings of the ASIANCON, pp. 1–6 (2021)

9. Chung, J., Gulcehre, C., Cho, K., Bengio, Y.: Empirical evaluation of gated recurrent neural networks on sequence modeling. arXiv preprint arXiv:1412.3555 (2014)

10. Dai, A.: Drought under global warming: a review. Wiley Interdiscip. Rev. Climate Change **2**(1), 45–65 (2011)

11. Danandeh Mehr, A., Rikhtehgar Ghiasi, A., Yaseen, Z.M., Sorman, A.U., Abualigah, L.: A novel intelligent deep learning predictive model for meteorological drought forecasting. J. Ambient Intell. Humaniz. Comput. 1–15 (2022)

12. Deo, R.C., Kisi, O., Singh, V.P.: Drought forecasting in eastern Australia using multivariate adaptive regression spline, least square support vector machine and M5Tree model. Atmos. Res. **184**, 149–175 (2017)

13. Eyring, V., et al.: Overview of the Coupled Model Intercomparison Project Phase 6 (CMIP6) experimental design and organization. Geosci. Model Dev. **9**(5), 1937–1958 (2016)

14. Hargreaves, G.H.: Defining and using reference evapotranspiration. J. Irrig. Drain. Eng. **120**(6), 1132–1139 (1994)

15. Jais, I.K.M., Ismail, A.R., Nisa, S.Q.: Adam optimization algorithm for wide and deep neural network. Knowl. Eng. Data Sci. **2**(1), 41–46 (2019)

16. Konda, G., Vissa, N.K.: Evaluation of CMIP6 models for simulations of surplus/deficit summer monsoon conditions over India. Clim. Dyn. **60**(3–4), 1023–1042 (2023)

17. LeCun, Y., Bottou, L., Bengio, Y., Haffner, P.: Gradient-based learning applied to document recognition. Proc. IEEE **86**(11), 2278–2324 (1998)

18. McKee, T.B., Doesken, N.J., Kleist, J., et al.: The relationship of drought frequency and duration to time scales. In: Proceedings of the Applied Climatology, vol. 17, pp. 179–183 (1993)

19. Mei, P., Liu, J., Liu, C., Liu, J.: A deep learning model and its application to predict the monthly MCI drought index in the Yunnan province of China. Atmosphere **13**(12), 1951 (2022)

20. Miao, T.: Research of regional drought forecasting based on phase space reconstruction and wavelet neural network model. In: Proceedings of the ISAM, pp. 1–4 (2018)

21. Mishra, V., Bhatia, U., Tiwari, A.D.: Bias Corrected Climate Projections from CMIP6 Models for South Asia, June 2020. https://doi.org/10.5281/zenodo.3873998

22. Nair, S.C., Mirajkar, A.: Drought vulnerability assessment across Vidarbha region, Maharashtra. India Arabian J. Geosci. **15**(4), 355 (2022)

23. Nandi, S., Patel, P., Swain, S.: IMDLIB: a python library for IMD gridded data, October 2022. https://doi.org/10.5281/zenodo.7205414

24. Rhee, J., Im, J.: Meteorological drought forecasting for ungauged areas based on machine learning: using long-range climate forecast and remote sensing data. Agric. For. Meteorol. **237**, 105–122 (2017)

25. Ruddiman, W.F.: The anthropogenic greenhouse era began thousands of years ago. Clim. Change **61**(3), 261–293 (2003)

26. Sardar, V.S., Yindumathi, K., Chaudhari, S.S., Ghosh, P.: Convolution neural network-based agriculture drought prediction using satellite images. In: Proceedings of the MysuruCon, pp. 601–607 (2021)
27. Team, R.: RStudio: Integrated Development Environment for R. RStudio, PBC., Boston, MA (2020). http://www.rstudio.com/. Accessed 19 Apr 2023
28. Thornthwaite, C.W.: An approach toward a rational classification of climate. Geogr. Rev. **38**(1), 55–94 (1948)
29. Vicente-Serrano, S.M., Beguería, S., López-Moreno, J.I.: A multiscalar drought index sensitive to global warming: the standardized precipitation evapotranspiration index. J. Clim. **23**(7), 1696–1718 (2010)
30. Yang, T.H., Liu, W.C.: A general overview of the risk-reduction strategies for floods and droughts. Sustainability **12**(7), 2687 (2020)
31. Yu, J., Zhang, X., Xu, L., Dong, J., Zhangzhong, L.: A hybrid CNN-GRU model for predicting soil moisture in maize root zone. Agric. Water Manag. **245**, 106649 (2021)
32. Zhao, Z., et al.: Hybrid VMD-CNN-GRU-based model for short-term forecasting of wind power considering spatio-temporal features. Eng. Appl. Artif. Intell. **121**, 105982 (2023)

Bilevel Optimization by Conditional Bayesian Optimization

Vedat Dogan[1]([⊠])[iD] and Steven Prestwich[2][iD]

[1] Confirm Centre for Smart Manufacturing, School of Computer Science
and Information Technology, University College Cork, Cork, Ireland
`vedat.dogan@cs.ucc.ie`
[2] Insight Centre for Data Analytics, School of Computer Science and Information
Technology, University College Cork, Cork, Ireland
`s.prestwich@cs.ucc.ie`

Abstract. Bilevel optimization problems have two decision-makers: a
leader and a follower (sometimes more than one of either, or both). The
leader must solve a constrained optimization problem in which some
decisions are made by the follower. These problems are much harder to
solve than those with a single decision-maker, and efficient optimal algorithms are known only for special cases. A recent heuristic approach is
to treat the leader as an expensive black-box function, to be estimated
by Bayesian optimization. We propose a novel approach called ConBaBo
to solve bilevel problems, using a new conditional Bayesian optimization
algorithm to condition previous decisions in the bilevel decision-making
process. This allows it to extract knowledge from earlier decisions by both
the leader and follower. We present empirical results showing that this
enhances search performance and that ConBaBo outperforms some top-
performing algorithms in the literature on two commonly used benchmark datasets.

Keywords: Bilevel Optimization · Conditional Bayesian
Optimization · Stackelberg Games · Gaussian Process

1 Introduction

Many real-world optimization and decision-making processes are hierarchical:
decisions taken by one decision-maker must consider the reaction of another
decision-maker with their own objective and constraints. In this work, we consider non-cooperative games called Stackelberg Games [37], which are sequential
non-zero-sum games with two players. The first player is called the *leader* and
the second player is called the *follower*. We shall represent the leader decision
by x_u and the follower response by x_l^*. A decision pair x_u, x_l^* represents a choice
by the leader and an optimal feasible solution of the follower. The structure of

---

This publication has emanated from research conducted with the financial support of
Science Foundation Ireland under Grant number 16/RC/3918.

G. Nicosia et al. (Eds.): LOD 2023, LNCS 14505, pp. 243–258, 2024.
https://doi.org/10.1007/978-3-031-53969-5_19

Stackelberg games is asymmetric: the leader has perfect knowledge about the follower's objective and constraints, while the follower must first observe the leader's decisions before making its own optimal decisions. Any feasible bilevel solution should contain an optimal solution for the follower problem. The mathematical modelling of these games leads to nested problems called *bilevel optimization problems*. In these problems, the lower-level (follower) optimization problem is a constraint of the upper-level (leader) problem. Because the lower-level solution must be optimal given the upper-level decisions, these optimization problems are hard to solve. Bilevel optimization is known to be strongly NP-hard and it is known that even evaluating a solution for optimality is NP-hard [39]. There are several approaches proposed in the literature for solving bilevel problems. Most focus on special cases, for example, a large set of exact methods was introduced to solve small linear bilevel optimization problems. Another approach is to replace the lower-level problem with its Karush-Kuhn-Tucker conditions, reducing the bilevel problem to a single-level optimization problem [4]. A popular approach is to use nested evolutionary search to explore both parts of the problem [32], but this is computationally expensive and does not scale well.

Bayesian optimization [12] is another possible approach. It is designed to solve (single-level) optimization problems in which the calculation of the objective is very expensive so the number of such calculations should be minimized. This is done via an acquisition function which is learned as optimization proceeds. This function approximates the upper-level objective with increasing accuracy and is also used to select each upper-level decision. The solution of a lower-level problem could be viewed as an expensive objective evaluation, making Bayesian optimization an interesting approach to bilevel optimization.

Several kinds of applications can be modelled as bilevel optimization problems. In the toll setting problem [9] the authority acts as a leader and the network users act as followers, and the authority aims to optimize the tolls for a network of roads. The authority's toll price decision can be improved by taking into account the network users' previous acts. In environmental economics [35], an authority might want to tax an organization or individual that is polluting the environment. The authority acts as the leader and the polluting entity acts as the follower. If the authority uses the knowledge of previous acts of the polluting entity, then the authority can better regulate its tax policy.

In this work we propose an improved Bayesian approach to bilevel optimization, using the Gaussian process surrogate model with *conditional setting*. In [25] the Conditional Bayesian Optimization (ConBO) algorithm improved the knowledge gradient (KG) acquisition function by using a conditional setting. During the bilevel optimization process, it is important to know how the follower reacts to the leader's decisions, and we improved the algorithm by conditioning the follower's decisions. If we approach the leader problem as a black box and use previous leader decisions with the follower's best responses and leader fitness, conditioning makes us more likely to choose the next leader decision wisely, thus reaching optimality more quickly. Bayesian optimization with the Gaussian process embeds an acquisition function for determining the most promising areas to explore during the optimization process. The benefit of the acquisition function is to reduce the number of function evaluations. The results show that conditioning speeds up the search for the optimal region during bilevel optimization.

The rest of the paper is organized as follows. Section 2 surveys related work on solution methods for bilevel optimization problems. The preliminaries for the proposed CONBABO algorithm are discussed in Sect. 3 The algorithm is explained in Sect. 4. In Sect. 5, we present the experimental details and results and compare them with the state-of-the-art. Finally, Sect. 6 is devoted to conclusions and future directions of research.

2   Background

A bilevel optimization problem is a sequential nested optimization process in which each level is controlled by a different decision-maker. In the context of Game Theory, bilevel problems are Stackelberg games [37]. They were introduced by J. Bracked and J. McGill, and a defence application was published by the same authors in the following year [7]. Bilevel problems were modelled as mathematical programs at this time.

A considerable number of exact approaches have been applied to bilevel problems. Karush-Kuhn-Tucker conditions [4] can be used to reformulate a bilevel problem to a single-level problem. Penalty functions compute the stationary points and local optima. Vertex enumeration has been used with a version of a Simplex method [6]. Gradient information for the follower problem can be extracted for use by the leader objective function [28]. In terms of integer and mixed integer bilevel problems, branch-and-bound [5] and parametric programming approaches have been applied to solve bilevel problems [19]. Because of the inefficiency of exact methods on complex bilevel problems, meta-heuristics have been considered solvers.

Several kinds of meta-heuristics have been applied to bilevel problems in the literature. Four existing categories have been published in [38]: the nested sequential approach [18], the single-level transformation approach, the multi-objective approach [27], and the co-evolutionary approach [21]. An algorithm based on a human evolutionary model for non-linear bilevel problems [23], and the Bilevel Evolutionary Algorithm based on Quadratic approximations (BLEAQ), have been proposed by [33]. This is another work that attempts to reduce the number of follower optimizations. The algorithm approximates the inducible region of the upper-level problem through the feasible region of the bilevel problem. In [26] they consider a single optimization problem at both levels. They proposed the Sequential Averaging Method (SAM) algorithm. In different recent works [29] they used a truncated back-propagation approach for approximating the (stochastic) gradient of the upper-level problem. Basically, they use a dynamical system to model an optimization algorithm that solves the lower-level problem and replaces the lower-level optimal solution. Another work [13] developed a two-timescale stochastic approximation algorithm (TTSA) for solving a bilevel problem, assuming the follower problem is unconstrained and strongly convex, and the leader is a smooth objective function.

Bayesian Optimization for Bilevel Problems (BOBP) [16] uses Bayesian optimization to approximate the inducible region. Multi-objective acquisition approach is presented in [11]. In this work, bilevel programming problems have been solved by using priors in Gaussian processes to approximate the upper-level objective functions. Gaussian processes in Bayesian optimization make it possible to choose the next point wisely and use the previous iterations as knowledge to guide the optimization process. We use a conditional Bayesian optimization algorithm for bilevel optimization. We show empirically that conditioning can dramatically reduce the number of function evaluations for the upper-level problem. Many practical problems can be modelled and solved as Stackelberg games. In the field of economics [35] these include principal agency problems and policy decisions. Hierarchical decision-making in management [3] and engineering and optimal structure design [17] are other practical applications. Network design and toll setting problems are the most popular applications in the field of transportation [24]. Finding optimal chemical equilibria, planning the pre-positioning of defensive missile interceptors to counter an attacking threat, and interdicting nuclear weapons are further applications [8].

3   Preliminaries

The description of the algorithm will be divided into three parts. Firstly, we explain bilevel programming problems and structure. Secondly, we discuss conditional Bayesian optimization and Gaussian process settings. Thirdly, we explain the proposed CONBABO algorithm.

Bilevel Optimization Problems. For the upper-level objective function $F : \mathbb{R}^n \times \mathbb{R}^m \rightarrow \mathbb{R}$ and lower-level objective function $f : \mathbb{R}^n \times \mathbb{R}^m \rightarrow \mathbb{R}$, bilevel optimization problem can be defined as;

$$\underset{x_u, x_l}{\text{Minimize }} F(x_u, x_l)$$

$$\text{s.t. } x_l \in \underset{x_l}{\text{argmin}} \{ f(x_u, x_l) : g_j(x_u, x_l) \le 0, \ j = 1, 2, \ldots, J \} \qquad (1)$$

$$G_k(x_u, x_l) \le 0, \ k = 1, 2, \ldots, K$$

where $x_u \in \mathcal{X}_U, x_l \in \mathcal{X}_L$ are vector-valued upper-level and lower-level decision variables in decision spaces. G_k and g_j represent the constraints of the bilevel problem. Because the lower-level decision maker depends on the upper-level variables, for every decision x_u there is a follower-optimal decision x_l^* [34]. In bilevel optimization, a decision point $x^* = (x_u, x_l^*)$ is feasible for the upper-level only if it satisfies all the upper-level constraints and the vector x^* is an optimal solution to the lower-level problem with the upper-level decision as a parameter.

Bayesian Optimization and Gaussian Process. Bayesian optimization [15] is a method used to optimize black-box functions that are expensive to evaluate. BO uses a probabilistic surrogate model, commonly Gaussian Process [36], to obtain a posterior distribution $\mathbb{P}(\mathbf{f}|\mathcal{D})$ over the objective function $\mathbf{f}$ given the observed data $\mathcal{D} = \{(x_i, y_i)\}_{i=1}^{n}$ where n represents the number of observed data. The surrogate model is assisted by an acquisition function to choose the next candidate or set of candidates $X = \{x_j\}_{j=1}^{q}$ where q is the batch size of promising points found during optimization. The objective function is expensive but the surrogate-based acquisition function is not, so it can be optimized much more quickly than the true function to yield X. Let us assume that we have a set of collection points $\{x_1, \ldots, x_n\} \in \mathbb{R}^d$ and objective function values of these points $\{f(x_1), \ldots, f(x_n)\}$. After we observe n points, the mean vector is obtained by evaluating a mean function μ_0 at each decision point x and the covariance matrix by evaluating a covariance function or kernel Σ_0 at each pair of point x_i and x_j. The resulting prior distribution on $\{f(x_1), \ldots, f(x_n)\}$ is,

$$f(x_{1:n}) \sim \mathcal{N}(\mu_0(x_{1:n}), \Sigma_0(x_{1:n}, x_{1:n})) \tag{2}$$

where $x_{1:n}$ indicates the sequence $x_1, \ldots, x_n$, $f(x_{1:n}) = \{f(x_1), \ldots, f(x_n)\}$, $\mu_0(x_{1:n}) = \{\mu_0(x_1), \ldots, \mu_0(x_n)\}$ and $\Sigma_0(x_{1:n}, x_{1:n}) = \{\Sigma_0(x_1, x_1), \ldots, \Sigma_0 (x_1, x_k); \ldots; \Sigma_0(x_n, x_1), \ldots, \Sigma_0(x_n, x_n)\}$. Let us suppose we wish to find a value of $f(X)$ at some new point X. Then we can compute the conditional distribution of $f(X)$ given the observations

$$f(X)|f(x_{1:n}) \sim \mathcal{N}(\mu_n(X), \sigma_n^2(X)) \tag{3}$$

$$\mu_n(X) = \Sigma_0(X, x_{1:n})\Sigma_0(x_{1:n}, x_{1:n})^{-1}(f(x_{1:n}) - \mu_0(x_{1:n})) + \mu_0(X) \tag{4}$$

$$\sigma_n^2(X) = \Sigma_0(X, X) - \Sigma_0(X, x_{1:n})\Sigma_0(x_{1:n}, x_{1:n})^{-1}\Sigma_0(x_{1:n}, X) \tag{5}$$

where $\mu_n(X)$ is the posterior mean and can be viewed as the prediction of the function value. $\sigma_n^2(X)$ is the posterior variance, and a measure of uncertainty of the prediction. The conditional distribution is called the posterior probability distribution in Bayesian statistics. It is very important during the Bayesian optimization and Gaussian process to carefully choose the next point to evaluate. During the likelihood optimization, acquisition functions are used to guide the search to a promising next point. Several acquisition functions have been published in the literature and nice categorization can be found in [30].

4 Method

Bilevel problems have two levels of optimization task, such that the lower-level problem is a constraint of the upper-level problem. In general bilevel problems, the follower depends on the leader's decisions x_u. The leader has no control over the follower's decision x_l. As declared in [34], for every leader's decision, there is an optimal follower decision, such that those members are considered feasible and also satisfy the upper-level constraints. Because the follower problem is a parametric optimization problem that depends on the leader decision x_u, it is very time-consuming to adopt a nested strategy approach that sequentially solves both levels. In the continuous domain, the computational cost is very high. During the optimization process, it is important to choose wisely the next leader decision x_u, to speed up the search. So the question is: *How to choose the next leader decision x_u^*?* Treating the leader problem as a black box, and fitting the previous leader's decisions and best follower reactions to the Gaussian process model makes it possible to estimate the next point to evaluate wisely.

Recently an algorithm has been proposed based on Bayesian optimization (BOBP) [16], using differential evolution to optimize the acquisition function. It has been shown to perform better than an evolutionary algorithm based on quadratic approximations (BLEAQ) [32]. In this work, we further improve the algorithm by using the previous iterations, including followers as priors and updating the acquisition function with conditioning on the follower's decisions to improve performance. Moreover, we investigated the contribution of follower decisions for optimizing the overall bilevel problem. The follower can observe the leader's decisions, but if we treat leader fitness as a black box then has been proposed then we also have available the best follower reactions to the leader's decisions during the optimization process. Using this data and extracting knowledge from previous iterations of the optimization process is the main idea of ConBaBo. The other question is: *Can the leader learn from the follower's decisions and use this knowledge for its own decision-making process?* Because of the structure of bilevel problems, the leader has no idea about the follower's reaction at the beginning of the optimization process. But in the following iterations, it can observe the optimum reaction of the follower without knowing the follower's objective.

Problem Statement. Following the idea above, let us assume that we have an expensive black-box function that takes leader decisions in leader decision space $x_u \in \mathcal{X}_u$ and the optimal follower's response $x_l^* \in \mathcal{X}_l$ as input. The upper-level function returns a scalar fitness score, $F(x_u, x_l^*) : \mathcal{X}_u \times \mathcal{X}_l \to \mathbb{R}$. Given a budget of N, the leader makes a decision and the follower responds to the leader accordingly. The leader can observe this information during the optimization process, learn how the follower reacts to the leader's decisions in every iteration, and choose the next leader's decision to optimize the fitness score.

Algorithm Description. First, we discuss fitting the decision data to the Gaussian process model, then the motivation behind conditioning. After observing n data points, that is leader decisions, follower responses, and leader's fitness score, $\{(x_{u_i}, x_{l_i}^*, y_i)\}_{i=1}^n$ where $y_i = F(x_{u_i}, x_{l_i}^*)$ upper-level objective, we fit the Gaussian process from $\mathcal{X}_u \times \mathcal{X}_l$ to $y \in \mathbb{R}$. After we have the vector-valued dataset $\mathbf{x}_{1:n} = \{(x_{u_1}, x_{l_1}^*), ..., (x_{u_n}, x_{l_n}^*)\}$ and $\mathbf{y}_{1:n} = \{y_1, ..., y_n\}$, then we construct the Gaussian process by a prior mean μ_0 and prior covariance function Σ_0. Let us suppose to find a value of $F(\mathbf{x})$ at some vector-valued decision $\mathbf{x}$. So, the conditional distribution of $F(\mathbf{x})$ for given observations is,

$$F(\mathbf{x})|F(\mathbf{x}_{1:n}) \sim \mathcal{N}(\mu_n(\mathbf{x}), \sigma_n^2(\mathbf{x})) \tag{6}$$

$$\mu_n(\mathbf{x}) = \Sigma_0(\mathbf{x}, \mathbf{x}_{1:n})\Sigma(\mathbf{x}, \mathbf{x})^{-1}(\mathbf{y}_{1:n} - \mu_0(\mathbf{x}_{1:n})) + \mu_0(\mathbf{x}) \tag{7}$$

$$\sigma_n^2(\mathbf{x}) = \Sigma_0(\mathbf{x}, \mathbf{x}) - \Sigma_0(\mathbf{x}, \mathbf{x}_{1:n})\Sigma_0(\mathbf{x}_{1:n}, \mathbf{x}_{1:n})^{-1}\Sigma_0(\mathbf{x}_{1:n}, \mathbf{x}) \tag{8}$$

Algorithm 1. CONBABO: Conditional Bayesian Optimization for Bilevel Optimization

Input: Random starting upper-level decisions with size of n

$\mathbf{x}_{u_{1:n}} : \{x_{u_1}, \ldots, x_{u_n}\}$

1: Find the best *lower-level* responses
   $\mathbf{x}_{l_{1:n}}^* : \{x_{l_1}^*, \ldots, x_{l_n}^*\}$ regarding $\mathbf{x}_{u_{1:n}}$ with *SLSQP* algorithm [20]
2: Evaluate *upper-level* fitness values : $\mathbf{y}_{1:n} = \{F(x_{u_1}, x_{l_1}^*), \ldots, F(x_{u_n}, x_{l_n}^*)\}$
3: Initialize Gaussian process ($\mathcal{GP}$) model and fit the observed data :
   $\mathcal{GP}(\mathbf{x}_{u_{1:n}}, \mathbf{x}_{l_{1:n}}^*, \mathbf{y}_{1:n})$
4: Update the $\mathcal{GP}$ model
5: **for** i = 0 : *Budget* **do**
6:     Evaluate the *conditional LCB acquisition function* value by using the Eq. 9
7:     Suggest the next candidate $x_{u_{n+i+1}}$ by optimizing *acquisition function*
8:     Evaluate the best *lower-level* response $(x_{l_{n+i+1}}^*)$ regarding $x_{u_{n+i+1}}$ using *SLSQP*
9:     Calculate the *upper-level* fitness:
       $F(x_{u_{n+i+1}}, x_{l_{n+i+1}}^*)$
10:    Update the $\mathcal{GP}$ model with new observations
11: **end for**
12: **return** the best decision values (x_u, x_l^*) with *upper-level* fitness value $F(x_u, x_l^*)$

Note that there is an interaction between decision-makers in bilevel problems because of their hierarchical structure. Thus, it is important not to violate the upper-level constraints during lower-level optimization. In this paper, we follow the bilevel definition in [34] to avoid violating the upper-level constraints. As declared in the paper, the optimal lower-level decision satisfies the upper-level constraints. Every optimal follower's decision is a reaction to the leader's decision, which makes possible the use of conditional settings.

After fitting the data to the model, we choose the next leader decision $x_{u_{n+1}}$. Then we find the optimal reaction $x_{l_{n+1}}^*$ and the fitness score of the leader function $y_{n+1} = F(x_{u_{n+1}}, x_{l_{n+1}}^*)$. We update the Gaussian process model with new data. Then we update the predicted optimal decisions for every iteration. After updating the model with the new follower optimal point at each iteration, we calculate the *conditional acquisition values*. For each follower optimal point x_l^* there is a global optimization problem over $x_u \in \mathcal{X}_u$. The lower confidence bound (LCB) acquisition function by [10] is used to choose the next point. We choose the exploration weight $w = 2$ for the LCB acquisition function, as it is the default setting. Acquisition function by conditioning on the follower's decision is defined as follows:

$$\alpha_{LCB}(x_u | x_l^*) = \hat{\mu}(x_u | x_l^*) - w\hat{\sigma}(x_u | x_l^*) \tag{9}$$

where w is a parameter that balances exploration and exploitation during optimization. $\hat{\mu}$ represents the conditional mean and $\hat{\sigma}$ represents the standard deviation. The conditional mean is calculated as follows:

$$\hat{\mu}(x_u | x_l^*) = \mu(x_u) + \Sigma(x_u, x_l^*)\Sigma(x_l^*, x_l^*)^{-1}(x_l^* - \mu(x_l^*)) \tag{10}$$

where $\mu(\cdot)$ is mean function and $\Sigma(\cdot)$ is covariance function. In Eq. 10 we can see that correlation between the conditioned data set and the leader decision data

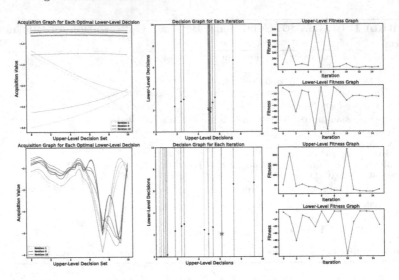

Fig. 1. Three graphs at the top show the results for the CONBABO algorithm, and the three graphs at the bottom show the results without conditioning the lower-level decisions. Left: Acquisition function graph with and without conditioning on lower-level decisions $LCB_{x_u|x_l=x_l^*}(x_{u_{n+1}})$ during upper-level optimization. The line's colour changes from light grey to dark grey in each iteration, and the red line is the shape of the acquisition function at the end. Middle: Decision space for upper-level decisions and best lower-level reaction at each iteration. Vertical lines show the upper-level decisions and black dots show the decision pair (x_u, x_l^*). The red line shows the last decision pair after 10 iterations with 5 training points. Right: Fitness graphs show CONBABO converges even after 10 iterations while not conditioning the lower-level decisions still searching for the optimal bilevel solution. (Color figure online)

set affects the mean function which represents the prediction. This change in posterior distribution is vital for the prediction of the Gaussian process. In this way, we used conditional setting on bilevel problems by improving the optimization with the follower's decisions.

We can see the observation for the first 10 iterations for the acquisition graph and how conditioning affects the optimization process in Fig. 1. In the figure, there are three different graphs. The left graph shows the acquisition function shape over 10 iterations. The light grey colour represents the previous acquisition graph and the red one represents the 10th iteration. During the Gaussian process, the acquisition function is the key point to select the next iteration and it is being done by optimizing the acquisition function. This is the reason for the importance of acquisition function during optimization. As we can observe from the Fig. 1, the conditional acquisition function shape (top-left in Fig. 1) is becoming convex even after 8 iterations while the standard acquisition function shape (bottom-left in Fig. 1) is maintaining non-convexity. Having convexity in acquisition optimization is vital in Bayesian optimization for improving the optimization performance in terms of computational cost. It shows the importance

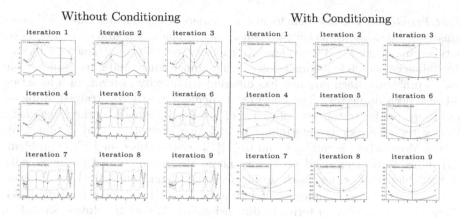

Fig. 2. First 9 iterations of the acquisition function graph with and without conditioning the follower's decision.

of gaining the lower-level optimal decision information for the upper-level decision makers. The graphs in the middle show the upper- and lower-level decision space. The vertical lines show the upper-level decisions and the stars show the lower-level responses to the upper-level decision-makers. We shared these graphs because of observing the behaviour of searching during the optimization process. We can see clearly that conditioning the lower-level decisions affects the choice of decisions at both levels and trying to converge at global optima even in 8 iterations while the standard acquisition function tries to minimize the uncertainty at different points in the search space. Practical problems in the bilevel structure are extremely hard and need time and consume computational power in most cases. So, we believe the improvement of CONBABO algorithm and the technique proposed provides a good technique for applying it to several practical problems.

We shared more acquisition graphs in a detailed way in Fig. 2. It shows the acquisition graph for each iteration for 9 iteration. As we can see from Fig. 2, the standard acquisition function behaves more explorative and tries different points at each step while trying to obtain a global optimal decision. As we can see from the graphs, the conditional acquisition function reached the near global optima even in 6 iterations. This fact makes us save lots of function evaluation during the bilevel optimization process. The acquisition function shape with conditioning becomes convex around global optima over the iterations quicker compared with the non-conditional setting. The CONBABO algorithm details are shown in Algorithm 1.

5   Numerical Experiments and Results

We evaluate ConBaBo on two test benchmarks including linear, non-linear, constrained, and unconstrained minimization test problems containing multiple dimensional decision variables.

Test Problems. The first benchmark is standard test problems, called TP problems, and contains 10 bilevel problems. Each problem has different interaction between upper-level and lower-level decision-makers. The benchmark includes 1 three-dimensional, 6 four-dimensional, 1 five-dimensional, and 2 ten-dimensional bilevel problems. Each problem has different complexity in terms of interaction between decision-makers. The benchmark includes mostly constrained problems and is created for testing the efficiency of proposed methods by the researchers. Mostly the efficiency is determined by function evaluations and obtained fitnesses. More details is can be found in [32]. The second is called SMD problems and it has 6 unconstrained problems. These are unconstrained problems with controllable complexities. Each problem has a different difficulty level in terms of convergence, the complexity of interaction, and lower-level multimodality. They are also scalable in terms of dimensionality. We set the dimensions of all problems as 4 for all experiments. The termination criteria is selected as the same as TP problems. More details about this benchmark can be found in [31].

Parameters. All experiments were conducted on a single core of 1.4 GHz Quad Core i5, 8Gb 2133 Mhz LPDDR3 RAM. The algorithm is executed 31 times for each test function and the results shown are medians. The termination criterion is selected as $\epsilon = 10^{-6}$ representing the difference between the obtained result and the best result of compared algorithms. Bayesian optimization and Gaussian processes are implemented in Python via the GPyOpt library [2] and *optimize_restarts* selected as 10 with the parameters of *verbose = False*, and *exact_feval = True*. The method initialized 5 Sobol points to construct the initial GP model. The LCB acquisition function has been selected for Gaussian Process and the Radial Basis Function (RBF) kernel is used. For optimizing the acquisition function, the L-BFGS method [22] has been used. The exploration-exploitation parameter for LCB, w is set to 2. After we set up the upper-level optimization method by conditional acquisition function, we used Sequential Least Squares Programming (SLSQP) [20] for solving the lower-level problem in Python programming language. For evaluating the CONBABO performance and making the comparison with the other selected algorithms, we set the termination criteria considering the best-obtained result of the other algorithms. The number of function evaluations is shared considering the Bayesian optimization iteration and compared with the selected algorithms median of 31 runs. We run the algorithm with different random seeds for making the comparison fair.

5.1   Results and the Discussion

We compared the experimental results with some state-of-art algorithms in the literature. To show the effectiveness of ConBaBo we compare it with the results in [16,32]. Sinha et al. present an improved evolutionary algorithm in [33] based on quadratic approximations of the lower-level problem, called BLEAQ. [16] propose an algorithm called BOBP with a Bayesian approach. Table 1 shows the median fitness values at both upper and lower-level for CONBABO and other algorithms. We also show the median function evaluations for upper-level and lower-level optimization in Table 2. We did not share the standard deviation over multiple runs because we could not compare it with the other algorithms as there is no information presented about the confidence interval. The primary purpose of bilevel optimization is to solve the leader problem, so in comparing the algorithms we focus on leader fitness and function evaluations at each level.

Table 1. Median Upper-level and lower-level fitness scores of ConBaBo, BOBP, BLEAQ algorithms for TP1-TP10.

| | Median Upper-level (UL) and Lower-level (LL) Fitness Scores | | | | | |
| | ConBaBo | | BOBP | | BLEAQ | |
	UL	LL	UL	LL	UL	LL
TP1	226.0223	98.0075	253.6155	70.3817	224.9989	99.9994
TP2	0.0000	200.0000	0.0007	183.871	2.4352	93.5484
TP3	−18.0155	−1.0155	−18.5579	−0.9493	−18.6787	−1.0156
TP4	−30.9680	3.1472	−27.6225	3.3012	−29.2	3.2
TP5	−4.3105	−1.5547	−3.8516	−2.2314	−3.4861	−2.569
TP6	−1.2223	7.6751	−1.2097	7.6168	−1.2099	7.6173
TP7	−1.9001	1.9001	−1.6747	1.6747	−1.9538	1.9538
TP8	0.0000	200.0000	0.0008	180.6452	1.1463	132.5594
TP9	0.00003	1.0000	0.0012	1.0000	1.2642	1.0000
TP10	0.000285	1.0000	0.0049	1.000	0.0001	1.0000

Table 2. Median function upper-level and lower-level function evaluations for ConBaBo and other known algorithms for TP1-TP10

| | Median Upper-level (UL) and Lower-level (LL) Function Evaluations | | | | | |
| | ConBaBo | | BOBP | | BLEAQ | |
	UL	LL	UL	LL	UL	LL
TP1	18.1333	190.6646	211.1333	1,558.8667	588.6129	1543.6129
TP2	17.6772	148.2746	35.2581	383.0645	366.8387	1,396.1935
TP3	10.1836	111.1374	89.6774	1,128.7097	290.6452	973.0000
TP4	14.6129	1,063.901	16.9677	334.6774	560.6452	2,937.3871
TP5	23.5164	222.2308	57.2258	319.7742	403.6452	1,605.9355
TP6	8.3333	323.1333	12.1935	182.3871	555.3226	1,689.5484
TP7	13.7092	274.9229	72.9615	320.2308	494.6129	26,682.4194
TP8	14.6066	114.9942	37.7097	413.7742	372.3226	1,418.1935
TP9	8.9215	1,020.8618	16.6875	396.3125	1,512.5161	141,303.7097
TP10	15.1935	3,689.6452	21.3226	974.0000	1,847.1000	245,157.9000

As can be seen in Table 1, ConBaBo achieved better upper-level results for 7 problems TP2, TP4, TP5, TP6, TP7, TP8, and TP9 in terms of fitness scores obtained. Moreover, the number of function calls by the leader decreases significantly as we encounter near-optimal solutions at both levels as we can see in Table 2. The test benchmark has constraint problems with various dimensional decision variables. The CONBABO results are promising when we consider the complexity of the problems in the benchmark. We can see that ConBaBo performs well on this standard test set.

Table 3. ConBaBo, CGA-BS [1], BLMA [14], NBLE [14], BIDE [14] and BLEAQ [33] upper-level fitness for SMD1-SMD6.

	Median Upper-level Fitness Scores					
	ConBaBo	CGA-BS	BLMA	NBLE	BIDE	BLEAQ
SMD1	1.8×10^{-6}	0	1.0×10^{-6}	5.03×10^{-6}	3.41×10^{-6}	1.0×10^{-6}
SMD2	9.09×10^{-6}	2.22×10^{-6}	1.0×10^{-6}	3.17×10^{-6}	1.29×10^{-6}	5.44×10^{-6}
SMD3	1.6×10^{-6}	0	1.0×10^{-6}	1.37×10^{-6}	4.1×10^{-6}	7.55×10^{-6}
SMD4	6.2×10^{-6}	3.41×10^{-11}	1.0×10^{-6}	9.29×10^{-6}	2.3×10^{-6}	1.0×10^{-6}
SMD5	1.49×10^{-6}	1.13×10^{-9}	1.0×10^{-6}	1.0×10^{-6}	1.58×10^{-6}	1.0×10^{-6}
SMD6	2.36×10^{-6}	9.34×10^{-11}	1.0×10^{-6}	1.0×10^{-6}	3.47×10^{-6}	1.0×10^{-6}

Table 4. Median function upper-level function evaluations for ConBaBo and other known algorithms for SMD1-SMD6

	Median Upper-level Function Evaluations					
	ConBaBo	CGA-BS	BLMA	NBLE	BIDE	BLEAQ
SMD1	1.1×10^{1}	1.01×10^{4}	1.19×10^{3}	1.52×10^{3}	6.0×10^{3}	1.19×10^{3}
SMD2	1.4×10^{1}	5.0×10^{4}	1.20×10^{3}	1.56×10^{3}	6.0×10^{3}	1.20×10^{3}
SMD3	2.3×10^{1}	1.0×10^{4}	1.29×10^{3}	1.56×10^{3}	6.0×10^{3}	1.29×10^{3}
SMD4	1.01×10^{1}	1.25×10^{5}	1.31×10^{3}	1.53×10^{3}	6.0×10^{3}	1.31×10^{3}
SMD5	2.37×10^{1}	1.0×10^{5}	2.06×10^{3}	3.40×10^{3}	6.0×10^{3}	2.06×10^{3}
SMD6	2.41×10^{1}	1.37×10^{5}	4.08×10^{3}	4.06×10^{3}	6.0×10^{3}	4.08×10^{3}

Table 5. Median function lower-level function evaluations for ConBaBo and other known algorithms for SMD1-SMD6

	Median Lower-level Function Evaluations					
	ConBaBo	CGA-BS	BLMA	NBLE	BIDE	BLEAQ
SMD1	3.8×10^{2}	1.5×10^{4}	2.37×10^{5}	9.520×10^{5}	1.8×10^{7}	2.37×10^{5}
SMD2	3.81×10^{2}	1.5×10^{5}	4.08×10^{5}	9.63×10^{5}	1.8×10^{7}	4.08×10^{5}
SMD3	7.7×10^{2}	2.0×10^{4}	3.02×10^{5}	1.04×10^{6}	1.8×10^{7}	3.02×10^{5}
SMD4	2.58×10^{2}	2.58×10^{5}	3.07×10^{5}	8.33×10^{5}	1.8×10^{7}	3.07×10^{5}
SMD5	9.05×10^{2}	2.58×10^{5}	8.42×10^{5}	2.22×10^{6}	1.8×10^{7}	8.42×10^{5}
SMD6	7.69×10^{2}	3.39×10^{5}	1.98×10^{4}	1.11×10^{5}	1.8×10^{7}	1.98×10^{4}

SMD problems are scalable in terms of the number of decision variables. We compared our results with CGA-BS, BLMA [14], NBLE [14], BIDE [14], and BLEAQ algorithms. We could not compare it with the BOBP algorithm because lack if information in the reference paper. Table 3 shows the upper-level fitness comparison with the other algorithms. We shared the upper-level and lower-level median function evaluations at Table 4 and Table 5 to show the effectiveness of the CONBABO algorithm. Also for SMD problems, we did not share the standard deviation over multiple runs because of the same reason with TP problems. We can observe that from Tables 4 and 5 that the upper- and lower-level function evaluations decrease significantly. These results show that ConBaBo can manage the difficulties resulting from the proposed SMD benchmark by [31]. Moreover, it converges faster to optimal solutions compared to other algorithms in the literature. The problems in the benchmark are scalable and have different difficulties in terms of the interaction between decision-makers. Considering this, CONBABO performs well on this unconstrained bilevel benchmark problem.

6 Conclusion

Bilevel optimization problems are a specific kind of problem that feature two decision makers, with a mathematical representation called "Stackelberg Games" in Game Theory. There are two optimization problems in these problems called the "upper-level" and "lower-level". Decisions in each level are affected by those made in the other, so the bilevel optimization scheme is time-consuming in terms of performance. We propose CONBABO, a conditional Bayesian optimization algorithm for bilevel optimization problems. The conditional Bayesian approach allows us to extract knowledge from previous upper- and lower-level decisions, leading to smarter choices and therefore fewer function evaluations. ConBaBo increases algorithm performance quality and dramatically accelerates the search for an optimal solution. We evaluate our method on two common benchmark sets from the bilevel literature, comparing results with those for top-performing algorithms in the literature. The CONBABO algorithm can be considered a powerful global technique for solving bilevel problems, which can handle the difficulties of non-linearity and conflict between decision-makers. Moreover, it can deal with constrained and unconstrained problems with multi-dimensional decision variables. In future work, the practical applications of bilevel problems and multi-objective bilevel problem adaptations will be researched.

Acknowledgements. This publication has emanated from research conducted with the financial support of Science Foundation Ireland under Grant number 16/RC/3918 which is co-funded under the European Regional Development Fund. For the purpose of Open Access, the author has applied a CC BY public copyright licence to any Author Accepted Manuscript version arising from this submission.

References

1. Abo-Elnaga, Y., Nasr, S.: Modified evolutionary algorithm and chaotic search for bilevel programming problems. Symmetry **12** (2020). https://doi.org/10.3390/SYM12050767
2. authors, T.G.: GPyOpt: a Bayesian optimization framework in python (2016). http://github.com/SheffieldML/GPyOpt
3. Bard, J.F.: Coordination of a multidivisional organization through two levels of management. Omega **11**(5), 457–468 (1983). https://doi.org/10.1016/0305-0483(83)90038-5
4. Bard, J.F., Falk, J.E.: An explicit solution to the multi-level programming problem. Comput. Oper. Res. **9**(1), 77–100 (1982). https://doi.org/10.1016/0305-0548(82)90007-7
5. Bard, J.F., Moore, J.T.: A branch and bound algorithm for the bilevel programming problem. SIAM J. Sci. Stat. Comput. **11**(2), 281–292 (1990). https://doi.org/10.1137/0911017
6. Bialas, W., Karwan, M.: On two-level optimization. IEEE Trans. Autom. Control **27**(1), 211–214 (1982). https://doi.org/10.1109/TAC.1982.1102880
7. Bracken, J., McGill, J.T.: Defense applications of mathematical programs with optimization problems in the constraints. Oper. Res. **22**(5), 1086–1096 (1974). https://doi.org/10.1287/opre.22.5.1086
8. Brown, G., Carlyle, M., Diehl, D., Kline, J., Wood, R.: A two-sided optimization for theater ballistic missile defense. Oper. Res. **53**, 745–763 (2005). https://doi.org/10.1287/opre.1050.0231
9. Constantin, I., Florian, M.: Optimizing frequencies in a transit network: a nonlinear bi-level programming approach. Int. Trans. Oper. Res. **2**(2), 149–164 (1995). https://doi.org/10.1016/0969-6016(94)00023-M
10. Cox, D.D., John, S.: SDO: a statistical method for global optimization. In: Multidisciplinary Design Optimization: State-of-the-Art, pp. 315–329 (1997)
11. Dogan, V., Prestwich, S.: Bayesian optimization with multi-objective acquisition function for bilevel problems. In: Longo, L., O'Reilly, R. (eds.) AICS 2022. CCIS, vol. 1662, pp. 409–422. Springer, Cham (2023). https://doi.org/10.1007/978-3-031-26438-2_32
12. Frazier, P.: A tutorial on Bayesian optimization. ArXiv abs/1807.02811 (2018)
13. Hong, M., Wai, H.T., Wang, Z., Yang, Z.: A two-timescale framework for bilevel optimization: complexity analysis and application to actor-critic. ArXiv abs/2007.05170 (2020)
14. Islam, M.M., Singh, H.K., Ray, T., Sinha, A.: An enhanced memetic algorithm for single-objective bilevel optimization problems. Evol. Comput. **25**, 607–642 (2017). https://doi.org/10.1162/EVCO_a_00198
15. Jones, D., Schonlau, M., Welch, W.: Efficient global optimization of expensive black-box functions. J. Glob. Optim. **13**, 455–492 (1998). https://doi.org/10.1023/A:1008306431147
16. Kieffer, E., Danoy, G., Bouvry, P., Nagih, A.: Bayesian optimization approach of general bi-level problems. In: Proceedings of the Genetic and Evolutionary Computation Conference Companion, GECCO 2017, pp. 1614–1621. Association for Computing Machinery (2017). https://doi.org/10.1145/3067695.3082537
17. Kirjner-Neto, C., Polak, E., Kiureghian, A.D.: An outer approximation approach to reliability-based optimal design of structures. J. Optim. Theory Appl. **98**(1), 1–16 (1998)

18. Koh, A.: Solving transportation bi-level programs with differential evolution. In: 2007 IEEE Congress on Evolutionary Computation, pp. 2243–2250 (2007). https:// doi.org/10.1109/CEC.2007.4424750
19. Koppe, M., Queyranne, M., Ryan, C.T.: Parametric integer programming algorithm for bilevel mixed integer programs. J. Optim. Theory Appl. **146**(1), 137–150 (2010). https://doi.org/10.1007/S10957-010-9668-3
20. Kraft, D.: A software package for sequential quadratic programming. Deutsche Forschungs- und Versuchsanstalt für Luft- und Raumfahrt Köln: Forschungs-bericht, Wiss. Berichtswesen d. DFVLR (1988)
21. Legillon, F., Liefooghe, A., Talbi, E.G.: CoBRA: a cooperative coevolutionary algorithm for bi-level optimization. In: 2012 IEEE Congress on Evolutionary Computation, pp. 1–8 (2012). https://doi.org/10.1109/CEC.2012.6256620
22. Liu, D.C., Nocedal, J.: On the limited memory BFGS method for large scale optimization. Math. Program. **45**(1), 503–528 (1989). https://doi.org/10.1007/ BF01589116
23. Ma, L., Wang, G.: A solving algorithm for nonlinear bilevel programing problems based on human evolutionary model. Algorithms **13**(10), 260 (2020)
24. Migdalas, A.: Bilevel programming in traffic planning: models, methods and challenge. J. Glob. Optim. **7**, 381–405 (1995). https://doi.org/10.1007/BF01099649
25. Pearce, M., Klaise, J., Groves, M.J.: Practical Bayesian optimization of objectives with conditioning variables. arXiv Machine Learning (2020)
26. Sabach, S., Shtern, S.: A first order method for solving convex bi-level optimization problems (2017). https://doi.org/10.48550/ARXIV.1702.03999
27. Sahin, K., Ciric, A.R.: A dual temperature simulated annealing approach for solving bilevel programming problems. Comput. Chem. Eng. **23**, 11–25 (1998)
28. Savard, G., Gauvin, J.: The steepest descent direction for the nonlinear bilevel programming problem. Oper. Res. Lett. **15**(5), 265–272 (1994). https://doi.org/ 10.1016/0167-6377(94)90086-8
29. Shaban, A., Cheng, C.A., Hatch, N., Boots, B.: Truncated back-propagation for bilevel optimization. CoRR abs/1810.10667 (2018)
30. Shahriari, B., Swersky, K., Wang, Z., Adams, R.P., Freitas, N.D.: Taking the human out of the loop: a review of Bayesian optimization. Proc. IEEE **104**, 148–175 (2016). https://doi.org/10.1109/JPROC.2015.2494218
31. Sinha, A., Malo, P., Deb, K.: Unconstrained scalable test problems for single-objective bilevel optimization. In: 2012 IEEE Congress on Evolutionary Computation, pp. 1–8 (2012). https://doi.org/10.1109/CEC.2012.6256557
32. Sinha, A., Malo, P., Deb, K.: Efficient evolutionary algorithm for single-objective bilevel optimization (2013). https://doi.org/10.48550/ARXIV.1303.3901
33. Sinha, A., Malo, P., Deb, K.: An improved bilevel evolutionary algorithm based on quadratic approximations. In: 2014 IEEE Congress on Evolutionary Computation (CEC), pp. 1870–1877 (2014). https://doi.org/10.1109/CEC.2014.6900391
34. Sinha, A., Malo, P., Deb, K.: A review on bilevel optimization: from classical to evolutionary approaches and applications. IEEE Trans. Evol. Comput. **22**(2), 276–295 (2018). https://doi.org/10.1109/TEVC.2017.2712906
35. Sinha, A., Malo, P., Frantsev, A., Deb, K.: Multi-objective Stackelberg game between a regulating authority and a mining company: a case study in environmental economics. In: 2013 IEEE Congress on Evolutionary Computation, pp. 478–485 (2013). https://doi.org/10.1109/CEC.2013.6557607
36. Srinivas, N., Krause, A., Kakade, S.M., Seeger, M.W.: Information-theoretic regret bounds for gaussian process optimization in the bandit setting. IEEE Trans. Inf. Theory **58**(5), 3250–3265 (2012). https://doi.org/10.1109/tit.2011.2182033

37. Stackelberg, H.V.: The theory of the market economy. William Hodge, London (1952)
38. Talbi, E.G.: A taxonomy of metaheuristics for bi-level optimization. In: Talbi, E.G. (ed.) Metaheuristics for Bi-level Optimization. Studies in Computational Intelligence, vol. 482, pp. 1–39. Springer, Heidelberg (2013). https://doi.org/10.1007/978-3-642-37838-6_1
39. Vicente, L., Savard, G., Júdice, J.: Descent approaches for quadratic bilevel programming. J. Optim. Theory Appl. 81(2), 379–399 (1994)

Few-Shot Learning for Character Recognition in Persian Historical Documents

Alireza Hajebrahimi(✉)📷, Michael Evan Santoso📷, Mate Kovacs(✉)📷,
and Victor V. Kryssanov📷

College of Information Science and Engineering, Ritsumeikan University,
1-1-1 Nojihigashi, Kusatsu, Shiga 525-8577, Japan
{is0591ki,is0534is}@ed.ritsumei.ac.jp, kovacsm@fc.ritsumei.ac.jp,
kvvictor@is.ritsumei.ac.jp

Abstract. Digitizing historical documents is crucial for the preservation of cultural heritage. The digitization of documents written in Perso-Arabic scripts, however, presents multiple challenges. The Nastaliq calligraphy can be difficult to read even for a native speaker, and the four contextual forms of alphabet letters pose a complex task to current optical character recognition systems. To address these challenges, the presented study develops an approach for character recognition in Persian historical documents using few-shot learning with Siamese Neural Networks. A small, novel dataset is created from Persian historical documents for training and testing purposes. Experiments on the dataset resulted in a 94.75% testing accuracy for the few-shot learning task, and a 67% character recognition accuracy was observed on unseen documents for 111 distinct character classes.

Keywords: Few-shot learning · Historical Documents · Character Recognition · Persian Language · Siamese Neural Networks

1 Introduction

Historical documents are essential for understanding the past, providing invaluable insights into a nation's or civilization's cultural, social, economic, and political background. Historical documents are sources of information that offer a unique view of past events, beliefs, and practices. In the case of Iran, the significance of historical documents cannot be overstated. Iran is home to one of the world's oldest continually significant civilizations, with historical and urban settlements dating back to 4000 B.C [18]. Preserving these documents, however, is a significant challenge because many were lost, destroyed, or damaged because of wars, invasions, and natural disasters. Therefore, it is crucial to digitize the documents to ensure their preservation and accessibility for future generations. One of the primary challenges in digitalizing historical documents is the difficulty in making them accessible to everyone in terms of readability. Handwriting can be a

G. Nicosia et al. (Eds.): LOD 2023, LNCS 14505, pp. 259–273, 2024.
https://doi.org/10.1007/978-3-031-53969-5_20

major obstacle to reading and understanding historical documents, especially for older scripts written in a style no longer or rarely used. This can make it difficult for researchers, historians, and the general public to access and understand the information contained in these documents. Thus, there is a need to extract and store the historical documents in a searchable and editable text format. Figure 1 shows example pages of three different Persian historical documents written in the Nastaliq style.

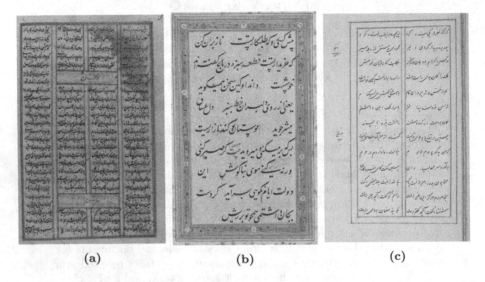

(a)   (b)   (c)

Fig. 1. Sample pages of Persian historical documents (a) Shahnameh by Firdausi, (b) Gulistan by Saadi, and (c) Qajar-era poetry. (The images were obtained from the Library of Congress, [5, 22, 25])

Optical character recognition (OCR) is a technology that converts scanned images of documents into editable and searchable texts that would offer a solution to the challenges of digitizing historical documents. Handwritten character recognition is a highly complex task that has been the subject of extensive research in computer vision. Despite the recent advances in deep learning, handwriting recognition remains a difficult task owing to variations in writing styles, font sizes, and layouts. Moreover, Perso-Arabic OCR must handle issues such as detecting and handling dots properly, dealing with different letter shapes based on position in a word, and managing the cursive connectivity between letters in addition to other factors such as unclear text and missing sections when dealing with historical documents [4]. One of the major factors is the use of the Nastaliq and Shekasteh Nastaliq calligraphy. Figure 2 compares the two calligraphy styles. Another issue that makes OCR challenging, in this case, is the presence of four contextual forms for almost all alphabet letters.

While different authors may have different handwritings, the contextual forms of the letters remain the primary obstacle for handwritten character recognition. These forms refer to the shape of the letter, depending on its position in a word, which is classified into four cases: initial, medial, final, and isolated. In the Persian alphabet, out of the 32 letters, 25 have four contextual forms, while the other 7 have two contextual forms.

Persian historical literature is mostly written in the Nastaliq or Shekasteh Nastaliq calligraphy. Multiple organizations, such as The Library of Congress[1] and Harvard Library's Islamic Heritage Project[2], created digitized copies of these documents. However, due to the fluid and cursive nature of the Nastaliq calligraphy and individual handwriting styles, only experts can read these documents easily. Machine learning models developed previously require large amounts of training data to predict characters correctly [2] but since not many people can easily read these documents, building a new large training dataset may not be practically feasible.

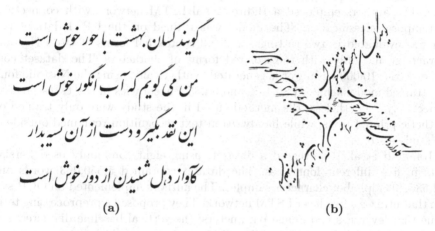

(a)   (b)

Fig. 2. Comparison of Nastaliq (a) and Shekasteh Nastaliq (b) text with the same content

In many machine learning applications, including character recognition from historical documents, the amount of labeled data available for training is limited. Few-shot learning (FSL) is a rapidly developing field of artificial intelligence that aims to create models that can learn from a small number of training samples available. This study deploys a few-shot learning approach, utilizing Siamese Neural Networks to recognize handwritten Nastaliq style characters from Persian historical documents.

The remainder of the paper is structured as follows: in Sect. 2, existing character recognition methods for Perso-Arabic scripts are surveyed. Section 3 presents the proposed approach, and Sect. 4 describes the dataset. The conducted experiments are presented in Sect. 5, where results obtained in the experiments are presented and discussed in Sect. 6. Finally, Sect. 7 summarizes the findings and outlines directions for future work.

---

[1] https://www.loc.gov.
[2] https://library.harvard.edu/collections/islamic-heritage-project.

2   Related Work

The authors of [1] presented an approach to recognizing Perso-Arabic handwritten digits, using a LeNet-5 CNN model [12] trained on the IFH-CDB dataset [14]. IFH-CDB contains 52,380 images of Persian alphabet letters in the isolated form and 17,740 images of digits. For handwritten digit recognition, the authors utilized an extended version of the dataset with 19,840 samples of digits. They implemented two rejection strategies to filter out samples with low confidence levels to improve the model accuracy. After applying the rejection strategies, the proposed method achieved an accuracy of over 99%. The study, however, only considered Persian digits by training the model on a total of 10,000 samples with 1,000 samples per class.

Ul-Hasan et al. [24] proposed an OCR system for the printed Urdu Nastaliq style. The authors employed a Bidirectional LSTM network with connectionist temporal classification. The model was trained on the UPTI dataset [21] and was evaluated in two settings: with all contextual forms of Urdu Nastaliq characters, and only with the isolated forms of characters. The dataset consisted of over 10,000 Urdu words generated synthetically, using the Nastaliq font. The trained model achieved 86.43% and 94.85% accuracies in the two scenarios, respectively. Nonetheless, the methods used in the study were only trained on synthetic printed texts, while handwritten text recognition remained outside of the study's scope.

Rahmati et al. [19] created a dataset, using eight commonly used Persian fonts in five different font sizes. The dataset contained 5,550,063 words and 102,536,493 alphabet character samples. The authors implemented an OCR system that utilizes a five-layer LSTM network. They proposed a preprocessing technique that leverages two image parameters: the vertical baseline diameter, and the line-height. The vertical baseline diameter is the vertical distance between the baseline of a line of text and the highest point of any glyph in that line. The line-height is the space between two connected letters, which can be lengthened or shortened, depending on the placement of the word in the paragraph. Experiments were conducted in the study with different depths for the LSTM network and various activation functions allowed for achieving a letter-level accuracy of 99.68% and a word-level accuracy of 99.08%. The developed model was trained on a large dataset consisting of 5.5 million word samples and 102 million alphabet character samples, all of which were in printed format.

Mohammed Aarif K.O. et al. [10] used transfer learning to develop an OCR system for the Urdu language. The authors used a variant of AlexNet and GoogleNet, which both were pre-trained [11,23]. For OCR-AlexNet, the weights of the initial layer were frozen, and the last three layers were fine-tuned on the IFH-CDB dataset. For the OCR-GoogleNet model, they retrained the network with only four inception modules. The last two layers were replaced, and all earlier layers were frozen to prevent overfitting. The authors achieved accuracies of 97.3% and 95.7% with OCR-AlexNet and OCR-GoogleNet, respectively. Although the proposed models were trained using different fonts and sizes, only printed Urdu characters were considered in the study.

Table 1. Summary of related work

Author	Method	Dataset	Total No. of Samples	No. of Classes	Accuracy
Ahranjany et al. [1]	LeNet-5	IFH-CDB	10000 handwritten digits	10	99.98%
Ul-Hasan et al. [24]	Bidirectional LSTM	UPTI	10063 synthetically generated text lines	191	86.42%
				99	94.85%
Rahmati et al. [19]	LSTM	Custom	5.5M words	110	99.08%
			102M alphabet characters		99.68%
Mohammed Aarif K.O et al. [10]	AlexNet	IFH-CDB	2380characters (isolated form) 17740 digits	54	97.3%
	GoogleNet				95.7%
Mushtaq et al. [15]	CNN	Custom	95508 handwritten characters	133	98.82%
Naseer & Zafar [17]	Siamese NN	CLETI	Not Specified	17	95.55%*
		UTG		17	98.72%*
		UOF		17	100%*

* These accuracies refer to the few-shot image differentiation task rather than the character recognition performance.

Mushtaq et al. [15] developed a CNN model for handwritten Urdu character recognition. A network with 10 layers was trained on 74,285 samples and tested on 21,223 samples collected from 750 individuals. The authors trained the model, considering all contextual forms of the characters, resulting in 133 classes. They observed a recognition accuracy of 98.82% on the dataset. It was argued that most misclassifications happened with letters having similar geometrical shapes, and were due to poorly written characters. While the study yielded a notable level of accuracy in recognizing the handwritten characters, it should be noted that the model was trained using approximately 500 samples per class, and the characters were written in a modern writing style.

Naseer and Zafar [17] used Siamese neural networks to develop a few-shot solution for optical character recognition of Nastaliq Urdu characters. Comparing the performances of the few-shot Siamese neural network and a standard CNN, the Siamese neural network performed better when trained with 5 samples per class. The Siamese network could recognize similarities and differences between character image pairs, achieving 95.55%, 98.72%, and 100% few-shot learning accuracies on the CLETI, UTG, and UOF corpora, respectively. The study, however, used printed texts with a uniform style of writing. Furthermore, only 17 classes were considered that simplified image differentiation and character recognition significantly.

Najam and Faizullah [16] provided a comprehensive review and critical analysis of deep learning techniques for Arabic handwritten text recognition and post-OCR text correction from 2020–2023. They analyzed numerous works utilizing CNN and LSTM with CTC loss function, Transformer, and GAN architectures for OCR, finding that CNN-LSTM-CTC is currently the standard approach, while Transformer shows promise given sufficient training data. The dataset sizes utilized in the analyzed studies ranged from thousands of images to hundreds of thousands of words of handwritten text. However, none of the datasets contained samples in Nastaliq calligraphy.

Although the previous character recognition studies on Perso-Arabic scripts delivered many positive results, historical documents have rarely been discussed in the related literature. Most of the proposed models were trained with a large number of samples per class. Since there is no available dataset for OCR on Persian historical documents, few-shot learning could be used to train a model for recognizing characters with a limited amount of training data. Few-shot algorithms are well-suited for recognizing differences between character images, such as typeface and size [26]. The latter is especially important in the case of historical documents, which may be in poor condition and would be difficult to process, using existing OCR methods. Table 1 gives a summary of the related work.

3   Proposed Approach

Figure 3 gives a general overview of the proposed approach. First, scanned historical documents are collected and preprocessed to create a small dataset for training a character recognition model. Next, a Siamese Neural Network is trained on the dataset to build a few-shot learning model that can be used for character recognition on unseen documents. Character recognition is performed using the trained Siamese model by doing one-shot prediction that is conducted by comparing a previously unseen character image with "anchor" images from the training dataset using the trained few-shot learning model. The input image is compared with the anchor images of each class, and the unseen character image is assigned to the most similar class.

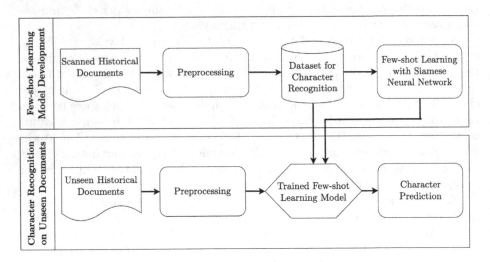

Fig. 3. Handwritten character recognition for Persian historical documents using few-shot learning

3.1   Siamese Neural Network

The Siamese network is a neural network architecture commonly used to assess the similarity between two image samples. In contrast to conventional machine learning methods, a Siamese network employs few-shot learning, so the model requires only a small number of training samples. Figure 4 illustrates the high-level architecture of the Siamese network used in this study. The network consists of two main CNN branches, sharing the same weights. Image A1 and Image A2 are the inputs of the network in the form of positive and negative sample pairs. In the present study, a positive pair means different images of the same character, and a negative sample pair means images of different characters. Sharing the same weights, the network layers act as feature extractors that output abstract representations $G_w(\cdot)$ of the image matrices. To assess the similarity between image pairs, the Euclidean distance is calculated for the image's internal representations. The network calculates the Contrastive loss to update the weights of the network layers. The Contrastive loss $L_{Contrastive}$ is defined as

$$L_{Contrastive} = \frac{1}{2\tau} \sum_{i=1}^{\tau} \left(y_i(d_i)^2 + (1 - y_i) \max((m - d_i), 0)^2 \right), \tag{1}$$

where y denotes the target label, which can take on binary values of either 0 or 1, where 0 implies dissimilar image pair and 1 suggests similar image pair. The variable τ represents the total number of classes encompassed within the dataset. The symbol d denotes the Euclidean distance calculated between the feature vectors of the two images under consideration, extracted from the CNN layers. This distance acts as a proxy for the predicted similarity between the input pair. The margin to prevent the model from being stuck at local optima is denoted by m. The model aims to maximize the distance for a negative input pair (representing different characters) and minimize it for positive pairs (representing the same character).

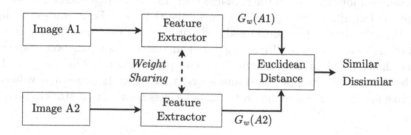

Fig. 4. The high-level architecture of the Siamese network

4   Dataset

In this study, a new dataset was built from Persian historical documents for character-level few-shot learning[3] The dataset was created to train models for predicting handwritten Persian letters from images. Figure 5 overviews the

---

[3] Available at: https://huggingface.co/datasets/iarata/PHCR-DB25.

dataset creation process. After scanned documents were collected from various sources, native speakers cropped and annotated the character images. Next, the preprocessing removed noise from the images and simplified them for model development. Finally, positive and negative image pairs were generated for the train and test datasets.

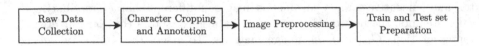

Fig. 5. Steps of the dataset creation process

In the data, all contextual forms of the letters are treated as separate classes. While characters آ, د, ذ, ر, ز, ژ and و have only two contextual forms, the rest of the letters have four contextual forms. Since the isolated form of the letter آ can be written as either آ or ا, these constitute distinct classes in the presented study.

The dataset was created from five historical Persian books, including Shahnameh by Firdausi (17th or 18th century) [5], Divan by Hafiz (16th century) [7], Kitab-i Rumi al-Mawlawi by Rumi Maulana (15th century) [13], Gulistan by Sa'di (18th century) [22], and Qajar-era by an unknown author (19th century) [25]. The main reason behind using these sources is that the authors' writing styles vary significantly, and training a model on a dataset with distinctive writing styles would contribute to the robustness of the resultant model. The dataset used thus contains 31 letters of the Persian alphabet and all their contextual forms. Four letters of the modern 32-letter Persian alphabet (گ, چ, پ and ژ) were seldom present in the classical Persian. In the five books, the letter گ was not used at all, so the dataset contains only 31 letters. Each contextual form of the Persian alphabet is considered a distinct class, and the dataset is comprised of 111 classes in total. Five images per historical document were extracted for each character class, resulting in 25 images per character. The images of Persian characters were cropped and annotated with their hexadecimal values by the native speakers. Each image in the dataset was resized to a size of 395×395 pixels.

4.1 Image Preprocessing

Images were first normalized to reduce noise from the background of the characters. The normalization filter is defined as

$$I_N = (I - min_0)\frac{max_1 - min_1}{max_0 - min_0} + min_1, \tag{2}$$

where I is the original image, and I_N is the image after normalization; max_0 and min_0 are the maximum and minimum pixel intensity values respectively, and max_1 and min_1 are the designated maximum and minimum pixel intensity values, respectively. The normalization scales the pixel values of the input image to be within a specified range, based on the minimum and maximum pixel values in the input image. The range chosen in this study is [0, 255], to improve color differentiation by using the full color spectrum. The normalized image is then converted to a single-channel grayscale image. Subsequently, the grayscale image undergoes a process of image thresholding, utilizing a binary inverse methodology with a value range of [100, 255]. This process is essential to the elimination of the background context associated with the individual character images. The thresholded image is binarized so that pixel values greater than 0 become 255 (white), and pixels with a value of 0 (black) remain unchanged. Finally, the binarized image is inversed. Figure 6 depicts the preprocessing steps applied to an example image.

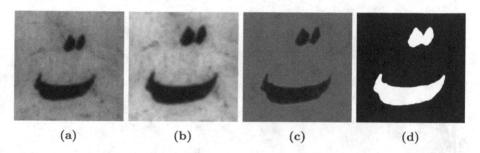

| (a) | (b) | (c) | (d) |

Fig. 6. An example of a preprocessed image: (a) original image, (b) normalized image, (c) thresholded image, and (d) inversed binarized image.

4.2 Train and Test Datasets

To build a dataset for the few-shot learning task, all combinations of the character images were generated. The image pairs were put into the train and test sets so that the numbers of positive and negative sample pairs were fixed equal. All classes of the test set have the same number of image pairs. Figure 7 illustrates two samples of the generated pairs.

| (a) | (b) |

Fig. 7. Examples of the generated pairs: (a) Positive pair, (b) Negative pair

5   Experiments

Two experiments were conducted, using the few-shot learning dataset created from the Persian historical documents. The first experiment involved a Siamese Network with a custom convolutional architecture. The second used transfer learning to leverage the feature-extracting ability of a robust pre-trained model for the few-shot learning task. Experiments were done to determine if the dataset necessitates a custom network, and if pre-trained models can be used with the same success. The networks were trained on 5, 10, and 15 samples per class in both experiments to assess the model performance for different training set sizes.

5.1   Custom Siamese Convolutional Network

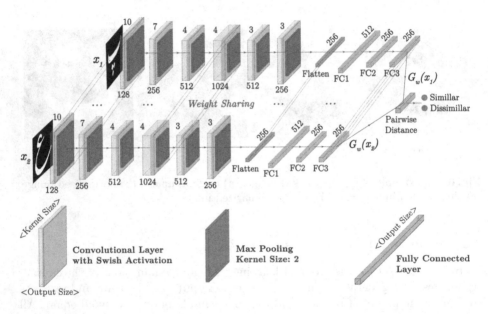

Fig. 8. The architecture of the custom Siamese network used in the study

Figure 8 depicts the detailed architecture and hyperparameters of the final custom Siamese network built after hyperparameter tuning and preliminary experiments. The network has six convolutional layers, followed by flattening, and three fully connected layers. The stride for the convolutional layers was fixed to 1×1. The kernel sizes for the first and second layers were set to 10×10 and 7×7, respectively. A kernel size of 4×4 was applied for the next two convolutional layers. The last two convolutional layers have kernel sizes of 3×3. After each convolutional layer, max pooling with a kernel size of 2×2 was applied. The three fully connected layers have 512, 256, and 256 units, respectively. The output of the last fully connected layer is used to calculate the similarity between the image pairs in the feature space. After preliminary experiments to find the optimal batch size, the model was trained on 32-sample batches with early stopping to avoid overfitting.

The Swish activation function [20] was used for neuron activation between the hidden layers. The function is defined as

$$Swish(x) = x \cdot \sigma(x\beta), \tag{3}$$

where β is a trainable parameter in the network that controls the gradient-preserving property for the weight updates, and σ denotes the sigmoid function.

The adaptive learning rate optimizer Adam [9] is used to update the weights of the network during training. To facilitate efficient learning, the Xavier uniform initializer [6] was used to set appropriate initial weight values which are neither too large nor too small.

5.2 Pre-trained Siamese Network

For the second experiment, a pre-trained ResNet101 model [8] was incorporated into both branches of the Siamese architecture. ResNet is a type of deep neural network introduced to address the problem of vanishing gradients during training by using shortcut connections between layers. ResNet101 has 101 layers, and it is pre-trained on the ImageNet-1k dataset [3]. In the experiment, the final layer of ResNet101 was used as the primary feature extractor of the Siamese network. To fine-tune the network for the target task, the ResNet layer is supplemented by three fully connected layers of 512, 256, and 256 units, respectively. While the fully connected layers were set as trainable, the weights of the ResNet layer were frozen to keep the feature extraction ability of the pre-trained network. Early stopping was deployed during training to avoid overfitting on the train set.

6   Results and Discussion

The custom and pre-trained models were trained on a range of sample sizes per class, and both were evaluated, using the same test dataset to ensure that the obtained results were comparable. Table 2 presents performance metrics obtained with the custom Siamese network for different training set sizes in the few-shot image differentiation task.

Table 2. Few-shot learning performance of the Custom Siamese CNN Network

Training Size (Image per class)	Testing Size (Image per class)	Accuracy (Test set)	Precision		Recall		F1-Score	
			Negative	Positive	Negative	Positive	Negative	Positive
5 Images	10 Images	77.74%	0.78	0.78	0.78	0.78	0.78	0.78
10 Images		87.97%	0.90	0.86	0.85	0.91	0.88	0.88
15 Images		94.75%	0.96	0.94	0.93	0.96	0.95	0.95

Table 3. One-shot character recognition performance of the custom Siamese model

Training Size (Image per class)	Testing Size (Image per class)	Accuracy (Test set)	Precision	Recall	F1-Score
15 Images	10 Images	67%	0.69	0.67	0.66

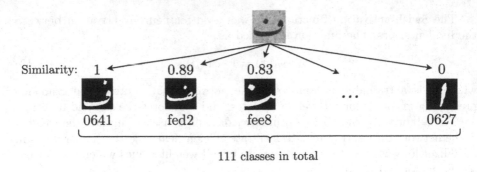

Fig. 9. An example of one-shot character recognition

The general trend observed is that precision, recall, and F1-score improve as the training set size grows. It is to note that balancing the training dataset is crucial for achieving a satisfactory performance. Preliminary experiments revealed that when the training set was imbalanced (i.e., the numbers of negative and positive sample pairs are significantly different), the model performed poorly, and was only able to achieve a 61.19% accuracy. However, when the positive/negative sample ratio was set to 1, the model could achieve 77.74%, 87.97%, and 94.75% accuracies in 10 epochs for 5, 10, and 15 training samples per class, respectively. Table 3 shows the one-shot character recognition performance of the custom Siamese model trained on 15 samples per class. The model achieved a 67% one-shot prediction accuracy for the 111 classes (random prediction would be 0.9%), in which a given input image was compared against the anchor images of all classes. The precision, recall, and F1-score values are 0.69, 0.67, and 0.66, respectively. Figure 9 illustrates the one-shot prediction of a sample. The trained few-shot learning model[4] was used to compare the new image with anchor images of all classes, and the label of the most similar class was assigned to the sample. It is important to acknowledge that the 67% accuracy derived from this one-shot prediction paradigm includes many occurrences where the input was mistakenly identified as an alternative contextual presentation of the same parent letter.

Table 4. Few-shot learning performance of the fine-tuned pre-trained network

Training Size (Image per class)	Testing Size (Image per class)	Accuracy (Test set)	Precision		Recall		F1-Score	
			Negative	Positive	Negative	Positive	Negative	Positive
5 Images	10 Images	73.70%	0.75	0.72	0.71	0.77	0.73	0.74
10 Images		79.17%	0.80	0.79	0.78	0.80	0.79	0.79
15 Images		82.81%	0.83	0.80	0.78	0.88	0.82	0.84

Table 4 presents the results of the fine-tuned pre-trained network. Although the two models resulted in comparable accuracies when trained on 5 samples per

---

[4] Available for academic purposes at https://huggingface.co/iarata/Few-Shot-PHCR.

class, the performance difference between the custom and pre-trained networks increased as the training set size grew. When the training size was increased to 10 samples per class, the pre-trained network could not learn as effectively as the custom network. With 15 samples per class, the difference between the two networks became approximately 12%. Thus, the one-shot prediction performance of the pre-trained network was not evaluated in the following experiments. There would be multiple reasons for why the fine-tuned pre-trained model performed worse than the custom network. For one, there is a possible complexity mismatch between the robust pre-trained model and the dataset: ResNet101's feature extraction capability may be excessive for the given character dataset. Another possible reason would be the domain mismatch. The ImageNet-1k dataset used to pre-train ResNet101 involves various image sources and mainly consists of photographs.

Most misclassifications observed in the experiments happened due to the detailed classification scheme, where all contextual forms of a letter were treated as separate classes. Given that there are only a few samples per class for training, the intraclass variations between the character images are relatively low. Also, the contextual forms can look extremely similar, depending on the writing style and the letter. For example, ﺱ and ﺳ belong to the same letter class but are different contextual forms. In most cases of misclassification, the model could predict the alphabet letter correctly but failed to identify the specific contextual form. When the evaluation of the model trained on all contextual forms was limited to the 31 alphabet letters without considering their contextual forms, the one-shot prediction accuracy increased to 84%.

Another reason for the misclassifications would be the quality of the images. Few-shot learning is known to be unreliable in the case of noisy and low-quality images, making it difficult to distinguish between meaningful patterns in the data. As a result, the model may inadvertently learn to prioritize irrelevant and unrepresentative features. Despite the high-quality images of the books used to make the dataset, the characters extracted involve only a small part of the images, often resulting in a relatively low character image quality.

Overall, however, the results obtained demonstrated the effectiveness of few-shot learning for recognizing Persian characters in historical documents. There are a few limitations of the presented work. Since the dataset built in this study is limited to only the five books by the five authors, it may not capture certain writing styles. It is also to note that annotation was a time-consuming process, even for such a small dataset. Despite being native speakers, the annotators often encountered difficulties when labeling characters, owing to the intricacies inherent to the Nastaliq calligraphy. Finally, this study focused on the Nastaliq style calligraphy, while the Shekasteh Nastaliq calligraphy was not considered.

7   Conclusions

This paper presented a successful application of few-shot learning for recognizing Persian handwritten characters from historical documents. Experiments were conducted on a new dataset, using two neural networks with different architectures: a custom convolutional Siamese network, and a pre-trained model that was fine-tuned for the few-shot learning task. Both models were trained with a different number of samples per class, and their performances were evaluated. The obtained results convincingly demonstrated the effectiveness of few-shot learning for character recognition from historical documents.

Future work would include building a large and diverse dataset with a broader range of historical documents from many authors. Also, as this study dealt with characters written in the Nastaliq style calligraphy, the model robustness could be improved by incorporating samples written in the Shekasteh Nastaliq style. Overall, the findings of this study would provide a basis for further research on character recognition of handwritten Perso-Arabic scripts using a limited amount of training data.

References

1. Ahranjany, S.S., Razzazi, F., Ghassemian, M.H.: A very high accuracy handwritten character recognition system for Farsi/Arabic digits using convolutional neural networks. In: 2010 IEEE Fifth International Conference on Bio-inspired Computing: Theories and Applications (BIC-TA), pp. 1585–1592. IEEE (2010)
2. Bonyani, M., Jahangard, S., Daneshmand, M.: Persian handwritten digit, character and word recognition using deep learning. Int. J. Doc. Anal. Recognit. (IJDAR) **24**(1–2), 133–143 (2021)
3. Deng, J., Dong, W., Socher, R., Li, L.J., Li, K., Fei-Fei, L.: Imagenet: a large-scale hierarchical image database. In: 2009 IEEE Conference on Computer Vision and Pattern Recognition, pp. 248–255. IEEE (2009)
4. Faizullah, S., Ayub, M.S., Hussain, S., Khan, M.A.: A survey of OCR in Arabic language: applications, techniques, and challenges. Appl. Sci. **13**(7), 4584 (2023)
5. Firdausi: Shah-Nameh by Firdausi. (1600). https://www.loc.gov/item/2012498868/
6. Glorot, X., Bengio, Y.: Understanding the difficulty of training deep feedforward neural networks. In: Proceedings of the Thirteenth International Conference on Artificial Intelligence and Statistics, pp. 249–256. JMLR Workshop and Conference Proceedings (2010)
7. Hafiz: Dīvān. (1517). https://www.loc.gov/item/2015481730/
8. He, K., Zhang, X., Ren, S., Sun, J.: Deep residual learning for image recognition. In: Proceedings of the IEEE Conference on Computer Vision and Pattern Recognition, pp. 770–778 (2016)
9. Kingma, D.P., Ba, J.: Adam: a method for stochastic optimization. arXiv preprint arXiv:1412.6980 (2014)
10. KO, M.A., Poruran, S.: OCR-nets: variants of pre-trained CNN for Urdu handwritten character recognition via transfer learning. Procedia Comput. Sci. **171**, 2294–2301 (2020)
11. Krizhevsky, A., Sutskever, I., Hinton, G.E.: Imagenet classification with deep convolutional neural networks. In: Proceedings of the 25th International Conference on Neural Information Processing Systems, vol. 1, pp. 1097–1105. NIPS'12, Curran Associates Inc., Red Hook, NY, USA (2012)

12. LeCun, Y., Bottou, L., Bengio, Y., Haffner, P.: Gradient-based learning applied to document recognition. Proc. IEEE **86**(11), 2278–2324 (1998)
13. Maulana, R.: Kitāb-i Rūmī al-Mawlawī. (1498). https://www.loc.gov/item/2016397707/
14. Mozaffari, S., Faez, K., Faradji, F., Ziaratban, M., Golzan, S.M.: A comprehensive isolated Farsi/Arabic character database for handwritten OCR research. In: Tenth International Workshop on Frontiers in Handwriting Recognition. Suvisoft (2006)
15. Mushtaq, F., Misgar, M.M., Kumar, M., Khurana, S.S.: Urdudeepnet: offline hand-written Urdu character recognition using deep neural network. Neural Comput. Appl. **33**(22), 15229–15252 (2021)
16. Najam, R., Faizullah, S.: Analysis of recent deep learning techniques for Arabic handwritten-text OCR and Post-OCR correction. Appl. Sci. **13**(13), 7568 (2023)
17. Naseer, A., Zafar, K.: Meta-feature based few-shot Siamese learning for urdu optical character recognition. Comput. Intell. **38**(5), 1707–1727 (2022). https://doi.org/10.1111/coin.12530, https://onlinelibrary.wiley.com/doi/abs/10.1111/coin.12530
18. Potts, D.T.: The Immediate Precursors of Elam, pp. 45–46. Cambridge Univ. Press, Cambridge (2004)
19. Rahmati, M., Fateh, M., Rezvani, M., Tajary, A., Abolghasemi, V.: Printed Persian OCR system using deep learning. IET Image Process. **14**(15), 3920–3931 (2020). https://doi.org/10.1049/iet-ipr.2019.0728, https://ietresearch.onlinelibrary.wiley.com/doi/abs/10.1049/iet-ipr.2019.0728
20. Ramachandran, P., Zoph, B., Le, Q.V.: Searching for activation functions. arXiv preprint arXiv:1710.05941 (2017)
21. Sabbour, N., Shafait, F.: A segmentation-free approach to Arabic and Urdu OCR. In: Document Recognition and Retrieval XX, vol. 8658, pp. 215–226. SPIE (2013)
22. Sa'dī: Gulistān (1593). https://www.loc.gov/item/2016503247/
23. Szegedy, C., et al.: Going deeper with convolutions. In: 2015 IEEE Conference on Computer Vision and Pattern Recognition (CVPR), pp. 1–9 (2015). https://doi.org/10.1109/CVPR.2015.7298594
24. Ul-Hasan, A., Ahmed, S.B., Rashid, F., Shafait, F., Breuel, T.M.: Offline printed Urdu nastaleeq script recognition with bidirectional LSTM networks. In: 2013 12th International Conference on Document Analysis and Recognition, pp. 1061–1065. IEEE (2013)
25. Unknown: Qajar-era poetry anthology (1800). https://www.loc.gov/item/2017498320/
26. Wang, Y., Yao, Q., Kwok, J.T., Ni, L.M.: Generalizing from a few examples: a survey on few-shot learning. ACM Comput. Surv. (CSUR) **53**(3), 1–34 (2020)

ProVolOne – Protein Volume Prediction Using a Multi-attention, Multi-resolution Deep Neural Network and Finite Element Analysis

Eric Paquet[1,2(✉)] , Herna Viktor[2] , Wojtek Michalowski[3] ,
and Gabriel St-Pierre-Lemieux[3]

[1] National Research Council of Canada, Ottawa, Canada
[2] School of Electrical Engineering and Computer Science, University of Ottawa,
Ottawa, Canada
hviktor@uottawa.ca
[3] Telfer School of Management, University of Ottawa, Ottawa, Canada

Abstract. Protein structural properties are often determined by experimental techniques such as X-ray crystallography and nuclear magnetic resonance. However, both approaches are time-consuming and expensive. Conversely, protein amino acid sequences may be readily obtained from inexpensive high-throughput techniques, although such sequences lack structural information, which is essential for numerous applications such as gene therapy, in which maximisation of the payload, or volume, is required. This paper proposes a novel solution to volume prediction, based on deep learning and finite element analysis. We introduce a multi-attention, multi-resolution deep learning architecture that predicts protein volumes from their amino acid sequences. Experimental results demonstrate the efficiency of the ProVolOne framework.

Keywords: Deep Learning · Multi-Attention Learning · Protein Volume Prediction · Multi-Resolution Learning

1 Introduction

Gene therapy [4] focuses on the genetic modification of cells to produce a therapeutic effect. To achieve this objective, a protein or a virus must carry a payload of genetic material inside a cell: the larger the payload, the greater the efficacy of the therapy [2]. The upper bound of the payload is closely related to the volume of the proteins. This volume corresponds, in turn, to the region enclosed by their macromolecular surfaces, which is a fundamental representation of their three-dimensional geometric shapes [20].

Proteins have highly complex shapes, which makes the evaluation and prediction of their volume challenging. This is illustrated in Fig. 1, showing the macromolecular surface of protein GalNAc-T7, which corresponds to entry 6IWR in the Protein Data Bank (PDB) [3]. To evaluate the macromolecular surface, it is necessary to know the positions of the atoms forming the protein [20]. Their positions may be obtained by X-ray crystallography, for large proteins, or by nuclear magnetic resonance, for smaller ones [20]. However, these techniques are expensive and time-consuming. In contrast,

G. Nicosia et al. (Eds.): LOD 2023, LNCS 14505, pp. 274–287, 2024.
https://doi.org/10.1007/978-3-031-53969-5_21

their amino acid sequences are readily available from non-expensive high-throughput techniques such as mass spectrometry [15]. In this paper, we employ deep learning to directly predict the volume from amino acid sequences, and we propose a new framework called Protein Volume One (ProVolOne), based on deep learning and finite element analysis (FEA).

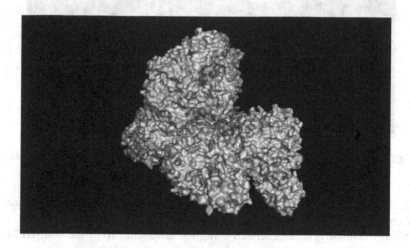

Fig. 1. Macromolecular surface of GalNAc-T7 (PDB entry 6IWR).

This paper is organised as follows. In Sect. 2, the calculation of the crystallised structure volume using FEA is addressed. In Sect. 3, a new multi-resolution, multi-attention convolutional neural network is proposed for volume prediction. The experimental evaluation is discussed in Sect. 4, and Sect. 5 concludes the paper.

2   Evaluation of the Payload from Crystallised Structures

X-ray crystallography can be used to determine atomic positions by studying the diffraction of an X-ray beam on a crystallised protein. The positions of the atoms are inferred by analysing the diffraction pattern resulting from the interaction of the beam with the crystallised structure. The atoms associated with the protein in Fig. 1 are represented by spheres in Fig. 2.

The macromolecular surface, also known as assessable surface area (ASA), is evaluated using the Shrake-Rupley algorithm, and implemented in PyMOL™ [14]. This algorithm is widely used in Proteomics. The Shrake-Rupley method is a numerical algorithm that generates spherical point clouds around each atom, each sphere having the same radius, which is chosen beyond the van der Waals radius. The points that are solvent accessible determine the surface. This is effectively similar to rolling a ball along the surface. Then, to determine the volume, the ASA is first tessellated into a triangular mesh, as illustrated in Fig. 3.

This mesh determines a three-dimensional boundary in which volume is partitioned into a grid of finite elements (tetrahedrons), for which the volume may be easily

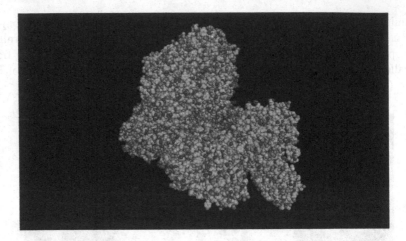

Fig. 2. Atomic structure of GalNAc-T7: each sphere corresponds to a particular atom.

evaluated [7]. Grid generation is challenging because it is intrinsically unstructured and highly complex.

The unstructured grid is generated using the advancing front method (AFM) [7]. The grid is constructed by progressively adding tetrahedral elements (tetrahedrons), starting at the boundary delimited by the ASA. This results in a propagation of a front that may be assimilated into the boundary, which divides the grid regions from the regions without the grid. The main difficulty of this approach resides in the merging of the advancing front. A tetrahedral element is added by inserting a new point, whose location is determined by the quality of the tetrahedron, the desired grid spacing, and neighbourhood constraints.

These constraints imply that none of the created edges intersect any edge/facet of the front, none of the created facets intersect any edge of the front, and the newly created tetrahedron does not engulf any other points. The tetrahedrons must always lie inside the boundary. The process is repeated until the grid is completely merged, such that there is no active front remaining. It should be noted that a high-quality boundary triangulation is required to generate a tetrahedral grid [7]. Once the grid is generated, the volume of the ASA is obtained by summing the volume of the individual tetrahedrons, which has a closed-form expression. The AFM was selected because of its ability to achieve a high-quality grid for complex boundaries [21].

3   Volume Prediction from Amino Acid Sequences

3.1   Amino Acid Sequence Representation

Twenty standard amino acids are found in protein sequences such as alanine, glycine, and tyrosine [16]. An additional unknown amino acid is added to account for unresolved material that may remain after the experimental procedure. The amino acids are characterised by their physicochemical properties, such as polarity, net charge, hydropathy,

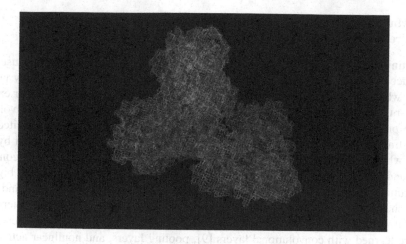

Fig. 3. Tessellation of the ASA associated with GalNAc-T7.

molar absorptivity, and molecular mass, just to mention a few [16]. They form oriented sequences [16], which determine their three-dimensional structure. The latter results from their interaction with their surrounding environment, which is essentially aqueous. Thus, hydropathy plays a key role in structure determination.

Amino acid types are categorical features that correspond to their names, e.g., leucine. As categorical features, they may be encoded by one-hot binary vectors in which all bits are zeros except the one corresponding to a given amino acid. This representation is not particularly informative because most entries are zeros [18], and it may lead to high dimensional sparse data. The amino acids may also be substituted by their corresponding physicochemical properties, which result in numerical features. In both cases, a variable-length descriptor is obtained, because there is a one-hot indicator/physicochemical properties vector associated with each amino acid. A fixed-length descriptor can be obtained by padding shorter sequences with zeros up to a maximum length, but the procedure is both arbitrary and non-informative because of the large number of arbitrary zero elements. For these reasons, we employ the unnormalised bigram frequency distribution for the description [6]. The bigrams consist of all permutations containing exactly two amino acids, which is equal to 420 in this case (permutations of 21 tokens). Therefore, the length of the description is 420, irrespective of the number of amino acids forming the amino acid sequence. The bigrams distribution is more informative because it provides the joint distribution of an amino acid, given the preceding one $p(a_{n-1}, a_n)$, in addition to being related to their conditional probability through Bayes' theorem:

$$p(a_n \mid a_{n-1}) = \frac{p(a_{n-1}, a_n)}{p(a_{n-1})} \tag{1}$$

where a_n is the amino acid occurring in position n. Experimental results demonstrate a posteriori that the bigrams distribution is well suited for characterising the volume of a protein.

3.2 Multi-attention, Multi-resolution Deep Neural Network for Volume Prediction

Predicting the volume from the amino acid sequence is challenging because the sequence does not contain any geometrical information per se. Yet, it is the only viable option when the volume of a large number of proteins must be determined. For example, X-ray crystallography is too expensive and time-consuming. Therefore, volume can be predicted with a novel convolutional deep neural network, whose architecture is illustrated in Fig. 13, shown in the Appendix. This network is characterised by the input, which is constantly reinjected into the output, both locally, with skip connections, and globally, with branching connections. The input to the network is the bigram distribution. The network consists of three similar modules referred to as 1, 2, and 3.

Each module consists of an attention mechanism [19], whose role is to determine the relative importance of the various elements forming the input. The value, query, and key are learned with convolutional layers [9], pooling layers, and nonlinear activation units, i.e., sigmoids. Each convolutional layer consists of 32 channels, and each channel corresponds to a kernel of size 7 (1-D convolutions). The pooling layers consist of a kernel of size 3, and the aggregation function is the maximum function to capture the most prominent value. All these hyperparameters are set by inspection. The role of the convolutional module is to summarise the information of the input while providing a lower-resolution output.

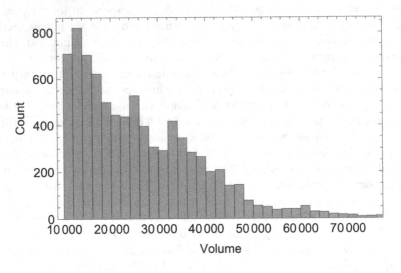

Fig. 4. Volume distribution for the dataset.

Skip connections are associated with each value, query, and key. They are combined at the output of the attention layer with a summation function (#). They make it possible to reinject the input information, which otherwise would be lost, while simultaneously smoothing the energy landscape of the loss function, thus allowing for more efficient training [13]. This module is repeated three times, allowing for a multi-resolution analysis resulting from the repeated application of the convolution modules.

To maintain the high-resolution information, the outputs of each module are combined with a catenation layer, thus taking advantage of low and high-resolution information, concurrently and on equal footing. The long-range connections required to reinject the multi-resolution information are called branching connections. The last part of the network consists of a batch normalisation layer, a dropout layer for regularisation, and two dense layers to extract the volume from the latent information.

The dropout probability was set to 0.1 by inspection, and the output is the predicted volume. The loss function is the mean-squared error function, which measures the discrepancy between the real or actual volume and the predicted one.

4   Experimental Results

4.1   Dataset

To train and validate the proposed approach, Homo sapiens (human) proteins were retrieved from the PDB, which is a curated database of three-dimensional structural data for large biomolecules, such as proteins and amino acids [3]. The process was entirely automated with a Python script filtering for human proteins and retrieving their corresponding crystallographic information file, which is a standard format for representing crystallographic information [1]. For each file, the macromolecular surface was evaluated with the method presented in Sect. 3, using the PyMOL [14] software. These surfaces were subsequently tessellated into a triangular mesh with the same software. Each mesh represents the boundary of the volume enclosed by the ASA. Within this boundary, an unstructured grid was generated with the AFM (Sect. 2), from which the volume enclosed by the boundary was evaluated by summing up the volumes of the individual elements. This was implemented using Mathematica^{TM}. Only the proteins with a volume of more than $10,000$ Å^3 and less than $110,000$ Å^3 were considered. Small proteins are less interesting because they cannot carry large payloads, and large proteins are scarce [4]. As illustrated by Fig. 4, large proteins are relatively rare, making the network training more challenging for that volume range. In total, 8,370 proteins were employed for training and testing.

For each protein, the corresponding bigram distribution was evaluated and an association list, consisting of 8,370 entries, was created from the bigrams distributions and their corresponding volumes. The histogram of the number of proteins given their amino acid count appears in Fig. 5. The number of amino acids varies between 2 and 4,646.

4.2   Results

The network was trained and tested with 10-fold cross-validation: 90% of the elements of the list were selected randomly for training while the rest of the elements, the complement, were employed for testing. This process was repeated ten times. The network was trained with stochastic optimisation using the Adam algorithm [11], whose parameters were determined by inspection. The network was implemented using Mathematica^{TM}.

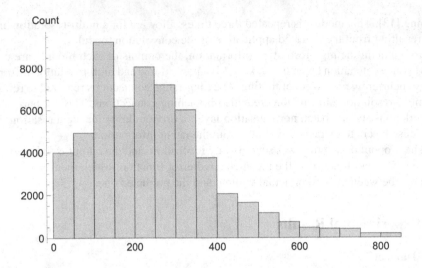

Fig. 5. Histogram of the number of proteins as a function of the number of their amino acids.

The network was trained and tested on a Linux computing node consisting of two Intel Xeon Gold 6130 CPUs at 2.1 GHz, with 16 cores each, two NVIDIA Tesla V100 GPUs with 32 GB of memory each, and 192 GB of RAM. The 10-fold cross-validation took 15 h and 30 min to complete.

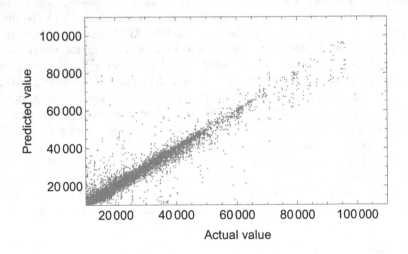

Fig. 6. Predicted volume in relation to the actual volume for all ten folds. The figure consists of 8,370 points.

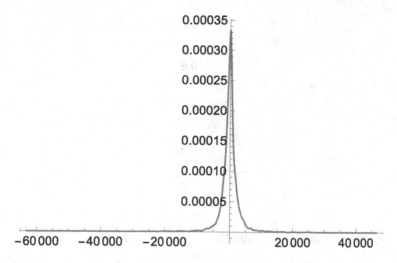

Fig. 7. Probability density function in relation to the residual for all ten folds for the whole domain.

The predicted volumes as a function of the actual volume, for all ten folds, are reported in Fig. 6. This figure consists of 8,370 points. Ideally, the points should be distributed along the main diagonal, with the predicted value being equal to the real one, which is the case for most of them. More often than not, outliers are associated with large volumes. This may be explained by the fact that large volumes are relatively scarce, which makes the training of the network for those values more challenging.

To further ascertain the quality of the results, the probability density function (PDF) of the residuals was evaluated by linearly interpolating smoothing kernels [8], with the residual corresponding to the difference between the actual (real) and predicted values. The corresponding PDF is reported in Fig. 7, which has a very sharp distribution centred on zero. This also demonstrates the high quality of the predictions.

This may be further emphasised by looking at the cumulative distribution function (CDF) [8], as illustrated in Fig. 8. The CDF may be assimilated to a Heaviside step function, which represents the perfect case in which all predicted values are equal to their actual counterparts.

Outliers may be detected, for instance, with the Studentised residual, which is the quotient resulting from the division of a residual by an estimate of its standard deviation [5]. The Studentised residuals for all ten folds are reported in Fig. 9.

The outliers typically have a standardised residual between 1.0 and 1.5. Outliers may be eliminated by learning the distribution of non-anomalous data to subsequently detect anomalies [17]. This involves an acceptance threshold that corresponds to the rarer probability threshold to consider a point anomalous. If the non-anomalous distribution is univariate, continuous, and symmetric, which is the case here, the rarer probability is equal to the two-tailed p-value [Goodman, 2008]. The predicted volumes in relation to the actual volumes, for all ten folds, with anomalies detection threshold set to 0.01, 0.1, and 0.2, are reported in Figs. 10, 11, and 12, respectively.

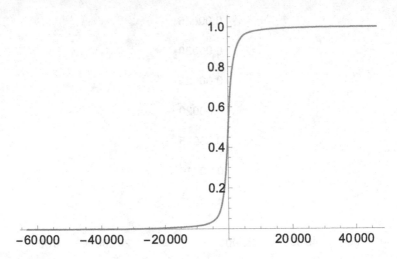

Fig. 8. Cumulative density function in relation to the residual for all ten folds for the entire domain.

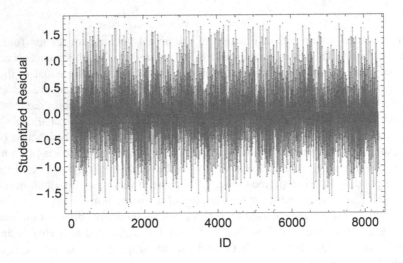

Fig. 9. Studentised residuals for ten folds for all 8,370 points.

These figures reveal that the network performs best for volumes less than 50,000 Å^3. Again, this is not surprising considering that large-volume proteins are relatively scarce. Beyond this point, there are not enough examples to adequately perform the learning task.

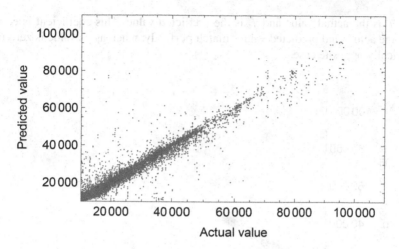

Fig. 10. Predicted volume in relation to the actual volume for all ten folds with anomalies detection with a threshold set to 0.01.

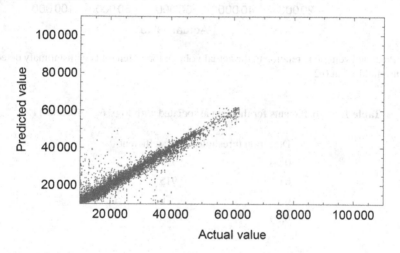

Fig. 11. Predicted volume in relation to the actual volume for all ten folds with anomaly detection, with a threshold set to 0.10.

These results may be summarised by employing the coefficient of determination, denoted by r^2, which is defined as

$$r^2 = 1 - \frac{\sum\limits_{i=1}^{n} (y_i - f_i)^2}{\sum\limits_{i=1}^{n} \left(y_i - \frac{1}{n} \sum\limits_{j=1}^{n} y_j\right)^2} \tag{2}$$

where y_i is the actual value and f_i is the predicted value. This coefficient is equal to 1 when the actual and predicted values match perfectly, whereas its value is zero in the diametrically opposite case.

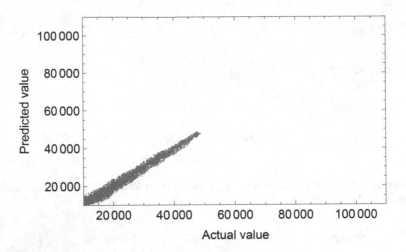

Fig. 12. Predicted volume in relation to the actual volume for all ten folds with anomaly detection, with a threshold set at 0.2.

Table 1. r^2 coefficients for the data associated with Figs. 6, 9, 10, and 11.

Detection threshold	r^2 coefficient
0.00	0.914
0.01	0.915
0.10	0.976
0.20	0.980

This coefficient was evaluated from the data of Figs. 6, 10, 11, and 12, and the results are reported in Table 1. Even without anomaly detection, and with volumes ranging from 10,000 to 110,000 Å^3, the coefficient has a value of 0.914. This value increases to 0.980 if a threshold of 0.2 is applied. Once more, Table 1 illustrates the efficacy of the proposed method.

5   Conclusions and Future Work

A new approach for protein volume prediction from amino acid sequences was proposed based on a multi-attention and multi-resolution deep neural network and FEA. The experimental results have demonstrated the efficiency of the proposed approach.

Volume prediction for very large proteins is more challenging because of data scarcity in that range, as clearly illustrated by Fig. 4. In future work, it is proposed to address this problem. It is also possible to extend the method to other structural properties [12], such as secondary structures.

The tessellated macromolecular surface often presents defects such as holes and inverted triangles. These are unavoidable given the geometrical complexity of the ASA. Small defects may, and were, repaired automatically by ProVolOne, but several surfaces had to be discarded because of the proportion of affected triangles. These defects are troublesome because the ASA acts as a boundary, and as such, it must be watertight (no holes or inverted triangles) for FEA to be applied successfully [7]. Correcting them automatically would allow for application of the approach to the entirety of the PDB, which contains 198,165 experimentally determined structures (22 November 2022). This will be the subject of future work.

Recently, great progress has been made in the prediction of protein three-dimensional structures from their amino acid sequences using frameworks such as AlphaFold [10] and ESMFold [15]. Given a predicted structure, it would be possible to infer its volume with FEA, but it is currently unclear whether the precision reached by these systems is adequate. Furthermore, these algorithms are extremely complex and require massive computing resources for training, whereas the proposed approach is computationally inexpensive and more efficient because the neural network only learns the property of interest, e.g., only the volume but not about the whole protein.

The accuracy of the experimental results has also demonstrated, a posteriori, the key role played by bigrams in volume determination. To the best of our knowledge, this is the first time this observation has been reported.

Acknowledgements. This research was funded by the Artificial Intelligence For Design (AI4Design) Challenge Program from the Digital Technologies Research Centre of the National Research Council (NRC) of Canada.

Appendix

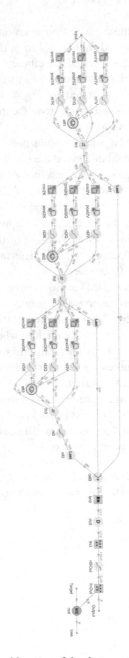

Fig. 13. Architecture of the deep neural network.

References

1. Adams, P.D., et al.: Announcing mandatory submission of PDBx/mmCIF format files for crystallographic depositions to the Protein Data Bank (PDB). Acta Crystallogr. Sect. D **75**(4), 451–454 (2019)
2. Aslanidi, G., et al.: Optimization of the capsid of recombinant adeno-associated virus 2 (aav2) vectors: the final threshold? PLoS One **8**(3) (2013). https://doi.org/10.1371/journal.pone.0059142
3. Burley, S., et al.: RCSB protein data bank: powerful new tools for exploring 3D structures of biological macromolecules for basic and applied research and education in fundamental biology, biomedicine, biotechnology, bioengineering and energy sciences. Nucleic Acids Res. **49**(1), D437–D451 (2021). https://doi.org/10.1093/nar/gkaa1038
4. Bylaklak, K., Charles, A.G.: The once and future gene therapy. Nat. Commun. **11**, 1–4 (2020)
5. Celik, R.: RCEV heteroscedasticity test based on the studentized residuals. Commun. Stat. Theory Methods **48**(13), 3258–268 (2019)
6. Chandra, A.A., Sharma, A., Dehganzi, A., Tsunoda, T.: Evolstruct-phogly: incorporating structural properties and evolutionary information from profile bigrams for the phosphoglycerylation prediction. BMC Genomics 984–992 (2019)
7. Chung, T.J.: Computational Fluid Dynamics. Cambridge University Press, Cambridge, UK (2010)
8. Gelman, A., Carlin, J.B., Stern, H.S., Dunson, D.B., Vehtari, A., Rubin, D.B.: Bayesian Data Analysis. CRC Press, Boca Raton, FL (2013)
9. Gu, J., et al.: Recent advances in convolutional neural networks. Pattern Recogn. **77**, 354–377 (2018)
10. Jumper, J., et al.: Highly accurate protein structure prediction with AlphaFold. Nature **596**, 583–589 (2021)
11. Kingma, D., Ba, J.: Adam: a method for stochastic optimization, December 2014. ArXiv: 1412.6980
12. Kuhlman, B., Bradley, P.: Advances in protein structure prediction and design. Nature **20**, 681–697 (2019)
13. Li, H., Xu, Z., Taylor, G., Studer, C., Goldstein, T.: Visualizing the loss land-scape of neural nets. In: 32nd Conference on Neural In-formation Processing Systems, p. 11, Montréal, Canada, December 2018
14. Lill, M.A., Danielson, M.L.: Computer-aided drug design platform using PyMOL. J. Comput. Aided Mol. Des. **25**, 13–19 (2011)
15. Lin, Z., et al.: Language models of protein sequences at the scale of evolution enable accurate structure prediction. BioRxiv 2022.07.20.500902 (October 2022)
16. Lovric, J.: Introducing Proteomics: From Concepts to Sample Separation, Mass Spectrometry and Data Analysis. John Wiley & Sons, Oxford, UK (2011)
17. Pang, G., Shen, C., Cao, L.: Deep learning for anomaly detection: a review. ACM Comput. Surv. **54**((2)38), 1–38 (2022)
18. Rodríguez, P., Bautista, M.A., Gonzàlez, J., Escalera, S.: Beyond one-hot encoding: Lower dimensional target embedding. Image Vis. Comput. **75**, 21–31 (2018)
19. Vaswani, A., et al.: Attention is all you need. In: 31st Conference on Neural Information Processing Systems (Neurips 2017), December 2017
20. Xu, D., Zhang, Y.: Generating triangulated macromolecular surfaces by Euclidean distance transform. PLoS ONE **4**(12) (2009)
21. Zienkiewicz, O.C., Taylor, R.L., Fox, D.D.: The Finite Element Method for Solid and Structural Mechanics. Elsevier, London, UK (2013)

A Data-Driven Monitoring Approach for Diagnosing Quality Degradation in a Glass Container Process

Maria Alexandra Oliveira[✉][ID], Luís Guimarães[ID], José Luís Borges[ID], and Bernardo Almada-Lobo[ID]

Faculty of Engineering, University of Porto, 4200-465 Porto, Portugal
{alexandra.oliveira,lguimaraes,jlborges,almada.lobo}@fe.up.pt

Abstract. Maintaining process quality is one of the biggest challenges manufacturing industries face, as production processes have become increasingly complex and difficult to monitor effectively in today's manufacturing contexts. Reliance on skilled operators can result in suboptimal solutions, impacting process quality. In doing so, the importance of quality monitoring and diagnosis methods cannot be undermined. Existing approaches have limitations, including assumptions, prior knowledge requirements, and unsuitability for certain data types. To address these challenges, we present a novel unsupervised monitoring and detection methodology to monitor and evaluate the evolution of a quality characteristic's degradation. To measure the degradation we created a condition index that effectively captures the quality characteristic's mean and scale shifts from the company's specification levels. No prior knowledge or data assumptions are required, making it highly flexible and adaptable. By transforming the unsupervised problem into a supervised one and utilising historical production data, we employ logistic regression to predict the quality characteristic's conditions and diagnose poor condition moments by taking advantage of the model's interpretability. We demonstrate the methodology's application in a glass container production process, specifically monitoring multiple defective rates. Nonetheless, our approach is versatile and can be applied to any quality characteristic. The ultimate goal is to provide decision-makers and operators with a comprehensive view of the production process, enabling better-informed decisions and overall product quality improvement.

Keywords: Quality Monitoring · Quality Diagnosis · Condition Index · Manufacturing

1 Introduction

In manufacturing industries today, maintaining process quality is a major challenge due to the increasing complexity of production processes and the difficulty

Supported by FEUP-PRIME program in collaboration with BA GLASS PORTUGAL.

G. Nicosia et al. (Eds.): LOD 2023, LNCS 14505, pp. 288–302, 2024.
https://doi.org/10.1007/978-3-031-53969-5_22

in effectively monitoring them. Despite having more data from the shop floor, relying on skilled operators to ensure product quality can lead to suboptimal solutions and ultimately poor process quality. This emphasizes the importance of quality monitoring and diagnosis methods. To improve process quality, measuring and assessing the process Quality Characteristic (QC) to monitor its progress is crucial. When the measured QC deviates from the desired levels, input process variables should be adjusted. Understanding how these input variables influence the output variable is essential for successfully achieving this task.

A process monitoring system, or Fault Detection and Diagnosis (FDD) system, employs data-driven methods to identify and diagnose anomalous deviations from normal operating conditions. The foundation of most basic monitoring schemes lies in control charts, which have evolved to develop more efficient and advanced monitoring schemes. However, there are still opportunities for improvement in this field. According to a review by [8], most FDD studies heavily rely on simulated data, disregarding important aspects of real-world problems and hindering their practical application. Additionally, many methods in this field rely on assumptions about data distribution, which are challenging to determine in practice. Moreover, collecting labelled In-Control (IC) data is often assumed as a prerequisite, adding to the difficulties. Furthermore, existing diagnosis approaches either lack interpretability or require prior knowledge of process variables.

To address these problems and driven by the challenges encountered in our application in a glass container production process, we propose a novel quality monitoring and diagnosis approach inspired by FDD methods. In this paper, we present a general framework to measure and monitor the condition degradation of any QC and detect when it deviates from the desired operation levels to identify the need for interventions. Deviations are signalled every time a scale and/or location shift are detected. Secondly, as the originally unsupervised problem was transformed into a supervised one, it is intended to correlate the process parameterisation with the condition assessment of the QC for workers also know where to intervene. This is achieved by training a Logistic Regression (LR) classification model and interpreting its results. The approach has the ultimate goal of supporting the operator's decision-making regarding when and where to intervene and standardising the method of quality assessment to avoid misjudgements and unnecessary adjustments. It is worth mentioning that the presented methodology is not a control scheme since the assessment does not account for statistical stability. Instead, we present a monitoring scheme that provides information on the current condition of the analysed signal and identifies its critical moments. Thereby, IC and Out-Of-Control (OOC) nomenclature will not be used while referring to our approach. The proposed methodology addresses the above-mentioned problems as follows:

- The monitoring and detection approach is based on a created condition measurement - the condition index (CI)- that considers user-defined targets so quality assessment is performed according to the company's goals. The CI reflects any QC's condition at two levels. In the worse condition, an alarm is

raised. In this way, the user is provided with a more straightforward interpretation of the current state and evolution of the QC. To the best of our
knowledge, there are no approaches for assessing quality in this way;

- The *CI* is simple to calculate and independent of the data type and underlying
  distribution;
- It is unnecessary to have high-quality labelled training data as the approach
  is unsupervised;
- The diagnosis method is enhanced by its reliance on a linear model, providing an increased level of interpretability that proves indispensable in various
  practical applications.

To the literature, our method's contributions can be summarised as follows:

- The application's QCs are defective rates of a glass container production
  process (i.e. the proportion of nonconforming). These QCs types are less likely
  to be found in a monitoring and control study. Nevertheless, the methodology
  can be replicated in another industry domain and for a different QC;
- We proposed a one-sided monitoring scheme instead of the usual two-sided
  approach since the approach was conceived for monitoring defective rates;
- The diagnosis methodology does not demand prior knowledge of process variables and their relationships;
- By adapting existing methods to this new problem setting, the study offers
  valuable contributions to the scientific community in practical applications
  within the quality monitoring and diagnosis field.

The rest of the paper is organised as follows. First, in Sect. 2, we formally
define the problem on hand and present its main challenges. Next, FDD techniques are briefly surveyed in Sect. 2.3. Then, in Sect. 3, the monitoring and
diagnosis approaches are described. Section 4 demonstrates an actual application on a glass container production process and its use cases. Finally, Sect. 5
ends with conclusions.

2   Problem Description

This section starts by describing the glass container production process, followed
by an exposition of the application's challenges.

2.1   The Glass Container Production Process

The glass container production begins with a mixture of raw materials, which is
transported to a furnace to melt. After leaving the furnace, the liquefied glass
is conducted to a distribution channel and subjected to a conditioning process
that ensures the glass paste's thermal homogeneity and consequently provides
the glass with the desired viscosity. At the end of the conditioning process, the
glass paste falls into a feeder mechanism, which aims to cut the glass into gobs
with the right shape and deliver them to a moulding machine. The cut gobs

are distributed to a set of parallel Independent Section (IS) machines, which transform them into containers. The IS machines typically have 6 to 20 sections working independently, achieving high production rates. The primary function of the IS machines is to mould the gob into a bottle. The IS machines' electric timing systems synchronously control servo-electric, pneumatic, and hydraulic movements. The timing and duration of all these movements are controlled by adjustable parameters that allow managing the moulding process in each section individually. Per section, there are approximately 80 adjustable variables. After being given a shape, the containers are submitted to a hot surface treatment to increase their mechanical resistance. The containers are then subjected to strict quality control, performed by automated Inspection Machines (IM) capable of rejecting defective containers by various defect types. The IM also provide a production report where indicators such as rejection and rejection rates per type of defect are presented. The containers approved in the quality control process are packed on pallets at the end of the production lines.

This manufacturing process can also be regarded as a system with inputs and outputs, as depicted in Fig. 1. The controllable inputs are usually process variables, such as temperatures, pressures, and speeds. In turn, the uncontrollable inputs, like the external temperature and humidity, are the parameters that can not be changed. The production process transforms the input raw materials into finished products with several QCs - the output variables.

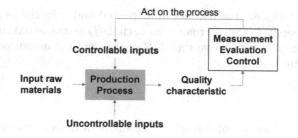

Fig. 1. Production process seen as a system, adapted from [13].

In general, every time the QC deviates from its desired levels, one should act on the process by adjusting the controllable variables while accounting for the influence of the uncontrollable ones. Yet, given that the stages of the glass container production process are sequential and have an associated duration, the value of the outputs at one time instant cannot be directly related to the value of the variables at the same time instant. This implies that the adjustments on the controllable variables do not immediately impact the process outputs as they depend on the stage where they were made. Figure 2 illustrates this situation. For example, suppose an adjustment is performed on the controllable variables of the first stage. In that case, its impact on the process outputs can only be evaluated after a time equivalent to at least the sum of all stages' duration.

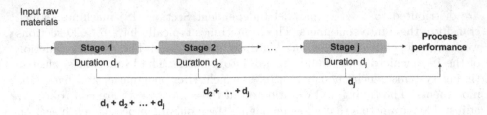

Fig. 2. Representation of the impact of the adjustments of the process variables on the system outputs

2.2   Challenges to Monitoring and Diagnose Glass Manufacturing Quality Output

The primary goal of the company is to minimise the company's quality production losses. As so, the system's QC, as a measure of the product quality, is the proportion of nonconforming containers, i.e. the defective rate of the process. However, as the IM detects several defect types, the system has multiple QCs, measured at each time instant t, represented by the defective rate DR_t of the defect d, DR_{dt}:

$$DR_{dt} = \frac{D_{dt}}{Insp_t} \tag{1}$$

where $Insp_t$ and D_{dt} are the number of inspected and defective containers measured on t, respectively. Furthermore, as each DR_d is measured in each section of the IS machines, the defective rate DR of the defect d measured on section s on t is defined by:

$$DR_{dst} = \frac{D_{dst}}{Insp_{st}} \tag{2}$$

where D_{dst} is the number of defective containers of defect type d measured on section s on t.

There are six macro categories of defects measured in each section. Table 1 displays such categories accompanied by an example. Figure 3 shows the nomenclature of the principal areas of a glass container to facilitate the interpretation of the defect types.

As mentioned earlier, the defective rates are the QCs for this production process, as quality cannot be evaluated solely based on a single indicator but rather a combination of defect types. However, these QCs are initially lacking in clear assessment or definition criteria that would unequivocally indicate a high or low-quality process. This lack of clarity and standardisation among operators with varying experience levels often leads to misjudgements and unnecessary adjustments. Furthermore, the adjustments made to the process parameterisation play a crucial role in determining the process quality. However, the quality of the process is also influenced by a dynamic and uncontrollable environment, which results in different outcomes despite using the same parameterisation at

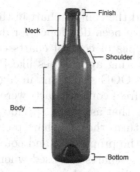

Fig. 3. Nomenclature of the glass container areas

Table 1. Groups of defects of the glass containers at the company

Defect group	Example
Checks (Cracks)	Check in the body
Thickness nonconformities	Uneven glass distribution
Dimensional nonconformities	Deformed neck
Top nonconformities	Broken finish
Bottom nonconformities	Flanged bottom
Sidewall nonconformities	Wrinkles

different moments. This inherent complexity also contributes to operators' difficulty in replicating the best historical production cycles. As a result, optimising the process often involves a trial-and-error approach, which incurs costs in terms of time and materials. Additionally, the impact of these adjustments may only become noticeable within one hour to one day later. Consequently, promptly responding to quality problems in real-time becomes challenging.

2.3 Related Work

An FDD system uses data-driven methods to identify and characterise abnormal deviations from normal operating conditions. These techniques rely on historical process data for creating monitoring systems, making them preferred in industrial applications due to their speed, ease of implementation, and minimal requirement for prior knowledge [14]. A common data-driven process monitoring scheme is the control chart, which aims to quickly detect shifts in location or scale on a QC, allowing for prompt remedial actions [13]. The control chart is typically implemented in two phases. Phase I involves retrospectively analysing historical data to assess process stability and IC model parameters for future production monitoring [13]. In Phase II, real-time data is used to detect process deviations from the estimated IC model [22] and proceed with fault diagnosis to identify and isolate the root cause [21].

Since Shewhart introduced the control chart in the 1900s, various univariate and multivariate methods have been developed for detecting process upsets. As research evolved, different types of control charts emerged to address specific needs. Latent variable monitoring approaches like Principal Component Analysis (PCA) effectively detect OOC situations in high-dimensional systems [9]. Nonparametric (distribution-free) control charts were developed to overcome the limitations of traditional ones that assume a specific probability distribution for process behaviour [3,11]. In turn, the Shewhart p-chart and np-chart are commonly employed to monitor the proportion and count of non-conforming products, respectively [4,15]. These charts are used when certain QCs of a product cannot be expressed numerically. In such cases, an attribute-based classification system can be used to determine if items are conforming or non-conforming [4]. Nevertheless, FDD applications with attribute data are less common.

Nevertheless, with vast quantities of data being collected in modern industrial processes, traditional control chart applications are often ineffective due to unknown data distributions and the difficulty in describing process variables and correlations among them. As a result, the popularity of Machine Learning (ML) techniques has led to the growth of FDD applications using these techniques. For the monitoring and detection, [5] and [25] used the K-Nearest Neighbour to monitor and detect faults in a refrigerant compression process and on an industrial pyrolysis furnace, respectively. Other authors applied the One-Class Support Vector Machines (SVM) [20] to detect typical chiller faults, and [18] used Neural Networks (NNs) to detect abnormal equipment performance. Deep learning approaches have also been successfully used for fault detection, as in [19] and [23].

Regarding the diagnostic task, the PCA-based monitoring method is very popular since it provides a contribution plot that indicates potential relationships between the OOC signal and the original variables that may have caused it. However, it suffers from the "smearing-out effect", where faulty variables may be omitted [16]. Supervised ML approaches like SVM [7], deep belief network [24], and Artifical NN [6] are used to overcome the difficulty of diagnosing faults in complex systems. They are often used to make more conclusive diagnoses when a sufficiently large history record with defect labels is available. Besides, classical multivariate methods also suffer from low interpretability and lack of causal relationships. In this sense, structured approaches have been proposed. They integrate the process-specific structure (causal connections) to improve diagnostic speed [17]. Examples of structured approaches include transfer entropy [10], Granger causality [12], causal maps [2], and Bayesian networks [21].

All revised approaches evaluate the QC's statistical stability but do not necessarily ensure adherence to customer standards. Process stability, achieved through control charts, aims to produce consistent and predictable output. However, a process can be IC while consistently producing nonconforming parts that do not meet customer requirements [1]. This is the philosophy on which process capability indexes are based since they measure a process's ability to meet specifications, focusing on customer needs.

Furthermore, according to [8], only 12% of the FDD studies from 2017–2021 applied the methods to real industrial processes. Monitoring industrial processes is challenging due to unique characteristics and varying operating conditions in industrial settings. While many studies rely on simulated cases and benchmark datasets, practical applications demand interpretable and actionable results, particularly in manufacturing. Often, practical applications offer new perspectives, while some works overlook crucial factors, rendering them inapplicable.

Finally, current approaches necessitate knowledge of the key process variables that significantly impact the QC under monitoring. This is essential not just for diagnostic purposes but also for effectively monitoring the QC. However, selecting these critical variables can be challenging, particularly in complex production processes with numerous variables. Moreover, when monitoring multiple QCs of the same type (e.g., various defective rates) where the same process variables are involved, linking their deviations directly with different system targets becomes increasingly difficult.

3 Methods

This section describes the monitoring and detection approach and the diagnosis methodology.

3.1 Monitoring and Detection Approach

The goal of this phase e to develop a monitoring scheme that measures the condition of a given QC and detects when such a condition is degrading. The proposed monitoring and detection technique employs a Condition Index (CI) to measure the QC's condition by locating the attained index on an ordinal scale that informs on criticality occurrence. The CI considers any period's variability and how it is positioned regarding two parameters. These parameters must be defined beforehand by the user. The two parameters are:

- Upper Limit (UL): represents the maximum acceptable value the QC can reach. For example, in a production process, the organisation must be capable of defining the maximum allowable value a defective rate may reach;
- Target (T): represents the desired value of the QC, the value around which it should vary. For instance, in a production process set the defective rate to an achievable value, as a zero defective rate is impractical to reach.

Once these two parameters are specified, the condition measurement formula of the CI, which was inspired by the one-sided C_p process capability index (see [1] Chap. 5), can be formally defined. Let $X_i = \{x_{i;t}, x_{i;t+1}, ..., x_{i;t+n-1}\} \ \forall i, n, t, \in \mathbb{N}$ denote the n observations of the i-th time-series sample of the monitored QC X. For each sample i of X, the CI is calculated as follows:

$$CI_i = \frac{UL - \bar{x}_i}{s_i} \times (1 - \frac{\bar{x}_i - T}{UL - T})^2 \tag{3}$$

where $\bar{x}_i$ and s_i are the sample mean and standard deviation values of X_i, respectively. Figure 4 illustrates an example, applied on a defective rate as a QC, to support the interpretation of (3). In period A, $\bar{x}_A < T, UL$. According to (3), the first term is positive, while the second term, which is always positive, is greater than one. As a result, even if the period's high variability s_A decreases the first term, the second term will increase $C!I_A$. In period B, $T < \bar{x}_B < UL$. Thus, the first term is positive, and the second is less than one. As so, the second term decreases CI_B, penalising the period for being above T. Finally, in period C, $\bar{x}_C > T, UL$. In doing so, CI_C becomes negative due to the first term. Additionally, as the second term will be less than one because the period's mean is very distant from T, CI_C will even decrease more. In summary, higher $C!I$ values indicate a better condition, while lower values indicate a deteriorating condition. As so, when $\bar{x}_i < T$, the CI only penalises for the variability shifts - scale shifts - since the sample is already located favourably. On the other hand, when $\bar{x}_i > T$, the CI also penalises for the departure of the sample mean from T - mean or location shifts.

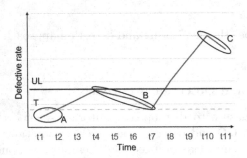

Fig. 4. Example to illustrate how the signalled periods are assessed by the CI formula

Yet, to map the CI values into an ordinal and more interpretable scale, one must define the CI value below which the condition is considered undesirable. In other words, defining a threshold th capable of detecting small to large shifts, both in scale and location, is necessary. The most straightforward assessment scale is composed of two levels:

$$CI_i = \begin{cases} Good & \text{if } CI_i \geq th \quad (4a) \\ Poor & \text{otherwise} \quad (4b) \end{cases}$$

A good condition represents a desirable state of the QC, and a poor condition expresses an alarming state where some intervention is required. The CI maps the QC's values into a condition measurement. Figure 5 exemplifies the result of such mapping with the previous example of the defective rate. The defective rate is the solid blue line, and the CI is the dashed orange line. As the defective rate increases, its corresponding condition starts to degrade (i.e. decrease) and vice-versa.

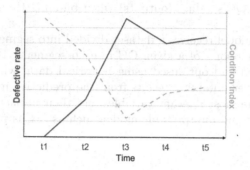

Fig. 5. Exemplifying image of the mapping of a QC's values to a measurement of its condition (Color figure online)

The threshold, th, is essential for correctly implementing the proposed monitoring and detection approach, as the methodology's performance depends on its accuracy. The method used to calculate th was inspired by the procedure used to design control charts. In practice, researchers resort to Monte Carlo simulations to search for the control limit(s) value(s) that best minimises the false alarm rate. We relied on the same principle for the presented approach to tuning th.

3.2 Diagnosis Approach

In this phase, the aim is to associate the production process parameterisation (controllable variables) with the condition attributed to a QC by the monitoring and detection methodology previously described. The goal is to learn about the combination of process parameters that usually result in poor or good conditions. We applied an LR model that leverages process parameterisation to predict and characterise the QC's condition to meet this objective. This is only possible because the monitoring and detection methodology provided labels to the QC values. The proposed approach is divided into three parts: i) data processing, ii) development and training of the forecasting model, and iii) characterisation of the QC's condition.

The data processing stage encompasses two tasks: feature selection and engineering and data transformation. The former task aims to reduce dimensionality and to identify context features to improve predictive performance. To reduce the problem's dimensionality, a Pareto analysis is used to select the process variables responsible for 80% of the variables' value changes. As the QC condition does not depend only on the process parameterisation, context variables (e.g. maintenance activities) were also included to explain the variation the parameterisation cannot capture. In turn, the data transformation task aggregates consecutive records with constant process parameter values into segments for predicting the QC condition. To each segment, one or two conditions may be associated. For this reason, it is important to include the context variables that will explain this variation. Moreover, one must consider the time lag between the inputs and the QC's values (outputs) when building the dataset, as shown in Fig. 2.

The predictive model receives a dataset divided into segments to predict the condition - good or poor - of a given QC in each segment. To achieve this, an LR model is trained and optimised using historical data, which is then tested on unseen data. After obtaining results from the predictive model, one can use them to identify the most relevant variables and their impact on the condition prediction. A linear and interpretable model, such as LR, is preferred whenever possible.

4   Results

This section illustrates the implementation of the proposed approach with actual data from the glass container production process described in Sect. 2. The methodology was applied to each DR_d on the 12 sections of the company's IS machine. In doing so, six defective rates were measured every hour in 12 sections, summing 72 QCs. The data of each of these 72 QCs were sampled every hour for 3 months. The value of th was calculated for each DR_d since, for the same defect, the sections share the same value of th.

As 72 QCs were monitored, we developed three use cases to give the operators a more straightforward interpretation of the results. The first use case gives the workers a global perspective of the system state. The colour map depicted in Fig. 6 concerns the condition of each DR_d (vertical axis) assessed in all sections every hour (horizontal axis) - DR_{dst}. At each hour t, for each DR_d, if the condition is classified as poor in i) all sections, the colour is red; ii) a subset of sections, the colour is yellow; iii) one section, the colour is light green. In turn, if a good condition is attributed to all sections of DR_{dt}, the colour is dark green. For simplicity, Fig. 6 only illustrates the system state for 13 h.

The second use case offers a more local perspective. The colour map represented in Fig. 7 shows the CI values (without discretising into poor and good) of each D_{dst}. Blue and orange tones are associated with good and poor conditions, respectively. For simplicity, Fig. 7 only illustrates the condition of one D_d in each section for 13 h.

Finally, in a more detailed perspective, Fig. 8 showcases the application of the proposed monitoring scheme in one of the six defective rates measured in one section, DR_{ds}. Figure 8(a) displays the evolution of the CI of DR_{ds}, where the horizontal borderline is the value of th. Above that line are the values of the CI for which the condition is good, and below are values of the CI for which condition is poor. In turn, Fig. 8(b) illustrates directly the evolution of DR_{ds} highlighted with colours according to the condition detected by our method: green for good and red for poor. For DR_{ds}, the company defined $UL = 0.005$ and $T = 0.003$. From the simulation, we obtained $th = 0.812$. The magenta

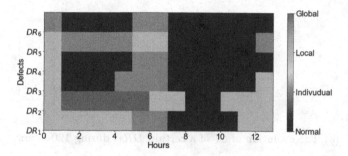

Fig. 6. Colour map of the global condition of each DR_d during 13 h (Color figure online)

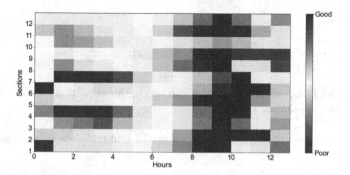

Fig. 7. Colour map of the CI values of one DR_d measured in each section during 13 h (Color figure online)

horizontal line is the UL, while the blue horizontal dashed line is the T. For simplicity, Fig. 8 only illustrates this scenario for 120 h.

Table 2 displays the prediction performance metrics achieved on the test data by applying the LR model to predict the condition of a given DR_d. To build the dataset, we combined the data from all sections of DR_d. However, the dataset suffered from a significant class imbalance, containing many segments associated with good conditions. To address this issue, we employed an undersampling strategy to balance the classes, ensuring fair representation during training. The training data comprised 80% of the balanced dataset.

Table 2. Results of LR applied on test data

Condition	Precision	Recall
Poor	0.95	0.76
Good	0.79	0.96

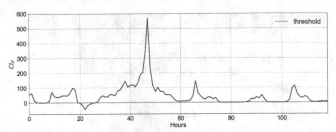

(a) The evolution of CI of a certain DR_{ds} during 120 hours

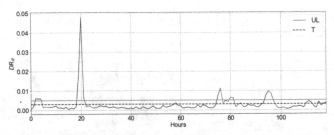

(b) DR_{ds} with the colours signalling the condition of the process during 120 hours

Fig. 8. The results of applying the proposed method on real data (Color figure online)

Figure 9 illustrates the impact's magnitude of the coefficients of the first ten most important variables for the LR prediction results. A positive coefficient indicates a variable that predicts the class "Good condition", whereas a negative coefficient indicates a variable that predicts the class "Poor condition". We do not show the variables' names and coefficient values for confidentiality reasons.

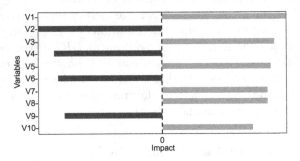

Fig. 9. Impact's magnitude of the coefficients of the first ten most important variables provided by LR

Relevant prediction results were obtained with a simple and interpretable model. This way, it is possible to clearly interpret the process variables' influence and combine this knowledge with skilled operators' experience to understand the relationships better.

5   Conclusions

This work presents a novel monitoring and diagnosis application of a glass container manufacturing process. By adapting existing methods to this new problem setting, the study offers valuable contributions to the scientific community in practical applications within the quality monitoring and diagnosis field. The methodology aims to measure and assess the condition degradation of any QC and provide timely signals when poor conditions are detected. Then, the process parameterisation is correlated with the most critical detected moments resorting to an LR model. Real data from the glass container production process was used to illustrate the methodology application, showcasing how the methodology results can be utilised to assess quality production losses at different levels. This method provides operators with a more straightforward evaluation of the most significant defective rates of the production process, along with a diagnosis of the condition degradation causes. Consequently, it enables informed intervention decisions, reducing process variability from operator experiments and minimising problem-solving time. Additionally, it generates actionable knowledge for further improvement actions and can be easily adapted for other QCs of any type.

References

1. Bothe, D.R.: Measuring Process Capability: Techniques and Calculations for Quality and Manufacturing Engineers. McGraw-Hill (1997)
2. Chiang, L.H., Jiang, B., Zhu, X., Huang, D., Braatz, R.D.: Diagnosis of multiple and unknown faults using the causal map and multivariate statistics. J. Process Control **28**, 27–39 (2015)
3. Chong, Z.L., Mukherjee, A., Khoo, M.B.: Some distribution-free Lepage-type schemes for simultaneous monitoring of one-sided shifts in location and scale. Comput. Ind. Eng. **115**, 653–669 (2018)
4. Chukhrova, N., Johannssen, A.: Improved control charts for fraction nonconforming based on hypergeometric distribution. Comput. Ind. Eng. **128**, 795–806 (2019)
5. Ha, D., Ahmed, U., Pyun, H., Lee, C.J., Baek, K.H., Han, C.: Multi-mode operation of principal component analysis with k-nearest neighbor algorithm to monitor compressors for liquefied natural gas mixed refrigerant processes. Comput. Chem. Eng. **106**, 96–105 (2017)
6. Heo, S., Lee, J.H.: Fault detection and classification using artificial neural networks. IFAC-PapersOnLine **51**(18), 470–475 (2018)
7. Hu, H., He, K., Zhong, T., Hong, Y.: Fault diagnosis of FDM process based on support vector machine (SVM). Rapid Prototyping J. **26**, 330–348 (2019)
8. Ji, C., Sun, W.: A review on data-driven process monitoring methods: characterization and mining of industrial data. Processes **10**(2), 335 (2022)

9. Kumar, A., Bhattacharya, A., Flores-Cerrillo, J.: Data-driven process monitoring and fault analysis of reformer units in hydrogen plants: industrial application and perspectives. Comput. Chem. Eng. **136**, 106756 (2020)

10. Lee, H., Kim, C., Lim, S., Lee, J.M.: Data-driven fault diagnosis for chemical processes using transfer entropy and graphical lasso. Comput. Chem. Eng. **142**, 107064 (2020)

11. Li, C., Mukherjee, A., Su, Q.: A distribution-free phase i monitoring scheme for subgroup location and scale based on the multi-sample Lepage statistic. Comput. Ind. Eng. **129**, 259–273 (2019)

12. Liu, Y., Chen, H.S., Wu, H., Dai, Y., Yao, Y., Yan, Z.: Simplified granger causality map for data-driven root cause diagnosis of process disturbances. J. Process Control **95**, 45–54 (2020)

13. Montgomery, D.C.: Introduction to Statistical Quality Control. Wiley, Hoboken (2020)

14. Nor, N.M., Hassan, C.R.C., Hussain, M.A.: A review of data-driven fault detection and diagnosis methods: applications in chemical process systems. Rev. Chem. Eng. **36**(4), 513–553 (2020)

15. Quinino, R.D.C., Cruz, F.R., Ho, L.L.: Attribute inspection control charts for the joint monitoring of mean and variance. Comput. Ind. Eng. **139**, 106131 (2020)

16. Reis, M.S., Gins, G.: Industrial process monitoring in the big data/industry 4.0 era: from detection, to diagnosis, to prognosis. Processes **5**(3), 35 (2017)

17. Reis, M.S., Gins, G., Rato, T.J.: Incorporation of process-specific structure in statistical process monitoring: a review. J. Qual. Technol. **51**(4), 407–421 (2019)

18. Sun, J., Zhou, S., Veeramani, D.: A neural network-based control chart for monitoring and interpreting autocorrelated multivariate processes using layer-wise relevance propagation. Qual. Eng. 1–15 (2022)

19. Sun, W., Paiva, A.R., Xu, P., Sundaram, A., Braatz, R.D.: Fault detection and identification using Bayesian recurrent neural networks. Comput. Chem. Eng. **141**, 106991 (2020)

20. Yan, K., Ji, Z., Shen, W.: Online fault detection methods for chillers combining extended Kalman filter and recursive one-class SVM. Neurocomputing **228**, 205–212 (2017)

21. Yang, W.T., Reis, M.S., Borodin, V., Juge, M., Roussy, A.: An interpretable unsupervised Bayesian network model for fault detection and diagnosis. Control. Eng. Pract. **127**, 105304 (2022)

22. Zhang, J., Li, E., Li, Z.: A cramér-von mises test-based distribution-free control chart for joint monitoring of location and scale. Comput. Ind. Eng. **110**, 484–497 (2017)

23. Zhang, Z., Jiang, T., Li, S., Yang, Y.: Automated feature learning for nonlinear process monitoring-an approach using stacked denoising autoencoder and k-nearest neighbor rule. J. Process Control **64**, 49–61 (2018)

24. Zhang, Z., Zhao, J.: A deep belief network based fault diagnosis model for complex chemical processes. Comput. Chem. Eng. **107**, 395–407 (2017)

25. Zhu, W., Sun, W., Romagnoli, J.: Adaptive k-nearest-neighbor method for process monitoring. Ind. Eng. Chem. Res. **57**(7), 2574–2586 (2018)

Exploring Emergent Properties of Recurrent Neural Networks Using a Novel Energy Function Formalism

Rakesh Sengupta[1]([✉])[ID], Surampudi Bapiraju[2][ID], and Anindya Pattanayak[3][ID]

[1] Center for Creative Cognition, SR University, Warangal, India
`rakesh.sengupta@sru.edu.in`
[2] Cognitive Science Lab, IIIT Hyderabad, Hyderabad, India
[3] School of Commerce, XIM University, Bhubaneswar, India

Abstract. The stability analysis of dynamical neural network systems typically involves finding a suitable Lyapunov function, as demonstrated in Hopfield's famous paper on content-addressable memory networks. Another approach is to identify conditions that prevent divergent solutions. In this study, we focus on biological recurrent neural networks (bRNNs), specifically the Cohen-Grossberg networks that require transient external inputs. We propose a general method for constructing Lyapunov functions for recurrent neural networks using physically meaningful energy functions. This approach allows us to investigate the emergent properties of the recurrent network, such as the parameter configuration required for winner-take-all competition in a leaky accumulator design, which extends beyond the scope of standard stability analysis. Furthermore, our method aligns well with standard stability analysis (ordinary differential equation approach), as it encompasses the general stability constraints derived from the energy function formulation. We demonstrate that the Cohen-Grossberg Lyapunov function can be naturally derived from the energy function formalism. Importantly, this construction proves to be a valuable tool for predicting the behavior of actual biological networks in certain cases.

Keywords: Cohen-Grossberg neural networks · Lyapunov function · Recurrent networks · Stability · Winner-take-all

1 Introduction

Recurrent neural networks (RNNs) are comprised of interconnected neurons that incorporate feedback loops between nodes. This feedback can originate from the same or different nodes at each time step, resulting in a network behavior that resembles that of nonlinear dynamical systems. RNNs have found applications in various domains, including the construction of neural models for memory [1], decision making [2], and the visual sense of numbers [3]. In dynamic vision algorithms, recurrent neurons are often employed as fundamental building blocks due

G. Nicosia et al. (Eds.): LOD 2023, LNCS 14505, pp. 303–317, 2024.
https://doi.org/10.1007/978-3-031-53969-5_23

to their ability to integrate and propagate local and global influences through-out the network [4]. Biological RNNs (bRNNs) are typically conceptualized as a single layer of neurons.

The utilization of on-center off-surround recurrent networks has gained sig-nificant popularity in the literature, particularly in the domains of short-term memory, decision making, contour enhancement, pattern recognition, and vari-ous other problem domains [21, 23]. These networks are valued for their versatil-ity and ability to generate self-organized outputs based on the inherent nonlinear mathematical properties of the network. Traditionally, ensuring stability in such networks has relied on the identification of a suitable Lyapunov function or the determination of conditions under which network trajectories do not diverge [5, 6]. A comprehensive examination of Lyapunov and other stability approaches can be found in [7].

In this paper, we present a novel and intuitive general-purpose method for constructing a Lyapunov function applicable to recurrent networks in gen-eral. We demonstrate the effectiveness of this formalism through several specific cases, highlighting its potential. Furthermore, we compare the stability criteria obtained from the energy function formalism with the conventional approach of ordinary differential equations. Towards the conclusion of the paper, we illustrate how this general-purpose framework can be employed to generate predictions in real-world biological systems, drawing inspiration from previous works [22, 24].

2   Methods

2.1   Cohen-Grossberg Lyapunov Function Derived from Energy Function Formalism

We begin by considering a single layer of fully connected recurrent neural nodes. The activation of each node is denoted by x_i. In this analysis, we focus on a general recurrent shunting network with dynamic behavior described by Eq. 1, where we neglect the noise terms for simplicity.

$$\dot{x}_i = -Ax_i + (B_i - x_i)(I_i + S(x_i)) - (x_i + C_i)\left(J_i + \sum_{j \neq i} w_{ji}S(x_j)\right) \quad (1)$$

In Eq. 1, the parameter A represents the decay constant, determining the rate at which the activation x_i decreases over time. A higher value of A leads to faster decay, while a lower value results in slower decay. I_i and J_i denote the excitatory and inhibitory inputs to node i, respectively. S denotes a sigmoid function. The constants B_i and $-C_i$ determine the upper and lower bounds for the network activation, respectively.

To simplify the equation, we introduce a variable transformation by letting $y_i = x_i + C_i$. This leads to the following form:

$$\dot{y}_i = y_i\left(b_i(y_i) - \sum_{j=1}^{n} w_{ji}S(y_j - C_j)\right) \quad (2)$$

Here, $b_i(y_i)$ is defined as:

$$b_i(y_i) = \frac{1}{y_i}[AC_i - (A + J_i)y_i + (B_i + C_i - y_i)(I_i + \mathcal{S}(y_i - C_i))] \tag{3}$$

By applying the variable transformation, we obtain a more simplified representation of the dynamics. The transformed equation allows us to study the network's behavior more effectively.

The activation of a specific node i in a recurrent network with n nodes is denoted as x_i. The general time evolution of all Cohen-Grossberg systems can be described by the following equation:

$$\frac{dx_i}{dt} = a_i(x_i)\left[b_i(x_i) - \sum_{j=1}^{n} c_{ij}d_j(x_j)\right] \tag{4}$$

In this equation, the coefficients c_{ij} are symmetric. It is worth noting that this formulation is quite general and applicable to various network models. It encompasses additive and shunting model networks, continuous-time McCulloch-Pitts models, Boltzmann machines, mean field models, and more.

For instance, a continuous-time Hopfield network [8] with the network model:

$$\frac{dx_i}{dt} = -\lambda_i x_i(t) + \sum_{j}^{n} c_{ij}d_j(x_j) + I_i \tag{5}$$

can be seen as a special case of Eq. 4. In this case, we have $a_i(x_i(t)) = 1$ and $b_i(x_i(t)) = -\lambda_i x_i(t) + I_i$.

The global Lyapunov function used by Cohen and Grossberg is given by [5,6]:

$$V = -\sum_{i=1}^{n} \int^{x_i} b_i(\xi_i)d_i'(\xi_i)d\xi_i + \frac{1}{2}\sum_{j,k=1}^{n} c_{jk}d_j(x_j)d_k(x_k) \tag{6}$$

In the context of the network's time evolution, the steady-state or equilibrium solution implies that the activations of the nodes will no longer influence each other in the long term. In other words, all local instabilities will decay, allowing us to consider the set x_i as a set of generalized coordinates describing the state of the network.

Drawing inspiration from the energy function principle in classical mechanics, we can express the relationships between the derivatives of the activations x_i and the corresponding energy function H_i as follows:

$$\frac{\partial x_i}{\partial t} \propto \frac{\partial H_i}{\partial \dot{x}_i} \tag{7}$$

$$\frac{\partial \dot{x}_i}{\partial t} \propto -\frac{\partial H_i}{\partial x_i} \tag{8}$$

Here, $\dot{x}_i = \frac{dx_i}{dt}$. If the proportionality constants in Eqs. 7 and 8 are unequal (equality leading to a trivial case), the energy function for a particular node i can be expressed as:

$$dH_i = \frac{\partial H}{\partial \dot{x}_i} d\dot{x}_i + \frac{\partial H}{\partial x_i} dx_i \propto \dot{x}_i d\dot{x}_i \tag{9}$$

In the above equation, we have utilized the identity $\dot{x}_i d\dot{x}_i = \ddot{x}_i dx_i$, anticipating the final derivation.

Using Eq. 4, we can derive the expression for $d\dot{x}_i$:

$$d\dot{x}_i = \left[a_i(x_i) \left[b'_i(x_i) - c_{ii} d'_i(x_i) \right] + \frac{a'_i(x_i)}{a_i(x_i)} \dot{x}_i \right] dx_i \tag{10}$$

Similarly, by considering the term $\dot{x}_i d\dot{x}_i$, we obtain:

$$\dot{x}_i d\dot{x}_i = (a_i(x_i))^2 \left[b'_i(x_i) b_i(x_i) \right.$$

$$- c_{ii} b_i(x_i) d'_i(x_i) + c_{ii} \sum^{n} c_{ij} d'_i(x_i) d_j(x_j) \tag{11}$$

$$\left. - \sum^{n} c_{ij} b'_i(x_i) d_j(x_j) \right] dx_i + \mathcal{O}(\dot{x}_i^3)$$

In Eq. 11, the term $\mathcal{O}(\dot{x}_i^3)$ represents terms that are of the order of $\dot{x}_i^3$.

Near equilibrium ($\dot{x}_i \rightarrow 0$), we can safely ignore the terms $\mathcal{O}(\dot{x}_i^3) = \left(\frac{a'_i(x_i)}{a_i(x_i)} \right) \dot{x}_i^3 dt$. Additionally, as $\dot{x}_i \rightarrow 0$, we have $b_i(x_i) \rightarrow \sum c_{ij} d_j(x_j)$, which means that the first and last terms in the sum within the parentheses cancel each other. Ignoring the multiplicative coefficients, we can say that near equilibrium:

$$\sum^{n}_{i} dH_i \propto - \sum^{n}_{i} b_i(x_i) d'_i(x_i) dx_i + \frac{1}{2} \sum^{n}_{i,j} c_{ij} d(d_i(x_i) d_j(x_j)) \tag{12}$$

Hence, the full energy function for the system can be written as (with some changes in the dummy indices):

$$H = \sum^{n} \int dH_i$$

$$\propto - \sum^{n} \int^{x_i} b_i(\xi_i) d'_i(\xi_i) d\xi_i + \frac{1}{2} \sum^{n}_{j,k=1} c_{jk} d_j(x_j) d_k(x_k) \tag{13}$$

Therefore, we can establish that the Cohen-Grossberg Lyapunov function is not only a special case but can also be derived from the more general energy function formalism. This realization carries profound implications for neural networks. By applying Eq. 4 to a wide range of neural networks, we are now able to derive energy functions based on fundamental principles rather than relying on inspired guesses.

In the upcoming section, we will explore specific instances of this general formalism and demonstrate how stability criteria can be derived for various types of recurrent networks. Additionally, we will compare the stability criterion obtained from the energy function with the criterion derived from divergence tests.

2.2 Coupled Oscillatory Brain Network from the Energy Function Formalism

Another intriguing contribution of the energy function formalism is its application to oscillatory neural theories. Based on Eq. 4, we can deduce the following expression:

$$\dot{x}_i^2 = (a_i(x_i))^2 [b_i(x_i)^2 - 2 \sum_{j}^{n} b_i(x_i) c_{ij} d_j(x_j) + \sum_{j}^{n} \sum_{k}^{n} c_{ij} c_{ik} d_j(x_j) d_k(x_k)] \tag{14}$$

As we approach equilibrium, where $b_i(x_i) \to \sum c_{ij} d_j(x_j)$, we can express it as:

$$(a_i(x_i))^2 \left(\sum_{j}^{n} \sum_{k}^{n} c_{ij} c_{ik} d_j(x_j) d_k(x_k) \right) = \dot{x}_i^2 + \mathcal{F}_1(x_1, \ldots, x_n) \tag{15}$$

If $C = ||c_{ij}||$ is a symmetric matrix ($c_{ij} = c_{ji}$), if the following property is satisfied

$$\sum_{i}^{n} c_{ij} c_{ik} = \sum_{i}^{n} c_{ji} c_{ik} = m_i c_{jk} \tag{16}$$

then m_i are the diagonal elements of the diagonal matrix D which satisfies the following matrix relation

$$C^2 = DC \tag{17}$$

So using the multiplicative coefficients in Eq. 11, and using Eq. 15, 16 and 13, we can write the full energy function as the following

$$H = \frac{1}{2} \sum_{i}^{n} m_i \dot{x}_i^2 + \mathcal{F}_2(x_1, \ldots, x_n) \tag{18}$$

Here, m_i represents proportionality constants. Remarkably, the form of the energy function presented in Eq. 18 resembles the energy function for multiple coupled nonlinear oscillators with equations of motion of the form [9]

$$m_i \ddot{x}_i + f_i(x_1, \ldots, x_n) = 0 \tag{19}$$

It is interesting to note that to generate meaningful oscillatory solutions, the energy function requires transient input to the system as well as a period where the network can settle or stabilize. Therefore, for oscillatory dynamics to be biologically feasible in the brain, a recurrent layer is needed to receive transient input from a feed-forward network. The aforementioned formulation provides crucial insights and constraints for deriving oscillatory models of brain function (this is further supported in our work [20], where we validated this conjecture through simulations of a complex-valued additive recurrent neural network).

3   Results

A recurrent shunting network with the range $[-D, B]$ follows the dynamics

$$\frac{dx_i}{dt} = -Ax_i + (B - x_i)f(x_i)$$
$$-(D + x_i) \sum_{k=1,k\neq i}^{N} f(x_k) + I_i \tag{20}$$

A general additive recurrent network is given by

$$\frac{dx_i}{dt} = -\lambda x_i + \alpha F(x_i) - \beta \sum_{j=1,j\neq i}^{N} F(x_j) + I_i \tag{21}$$

For the additive network we assume a decay constant of λ.

3.1   Additive Recurrent Network with Slower Than Linear Activation Function

Let us assume that in Eq. 21,

$$F(x) = \begin{cases} 0 & \text{for } x \leq 0 \\ \frac{x}{1+x} & \text{for } x > 0 \end{cases} \tag{22}$$

The network should reach steady state activity when the external input is taken away. If we disregard noise, at steady state, i.e. when, $\frac{dx_i}{dt} = 0$,

$$\lambda x_i = \alpha F(x_i) - \beta \sum_{j=1,j\neq i}^{N} F(x_j) \tag{23}$$

As the equation is symmetric under permutation of units, the system should have symmetric solutions characterized by number of active units n, and their activation $x(n)$, all other units having 0 activation.

$$x(n) = \left(\frac{\alpha - (n-1)\beta}{\lambda} \right) F(x(n)) \tag{24}$$

Using Eq. 22, we get

$$x(n) = \left(\frac{\alpha - (n-1)\beta}{\lambda}\right) - 1 \tag{25}$$

Noise can bring in additional fluctuation that can destabilize the solution for a pair of active modes (with equal activation according to Eq. 25), unless the the difference of activations between the said nodes $\Delta x = x_i - x_j$ decays. Using Eq. 21 & 25 we get

$$\frac{d\triangle x}{dt} = \triangle x \left[-\lambda + \lambda^2 \left(\frac{\alpha + \beta}{(\alpha - (n-1)\beta)^2}\right)\right] \tag{26}$$

Thus the fluctuation decays only if $\frac{d\triangle x}{dt} \leq 0$, i.e.,

$$\frac{\alpha + \beta}{(\alpha - (n-1)\beta)^2} \leq \frac{1}{\lambda} \tag{27}$$

As we can see that the decay parameter, excitation parameter and inhibition parameter are not completely independent for stable solutions. For the present purposes we use $\lambda = 1$ (in line with [1] in order to set the time scale to synaptic currents).

From Eq. 21 and 22 we have

$$d\dot{x}_i = -dx_i + \alpha \left(\frac{F(x_i)}{x_i}\right)^2 dx_i \tag{28}$$

It is easy to show that from the definition given in 9,

$$H = \sum_i H_i \propto -\sum_i \int \left(1 - \alpha \left(\frac{F(x_i)}{x_i}\right)^2\right) \dot{x}_i{}^2 dt \tag{29}$$

From Eq. 25 we can substitute terms in steady state to get

$$H \propto -\sum_i \int \left(1 - \frac{\alpha}{(\alpha - (n-1)\beta)^2}\right) \dot{x}_i{}^2 dt \tag{30}$$

In our previous work [25] using the energy function we derived an expression of reaction time in enumeration tasks as being proportional to maximum allowed fluctuation in energy (as the network needs to reset to continue enumeration). This allowed us to derive meaningful predictions that were corroborated by experiments on human subjects.

Now if $dH < 0$ and thus a monotonically decreasing Lyapunov type function in absence of external input, we have the stability condition as

$$\frac{\alpha}{(\alpha - (n-1)\beta)^2} \leq 1 \tag{31}$$

Comparing this to Eq. 27, we can see that the conditions derived from the energy value is slightly different and diverges greatly for higher β. This is due to the fact that Eq. 27 excludes winner-take-all mechanisms operating at higher inhibition, whereas the energy function does not. And it is evident that for all β,

$$\frac{\alpha}{(\alpha - (n-1)\beta)^2} \leq \frac{\alpha + \beta}{(\alpha - (n-1)\beta)^2} \leq 1 \tag{32}$$

and thus the energy function is a very suitable candidate for the network as it is in line with the stability analysis derived from the dynamics of the network.

3.2   Shunting Recurrent Network with Constant Activation Function

Here we assume in Eq. 20,

$$f(x) = \begin{cases} 0 & x \leq 0 \\ k & x > 0 \end{cases} \forall k \epsilon \mathbb{R} \tag{33}$$

A search for a symmetric solution at steady state leads to steady state activation value

$$x(n) = \frac{(B - D(n-1))f(x(n))}{A + nf(x(n))} = \frac{(B - D(n-1))k}{A + nk} \tag{34}$$

The stability criterion is determined considering the decay of $\Delta x = x_i - x_j$ and is calculated to the following condition

$$nk \leq -A \tag{35}$$

Constructing the energy function using

$$d\dot{x}_i = \left(-A - \sum_{j=1}^{N} f(x_j)\right) dx_i \tag{36}$$

and Eq. 9, we have

$$H = \sum_i H_i \propto -\sum_i \int \left(A + \sum_{j=1}^{N} f(x_j)\right) \dot{x}_i^2 dt \tag{37}$$

As $\sum_{j=1}^{N} f(x_j) = nk$ near steady state, we have the same criterion for stability from the energy function as Eq. 35.

3.3   Shunting Network with Linear Activation Function

The activation function for such networks is given by

$$f(x) = \begin{cases} 0 & x \leq 0 \\ x & x > 0 \end{cases} \tag{38}$$

Using the above activation function in Eq. 20 we have at steady state condition,

$$x(n) = \frac{(B - D(n-1))f(x(n))}{A + nf(x(n))} = \frac{(B - D(n-1))x(n)}{A + nx(n)} \tag{39}$$

This leads to the non-trivial solution

$$x(n) = \frac{B - A - D(n-1)}{n} \tag{40}$$

Looking at the decay of $\Delta x = x_i - x_j$ leads to the stability criterion,

$$D \leq 0 \tag{41}$$

The energy function constructed using

$$d\dot{x}_i = -dx_i + (B - x_i)dx_i - \sum_{j=1}^{N} f(x_j)dx_i \tag{42}$$

and Eq. 9 leads to,

$$B - A - D(n^2 - 1) < 0 \tag{43}$$

which upon substituting $B - A = nx(n) + D(n-1)$ (from Eq. 40) yields $nx(n) - nD(n-1) < 0$. For the situation where total network activation $nx(n) \geq 0$, we have $D \geq 0$ for $n > 1$. In fact this leads to a better stability criterion combining the stability criterion given in Eq. 41 with the condition $D \geq 0$ obtained from the energy function

$$D = 0 \tag{44}$$

This is a sensible result as by definition of shunting networks B and D are positive real quantities.

3.4   Shunting Network Using Reciprocal Activation Function

The activation function for such networks is given by

$$f(x) = \begin{cases} 0 & x \leq 0 \\ \frac{1}{x} & x > 0 \end{cases} \tag{45}$$

The steady state solution (assuming the positive root of a quadratic equation) is given by,

$$x(n) = \frac{-n + k}{2A} \tag{46}$$

where $k = \sqrt{n^2 + 4A(B - D(n-1))}$.

The stability criterion from decay condition is derived to be

$$\frac{4(B+D) + 2n(-n+k)}{(-n+k)^2} \geq -1 \tag{47}$$

Constructing the energy function in the now familiar way, we get

$$\frac{(2B+n-k)2 + 2n(-n+k)}{(-n+k)^2} \geq -1 \tag{48}$$

It can be shown that both the criteria given by Eq. 47 and 48 are same if $n - k = 2D$.

3.5   Analytical Prediction of Winner-take-all

In Sect. 3.1, we have discussed the constraints on the stability of the network response in an additive recurrent model. These constraints ensure that the activation of two nodes does not diverge during simulation. The stability condition can be expressed as:

$$\frac{\alpha + \beta}{(\alpha - (n-1)\beta)^2} < 1 \tag{49}$$

Here, n represents the number of nodes active after the simulation. By considering the novel energy function formulation, we derived another stability condition:

$$\frac{\alpha}{(\alpha - (n-1)\beta)^2} < 1 \tag{50}$$

However, for the important winner-take-all (WTA) dynamics, it is necessary to amplify activation differences between nodes until only one node emerges as the winner. When $n = 1$, corresponding to the standard WTA interaction, Eq. 50 simplifies to:

$$\alpha - 1 > 0 \tag{51}$$

This inequality provides the lower limit for α in WTA behavior. The upper limit is determined by Eq. 49 when $n = 2$ reaches stability:

$$\alpha^2 - \alpha - 2\beta > 0 \tag{52}$$

Hence, the WTA behavior is supported by a range of α that satisfies the conditions $\alpha \geq 1$ and $\alpha^2 - \alpha - 2\beta \leq 0$. Figure 1 illustrates how the ranges of α should be calculated for different β values. For instance, for $\beta = 0.25$, the range is $1 \leq \alpha \leq 1.37$, for $\beta = 0.3$, it is $1 \leq \alpha \leq 1.42$, and for $\beta = 0.35$, it is $1 \leq \alpha \leq 1.47$.

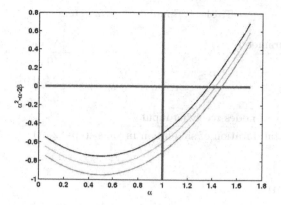

Fig. 1. A plot of $\alpha^2 - \alpha - 2\beta$ for different β values. Blue line for $\beta = 0.25$, green for $\beta = 0.3$ and magenta for $\beta = 0.35$. In the bottom right quadrant of the plot bound by the lines $\alpha^2 - \alpha - 2\beta = 0$ and $\alpha = 1$ we get the α values desired for WTA interaction, mainly $\alpha \geq 1$ and $\alpha^2 - \alpha - 2\beta \leq 0$. For $\beta = 0.25$ the range is $1 \leq \alpha \leq 1.37$, for $\beta = 0.3$, $1 \leq \alpha \leq 1.42$, for $\beta = 0.35$, $1 \leq \alpha \leq 1.47$ (Color figure online).

We also validated the parameters through actual simulations. For each simulation, we considered a network consisting of 10 nodes, with inputs given to 2 nodes. The input level was fixed at 0.3. To calculate the probability of WTA interaction, we determined the fraction of simulations (out of 1000) in which only one node survived. We varied the value of α in the range of 0.5 to 1.5. Each stimulus was presented for 255 time steps, and the total simulation duration was set to 2500 time steps. Gaussian noise with a mean of 0 and a standard deviation of 0.1 was added to the system. The results, depicted in Fig. 2, closely aligned with the analytic limits obtained in Fig. 1. Detailed parameter values for the simulation can be found in Table 1.

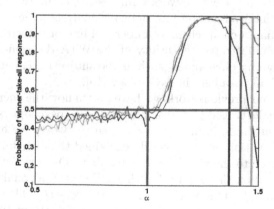

Fig. 2. Probability of winner take all interaction plotted against α for three different β values 0.25 (blue line), 0.30 (green line) and 0.35 (magenta line). For $\alpha > 1$, the WTA probability increases beyond chance level (0.5). The analytical limits on α obtained in Fig. 1 are shown in correspondingly colored lines. (Color figure online)

Table 1. Simulation parameters

Parameter	Value
N	10
α	0.5:0.01:1.5
β	0.3
No. of nodes receiving input	2
Total duration of simulation in time steps	2500

4   Discussion

In this study, we have demonstrated the effectiveness of the analytical energy function formalism developed by the authors in deriving the Cohen-Grossberg Lyapunov function for general single-layer shunting recurrent neural networks. This has significant implications as it allows the dynamics of such networks to be effectively modeled across a wide range of neural network architectures, including additive recurrent networks, continuous-time McCulloch-Pitts neurons, Boltzmann machines, and Mean field models, among others.

Furthermore, we have compared the stability criteria obtained from the energy function formulation with the stability criterion derived from network analysis. Our results indicate a remarkably close agreement between the two stability criteria across various activation functions and for both additive and shunting networks. This finding highlights the robustness and applicability of the energy function-based stability analysis in capturing the dynamics of recurrent neural networks.

For the additive variant, the stability criterion derived from the energy function offers the advantage of predicting the onset of Winner-take-all (WTA) behavior in additive recurrent networks. Interestingly, the two stability criteria derived from the energy function and network analysis appear to complement each other, providing a comprehensive understanding of network dynamics. We have previously demonstrated the utility of the WTA decision-making process in predicting intriguing phenomena, such as the subjective expansion of time in the temporal oddball paradigm in psychology [10].

In addition, in our previous work, we have shown how the energy function for the network can be leveraged to derive predictions for psychophysical attributes, specifically reaction times, in biological recurrent networks for humans. By utilizing the energy function, we successfully explained the distribution of reaction times in tasks related to the visual sense of numbers. Our prediction for reaction time (RT) was based on the assumption that RT is proportional to the negative of the energy required for network resetting ($RT \propto -H$). This prediction was employed to anticipate and interpret the patterns of reaction times in enumeration experiments involving human participants [25]. Furthermore, the fMRI activation patterns associated with enumeration and visual working memory tasks were predicted using the model and subsequently verified through human experimentation [11].

In recent years, the understanding of neural codes has advanced beyond traditional rate coding, with increasing recognition of the importance of temporal codes involving spike timing and phase information for reliable information transmission and processing [12–15]. However, most investigations have focused on neural codes following stimulus presentation. Recently, there has been a growing interest in pre-stimulus brain states, as demonstrated in MEG and EEG studies, which have shown the potential to predict conscious detection of stimuli based on oscillatory brain activity prior to stimulus onset [16–19]. For example, the modulation of the pre-stimulus α frequency band has been found to be important for near-threshold stimuli [19]. While these studies have garnered significant interest, they have offered limited theoretical or physiological insights into these phenomena. In Sect. 2.2, we explore the possibility of oscillatory brain states under specific equilibrium conditions in a neural assembly comprising both feed-forward and recurrent connections. Notably, we highlight the crucial role of the delay between feed-forward and recurrent connections in shaping the distribution of oscillatory brain states. This approach to understanding oscillatory brain states has been explored in [20]. However, how the formalism fares in applying it to Cohen-Grossberg networks with multiple delays and corresponding improved Lyapunov function (as seen in [26]) needs to be explored in future work.

5   Conclusion

In this study, we've showcased the potency of our analytical energy function formalism in deriving the Cohen-Grossberg Lyapunov function for general single-layer shunting recurrent neural networks. This has broad-reaching implications, enabling effective modeling of network dynamics across a spectrum of architectures, from additive recurrent networks to continuous-time McCulloch-Pitts neurons and beyond. Our comparison of stability criteria derived from the energy function with conventional network analysis reveals a remarkable alignment, underscoring the robustness of our approach in capturing recurrent neural network dynamics. Notably, for additive networks, the energy function-derived stability criterion predicts the onset of Winner-take-all behavior, complementing traditional analyses. Our past work has demonstrated the utility of this formalism in predicting psychophysical attributes, from reaction times to fMRI activation patterns, showcasing its application in diverse domains. As we delve into the exploration of oscillatory brain states, particularly under equilibrium conditions with feed-forward and recurrent connections, we illuminate the crucial role of delays in shaping these states. While this study sets a solid foundation, future endeavors will unravel the formalism's potential in networks with multiple delays.

References

1. Usher, M., Cohen, J.D. : Short term memory and selection processes in a frontal-lobe model. In: Heinke, D., Humphreys, G.W., Olson, A. (eds.) Connectionist Models in Cognitive Neuroscience, pp. 78–91 (1999)
2. Bogacz, R., Usher, M., Zhang, J., McClelland, J.L.: Extending a biologically inspired model of choice: multi-alternatives, nonlinearity and value-based multi-dimensional choice. Philos. Trans. Roy. Soc. B: Biol. Sci. **362**(1655), 1655–1670 (2007)
3. Sengupta, R., Bapiraju, S., Melcher, D.: A visual sense of number emerges from the dynamics of a recurrent on-center off-surround neural network. Brain Res. **1582**, 114–124 (2014)
4. Andreopoulos, A., Tsotsos, J.K.: 50 Years of object recognition: directions forward. Comput. Vis. Image Underst. **117**(8), 827–891 (2013)
5. Grossberg, S.: Nonlinear neural networks: principles, mechanisms, and architectures. Neural Netw. **1**(1), 17–61 (1988)
6. Cohen, M.A., Grossberg, S.: Absolute stability of global pattern formation and parallel memory storage by competitive neural networks. IEEE Trans. Syst. Man Cybern. **SMC-13**, 815–826 (1983)
7. Zhang, H., Wang, Z., Liu, D.: A comprehensive review of stability analysis of continuous-time recurrent neural networks. IEEE Trans. Neural Netw. Learn. Syst. **25**, 1229–1262 (2014)
8. Hopfield, J.J.: Neurons with graded response have collective computational properties like those of two-state neurons. Proc. Natl. Acad. Sci. **81**, 3088–3092 (1984)
9. Durmaz, S., Altay Demirbag, S., Kaya, M.O.: Energy function approach to multiple coupled nonlinear oscillators. Acta Phys. Polonica-Ser. A Gener. Phys. **121**, 47–49 (2012)
10. Sengupta, R., Bapiraju, S., Basu, P., Melcher, D.: Accounting for subjective time expansion based on a decision, rather than perceptual, mechanism. J. Vis. **14**, 1150 (2014)
11. Knops, A., Piazza, M., Sengupta, R., Eger, E., Melcher, D.: A shared, flexible neural map architecture reflects capacity limits in both visual short term memory and enumeration. J. Neurosci. **34**, 9857–9866 (2014)
12. Stanley, G.B.: Reading and writing the neural code. Nat. Neurosci. **16**(3), 259–263 (2013)
13. Van Rullen, R., Thorpe, S.J.: Rate coding versus temporal order coding: what the retinal ganglion cells tell the visual cortex. Neural Comput. **13**(6), 1255–1283 (2001)
14. Gautrais, J., Thorpe, S.: Rate coding versus temporal order coding: a theoretical approach. Biosystems **48**(1), 57–65 (1998)
15. Masquelier, T.: Relative spike time coding and STDP-based orientation selectivity in the early visual system in natural continuous and saccadic vision: a computational model. J. Comput. Neurosci. **32**(3), 425–441 (2012)
16. Mathewson, K.E., Gratton, G., Fabiani, M., Beck, D.M., Ro, T.: To see or not to see: prestimulus α phase predicts visual awareness. J. Neurosci. **29**(9), 2725–2732 (2009)
17. Keil, J., Müller, N., Ihssen, N., Weisz, N.: On the variability of the McGurk effect: audiovisual integration depends on prestimulus brain states. Cereb. Cortex **22**(1), 221–231 (2012)

18. May, E.S., Butz, M., Kahlbrock, N., Hoogenboom, N., Brenner, M., Schnitzler, A.: Pre- and post-stimulus alpha activity shows differential modulation with spatial attention during the processing of pain. Neuroimage **62**(3), 1965–1974 (2012)
19. Weisz, N., et al.: Prestimulus oscillatory power and connectivity patterns predispose conscious somatosensory perception. Proc. Natl. Acad. Sci. **111**(4), E417–E425 (2014)
20. Sengupta, R., Raja Shekar, P.V.: Oscillatory dynamics in complex recurrent neural networks. Biophys. Rev. Lett. **17**(1), 75–85 (2022)
21. Hopfield, J.J.: Neural networks and physical systems with emergent collective computational abilities. Proc. Natl. Acad. Sci. **79**(8), 2554–2558 (1982)
22. Hopfield, J.J., Brody, C.D.: Pattern recognition computation using action potential timing for stimulus representation. Nature **376**(3535), 33–36 (1995)
23. Amari, S.-I.: Dynamics of pattern formation in lateral-inhibition type neural fields. Biol. Cybern. **27**(2), 77–87 (1977)
24. Izhikevich, E.M.: Simple model of spiking neurons. IEEE Trans. Neural Netw. **14**(6), 1569–1572 (2003)
25. Sengupta, R., Bapiraju, S., Melcher, D.: Big and small numbers: empirical support for a single, flexible mechanism for numerosity perception. Attent. Percept. Psychophys. **79**, 253–266 (2017)
26. Faydasicok, O.: An improved Lyapunov functional with application to stability of Cohen-Grossberg neural networks of neutral-type with multiple delays. Neural Netw. **132**, 532–539 (2020)

Co-imagination of Behaviour and Morphology of Agents

Maria Sliacka[1], Michael Mistry[2] (iD), Roberto Calandra[3,4] (iD), Ville Kyrki[1] (iD),
and Kevin Sebastian Luck[1,5,6(✉)] (iD)

[1] Department of Electrical Engineering and Automation (EEA), Aalto University,
Espoo, Finland
`ville.kyrki@aalto.fi`
[2] University of Edinburgh, Edinburgh, UK
`mmistry@ed.ac.uk`
[3] Learning, Adaptive Systems, and Robotics (LASR) Lab, TU Dresden, Dresden, Germany
`roberto.calandra@tu-dresden.de`
[4] The Centre for Tactile Internet with Human-in-the-Loop (CeTI), Dresden, Germany
[5] Finnish Center for Artificial Intelligence, Espoo, Finland
[6] Vrije Universiteit Amsterdam, Amsterdam, Netherlands
`k.s.luck@vu.nl`

Abstract. The field of robot learning has made great advances in developing behaviour learning methodologies capable of learning policies for tasks ranging from manipulation to locomotion. However, the problem of combined learning of behaviour and robot structure, here called co-adaptation, is less studied. Most of the current co-adapting robot learning approaches rely on model-free algorithms or assume to have access to an a-priori known dynamics model, which requires considerable human engineering. In this work, we investigate the potential of combining model-free and model-based reinforcement learning algorithms for their application on co-adaptation problems with unknown dynamics functions. Classical model-based reinforcement learning is concerned with learning the forward dynamics of a specific agent or robot in its environment. However, in the case of jointly learning the behaviour and morphology of agents, each individual agent-design implies its own specific dynamics function. Here, the challenge is to learn a dynamics model capable of generalising between the different individual dynamics functions or designs. In other words, the learned dynamics model approximates a multi-dynamics function with the goal to generalise between different agent designs. We present a reinforcement learning algorithm that uses a learned multi-dynamics model for co-adapting robot's behaviour and morphology using imagined rollouts. We show that using a multi-dynamics model for imagining transitions can lead to better performance for model-free co-adaptation, but open challenges remain.

Keywords: Evolutionary Robotics · Co-Adaptation · Co-Design · Reinforcement Learning

G. Nicosia et al. (Eds.): LOD 2023, LNCS 14505, pp. 318–332, 2024.
https://doi.org/10.1007/978-3-031-53969-5_24

1   Introduction

Co-adaptation is a process that is present everywhere on Earth. From tiny insects adapting to human houses [13] to rats adapting their diets in cities [11], to the large whales that adapt to fight human-produced noise in the ocean [18], it has proven to be crucial for survival in a changing environment. The idea of co-adaptation brings together the two different ways that organisms adapt – behavioural and morphological. Behavioural adaptation to new tasks happens on short timescales and is not difficult for humans and animals, whereas, adapting morphological traits is often not possible and is a process operating on long timescales [17, 19].

When it comes to robotics, we usually only optimise the behaviour of our robots given a new task. This leads to robots being designed by human engineers to be multipurpose and easy to control, such that they can be used for a wide range of tasks. However, as nature has shown repeatedly, having evolved a specialised body morphology can lead to vastly improved behavioural policy, performance and excellence in a low number of essential tasks. Even humans tend to complement or change their morphology for improving their performance, for example, by using artificial modifications to their bodies, such as bodysuits for diving.

This leads to the idea of co-adaptation of the behaviour and morphology in robots [6, 15, 16, 21, 22, 24–26]. The goal is to jointly optimise the control and design parameters of a robot given its task. The challenge of co-adaptation comes primarily from the design search space: especially with a high number of continuous design parameters, it is impossible to evaluate all possible body shapes. One possible alternative option to reduce these costs drastically is by utilising simulations to build and evaluate robots' morphologies [6, 21, 25]. However, creating, evaluating and mutating possible robot candidates in a simulation is not only computationally demanding but also suffers from the simulation-to-reality gap. Designs and behaviours found to be optimal in simulation may not be optimal in the real world [15, 23]. However, to allow for the co-adaptation of robots in the real world, data-efficient co-adaptation methodologies are required. Especially methods, which are able to optimise robot morphologies within a low number of design iterations, as manufacturing robots in the real world requires many resources and person-hours. Prior work tackling this problem has largely focused on model-free reinforcement learning approaches [8, 16, 21, 24]. In this work, we will explore the possibility to utilise model-based reinforcement learning techniques [10, 20] to further improve the performance and sample-efficiency of model-free co-adaptation algorithms [16]. This requires us to learn not only one single forward-dynamics function for a specific agent morphology, but also the dynamics of multiple, if not infinitely many, agents. In our work, we will investigate the benefit of training a forward-dynamics function parameterised by the known design parameters of agents, and evaluate their generalisation ability across known and unseen agent morphologies. Furthermore, we will incorporate this model-based learning approach into state-of-the-art model-free co-adaptation methods to further increase the data- and training-efficiency of the co-adaptation using imagination-based [10, 20] training data augmentation for a co-adapting deep reinforcement learning method. We show that by adding artificial data in known design space, we are able to improve the performance of the co-adaptation algorithm in terms of total cumulative rewards collected.

2   Related Work

Co-adaptation of control and morphology of an agent or robot has a long history, including the usage of evolutionary algorithms and gradient-based optimisation methods, such as reinforcement learning. Evolutionary algorithms have been used to evolve populations of agent morphologies via mutations [6,25]. Gupta et al. [6] also added a reinforcement learning loop to optimise the controller directly. Evolutionary approaches require populations of solutions for each optimisation iteration, which often requires large amounts of data especially in the case of a high-dimensional space of morphologies. The need for such large populations means that these approaches have to rely primarily on simulations due to their high cost in real-world tasks, which makes them prone to suffer from the so-called simulation-to-reality gap.

There are multiple approaches for solving the co-adaptation problem using model-free reinforcement learning [2,7,16,24]. Schaff et al. [24] use a distribution of designs that is shifted towards better-performing morphologies with a policy using the design as context. Whereas, [7] directly considers the design parameters to be learnable but also requires keeping a population of morphologies to compute the policy gradient. Similarly to the evolutionary methods, both of these model-free approaches have the same drawback of requiring a population or distribution of morphologies during the entire algorithm run, thus being limited in their real-world applicability. A different approach was proposed in [2], where the design or hardware is considered as part of the policy parameters and jointly optimised. However, this requires hand-engineering and prior knowledge as it relies on simulating the hardware with an auto-differentiable computational graph. ORCHID [12] uses model-based reinforcement learning to simultaneously optimise the hardware and control parameters of the agent. It also uses an actor-critic algorithm similarly to the proposed approach, however, it relies on a differentiable transition function which is assumed to be known a-priori. Closest to our method is [16], which is a model-free co-adaptation approach utilising the Q-value function for data-efficient evaluation of design candidates. However, it only uses model-free reinforcement learning while our method uses a learned dynamics model in addition to generate artificial data, thus combining model-free and model-based reinforcement learning. We use model-free co-adaptation [16] as a starting point and baseline for evaluating the potential of joint co-imagination of agent behaviour and design as augmented training data throughout the learning process. Another approach [5,8] is to use function and control theory to co-optimise the morphology and control, both these methods rely on the accuracy of their models and equations, where [5] also require differentiable control planner and simulation. [14] use Bayesian optimisation to first optimise the morphology and then learn the corresponding controller. This method, however, requires a parameterised controller and would struggle to scale to high-dimensional spaces.

3   Problem Statement

We assume a Markov Decision Process (MDP) for solving the co-adaptation problem, extended with a design context ξ. The MDP is denoted by $(S, A, p, r, \gamma, \Xi)$, with state space $S \subseteq R^d$, action space $A \subseteq R^n$, design parameter space $\Xi \subseteq \mathbb{R}^i$ and a given reward function $r(s, a)$, with $r : S \times A \mapsto \mathbb{R}$. Without loss of generalisation we will assume for the reminder of the paper a continuous but bounded design space $\Xi \subseteq \mathbb{R}^i$, with i dimensions. However, Ξ could be in principle also be a set of discrete designs. In this MDP, the underlying transition probability $p(s_{t+1}|s_t, a_t, \xi) : S \times A \times S \times \Xi \mapsto \mathbb{R}^+$ maps the current state s_t and an action a_t to the next state s_{t+1}. Importantly, the transition probability does not only depend on the current state s_t and the agent's action a_t, but also on the agent's design variable ξ which parameterises the morphology of the agent such as lengths of legs, their shape or diameter.

The general problem of co-adapting behaviour and morphology of agents with reinforcement learning is to find a policy $\pi : S \times A \mapsto \mathbb{R}^+$ and designs $\xi \in \Xi$ such that the expected discounted return is maximised as

$$\max_{\pi, \xi} \mathbb{E}_{\substack{a_t \sim \pi(s_t) \\ s_{t+1} \sim p(s_{t+1}|s_t, a_t, \xi)}} \left[\sum_{t=0}^{\infty} \gamma^t r(s_t, a_t) \right], \tag{1}$$

given a specific reward function $r(s, a)$ and discount $\gamma \in [0, 1]$. While not required, we will also assume for the reminder of the paper that the policy π and reward r are parameterised by the design variable ξ, i.e. $\pi(a_t|s_t, \xi)$ and $r(s_t, a_t, \xi)$. This allows to learn a single policy π capable of adapting to specific agent designs. We will furthermore assume that design parameters ξ are observed, which is generally true as these parameters are required for manufacturing or simulating (e.g. via urdf files) specific agents.

4   Learning Behaviour, Design and Dynamics Across the Design-Space

To find an optimal combination of behaviour and design with respect to a given reward function as defined in Eq. (1) using model-based reinforcement learning we need to learn three components:

$$
\begin{aligned}
\text{Policy:} & \quad \pi(s_t, \xi), \\
\text{Design:} & \quad \xi, \\
\text{Transition model:} & \quad p(s_{t+1}|s_t, a_t, \xi).
\end{aligned}
\tag{2}
$$

The first two, policy and design, correspond to the core ideas in the co-adaptation framework, behaviour learning and morphology optimisation, which also appear in existing model-free co-adaptation approaches. The third corresponds to our main contribution: Learning a forward-dynamics model across designs. In the following, we will discuss how we will learn each component.

Behaviour Learning: A central component of co-adaptation is the learning of an optimal behavioural policy given a design and task. As discussed in the problem statement, we will operate with an extended MDP formulation considering the effect a parameterised design has on the reward function $r(s_t, a_t, \xi)$ and dynamics $p(s_{t+1}|s_t, a_t, \xi)$. In effect, we consider an extension of the standard reinforcement learning approach optimising for Eq. (1) in which both the policy π and value $V : S \mapsto \mathbb{R}$, or Q-value $Q :$ $S \times A \mapsto \mathbb{R}$, functions depend on the design variable ξ, i.e. $\pi(s_t, \xi) : S \times \Xi \times A \mapsto \mathbb{R}^+$ and $Q(s_t, a_t, \xi) : S \times A \times \Xi \mapsto \mathbb{R}$. For learning a policy π we employ Soft-Actor-Critic algorithm [9] with the double-Q-network approach for learning a probabilistic policy π. Given a set of training experience $\mathcal{D}$, the Q-value function is trained with the altered loss

$$J_Q(\theta) = \mathbb{E}_{(s_t, a_t, s_{t+1}, \xi) \sim \mathcal{D}} \left[\frac{1}{2} \left(Q_\theta(s_t, a_t, \xi) - (r(s_t, a_t, \xi) + \gamma V_{\bar{\theta}}(s_{t+1}, \xi)) \right)^2 \right], \quad (3)$$

where $V_{\bar{\theta}}$ is defined using the target Q-value function $Q_{\bar{\theta}}(s_t, a_t, \xi)$ parametrised by the target network parameters $\bar{\theta}$, given the transition and design (s_t, a_t, s_{t+1}, ξ). Similarly, we use the modified loss

$$J_\pi(\phi) = \mathbb{E}_{(s_t, \xi) \sim \mathcal{D}, \epsilon_t \sim \mathcal{N}} \left[\alpha \log \pi_\phi \left(f_\phi \left(\epsilon_t; s_t, \xi \right) \mid s_t, \xi \right) - Q_\theta \left(s_t, f_\phi \left(\epsilon_t; s_t, \xi \right), \xi \right) \right], \quad (4)$$

for training the parameters of the policy π. Both these equations are modified to include the design parameter ξ.

Furthermore, we train two separate sets of policies and value networks: The population networks (*Pop*), which train on all designs seen thus far, and the individual networks (*Ind*) which are trained only on current design. This is facilitated by using a replay buffer $\mathcal{D}_{\text{Pop}}$. containing the collected experience of all designs seen so far, i.e. $(s_t, a_t, s_{t+1}, \xi_s)$ with $\xi_s \in \Xi_{\text{seen}}$. The individual networks have its weights initialised by the *pop* networks each time we start training on a new design and primarily train with experience from the current design to facilitate fast and sample-efficient reinforcement learning. While training on an agent of design ξ_{current}, these individual networks will utilise the replay buffer $\mathcal{D}_{\text{Ind}}$. containing only experience of the current agent, i.e. $(s_t, a_t, s_{t+1}, \xi_{\text{current}})$.

Design Optimisation: During the design optimisation stage, we aim to identify a design variable maximising the expected return given the data collected thus far, which is then synthesised and a new round of behavioural learning is executed. While previous approaches have utilised simulations to gauge the potential performance of design candidates, [16] have pointed out that the value function can be utilised as a data-efficient and computationally more efficient alternative. By using the Q-value function as the fitness function, Particle Swarm Optimisation (PSO) [1] is used to quickly find the best next design parameter by optimising the objective

$$\max_\xi \mathbb{E}_{(s_0, a_0, s_1, \xi_{\text{orig}}) \sim \mathcal{D}} \left[\mathbb{E}_\pi \left[Q(s_0, a, \xi) \mid a = \pi(s_0, \xi) \right] \right]. \quad (5)$$

The expected return (see Eq. (1)) is estimated by evaluating the Q-function for start states s_0 sampled from a separate replay buffer $\mathcal{D} = Replay_{s_0}$, and replacing the original design variable ξ_{orig} with the design query ξ.

Model Learning: When learning a forward dynamics function in the context of co-adaptation, we have in principle two choices: we either learn a design-specific, individual, dynamics function or learn one multi-dynamics function that captures all possible designs. In the following, we will concentrate on the latter due to its potential to allow for generalisation between designs, and allow for queries of transitions for unseen designs. For learning the multi-dynamics function we will follow the probabilistic ensemble approach proposed by [3]. The ensemble consists of 3 probabilistic neural networks with outputs parametrising a Gaussian distribution that estimates the next state given the current state, taken action and design, defined as $Pr^i(s_{t+1}|s_t, a_t, \xi) = \mathcal{N}(\mu_{\psi^i}(s_t, a_t, \xi), \Sigma_{\psi^i}(s_t, a_t, \xi))$, where ψ^i corresponds to the i-th network's parameters. The ensemble is trained using the Negative Log Likelihood loss calculated for each network

$$J_{\tilde{p}}(\psi^i) = -\mathbb{E}_{(s_t, a_t, s_{t+1}, \xi) \sim Replay_{\text{Pop}}} \log \mathcal{N}(s_{t+1}|\mu_{\psi^i}(s_t, a_t, \xi), \Sigma_{\psi^i}(s_t, a_t, \xi))$$

$$= \mathbb{E}_{(s_t, a_t, s_{t+1}, \xi) \sim Replay_{\text{Pop}}} [\mu_{\psi^i}(s_t, a_t, \xi) - s_{t+1}]^\top \Sigma_{\psi^i}^{-1}(s_t, a_t, \xi)[\mu_{\psi^i}(s_t, a_t, \xi) - s_{t+1}]$$

$$+ \log \det \Sigma_{\psi^i}(s_t, a_t, \xi), \tag{6}$$

where we use transitions (s_t, a_t, s_{t+1}, ξ) sampled from the replay buffer $Replay_{\text{Pop}}$, containing experience from all designs seen thus far. The learned forward dynamics function $h(s_t, a_t, \xi)$ is then defined using the TS1 propagation method [3], where at each time step the network is uniformly sampled from the ensemble to produce an output. Importantly, we will consider the case where the dynamics functions is learned online, i.e. without pre-training on pre-collected datasets. At the start of the co-adaptation process, the dynamics network will be initialised with random weight initialisation and thereafter trained with the process described above. The number of designs the dynamics networks are trained on will increase with each new design selected (see Algorithm 1).

5   Co-adaptation by Model-Based Imagination

In this section, we will discuss the changes made to the behavioural learning process necessary to utilise a learned dynamics model in the context of co-adaptation. Specifically, we will propose and investigate the use of a learned forward dynamics model for supporting the model-free reinforcement learning process by supplying artificial, i.e. imagined, transitions on known or unseen designs. Assuming a forward dynamics function $h(s_t, a_t, \xi)$ learned in the manner described above, we re-formulate the model-free loss functions of the value function loss

$$J_Q(\theta) = \mathbb{E}_{(s_t, a_t, s_{t+1}, \xi) \sim \mathcal{D}} \left[\frac{1}{2} \left(Q_\theta(s_t, a_t, \xi) - (r(s_t, a_t, \xi) + \gamma V_{\bar{\theta}}(s_{t+1}, \xi)) \right)^2 \right]$$

$$+ \mathbb{E}_{\substack{(s_t, a_t, s_{t+1}, \xi) \sim \mathcal{D} \\ \xi_g \sim G(\xi) \\ a_g \sim \pi(s_t, \xi_g)}} \left[\frac{1}{2} \left(Q_\theta(s_t, a_g, \xi_g) - (r(s_t, a_g, \xi_g) + \gamma V_{\bar{\theta}}(h(s_t, a_g, \xi_g))) \right)^2 \right] \tag{7}$$

and for the policy loss as

$$J_\pi(\phi) = \mathbb{E}_{(s_t, \xi) \sim \mathcal{D}, \epsilon_t \sim \mathcal{N}} [\alpha \log \pi_\phi (f_\phi(\epsilon_t; s_t, \xi) \mid s_t, \xi) - Q_\theta(s_t, f_\phi(\epsilon_t; s_t, \xi), \xi)]$$

$$+ \mathbb{E}_{\substack{(s_t, \xi) \sim \mathcal{D}, \epsilon_t \sim \mathcal{N} \\ \xi_g \sim G(\xi)}} [\alpha \log \pi_\phi (f_\phi(\epsilon_t; s_t, \xi_g) \mid s_t, \xi_g) - Q_\theta(s_t, f_\phi(\epsilon_t; s_t, \xi_g), \xi_g)],$$

$$\tag{8}$$

where the design variable ξ_g is sampled from a generator $G(\xi)$. These losses correspond to a one-step imagination-based approach similar to [10]. Many potential choices exist for the design generator $G(\xi)$. For our experimental evaluation we will consider $G(\xi) = \xi$, $G(\xi) = N(\xi, \sigma^2)$ and $G(\xi) = \text{uniform}(\Xi)$. These correspond to (a) using the identity function, i.e. using the original design from the replay buffer, (b) adding Gaussian noise to ξ, and (c) using a randomly sampled design $\xi \in \Xi$. The action a_g is sampled from the policy with $\pi(s_t, \xi_g)$.

In practice, to increase the computational efficiency during training, we pre-compute a sufficiently large batch of $(s_t, a_g, h(s_t, a_g, \xi_g), \xi_g)$ after updating the forward dynamics model and store them in a separate replay buffer $Replay_{\text{Art}}$. after each episode. During the SAC networks training, the population networks are then trained on data that comes from both real and artificial replay buffers. This is done by taking 80% of the batch from the real experience buffer $Replay_{\text{Pop.}}$, and 20% from the artificial experience $Replay_{\text{Art}}$.. The final algorithm is shown in Algorithm 1: The algorithm presents a version of the algorithm in which agent designs, and behaviours are updated sequentially, i.e. no population of agents is maintained. This is to emulate the constraints experienced when co-adapting systems in the real world, where mass-parallelization is rarely possible [16]. The algorithm features two learning loops: The first is training and updating the behaviour and neural networks; the second is optimizing the agent design via the neural network surrogates.

6 Experiments

Using the developed data augmentation techniques to imagine transitions of unseen behaviour and designs, we will now investigate whether these techniques can improve the performance of model-free co-adaptation. To this end, we will train a multi-dynamics model on three continuous control tasks: Half-Cheetah, Hopper and Walker. In these environments, the co-adaptation algorithm can change design variables such as the limb lengths every hundred episodes, and the goal is to maximize the performance given a reward function. The trained models are then used to augment the training data with imagined trajectories from (a) previously seen designs, (b) previous designs with noise and (c) randomly selected designs. We will focus in our evaluation specifically on the data efficiency of the developed algorithms in a scenario where each agent is sequentially updated in behaviour and design, as it would be the case in real-world co-adaptation. We will conclude this section by discussing open challenges and the apparent non-linearities in the dynamics predictions when varying the design variables.

6.1 Experimental Setup

During co-adaptation, each experiment starts on the same initial five agent designs that were selected randomly but are kept constant for all experiments, then the algorithm runs for 50 more design iterations which are optimised (exploitation) or randomly chosen (exploration) in alternating fashion. For each of the initial designs, 300 episodes are being executed, and thereafter 100 episodes. For all environments, we execute 1000 steps per episode. During design optimisation, we use a batch size of 36 initial states

---

Algorithm 1. CoIm: Co-Imagination

Initialise replay buffers: $Replay_{\text{Pop.}}$, $Replay_{\text{Ind.}}$, $Replay_{s_0}$, $Replay_{\text{Art.}}$.
Initialise first design ξ
for $i \in (1, 2, ..., M)$ **do**
  $\pi_{\text{Ind.}} = \pi_{\text{Pop.}}$.
  $Q_{\text{Ind.}} = Q_{\text{Pop.}}$.
  Initialise an empty $Replay_{\text{Ind.}}$.
  Initialise an empty $Replay_{\text{Art.}}$.
  Fill $Replay_{\text{Art.}}$ with random batches from $Replay_{\text{Pop.}}$ based on design strategy
  while *Not finished optimising local policy* **do**
    Collect experience $(s_0, a_0, r_1, s_1, ..., s_T, r_T)$ for current design ξ with policy $\pi_{\text{Ind.}}$.
    Add quintuples $(s_t, a_t, r_{t+1}, s_{t+1})$ to $Replay_{\text{Ind.}}$.
    Add quintuples $(s_t, a_t, r_{t+1}, s_{t+1}, \xi)$ to $Replay_{\text{Pop.}}$.
    After each episode refresh $Replay_{\text{Art.}}$. experience using $\widetilde{p}$ and $\pi_{\text{Pop.}}$..
    Add start state s_0 to $Replay_{s_0}$.
    Train networks $\pi_{\text{Ind.}}$ and $Q_{\text{Ind.}}$ with random batches from $Replay_{\text{Ind.}}$ and Eq. (7–8)
    Train $\pi_{\text{Pop.}}$ and $Q_{\text{Pop.}}$ with batches from $Replay_{\text{Pop.}}$ and $Replay_{\text{Art.}}$ via Eq. (7–8)
    Train Gaussian Dynamics Ensemble model $\widetilde{p}$ with batches from $Replay_{\text{Pop.}}$ with Eq. (6)
  end
  if i *is even* **then**
    Sample batch of start states $s_b = (s_0^1, s_0^2, ..., s_0^n)$ from $Replay_{s_0}$
    Find optimal design ξ with objective function $\max_\xi \frac{1}{n} \sum_{s \in s_b} Q_{\text{Pop.}}(s, \pi_{\text{Pop.}}(s, \xi), \xi)$
  else
    Sample design ξ with exploration strategy
  end
end

---

to estimate the fitness. SAC parameters are: a discount of 0.99, a tau of 0.005, 3E$-$4 for policy and Q function learning rates, $\alpha = 0.01$, and 3 hidden layers with 200 neurons each for policy and q-value networks. Individual networks have 1000 updates per episode, population networks have 250, and the batch size is 256. The probabilistic dynamics ensemble [3] with 3 networks has 3 hidden layers with 500 neurons, the propagation method is the random model, the learning rate is 3E$-$4 and the weight decay is 1E$-$5.

6.2 Environments

Experiments are performed both in PyBullets' [4, 16] Half-Cheetah and MuJoCos' [27] Walker and Hopper continuous control environments:

Half Cheetah: Half Cheetah has a 17-dimensional state space that contain joint positions and velocities, angular velocity, the horizontal and vertical speed of the centre of mass. Morphological parameters of Half Cheetah are described by continuous design parameters $\xi \in \mathbb{R}^6$, where ξ is used to calculate the scaled version of the original Half-Cheetah leg lengths as $(0.29 \times \xi_1, 0.3 \times \xi_2, 0.188 \times \xi_3, 0.29 \times \xi_4, 0.3 \times \xi_5, 0.188 \times \xi_6)$, where each $\xi_i \in [0.8, 2.0]$. Design bounds are defined as 0.8 for lower and 2 for upper bound on every ξ_i. Action $a \in \mathbb{R}^6$ defines the joint acceleration and the reward function r is given by $r(s) = max(\frac{\Delta x}{10}, 10)$, where Δx is the horizontal speed defining the forward motion of the agent.

Walker: Similarly, Walker's state space is 18-dimensional, with an action space of $a \in \mathbb{R}^6$ and reward function of $r(s) = \frac{1}{10}((h_{torso} > 0.8) \times (max(\Delta x, 0) + 1) - \|y_{rot}\|_2 \times 0.1)$, where h_{torso} is the height of the torso and y_{rot} is the vertical orientation of the torso. The morphology parameters of Walker scale the agent's leg and foot lengths calculated as $(0.4 \times \xi_1, 0.45 \times \xi_2, 0.6 \times \xi_3, 0.2 \times \xi_4, 0.45 \times \xi_5, 0.6 \times \xi_6, 0.2 \times \xi_7)$.

Hopper: Hopper's state space is 11-dimensional, action space is 3-dimensional and reward function is similarly to Walker given by $r(s) = (max(h_{torso} > 0.5, 0.1) \times (max(\Delta x, 0) + 1) - \|y_{rot}\|_2 \times 0.1)$. The morphology of Hopper scales the original leg segments and is defined as $(0.4 \times \xi_1, 0.45 \times \xi_2, 0.5 \times \xi_3, 0.39 \times \xi_4)$. Design bounds for both Walker and Hopper are defined as 0.5 for lower and 2 for upper bound on every ξ_i.

6.3   Co-imagination of Unaltered Designs ($G(\xi) = \xi$)

First, we study how our method performs when the artificial experience is created without disturbance to experienced design parameters, using the design generator $G(\xi) = \xi$. This way the learned dynamics model imagines the transitions only for designs seen in the past or the current design. Figure 1 shows the maximum cumulative episodic reward achieved for each design, with the performance of the model-free baseline shown in blue and our proposed model-based approach shown in red. We are able to show that using model-based imagination leads to at least a similar performance in Half-Cheetah (Fig. 1(a)), but leads to an increase in data efficiency for Walker and Hopper (Fig. 1(b-c)). When it comes to the Walker task in 1(c), which is more difficult compared to Half-Cheetah, the outcome of using our artificial experience is more significant with a smaller variance indicating a more stable and reliable increase in performance. This shows that model-free co-adaptation can benefit from model-based data augmentation, or at least reach similar performance in the case of Half-Cheetah. To better compare both methods and investigate the performance of designs uncovered for Walker task, we train SAC from scratch on the final designs found by model-free and model-based co-adaptation. Figure 2 shows the maximum cumulative rewards achieved by the top 50% performing designs for both model-free and model-based co-adaptation, demonstrating the ability of the proposed method to outperform model-free co-adaptation (Welch's t-test: $p = 0.024$) and uncover better performing designs.

6.4   Co-imagination of Unknown Designs ($G(\xi) \sim N(\xi, \sigma^2)$ or uniform(Ξ))

Given the previous result, we hypothesise that adding even more diverse imagined experience should further improve our co-adaptation method's performance. We use the design generator $G(\xi) \sim N(\xi, \sigma^2)$, with $\sigma = 0.1$, and $G(\xi) \sim$ uniform(Ξ), meaning adding noise or selecting random designs respectively, to create the imagined rollouts with our dynamics model. Figure 3 shows the results of imagining transitions for

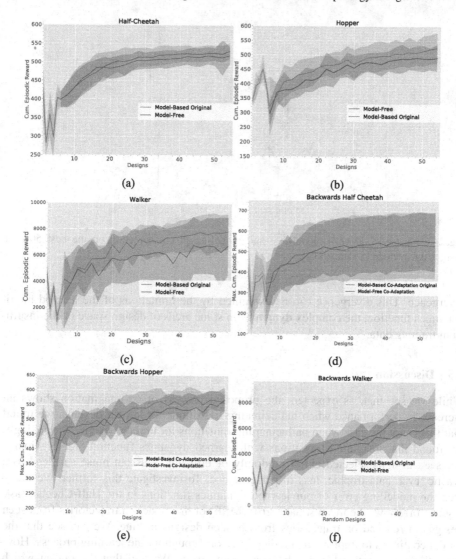

Fig. 1. Cumulative episodic rewards collected by our method with known designs in red and the model-free co-adaptation in blue. The first row shows forward walking and second row showing backwards walking task. Half-Cheetah includes 30 seeds, Hopper 10 and Walker 15. For each design the best performance is reported after 100 episodes. (Color figure online)

unknown designs on the co-adaptation performance with the random selection strategy in black and noisy in yellow. When it comes to Half-Cheetah in 3(a) which is easier to learn than others, the noisy strategy is able to continuously learn in it, however, is still not performing as well as the original strategy where the design parameter is unchanged. However, in the more difficult Hopper in 3(b) and Walker in 3(c) where falling over is possible, we can see that noisy and sometimes also random strategy quickly leads to

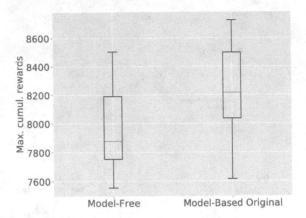

Fig. 2. Comparison between model-free and model-based co-adaptation on the best 50% of designs uncovered (18 samples).

divergence. This divergence can be explained by the limitations of the learned multi-dynamics function, the complex dynamics in some areas of design space or the insufficient training data.

6.5 Discussion

While we see that, as expected, the proposed process of co-imagination shows an increase in performance when hallucinating transitions for known designs, we found that this is not the case when hallucinating transitions for unseen designs. Surprisingly, we found that imagining transitions with learned dynamics functions for previously not seen designs selected either randomly or utilising Gaussian noise can lead to a drastic, even catastrophic, loss in performance. To investigate this further, we evaluated the modelling error of our learned dynamics functions in the Half-Cheetah task. Figure 4(a) shows the mean square error (MSE) of the model on trajectories from seen designs in red and on trajectories from unseen designs in blue. We can see that the model continues to improve its prediction error throughout the learning process. However, there is a noticeable gap between both errors. We find that the rate at which the prediction error slows decreasing for designs not yet experienced is much higher. This seems to hint at the limited capability of the dynamics model to generalise across the design space. Further analysis of the dynamics of the environment provides additional insight: Given a randomly selected state-action pair, Fig. 4(b) shows the magnitude of change in the predicted next state when changing the design variables only, i.e. $\| h(s_t, a_t, \xi_{original}) - h(s_t, a_t, \xi_{new}) \|^2$ with the original design $\xi_{original}$ and the adapted design ξ_{new}. The magenta circle shows where the original state s_t lies in the design space. Due to the high dimensionality of the design variable, the plot shows the resulting change across a two-dimensional manifold in the design space with two principal

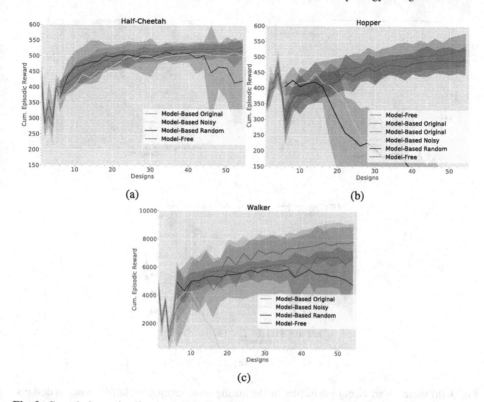

Fig. 3. Cumulative episodic rewards collected by our model-based co-adaptation method with known designs in red, noisy designs in yellow and random designs in black. The model-free co-adaptation baseline [16] in blue. Three agents from left are Half-Cheetah, Hopper and Walker with forward walking task. For each design the best performance is reported after 100 episodes. (Color figure online)

components. We can see that the change in the predicted next state is highly non-linear and may explain the inability of the dynamics function to accurately predict transitions for designs not yet experienced. This effect can also be found in much simpler dynamical systems, such as the cartpole system, where changes to the pole length and mass impact the x acceleration (Fig. 4(c)), showing non-linear changes to acceleration along the x-axis in certain states.

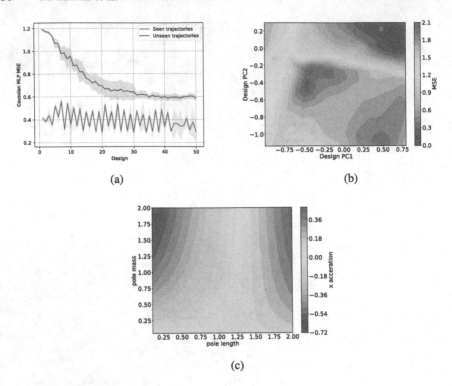

(a)   (b)

(c)

Fig. 4. (a) Mean square error of dynamics model during co-adaptation in Half-Cheetah on designs seen (red) vs an unknown design (blue). (b) Difference in dynamics model over changes of design in Half-Cheetah. Axes are the principal components of the design space. Colours indicate the magnitude of change when predicting the next state, using one design (purple) as reference point. (c) Difference in dynamics when changing pole length and mass of the classic CartPole task. (Color figure online)

7   Conclusion

The problem of co-adapting agents' behaviour and morphology has been studied primarily using model-free learning frameworks or algorithms utilising known dynamics models and simulators. We investigated and proposed using a model-based learning approach to improve the data-efficiency of model-free co-adaptation by co-imagining transitions for unseen behaviour and designs. We showed a performance improvement when co-adapting the design and behaviour in two out of three continuous control tasks. However, we also found limitations of using model-learning approaches in the context of co-adaptation: Using learned models to predict transitions of not yet experienced agent designs and augmenting the training process with the same leads to a severe and sometimes catastrophic deterioration in learning performance, impacting both the policy learning and design optimization process. This uncovered limitation of current approaches provides an interesting avenue for future research into robust and generalizable model-learning approaches, suitable for their use in co-adaptation problems, where we face an infinite number of design variations.

Acknowledgments. This work was supported by the Research Council of Finland Flagship programme: Finnish Center for Artificial Intelligence FCAI and by the German Research Foundation (DFG, Deutsche Forschungsgemeinschaft) as part of Germany's Excellence Strategy – EXC 2050/1 – Project ID 390696704 – Cluster of Excellence "Centre for Tactile Internet with Human-in-the-Loop" (CeTI) of Technische Universität Dresden.

The authors wish to acknowledge the generous computational resources provided by the Aalto Science-IT project and the CSC – IT Center for Science, Finland.

We thank the reviewers for their insightful comments and help for improving the manuscript.

References

1. Bonyadi, M.R., Michalewicz, Z.: Particle swarm optimization for single objective continuous space problems: a review. Evol. Comput. **25**(1), 1–54 (2017). https://doi.org/10.1162/EVCO_r_00180

2. Chen, T., He, Z., Ciocarlie, M.: Hardware as policy: mechanical and computational co-optimization using deep reinforcement learning (CoRL) (2020). http://arxiv.org/abs/2008.04460

3. Chua, K., Calandra, R., McAllister, R., Levine, S.: Deep reinforcement learning in a handful of trials using probabilistic dynamics models. In: Advances in Neural Information Processing Systems 2018-Decem (Nips), pp. 4754–4765 (2018)

4. Coumans, E., Bai, Y.: PyBullet, a python module for physics simulation for games, robotics and machine learning (2016-2021). http://pybullet.org

5. Dinev, T., Mastalli, C., Ivan, V., Tonneau, S., Vijayakumar, S.: Co-designing robots by differentiating motion solvers. arXiv preprint arXiv:2103.04660 (2021)

6. Gupta, A., Savarese, S., Ganguli, S., Fei-Fei, L.: Embodied intelligence via learning and evolution. Nat. Commun. **12**(1) (2021). https://doi.org/10.1038/s41467-021-25874-z, http://dx.doi.org/10.1038/s41467-021-25874-z

7. Ha, D.: Reinforcement learning for improving agent design. Artif. Life **25**(4), 352–365 (2019). https://doi.org/10.1162/artl_a_00301

8. Ha, S., Coros, S., Alspach, A., Kim, J., Yamane, K.: Computational co-optimization of design parameters and motion trajectories for robotic systems. Int. J. Robot. Res. **37**(13–14), 1521–1536 (2018)

9. Haarnoja, T., et al.: Soft actor-critic algorithms and applications (2018). https://doi.org/10.48550/ARXIV.1812.05905, https://arxiv.org/abs/1812.05905

10. Hafner, D., Lillicrap, T., Ba, J., Norouzi, M.: Dream to control: learning behaviors by latent imagination, pp. 1–20 (2019). http://arxiv.org/abs/1912.01603

11. Harpak, A., et al.: Genetic adaptation in New York City rats. Genome Biol. Evol. **13**(1) (2021). https://doi.org/10.1093/gbe/evaa247

12. Jackson, L., Walters, C., Eckersley, S., Senior, P., Hadfield, S.: ORCHID: optimisation of robotic control and hardware in design using reinforcement learning. In: 2021 IEEE/RSJ International Conference on Intelligent Robots and Systems (IROS), pp. 4911–4917 (2021). https://doi.org/10.1109/IROS51168.2021.9635865

13. Leong, M., Bertone, M.A., Savage, A.M., Bayless, K.M., Dunn, R.R., Trautwein, M.D.: The habitats humans provide: factors affecting the diversity and composition of arthropods in houses. Sci. Rep. **7**(1), 15347 (2017). https://doi.org/10.1038/s41598-017-15584-2

14. Liao, T., et al.: Data-efficient learning of morphology and controller for a microrobot. In: IEEE International Conference on Robotics and Automation (ICRA), pp. 2488–2494 (2019). https://doi.org/10.1109/ICRA.2019.8793802

15. Lipson, H., Pollack, J.B.: Automatic design and manufacture of robotic lifeforms. Nature **406**(6799), 974–978 (2000). https://doi.org/10.1038/35023115

16. Luck, K.S., Amor, H.B., Calandra, R.: Data-efficient co-adaptation of morphology and behaviour with deep reinforcement learning. In: Kaelbling, L.P., Kragic, D., Sugiura, K. (eds.) Proceedings of the Conference on Robot Learning. Proceedings of Machine Learning Research, vol. 100, pp. 854–869. PMLR (2020). https://proceedings.mlr.press/v100/luck20a. html

17. Mitteroecker, P.: How human bodies are evolving in modern societies. Nat. Ecol. Evol. **3**(3), 324–326 (2019). https://doi.org/10.1038/s41559-018-0773-2

18. Parks, S.E., Johnson, M., Nowacek, D., Tyack, P.L.: Individual right whales call louder in increased environmental noise. Biol. Let. **7**(1), 33–35 (2011)

19. Potts, R.: Evolution and environmental change in early human prehistory. Annu. Rev. Anthropol. **41**(1), 151–167 (2012). https://doi.org/10.1146/annurev-anthro-092611-145754

20. Racanière, S., et al.: Imagination-augmented agents for deep reinforcement learning. In: Advances in Neural Information Processing Systems, vol. 30 (2017)

21. Rajani, C., Arndt, K., Blanco-Mulero, D., Luck, K.S., Kyrki, V.: Co-imitation: learning design and behaviour by imitation. In: Proceedings of the AAAI Conference on Artificial Intelligence, vol. 37, no. 5, pp. 6200–6208 (2023). https://doi.org/10.1609/aaai.v37i5.25764, https://ojs.aaai.org/index.php/AAAI/article/view/25764

22. Reil, T., Husbands, P.: Evolution of central pattern generators for bipedal walking in a real-time physics environment. IEEE Trans. Evol. Comput. **6**(2), 159–168 (2002). https://doi.org/10.1109/4235.996015

23. Rosser, K., Kok, J., Chahl, J., Bongard, J.: Sim2real gap is non-monotonic with robot complexity for morphology-in-the-loop flapping wing design. In: 2020 IEEE International Conference on Robotics and Automation (ICRA), pp. 7001–7007. IEEE (2020)

24. Schaff, C., Yunis, D., Chakrabarti, A., Walter, M.R.: Jointly learning to construct and control agents using deep reinforcement learning. Proceedings - IEEE International Conference on Robotics and Automation, vol. 2019-May, pp. 9798–9805 (2019). https://doi.org/10.1109/ICRA.2019.8793537

25. Sims, K.: Evolving 3D morphology and behavior by competition. Artif. Life **1**(4), 353–372 (1994). https://doi.org/10.1162/artl.1994.1.4.353

26. Stanley, K.O., Miikkulainen, R.: Competitive coevolution through evolutionary complexification. J. Artif. Intell. Res. **21**, 63–100 (2004)

27. Todorov, E., Erez, T., Tassa, Y.: MuJoCo: a physics engine for model-based control. In: 2012 IEEE/RSJ International Conference on Intelligent Robots and Systems, pp. 5026–5033 (2012). https://doi.org/10.1109/IROS.2012.6386109

An Evolutionary Approach to Feature Selection and Classification

Rodica Ioana Lung[1] and Mihai-Alexandru Suciu[1,2]

[1] Centre for the Study of Complexity, Babeş-Bolyai University,
Cluj-Napoca, Romania
{rodica.lung,mihai.suciu}@ubbcluj.ro
[2] Faculty of Mathematics and Computer Science, Babeş-Bolyai University,
Cluj-Napoca, Romania

Abstract. The feature selection problem has become a key undertaking within machine learning. For classification problems, it is known to reduce the computational complexity of parameter estimation, but it also adds an important contribution to the explainability aspects of the results. An evolution strategy for feature selection is proposed in this paper. Feature weights are evolved with decision trees that use the Nash equilibrium concept to split node data. Trees are maintained until the variation in probabilities induced by feature weights stagnates. Predictions are made based on the information provided by all the trees. Numerical experiments illustrate the performance of the approach compared to other classification methods.

Keywords: feature selection · evolutionary strategy · decision tree · random forest · game theory

1 Introduction

Evolutionary algorithms (EAs) have been widely used for feature selection and classification purposes [6,23,27], as they are flexible and adaptable to different optimization environments. Genetic algorithms (GAs) have been the first natural choice, since the binary representation fits within this problem naturally [9]. Many examples of genetic algorithms are mentioned in [27], and in particular, the combination with Decision Trees (DT) has been appealing from the start [2], with many variants following, expanding to random forests [10] or multi-objective optimization [26]. Examples of EAs for feature selection and decision trees can be found in [13,14], and applications in network intrusion detection [21], chemistry [10], speech recognition [15], etc.

However, the efficiency of any evolutionary approach depends on proper parameter tuning and fitness evaluation mechanisms. Within EAs, the selection is mainly responsible for guiding the search, as the survival of newly created individuals ultimately relies on their fitness. When the fitness is associated with the results of classification tasks and is based on some performance indicator reported on data samples, we find a high variability between different samples, which makes comparisons of results irrelevant for selection purposes.

The fighting for survival paradigm is usually implemented within EAs by comparing individuals using their fitness values and deciding, depending on the selection mechanism used, which ones are preserved and which are discarded. It is considered that individuals compete for resources and the fittest will survive, to further access, exploit, and explore them.

In this paper, we introduce a feature selection-based classification method that evolves feature weights by using game-theoretic decision trees. Individuals represent vectors of feature importances, evolved with the purpose of identifying the most relevant features in the data set that may explain the classification problem. A game-theoretic decision tree is used for classification and evaluation purposes. However, there is no selection involved, trees are grown together, and they form an ecosystem in which all of them are involved in the prediction task. An individual stops evolving when there is no more variation in the probabilities that it provides for selecting features for inducting its tree. A practical application that analyses countries' income group classification based on world development indicators presents an interpretation of the feature selection approach.

2   Evolution Strategy Decision Forest (ESDF)

ESDF evolves individuals representing feature weights to identify those that most explain the data. The evolution strategy mechanism, as well as the decision trees used for classification, are presented in what follows.

Decision Trees and Random Forests. Decision trees were some of the most popular machine learning techniques [19,25] due to their efficiency and explainability. They recursively split the data space into separate regions aiming to find areas as pure as possible. Large trees tend to overfit the data, and small trees may not split it enough. One way to overcome these drawbacks is to use ensembles of trees, e.g. in the form of random forests [4], and aggregate their results in some form. Decision trees can also be used to assess feature importances based on the tree structure, and the purity of split data in each node [24].

In what follows, we consider the binary classification problem: given a data set $\mathcal{X} \subset \mathbb{R}^{N \times d}$ containing N instances $x_i \in \mathbb{R}^d$, $i = 1, \ldots, N$ and d and $\mathcal{Y}$ their corresponding labels, with $y_i \in \{0, 1\}$ the label of x_i, the goal is to find a rule that best predicts the labels $\hat{y}$ for instances x that come from the same distribution as $\mathcal{X}$.

Equilibrium-Based Decision Tree. Most decision trees are built top-down starting with the entire data set at the root level. Different trees split data in different manners, by using either axes parallel, oblique, or non-linear hyperplanes [1,12,16], computing their parameters by using some purity indicators that evaluate sub-nodes data, e.g. gini index, entropy, etc. [28]. At each node level, some optimisation process takes place involving either hyperplane parameters, the attributes to use for the split, or both.

In this paper, we propose the use of a decision tree that computes hyperplane parameters by approximating the equilibrium of a non-cooperative game [22]. The equilibrium of the game aims to find parameters such that each sub-node 'receives' data as pure as possible by shifting instances with different labels to the left/right of the hyperplane. Thus, in order to split node data X, Y based on an attribute j, we use the following non-cooperative game $\Gamma(X, Y | j)$:

- the players, L and R correspond to the two sub-nodes and the two classes, respectively;
- the strategy of each player is to choose a hyperplane parameter: β_L and β_R, respectively;
- the payoff of each player is computed in the following manner:

$$u_L(\beta_L, \beta_R | j) = -n_0 \sum_{i=1}^{n} (\beta_{1|j} x_{ij} + \beta_{0|j})(1 - y_i),$$

and

$$u_R(\beta_L, \beta_R | j) = n_1 \sum_{i=1}^{n} (\beta_{1|j} x_{ij} + \beta_{0|j}) y_i,$$

where

$$\beta = \frac{1}{2}(\beta_L + \beta_R)$$

and n_0 and n_1 represent number of instances having labels 0 and 1, respectively.

The concept of Nash equilibrium for this game represents a solution such that none of the players can find a unilateral deviation that would improve its payoff, i.e., none of the players can shift data more to obtain a better payoff. An approximation of an equilibrium can be obtained by using a stylized version of fictitious play [5] in the following manner. For a number of iterations (η), the best response of each player against the strategy of the other player is computed using some optimization algorithm. As we only aim to approximate β values that reasonably split the data, the search stops after the number of iterations has elapsed. In each iteration, the best response to the average of the other player's strategies in the previous ones is considered the fixed one. The procedure is outlined in Algorithm 1.

For each attribute $j \in \{1, \ldots, d\}$, data is split using Algorithm 1; the attribute that is actually used to split the data is chosen based on the Gini index [28]. The game theoretic decision tree splits data in this manner, recursively, until node data becomes pure (all instances have the same label) or a maximum tree depth has been achieved.

Prediction. A DT provides a partition for the training data. To predict the label for a tested instance x, the corresponding region of the space, i.e., its leaf, is identified. The decision is made based on the proportion of labels in that leaf. Let DT be a decision tree based on a data set $\mathcal{X}$ and x a tested value. Then

---

Algorithm 1. Approximation of Nash equilibrium

Input: X, Y - data to be split by the node; j - attribute evaluated
Output: $X_{L|j}, y_{L|j}, X_{R|j}, y_{R|j}$, and β_j to define the split rule for the node based on attribute j;
Initialize β_L, β_R at random (standard normal distribution)
for η iterations: **do**
  Find $\beta_L = \operatorname*{argmin}_b u_L(b, \beta_R)$;
  Find $\beta_R = \operatorname*{argmin}_b u_R(\beta_L, b)$;
end for
$\beta_j = \frac{1}{2}(\beta_L + \beta_R)$
$X_{L|j} = \{x \in X | x_j^T \beta \leq 0\}, \; y_{L|j} = \{y_i \in y | x_i \in X_{L|j}\}$
$X_{R|j} = \{x \in X | x_j^T \beta > 0\}, \; y_R = \{y_i \in y | x_i \in X_{R|j}\}$

---

the decision tree DT has partitioned $\mathcal{X}$ into data found in its leaves, denoted by $DT_1, \ldots, DT_m$, where m is the number of leaves of DT. Let $DT(x)$ be the data set corresponding to the leaf region of x, $DT(x) \subset \mathcal{X}$. Typically, the model would assign to x label y with a probability equal to the proportion of elements with class y in $DT(x)$.

Feature Importance. The game-based splitting mechanism of the tree indicates for each node the attribute that 'best' splits the data. It is reasonable to assume that the position of the node in the tree indicates also the importance of the attribute in classifying the data and an importance measure can be derived based on the structure of the tree. Thus, for each feature $j \in \{1 \ldots d\}$ we denote by

$$\nu_j = \{\nu_{jl}\}_{l \in I_j},$$

the set containing the nodes that split data based on attribute j, with I_j the set of corresponding indexes in the tree and let $\delta(\nu_{jl})$ be the depth of the node ν_{jl} in the decision tree, with values starting at one at the root node. Then the importance $\phi(j)$ of attribute j can be computed as:

$$\phi(j) = \begin{cases} \sum_{l \in I_j} \dfrac{1}{\delta(\nu_{jl})}, & I_j \neq \emptyset \\ 0, & I_j = \emptyset \end{cases}. \tag{1}$$

The formula (1) is based on the assumption that attributes that split data at the first levels of the tree may be more influential. Also, multiple appearances of an attribute in nodes with higher depths may indicate its importance and are counted in $\phi()$.

Evolution Strategy Decision Forest. The Evolution Strategy decision forest evolves a population of feature weights in order to identify their importance for classification. Individuals in the final population indicate feature importances while, overall, the evolution strategy performs classification using the evolved feature weights.

Encoding. Individuals w are encoded as real, positive valued vectors of length d, where w_j represents the importance of feature j, $j = 1, \ldots, d$.

Initialization. All individuals are initialized with equal weights of $1/d$. ESDF maintains a population of *pop_size* individuals.

Evaluation. There is no explicit fitness assignment mechanism within ESDF. Individuals are evolved regardless of their performance based on the information received from the environment. The motivation behind this approach can be expressed in two ways: on the one hand, the evaluation of feature weights may be performed using some classification algorithm based on its performance. However, there is no universally accepted performance indicator that can be used to compare results in a reliable manner. From a nature-inspired point of view, on the other hand, a forest paradigm does not require direct competition for resources. Trees grow in forests and adapt to each other. Some may stop growing due to a lack of resources, but they do not replace each other every generation. Thus, all trees are added to the forest and evaluation takes place on the entire forest at the end of the search.

Evolution. The evolution process takes place iteratively until a maximum number of generations is reached, or until all individuals have achieved maturity.

Updating Mechanism. In each iteration, a bootstrapped sample from the data is used to induct a game theoretic decision tree for each individual in the population. Attributes for each tree are selected with a probability proportional to their corresponding weights in the individual. Thus, for individual w representing feature weights, the probability to select feature j is:

$$P(j|w) = \frac{w_j}{\sum_{k=1}^{d} w_k}. \tag{2}$$

In the first iteration, probabilities are all equal. However, in subsequent generations, feature importances reported by each tree are used to update the corresponding individuals:

$$w_j \leftarrow w_j + \phi(j) \cdot \alpha, j = 1, \ldots, d, \tag{3}$$

where $\phi(j)$ represents the importance of feature j reported by the tree inducted by using individual w (Eq. (1)), and α is a parameter controlling the magnitude of the update.

In this manner, the weights of attributes that are deemed important by the tree are increased, also increasing the probability that they are selected in the next iterations. While apparently, this may lead to overfitting features, the fact that in each iteration, a different sample from the data is used for inducting trees, that the search of an individual stops when it reaches maturity, and also that there are several individuals maintained on the same data ensures diversity preservation.

Thus, the role of the tree is to assign feature importances for the updating mechanism. Apart from that, each tree is also preserved and further used in prediction for classification.

Maturity. Individuals are used to select attributes for training using probabilities in Eq. (2). The goal of the search is to find a distribution over the feature set: if several iterations of applying Eq. (3) feature importances reported by the tree do not change significantly, the standard deviation of the probabilities $P()$ will not vary. ESDF considers that an individual has reached maturity and stops evolving and inducting trees using it if there is no change in the variation of the corresponding probabilities. The following condition is used to compare the evolution of an individual from generation t to $t + 1$:

$$\frac{\sigma(P(\cdot|w_t))}{\sigma(P(\cdot|w_{t+1}))} < \epsilon \tag{4}$$

where σ denotes the standard deviation and $P(\cdot|w)$ the vector of probabilities in Eq. (2) taken for all attributes j. If condition 4 holds, the individual is considered to have reached maturity and is no longer updated,i.e., and no more trees are inducted based on him. Not all individuals reach maturity at the same time, which means that the size of the population decreases during the search, reducing the complexity of the method.

Classification. Each individual inducts several trees until reaching maturity. All these trees form a forest that can be used to make predictions for the classification problem. This is the last step of the algorithm, and it can be used to validate results. Trees are inducted by using different data samples and different attributes. Selecting attributes without any fitness measure may provide (or not) good classification trees. To avoid overfitting or using misleading trees, prediction is not made by considering labels in the trees' leaves but by aggregating data from the leaves corresponding to tested instances and further applying logistic regression (LR) to make predictions. Each tree offers a neighborhood for the tested instance. Aggregating all these regions will provide a set of relevant instances, allowing the algorithm to make an informed prediction.

Outline of ESDF. ESDF has two main steps: an evolution step (Algorithm 2, line 6) and a prediction step (Algorithm 2, line 17).

During the evolution step, a population of weights is updated several iterations until there is no variation in the probabilities they provide for selecting attributes for tree induction. Prediction is performed for each tested instance by aggregating data corresponding to its leaves in all inducted trees and applying logistic regression on the resulting data set.

The output of ESDF consists of prediction probabilities for the tested data that can be used to evaluate the entire approach and the average feature weights over the entire population.

Algorithm 2. ES-DF: Evolution Strategy Decision Forest

1: **input**: training set $\mathcal{X}, \mathcal{Y}$,
2: **parameters**:: - *pop_size* - population size;
   - p - the proportion of attributes used for a tree;
   - μ - maximum tree depth;
   - $MaxGen$ - maximum number of generations;
3: **output**: predictions C for a (test) set T; Feature weights ω;
4: $t = 0$;
5: Initialize population W_0 with $w_{0,ij} = 1/d$, $i = 1, \ldots, pop\_size$, $j = 1, \ldots, d$;
6: *Step 1: Evolution*
7: **while** $t < MaxGen$ **or not** all trees have reached maturity **do**
8:    $X_t \leftarrow$ sample of size N with replacement from $\mathcal{X}$;
9:    **for** each individual w_t **do**
10:      $\overline{X}_{t,w} \leftarrow$ sample proportion p of attributes from X_t using probabilities P in Eq. (2);
11:      $DT_{t,w} \leftarrow$ game based decision tree based on $\overline{X}_{t,w}, \mu$;
12:      Update w_{t+1} using Eq. (3);
13:      Check maturity using condition (4); if (4) holds, mark individual as mature and stop its update;
14:    **end for**
15:    $t \leftarrow t + 1$;
16: **end while**
17: *Step 2: Prediction*
18: **for** each $x_t \in T$ **do**
19:    $RF(x_t) = \cup_{w,t} DT(x_t)$;
20:    Fit LR on $RF(x_t)$;
21:    Assign c_t to x_t - probability that x_t has class 1, based on LR;
22: **end for**
23: **return**
   - $C = (c_1, c_2, \ldots, c(x_{|T|^1}))$
   - ω - average feature weights over the entire population.

[1] $|\cdot|$ denotes the cardinality of a set.

3   Numerical Experiments

Numerical experiments are used to test and illustrate the performance of ESDF and compare its results to other state-of-the-art classification models. This section is divided into two main parts: the first one presents results obtained on synthetic and real-world benchmarks with various degrees of difficulty used for classification, and the second part is a real data application involving the classification of countries' income groups.

Synthetic and Real-World Benchmarks

For synthetically generated test data, to ensure reproducibility and control the difficulty of the resulting data set, we use the `make_classification` function from the `scikit-learn`[1] Python library [18]. The degree of difficulty is con-

---

[1] version 1.1.1.

trolled by the generating function parameters: number of instances, number of attributes/features, degree of overlap between instances of different classes, the seed used to generate the test data, and class imbalance. For our experiments, we use the following: number of instances $(100, 200, 500, 1000, 2500)$, number of attributes $(20, 50)$, the seed used to generate the data (500), degree of overlap between instances of different classes $(0.1, 0.5)$, and all data sets generated are balanced. We generate test data sets for all combinations of the above parameters. In order to evaluate the feature selection mechanism, only half of the features in each data set are generated using the `make_classification` function, and the other half at random following a uniform distribution.

For real-world benchmarks, we use the following data sets from the UCI Machine Learning Repository [7]: iris data set (R1) from which we removed the *setosa* instances to obtain a linear non-separable binary classification problem, Pima Indians Diabetes (R2), Connectionist Bench (Sonar, Mines vs. Rocks) (R3), acute inflamations (R4), heart disease (R5), Somerville Happiness Survey (R6), appendicitis (R7), blogger (R8), bupa (R9), monks (R10), thoracic-surgery(R11), vertebra-column-2c (R12), wholesale-channel (R13), and the wdbc (R14) data set.

Experimental Set-Up. A *Stratified k-fold Cross-Validation* strategy [11] is used to estimate the expected prediction error. The data set is divided into $k = 10$ balanced folds, of which nine are used to train the model, and the tenth fold (the test fold) is used to evaluate the model. The train and test part are repeated $k = 10$ times, each time a different fold is used as a test fold. We repeat the k-fold cross-validation four times, each time a different seed is used to split the data (we use as seed the values $1, 2, 3, 4$), resulting in 40 indicator values that are compared.

For each test fold of a data set, we report performance metrics based on which we compare the performance of ESDF with other state-of-the-art classifiers. We train each compared classifier on the same train data as ES-DF for each fold and compare the results of ES-DF to those reported by the compared models on the test fold.

The performance metrics used for comparison are: AUC (area under the ROC curve) [8,20], the F_1 score [29], the accuracy ACC and the log-loss score [11].

ESDF results are compared to other decision tree-based classifiers, and because it uses Logistic Regression in the prediction step, we also compare results with this method. We also compare the performance of ES-DF to other well-known classifiers. The list of compared classifiers is: M0 - Support Vector Machine with a linear kernel, M1 - Support Vector Machine with a radial kernel, M2 - $k-$nearest-neighbour classifier with $k = 3$, M3 - AdaBoost classifier, M4 - Gaussian Naive Bayes, M5 - stochastic gradient descent, M6 - Gaussian process classification, M7 - decision tree classifier which splits nodes until its leaves contain only instances of one class, M8 - a decision tree with maximum depth equal to that of ESDF, M9 - a random forest classifier for which each estimator splits nodes until its leaves are pure, M10 - a random forest classifier with 10 estimators, M11 - a random forest classifier with 50 estimators, M12 - a random

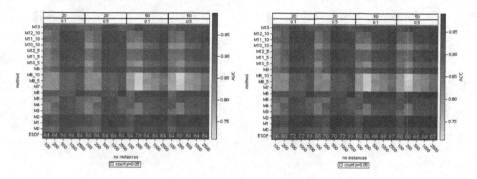

Fig. 1. Heatmaps for the AUC and ACC indicators for the synthetic data sets. A higher (darker) value is desirable. The last line also presents the number of times ESDF results were significantly better than the other method based on the p values of the t test comparing results reported for all folds (out of 120). The first line of the heading indicates the number of attributes and the second line the class separator for the data sets.

forest classifier with 100 estimators (for M10, M11 and M12 each estimator has a maximum depth equal to ESDF), and M13 - logistic regression classifier. For reproducibility and control, we use their implementation from the *scikit-learn* software library [18].

For each data set, each method reports 40 performance indicator values corresponding to the ten folds generated four times with different random number seed generators. To assess the difference in results, these values are compared using a paired t-test, with the null hypothesis that results provided by ESDF are worse than those reported by the other method. Rejecting this hypothesis, with a p-value smaller than 0.05, indicates that we can consider ESDF results significantly better than the others.

ESDF tested parameters are: population size 5, maximum number of generations 20, maximum tree depth 5 and 10, and α values 0.3, 0.8, and 1. All experiments are run with combination of these parameters.

Results. Figures 1 and 2 present the performance indicators obtained by all methods on the synthetic data sets. Figure 1 presents the AUC and accuracy indicators, and Fig. 2 the F1 and Log-loss indicators. A square in the heatmap shows the value obtained by a classifier for a specific set of parameters used to generate the data set (no instances - 100, 200, 500, 1000, 2500, no attributes - 20, 50, and overlap degree - 0.1, 0.5). Each row presents results obtained by a different classifier. For our approach, we also report the number of cases in which ESDF obtains significantly better results than the compared methods according to a t-test. In the case of AUC, ESDF obtains the best results for all data sets. For the accuracy indicator, ESDF consistently gives better results, and when more instances are available in the data set, the ensemble and logistic regression classifiers report, for a few data sets, results as good as the ones reported by ESDF. This is also the case for the F1 and Log-loss indicators (Fig. 2).

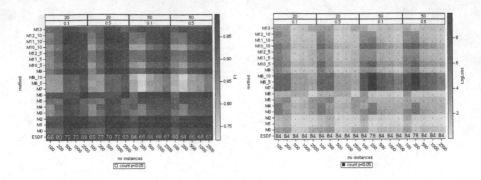

Fig. 2. Heatmaps for the F1 and LogLoss indicators for the synthetic data sets. For F1, a higher (darker) value is desirable, while for the LogLoss a smaller (lighter) value is better. The last line also presents the number of times ESDF results were significantly better than the other method based on the p values of the t test comparing results reported for all folds (out of 120). The first line in the heading indicates the number of attributes and the second line the class separator for the data sets.

Feature Selection. One possible manner to evaluate the efficiency of the feature selection mechanism is to compute the stability indicator SC [3,17] over the ten folds. Values of the stability indicator are based on average correlations: an SC close to 1 indicates that the feature selection method selects the same features in several runs on different samples of the data set, indicating stability. The values of the SC score when selecting half of the features based on their weights for the synthetic data sets vary between 0.7 and 1, indicating the stability of the approach. For data sets with 20 attributes the confidence interval for SC is (0.97, 0.99), and for those with 50 attributes is (0.88, 0.93).

Table 1 presents the results obtained by ESDF against the best result reported by the compared methods on the real-world data sets. The mean and standard deviation for the AUC and Log-loss indicators are presented. The statistically better results are highlighted. It can be seen that ESDF consistently gives better results. When comparing AUC values it can be seen that ESDF either gives statistically better results or is indifferent when compared to the best performing compared classifier.

Regarding different ESDF parameter settings, we found no significant differences among different values, either for synthetic or real-world benchmarks. The value of α does not influence the results because it is not directly used for the induction of trees.

Classification of low-income countries based on world development indicators: an application

The World Bank classifies countries into five income groups: high income, upper middle income, lower middle income, and low income yearly, based on gross national income (GNI) per capita in USD values, using the Atlas methodology[2]. The classification list for 2022 is based on 2021 data. In 2022, the GNI

---
[2] https://datahelpdesk.worldbank.org/knowledgebase/articles/378832-what-is-the-world-bank-atlas-method, last accessed January 2023.

Table 1. Mean and standard deviation for the AUC and Log-loss indicators in the case of real-world data sets for ESDF and the best performing compared classifier (best M). A (⋆) symbol highlights the ESDF results that can be considered statistically better than the other method.

data set	AUC (ESDF)	AUC (best M)	Log-loss (ESDF)	Log-loss (best M)
R1	0.99 ± 0.03⋆	M0: 0.95 ± 0.06	0.15 ± 0.14⋆	M3: 0.89 ± 0.08
R2	0.83 ± 0.05⋆	M6: 0.73 ± 0.05	0.65 ± 0.17	M5: 0.69 ± 0.06
R3	0.92 ± 0.06⋆	M2: 0.86 ± 0.08	0.56 ± 0.35⋆	M8: 0.72 ± 0.10
R4	1.00 ± 0.00	M0: 1.00 ± 0.00	0.00 ± 0.00⋆	M4: 0.83 ± 0.08
R5	0.88 ± 0.07⋆	M4: 0.84 ± 0.07	0.73 ± 0.53	M7: 0.72 ± 0.07
R6	0.66 ± 0.13⋆	M11: 0.62 ± 0.12	0.94 ± 0.30	M5: 0.53 ± 0.10⋆
R7	0.83 ± 0.16⋆	M3: 0.79 ± 0.16	1.34 ± 1.64	M8: 0.69 ± 0.18⋆
R8	0.92 ± 0.11⋆	M1: 0.82 ± 0.14	0.62 ± 1.05⋆	M3: 0.64 ± 0.15
R9	0.77 ± 0.08⋆	M9: 0.72 ± 0.08	0.66 ± 0.14⋆	M3: 0.70 ± 0.07
R10	1.00 ± 0.01	M8: 0.99 ± 0.02	0.13 ± 0.05⋆	M13: 0.76 ± 0.05
R11	0.63 ± 0.09⋆	M7: 0.56 ± 0.09	0.64 ± 0.22	M7: 0.56 ± 0.09⋆
R12	0.94 ± 0.04⋆	M6: 0.84 ± 0.07	0.39 ± 0.25⋆	M4: 0.80 ± 0.06
R13	0.96 ± 0.03⋆	M9: 0.91 ± 0.05	0.36 ± 0.28⋆	M8: 0.85 ± 0.05
R14	1.00 ± 0.01⋆	M1: 0.98 ± 0.02	0.12 ± 0.20⋆	M7: 0.92 ± 0.04

per capita is influenced by factors such as economic growth, inflation, exchange rates, and population growth. The classification is based on GNI intervals[3]. The World Bank also offers data related to a variety of other indicators. The World Development Indicator data-set contains information regarding various financial indicators that may be used to explain a country's income group classification. To test this assumption, as well as the efficiency of ESDF on a real-world application, we used these data to classify low (low and low-middle) income countries and identify features in the world development indicators list that most explain the classification.

Data Processing. The world development indicators data set (for the year 2021) contains 108 indicators for 218 countries for which an income category is also assigned. However, not all indicators have values for all countries. All indicators with values for less than half the number of countries were removed, resulting in a data set with 218 countries and 40 indicators. Further, removing all countries with less than half indicator values resulted in a data set containing 138 countries and 40 indicators. In this data set, we found 13.35% missing values that were replaced, for each indicator, with the average value of its country's region, which is part of the data set. Countries with lower and lower middle income were assigned the label 1, and the others 0, resulting in a slightly imbalanced data

---
[3] https://blogs.worldbank.org/opendata/new-world-bank-country-classifications-income-level-2022-2023, accessed Jan. 2023.

set with 37% instances having class 1. In what follows, we will call this data set the World Bank Income indicators (WBII) data set.

Experimental Set-Up. The same methodology used to test ESDF on the synthetic and real-world benchmarks was also used for the WBII data set. 10-fold cross-validation was applied four times with different seeds for the random number generator, and the four indicators were used to evaluate the results. ESDF parameters were $\alpha = 0.8$, the maximum tree depth used was 10, and the population size was 5.

Results - Classification. Numerical results for classification reported by all methods for the WBII data set are presented in Table 2. We find that results reported by ESDF are as good as or even better than those reported by the other methods. Particularly the Log Loss values are significantly better than all the other methods.

Table 2. WBII data set: mean and standard deviation values for the four indicators reported by all methods. We find results reported by ESDF better or as good as the others for all indicator values.

Method	AUC	ACC	F1	Log-loss
ESDF	0.90 ± 0.08	0.85 ± 0.09	0.79 ± 0.12	0.99 ± 1.28
M0	0.78 ± 0.11	0.80 ± 0.11	0.72 ± 0.16	7.02 ± 3.67
M1	0.85 ± 0.09	0.86 ± 0.08	0.81 ± 0.12	4.77 ± 2.93
M2	0.81 ± 0.09	0.84 ± 0.08	0.75 ± 0.13	5.40 ± 2.71
M3	0.81 ± 0.10	0.83 ± 0.10	0.76 ± 0.16	5.71 ± 3.32
M4	0.78 ± 0.11	0.76 ± 0.11	0.72 ± 0.13	8.19 ± 3.75
M5	0.75 ± 0.12	0.76 ± 0.11	0.67 ± 0.17	8.20 ± 3.97
M6	0.84 ± 0.10	0.85 ± 0.10	0.79 ± 0.13	5.20 ± 3.30
M7	0.81 ± 0.08	0.83 ± 0.08	0.76 ± 0.12	6.06 ± 2.73
M8_5	0.80 ± 0.08	0.82 ± 0.07	0.74 ± 0.11	6.38 ± 2.62
M8_10	0.79 ± 0.09	0.81 ± 0.08	0.72 ± 0.13	6.90 ± 2.91
M9	0.86 ± 0.09	0.87 ± 0.08	0.82 ± 0.11	4.58 ± 2.81
M10_5	0.85 ± 0.10	0.87 ± 0.09	0.81 ± 0.13	4.76 ± 3.12
M11_5	0.87 ± 0.09	0.88 ± 0.08	0.83 ± 0.11	4.37 ± 2.93
M12_5	0.87 ± 0.09	0.88 ± 0.08	0.83 ± 0.11	4.44 ± 2.80
M10_10	0.84 ± 0.09	0.86 ± 0.08	0.79 ± 0.12	5.16 ± 2.82
M11_10	0.86 ± 0.08	0.88 ± 0.08	0.83 ± 0.11	4.44 ± 2.80
M12_10	0.86 ± 0.08	0.88 ± 0.08	0.83 ± 0.11	4.44 ± 2.75
M13	0.81 ± 0.10	0.82 ± 0.10	0.75 ± 0.14	6.41 ± 3.54

Results - Feature Selection. To illustrate a possible practical interpretation of the selected features, Fig. 3 represents feature weights reported by ESDF on the 10 folds used for cross-validation. The corresponding stability score is 0.74 indicating a strong correlation between selected features (when half of them are chosen based on the value of their weights). The features with the highest weights are:

1. GFDD.AI.11: Received wages: into a financial institution account (% age 15+)
2. GFDD.AI.05: Financial institution account (% age 15+)
3. GFDD.AI.21: Debit card ownership (% age 15+) and
   GFDD.AI.20: Credit card ownership (% age 15+)
4. GFDD.EI.01: Bank net interest margin (%)
5. GFDD.AI.06: Saved at a financial institution (% age 15+)
6. GFDD.AI.10: Received domestic remittances: through a financial institution (% age 15+)

This list indicates that individual banking activities may be considered as indicators for a country's income group. While there is no causation involved here, results indicate a relationship between these indicators and the income group.

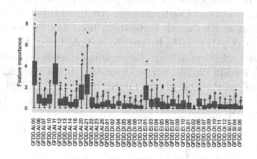

Fig. 3. Example of distribution of feature weights for one run on the WBII data set.

4 Conclusions and Further Work

The evolution strategy random forest for feature selection and classification proposed in this paper presents several original aspects: individuals are evolved without an explicit fitness; an updating mechanism, imitating mutation, always increases each component; converting values to probabilities when necessary decreases the additive effect; the search stops for each individual when there is no more variation in the probability values; the evaluation takes place at the end of the search, during the prediction phase for classification when data is gathered from the leaves in which tested instances are found from all trees, and logistic regression (but any classification method can be used) is applied for prediction.

Numerical experiments performed on synthetic and real-world benchmarks illustrate the efficiency of the approach for classification compared to other standard methods. A stability measure for feature selection indicates the potential of the approach to identify relevant feature sets. Furthermore, the method is used to analyse the classification of low-income countries based on several world development indicators. Classification results and most popular features are discussed.

Acknowledgements. This work was supported by a grant of the Ministry of Research, Innovation and Digitization, CNCS - UEFISCDI, project number PN-III-P1-1.1-TE-2021-1374, within PNCDI III.

References

1. Aich, S., Younga, K., Hui, K.L., Al-Absi, A.A., Sain, M.: A nonlinear decision tree based classification approach to predict the Parkinson's disease using different feature sets of voice data. In: 2018 20th International Conference on Advanced Communication Technology (ICACT), pp. 638–642 (2018)
2. Bala, J., Huang, J., Vafaie, H., Dejong, K., Wechsler, H.: Hybrid learning using genetic algorithms and decision trees for pattern classification. In: Proceedings of the 14th International Joint Conference on Artificial Intelligence, IJCAI 1995, vol. 1, p. 719–724. Morgan Kaufmann Publishers Inc., San Francisco (1995)
3. Bommert, A., Sun, X., Bischl, B., Rahnenführer, J., Lang, M.: Benchmark for filter methods for feature selection in high-dimensional classification data. Comput. Stat. Data Anal. **143**, 106839 (2020). https://doi.org/10.1016/j.csda.2019.106839
4. Breiman, L.: Random Forests. Mach. Learn. **45**(1), 5–32 (2001)
5. Brown, G.W.: Iterative solution of games by fictitious play. Act. Anal. Prod. Allocation **13**(1), 374–376 (1951)
6. Cai, J., Luo, J., Wang, S., Yang, S.: Feature selection in machine learning: a new perspective. Neurocomputing **300**, 70–79 (2018)
7. Dua, D., Graff, C.: UCI machine learning repository (2017)
8. Fawcett, T.: An introduction to ROC analysis. Pattern Recogn. Lett. **27**(8), 861–874 (2006). https://doi.org/10.1016/j.patrec.2005.10.010
9. Goldberg, D.E.: Genetic Algorithms in Search, Optimization and Machine Learning, 1st edn. Addison-Wesley Longman Publishing Co., Inc., USA (1989)
10. Hansen, L., Lee, E.A., Hestir, K., Williams, L.T., Farrelly, D.: Controlling feature selection in random forests of decision trees using a genetic algorithm: classification of class I MHC peptides. Combin. Chem. High Throughput Screen. **12**(5), 514–519 (2009). https://doi.org/10.2174/138620709788488984
11. Hastie, T., Tibshirani, R., Friedman, J.: The Elements of Statistical Learning: Data Mining, Inference and Prediction, 2nd edn. Springer, Heidelberg (2009). https://doi.org/10.1007/978-0-387-84858-7
12. Irsoy, O., Yıldız, O.T., Alpaydın, E.: Soft decision trees. In: Proceedings of the 21st International Conference on Pattern Recognition (ICPR2012), pp. 1819–1822. IEEE (2012)
13. Jovanovic, M., Delibasic, B., Vukicevic, M., Suknović, M., Martic, M.: Evolutionary approach for automated component-based decision tree algorithm design. Intell. Data Anal. (2014). https://doi.org/10.3233/ida-130628

14. Krętowski, M., Grześ, M.: Evolutionary learning of linear trees with embedded feature selection. In: Rutkowski, L., Tadeusiewicz, R., Zadeh, L.A., Żurada, J.M. (eds.) ICAISC 2006. LNCS (LNAI), vol. 4029, pp. 400–409. Springer, Heidelberg (2006). https://doi.org/10.1007/11785231_43

15. Mao, Q., Wang, X., Zhan, Y.: Speech emotion recognition method based on improved decision tree and layered feature selection. Int. J. Humanoid Rob. (2010). https://doi.org/10.1142/s0219843610002088

16. Murthy, S.K., Kasif, S., Salzberg, S.: A system for induction of oblique decision trees. J. Artif. Intell. Res. **2**, 1–32 (1994)

17. Nogueira, S., Brown, G.: Measuring the stability of feature selection. In: Frasconi, P., Landwehr, N., Manco, G., Vreeken, J. (eds.) ECML PKDD 2016. LNCS (LNAI), vol. 9852, pp. 442–457. Springer, Cham (2016). https://doi.org/10.1007/978-3-319-46227-1_28

18. Pedregosa, F., et al.: Scikit-learn: machine learning in Python. J. Mach. Learn. Res. **12**, 2825–2830 (2011)

19. Quinlan, J.R.: Induction of decision trees. Mach. Learn. (1986). https://doi.org/10.1007/bf00116251

20. Rosset, S.: Model selection via the AUC. In: Proceedings of the Twenty-First International Conference on Machine Learning, ICML 2004, p. 89. Association for Computing Machinery, New York (2004). https://doi.org/10.1145/1015330.1015400

21. Stein, G., Chen, B., Wu, A.S., Hua, K.A.: Decision tree classifier for network intrusion detection with GA-based feature selection. In: Proceedings of the 43rd Annual Southeast Regional Conference, vol. 2, p. 136–141. ACM-SE 43, Association for Computing Machinery, New York (2005). https://doi.org/10.1145/1167253.1167288

22. Suciu, M.A., Lung, R.: A new filter feature selection method based on a game theoretic decision tree. In: Abraham, A., Hong, T.P., Kotecha, K., Ma, K., Manghirmalani Mishra, P., Gandhi, N. (eds.) HIS 2022. LNNS, vol. 647, pp. 556–565. Springer, Cham (2022). https://doi.org/10.1007/978-3-031-27409-1_50

23. Vafaie, H., De Jong, K.: Genetic algorithms as a tool for feature selection in machine learning. In: Proceedings Fourth International Conference on Tools with Artificial Intelligence, TAI 1992, pp. 200–203 (1992). https://doi.org/10.1109/TAI.1992.246402

24. Wang, S., Tang, J., Liu, H.: Embedded unsupervised feature selection. In: Proceedings of the AAAI Conference on Artificial Intelligence, vol. 29, no. 1 (2015)

25. Wu, X., et al.: Top 10 algorithms in data mining. Knowl. Inf. Syst. **14**(1), 1–37 (2008). https://doi.org/10.1007/s10115-007-0114-2

26. Xue, B., Cervante, L., Shang, L., Browne, W.N., Zhang, M.: Multi-objective evolutionary algorithms for filter based feature selection in classification. Int. J. Artif. Intell. Tools **22**(04), 1350024 (2013)

27. Xue, B., Zhang, M., Browne, W.N., Yao, X.: A survey on evolutionary computation approaches to feature selection. IEEE Trans. Evol. Comput. **20**(4), 606–626 (2016). https://doi.org/10.1109/TEVC.2015.2504420

28. Zaki, M.J., Meira, W., Jr.: Data Mining and Machine Learning: Fundamental Concepts and Algorithms, 2nd edn. Cambridge University Press, Cambridge (2020)

29. Zijdenbos, A., Dawant, B., Margolin, R., Palmer, A.: Morphometric analysis of white matter lesions in MR images: method and validation. IEEE Trans. Med. Imaging **13**(4), 716–724 (1994). https://doi.org/10.1109/42.363096

"It Looks All the Same to Me": Cross-Index Training for Long-Term Financial Series Prediction

Stanislav Selitskiy$^{(\boxtimes)}$ (iD)

University of Bedfordshire, Park Square, Luton LU1 3JU, UK
stanislav.selitskiy@study.beds.ac.uk

Abstract. We investigate a number of Artificial Neural Network architectures (well-known and more "exotic") in application to the long-term financial time-series forecasts of indexes on different global markets. The particular area of interest of this research is to examine the correlation of these indexes' behaviour in terms of Machine Learning algorithms cross-training. Would training an algorithm on an index from one global market produce similar or even better accuracy when such a model is applied for predicting another index from a different market? The demonstrated predominately positive answer to this question is another argument in favour of the long-debated Efficient Market Hypothesis of Eugene Fama.

Keywords: Efficient Market Hypothesis · neural networks · cross-training

1 Introduction

The Efficient Market Hypothesis (EMH) in the well-known form was popularized by Eugene Fama [4] in the late 60's - the early '70s, though the earliest documented high-level formulation was known since 16'th century [28]. In various forms of strictness, it postulates that all potentially available information is immediately incorporated into market prices. This quite radical and superficial (and, in a way, superstitious and superfluous) proposition was fiercely debated over time. The early discussion of the EMH hypothesis was primarily conducted by proponents of the hypothesis on the material of the developed markets, which led to over-confident conclusions. In the '80s, the application of the theory was tested on emergent markets, which brought mixed results. In the '90s, the challenge of the EMH became widespread, though, even in the 21st century, the discussion still continues [35].

The economic [21], environment [20], logical paradoxes [10], and psychological [29] aspects of the debate are out of the scope of this paper. Machine Learning (ML) and statistical market prediction methods based on the previous historical data are the "bread and butter" of the practising traders and one of the arguments favouring the EMH [33]. However, even though these algorithms have

G. Nicosia et al. (Eds.): LOD 2023, LNCS 14505, pp. 348–363, 2024.
https://doi.org/10.1007/978-3-031-53969-5_26

some limited predicted power in plain form, they do not strictly show the use of the information in the predictions but possibly stochastic correlations. Research on tracing how publicly unknown information finds its way to the public and into market prices, the rate and cost of its dissipation is still to be conducted.

In this research, we still use the indirect method. We investigate the hypothesis that if today's global markets are affected by information, then their behaviour is correlated long-term. It is a trivial proposition, which has been debated for decades [3], but we give it a practical spin in our applied research, going beyond usual statistics correlation analysis [18]. If we train an ML model on one index (potentially on one market) and then use it for index prediction of another index on another global market, and those predictions have similar or better accuracy (due to less over-fitting), then this correlation would be an argument in favour to EMH.

In the study, we experiment with various Artificial Neural Network (ANN) architectures, the well-known easily assembled from the layers available in many ML libraries, more "exotic" that require additional manual coding, and developed from scratch by the authors, to ensure that cross-training effects are common and not bound to one particular ANN architecture. We show how successfully those ANN models work in long-term forecasting for 30 days when trained on the same index data and then when they are cross-trained. We use historical data from 2005 to the beginning of 2022 for the NASDAQ, Dow Jones, NIKKEI and DAX indexes to cover intra-market and cross-market effects.

The contribution is organized in the following way: Sect. 2 presents closely related to the study's existing work, Sect. 3 high-level describes ML algorithms chosen for the study, data sets and their partition, which were used in computational experiments, as well as accuracy metrics for algorithms' evaluation. Section 4 lists hardware parameters and model configurable parameters, as well as details of the less-popular ANN architectures implemented from scratch. Section 5 shows the results of the experiments in diagram and table form, discusses these results and study limitations, draws conclusions and outlines future research directions.

2   Related Work

ML algorithms of different types were used for various markets, stock types, observation periods, and prediction intervals.

Two classes of the ML algorithms, Decision Tree (DT) based and ANN-based, were used on the 10 years interval for the Tehran Stock Exchange data (financial, petroleum, mineral, and metal indexes) in [14]. Various prediction intervals ranging from 1 to 30 days in the future were used, measuring accuracy with four metrics. Unfortunately, the best results were reported over the whole time span and prediction intervals, which makes the research of limited interest. However, systematically, Long Short-term Memory (LSTM) ANN was reported as superior to other ML models.

In [30], not just composite indexes, but instead, stocks of the 25 individual companies on the Bucharest Stock Exchange were experimented with on the span of over 21 years. Two ANN models, LSTM and 1-dimensional temporal Convolutional Neural Network (CNN), were used not just for accuracy calculation but also for the trading gain simulation in comparison with traditional simple trading tactics. Predictions were run on training intervals from 30 to 120 days, but with only 1 day prediction in the future at a time. Both architectures demonstrated superior performance compared to the simple tactics specialising in the window frequency for CNN or gain amounts for LSTM.

Traditionally ML models are trained on the same indexes but on the previous chronological data. Of our particular interests are approaches of the ML models training on the different time series. "Cross-training" is not a well-established term; therefore, such methods come under various rubrics, such as ensembles, pre-training, and transfer learning. They include noise-augmented training data [36], contrastive self-supervised learning targeting on the extraction of the augmented components [34], averaged LSTM training on multiple index data [32], or averaging of the ensemble of the models trained on the multiple indexes [7].

Although some similarities could be seen between the above-mentioned and presented research, it is worth pointing out that the aim of the existing research is the enriching training data of the pre-trained transfer-learning models with the superset of the patterns found either in other indexes (or domain) or synthetic time series. In our study, ANN models are trained on the index of one single market and then tested on the other three markets, and such experiments are applied to all four markets in a circular fashion. No averaging or other use of data from multiple markets for training is done because the study aims to experimentally confirm the alleged by EMH synchronicity of the markets, strong enough to have no need for averaging or multiple data sets pre-training.

3   Methods

It could be easily seen in Fig. 1, Table 1, the stock indexes used in this study and described in more detail in Sect. 3.2, are correlated.

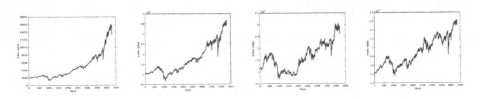

Fig. 1. Stock indexes NASDAQ, DOW, NIKKEI, DAX on intervals of the experiments.

However, how much of this correlation can be used for practical purposes of cross-training and cross-forecasting? Furthermore, how much of theoretical conclusions in the context of EMH can we extract? Remaining inside statistical methods, such cross-training is not intuitive, especially from the Frequentist's

Table 1. Pearson correlation coefficient, applied pairwise to the stock indexes under experimentation

Index 1	Index 2	Correlation
NASDAQ	NIKKEI	0.8879
NASDAQ	DOW	0.9750
NASDAQ	DAX	0.8975
DOW	DAX	0.9456
DOW	NIKKEI	0.9053
NIKKEI	DAX	0.8740

point of view, though the Bayesian can look at statistics obtained from another index as a starting belief to polish them on the target stock statistical forecasts using statistical [9, 22, 26] or ML methods [25]. Also, choosing particular statistics requires prior assumptions about patterns that can embed information the stocks are supposed to absorb by EMH.

ML methods, and especially ANNs', strength is freedom of necessity for such *a priory* assumptions. ANN can learn unexpected patterns, and if there are such patterns in the stock series which can be cross-learned and successfully used for cross-forecasting, that is an argument for EMH, considering the ubiquity of the information in global markets, economy, and the world itself. To leverage the best models mentioned in Sect. 2 and enhance them, we concentrated on ANN architectures, significantly increasing the variety of the being studied models and applying them for the extreme for the cited works prediction interval of 30 days.

Accuracy metrics (described below in Subsect. 3.1) and their distribution over session partitions (also described below in Subsect. 3.2), are collected for all being investigated ANN models. As ablation testing, prediction accuracies are calculated in the absence of cross-training (the same index is used for training and testing, but on different time intervals). Acquired accuracy distributions for cross-training are subjected to the non-parametric hypothesis testing Wilcoxon signed rank algorithm to verify that non-cross-training and cross-training accuracy distribution either have no shift or "right" (greater) shift for non-cross-training distributions, i.e. cross-training accuracy is at least no worse than the non-cross-training.

3.1   Accuracy Metrics

As an accuracy metrics, we use the Mean Absolute Percentage Error (MAPE), defined as follows:

$$MAPE = \frac{1}{n} \sum_{t=1}^{n} |\frac{A_t - F_t}{A_t}| \tag{1}$$

And Root Mean Square Error (RMSE):

$$RMSE = (\frac{1}{n}\sum_{t=1}^{n}(A_t - F_t)^2)^{\frac{1}{2}} \tag{2}$$

where A_t and F_t are the actual and predicted indexes at a given day t, respectively, and n is the number of test observations.

3.2 Data Sets

The data for NASDAQ, Dow Jones, NIKKEI, and DAX indexes for the beginning of 2005 year up to the end of January 2022 were used in computational experiments. The data were downloaded from https://tradingeconomics.com. Detailed statistical analysis of the data set is out of the scope of the paper; however, general intuition about this strongly non-stationary time series can be obtained from general Fig. 1, and simple Pearson correlation coefficients Table 1. The data represent a broad spectrum of behavioural tendencies: oscillations of the stagnant markets, explosive growth and fall during bubble bursts and busts in American, European, and Asian markets. The difference in stock exchanges' working days schedules made the data slightly (a few days) asynchronous. The reason for the particular indexes selection is an attempt to limit the combinatorial complexity of the experiments (only four indexes), still preserving geographical representability (at least for the most economically important markets: DOW, NIKKEI, DAX), and general vs. industry-specific markets (DOW and NASDAQ).

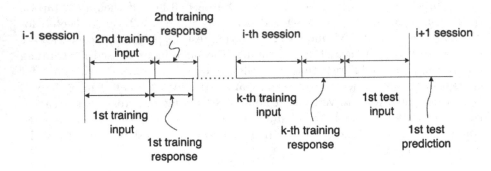

Fig. 2. Session partition schema.

To accommodate long-term prediction of 30 days forward, based on the past 30 days indexes performance, the data were divided into 35 subsets, each consisting of 30 observations used to train the sequence-to-sequence models. Each "observation" was comprised of 30 days values and the output "label" values were the 31th–60th days. Each following observation starts with a 1 day shift. The number of training sessions is first 34, and the whole session is 120 days. It was defined so that none of the training data (including the label days) would touch the following test session data. The number of test sessions is last 34. The last 30 days of the preceding training session (not used in the training process) were used to predict the first 30 days of the following test session, with step 1 until the end of the test session. Models' parameters were reset for each session, and training was done anew (see Fig. 2). For sequence-to-value models, such as LSTM, the whole 119 + 1 last label day was used in the training session.

4   Experiments

The experiments were run on the Linux (Ubuntu 20.04.3 LTS) operating system with two dual Tesla K80 GPUs (with 2×12 GB GDDR5 memory each) and one QuadroPro K6000 (with 12 GB GDDR5 memory, as well), X299 chipset motherboard, 256 GB DDR4 RAM, and i9-10900X CPU. Experiments were run using MATLAB 2022a with Deep Learning Toolbox. For inferential statistics, built-in R 4.2.1 implementations were used with default parameters unless mentioned otherwise.

Table 2. Accuracy of various ML models for NASDAQ index

Model	MAPE	RMSE
AR	0.0521 ± 0.0428	336.79 ± 471.61
ANN reg	0.0504 ± 0.0401	320.75 ± 405.89
Logistic	0.0541 ± 0.0417	313.21 ± 315.77
ReLU	0.0560 ± 0.0421	361.88 ± 432.68
LSTM vec	0.0506 ± 0.0273	283.15 ± 229.51
LSTM	0.0615 ± 0.0590	316.60 ± 274.46
SCNN	0.0643 ± 0.0590	360.77 ± 358.64
RBF	0.0588 ± 0.0418	343.20 ± 336.99
KGate	0.0623 ± 0.0555	363.56 ± 351.92
GMDH	0.0549 ± 0.0505	316.23 ± 321.70

The following models were experimented with: analytically-solved multivariate linear Auto-regression, ANN Regression (no activation functions), ANN with ReLU (Rectified Linear Unit), Logistic, Hyperbolic tangent activations, sequence-to-sequence and sequence-to-value LSTM and GRU (Gated Recurrent Unit), simple sequential and spectral cascade CNN, RBF (Radial Basis Function), KGate (Kolmogorov's Gate), and finally GMDH (Group Method of Data Handling) ANNs. Because ANN regression, ANN with ReLU, KGate activations and CNNs are tolerable to non-normalized input data and frequently produce better accuracy, for these models, experiments are done with non-normalized input. For other models sensitive to normalization, a min-max normalization is applied in a strict mode, calculated only for a given observation data without looking ahead of time or in the past. A mean squared error was used as a loss function for all ANNs.

4.1   Ready-to-Use ANN Layers

Well-known ANN layers and architectures, readily available in MATLAB or its Toolboxes, are not described here in detail - only non-default configuration parameters are presented below. Non-standard implementations and the whole source code used in experiments can be found on GitHub (https://github.com/Selitskiy/LTTS) and are implemented as follows.

If we look at linear regression, or the linear part of a layer transformation of an ANN, as a transformation from a higher dimensional space into the lower dimensional space f, $n < m$:

$$f : \mathcal{X} \subset \mathbb{R}^m \mapsto \mathcal{Y} \subset \mathbb{R}^n \tag{3}$$

Linear regression can be represented as a matrix multiplication:

$$\mathbf{y} = f(\mathbf{x}) = \mathsf{W}\mathbf{x}, \forall \mathbf{x} \in \mathcal{X} \subset \mathbb{R}^m, \forall \mathbf{y} \in \mathcal{Y} \subset \mathbb{R}^n \tag{4}$$

where $\mathsf{W} \in \mathcal{W} \subset \mathbb{R}^{n \times m}$ is the adjustable coefficient matrix that, using minimization of the sum of squared errors, can be found as [6]:

$$\mathbf{W} = (\hat{\mathbf{X}}^T \hat{\mathbf{X}})^{-1} \hat{\mathbf{X}}^T \hat{\mathbf{Y}} \tag{5}$$

where $\hat{\mathbf{X}}, \hat{\mathbf{Y}}$ are matrices of the observations of the input and output observations, respectively [23].

All other ANN architectures, to compare them on a similar level of complexity (number of learnable parameters), were designed to have two hidden layers with a number of neurons m and $2m + 1$ on the first and second hidden layer, respectively, where m is an input dimensionality, Formula 3. The reason to limit ANN models to two layers was if one looks at ANN as a Universal Approximation according to the Kolmogorov-Arnold superposition theorem [11], for general emulation of the $f : \mathcal{X} \subset \mathbb{R}^m \mapsto \mathcal{Y} \subset \mathbb{R}$ process, the 2-layer is a minimal ANN configuration needed (given that activation functions are complex enough):

$$f(\mathbf{x}) = f(x_1, \ldots, x_m) = \sum_{q=0}^{2m} \Phi_q(\sum_{p=1}^{m} \phi_{qp}(x_p)) \tag{6}$$

where Φ_q and ϕ_{qp} are continuous $\mathbb{R} \mapsto \mathbb{R}$ functions.

The practicality of such ANN, as a Universal Approximator, was disputed in [5], particularly because of the non-smoothness, hence non-practicality, of the inner ϕ_{qp} functions. However, these objections were rebutted in [13]. In [17] ϕ_{qp} activation functions are even called "pathological".

Therefore, in addition to the usual activation functions, much less frequently used architectures and activation functions were experimented with.

4.2   Custom-Coded and Originally-Developed ANN Layers

Radial Basis Functions (RBF) ANN were proposed at the end of the 80s [2]. It could be viewed as a "soft gate" which activates the transformation matrix coefficients in Gaussian proportion to the proximity of the test signal to the training signals, this transformation matrix's coefficients were trained at [16].

Table 3. Accuracy of various ML models for NIKKEI index

Model	MAPE	RMSE
AR	0.0808 ± 0.0723	1919.10 ± 2707.51
ANN reg	0.0799 ± 0.0723	1877.54 ± 2630.55
Logistic	0.0890 ± 0.0798	1963.34 ± 2596.88
ReLU	0.0904 ± 0.0989	2256.28 ± 3905.08
LSTM vec	0.0598 ± 0.0248	1165.07 ± 467.01
LSTM	0.0587 ± 0.0286	1120.97 ± 519.30
SCNN	0.0932 ± 0.0924	2122.16 ± 3200.82
RBF	0.0923 ± 0.0818	2105.33 ± 3015.20
KGate	0.1008 ± 0.1086	2386.27 ± 3854.98
GMDH	0.0932 ± 0.0926	2087.34 ± 2806.60

$$f(\mathbf{x}) = \mathbf{a}_k e^{-\mathbf{b}_k(\mathbf{x}-\mathbf{c}_k)^2} \tag{7}$$

where $k \in \{1 \ldots n\}$.

An apparent drawback of the architecture is its "fluffiness" due to the non-reuse of the neurons for the "missed" test-time data input, making the RBF ANNs less dense or compact than Deep ReLU ANNs. Still, RBF is a viable architecture used in niche applications [1,12].

Another ANN architecture [24] can be seen as a part of the Gated Linear Unit (GLU) family of activations. Using Directed Acyclic Graph (DAG) ANN, one can implement a cell (let us call it Kolmogorov's Gate or KGate for short) of perceptrons with logistic sigmoid activations that would work as allow or do not allow gates at saturation domain or multiplicative scaling of the main trunk of ANN, in the non-saturation domain of input values. Perceptrons with hyperbolic tangent activation would work as update/forget or the mean shift gates on the main ANN trunk, working together with the linear input transformation through the multiplication gate, Formula 8.

$$\mathbf{z}_i = (\mathbf{W}_i\mathbf{x}_i + (\tau \circ \mathbf{W}_{ti}\mathbf{x}_0) \odot (\mathbf{W}_{ai}\mathbf{x}_0)) \odot \sigma \circ \mathbf{W}_{si}\mathbf{x}_0,$$
$$\forall \mathbf{x}_0 \in \mathcal{X}_0 \subset \mathbb{R}^m, \forall \mathbf{x}_i \in \mathcal{X}_i \subset \mathbb{R}^{m_i} \tag{8}$$

where $\mathbf{x}_0$ is an ANN input, $\mathbf{x}_i$ is an input of the ith layer, $\mathbf{W}_i\mathbf{x}_i$ is the linear transformation of the main trunk, $\mathbf{W}_{ti}\mathbf{x}_0, \mathbf{W}_{ai}\mathbf{x}_0, \mathbf{W}_{si}\mathbf{x}_0$ are linear transformations inside the KGate cell, and τ, σ are hyperbolic tangent and logistic sigmoid activation functions, respectively.

Following Ivakhnenko [8], the multi-layer neural-network models could be grown by the Group Method of Data Handling (GMDH) using a neuron activation function defined by a short-term polynomial, in our case - the polynomial of the second degree of the pairwise connected neurons, which ensures linear gradient optimization hyperplane on each layer generation step [15].

$$f(x_i, x_j) = (w_{ki}x_i + w_{kj}x_j + w_{k0})^2 \qquad (9)$$

The GMDH is capable of generating new layers capable of predicting new data most accurately. The GMDH generates new neurons to be fitted to the training data in each layer. A given number of the best-fitted neurons are selected for the next layer.

Spectral cascade CNN (SCNN) was organized as parallel Directed Acyclic Graph (DAG) cells, similar to Inception or ResNet type cells [19,31], of 3×1, 5×1, 7×1, 11×1, 13×1 dimensions, packets of 16 of each [27].

ANN models were trained using the "adam" learning algorithm with 0.01 initial learning coefficient, mini-batch size 32, and 1000 epochs.

5   Results

As mentioned above, computational experiments were conducted on NASDAQ, Dow Jones, NIKKEI, and DAX indexes partitioned as described in Sect. 3.2 using linear Auto-regression, ANN Regression, ANN with ReLU, Logistic, Hyperbolic tangent activations, sequence-to-sequence and sequence-to-value LSTM and GRU, simple sequential and spectral cascade CNN, RBF, KGate, Transformer-like non-linearity, and GMDH ANNs.

Table 4. MAPE of cross-trained LSTM sequence-to-vector model (rows - trained on the index, columns - tested on the index).

Index	NASDAQ	DOW	NIKKEI	DAX
NASDAQ	0.0506 ± 0.0273	0.0409 ± 0.0269	0.0574 ± 0.0258	0.0523 ± 0.0239
DOW	0.0501 ± 0.0261	0.0417 ± 0.0270	0.0589 ± 0.0280	0.0546 ± 0.0268
NIKKEI	0.0521 ± 0.0284	0.0416 ± 0.0243	0.0598 ± 0.0248	0.0551 ± 0.0243
DAX	0.0522 ± 0.0291	0.0425 ± 0.0283	0.0615 ± 0.0291	0.0589 ± 0.0377

Logistic and Hyperbolic tangent activation ANNs, LSTM and GRU, KGate and Transformer, CNN and SCNN produced similar results; therefore, only one of them is shown. For ablation non-cross-training accuracy data, only NASDAQ and NIKKEI index results are shown as extremes. To save space, Dow Jones and DAX results are not shown here Table 2, Table 3), but can be downloaded from the same GitHub link as the source code.

Table 5. MAPE of cross-trained RBF model (rows - trained on the index, columns - tested on the index).

Index	NASDAQ	DOW	NIKKEI	DAX
NASDAQ	0.0588 ± 0.0418	0.0485 ± 0.0393	0.0756 ± 0.0397	0.0590 ± 0.0316
DOW	0.1328 ± 0.4250	0.1020 ± 0.3150	0.0709 ± 0.0344	0.4299 ± 2.1876
NIKKEI	0.0698 ± 0.0680	0.0544 ± 0.0619	0.0923 ± 0.0818	0.0592 ± 0.0354
DAX	0.0649 ± 0.0503	0.0542 ± 0.0521	0.0702 ± 0.0339	0.0794 ± 0.1053

Results of the cross-training experiments are shown for all indexes, but also for extremes of the models - LSTM sequence-to-vector being most accurate, and RBF activation being most vulnerable for volatility: Table 4, Table 5.

P-values of the paired Wilcoxon signed rank test between non-cross-trained and cross-trained accuracy distributions for various ANN architectures for NASDAQ, Dow Jones, NIKKEI indexes, Table 6, 7, 8. Again, to save space, DAX results, which are similar to NASDAQ, are not shown.

Table 6. Wilcoxon signed rank test p-values on accuracy distribution over 34 sessions for models trained on NASDAQ data and tested on other indexes. P-values for alternative hypotheses of distribution shift for the not-cross-trained relative to cross-trained to the right, left, and two-sided.

Model	Index	Greater	Less	Two-side
Reg	DOW	0.00003	0.99997	0.00007
Reg	NIKKEI	0.97359	0.02757	0.05513
Reg	DAX	0.89874	0.10434	0.20869
ANN	DOW	0.00003	0.99997	0.00006
ANN	NIKKEI	0.97683	0.02421	0.04843
ANN	DAX	0.90761	0.09528	0.19056
ReLU	DOW	0.00001	0.99999	0.00001
ReLU	NIKKEI	0.93946	0.06276	0.12552
ReLU	DAX	0.53362	0.47310	0.94619
Sig	DOW	0.00160	0.99850	0.00319
Sig	NIKKEI	0.85610	0.14799	0.29599
Sig	DAX	0.81975	0.18476	0.36953
LSTMv	DOW	0.00000	0.99999	0.00000
LSTMv	NIKKEI	0.86407	0.13988	0.27976
LSTMv	DAX	0.68786	0.31815	0.63630
LSTM	DOW	0.00004	0.99996	0.00009
LSTM	NIKKEI	0.99013	0.01038	0.02077
LSTM	DAX	0.49327	0.51346	0.98654
GMDH	DOW	0.00004	0.99996	0.00009
GMDH	NIKKEI	0.79701	0.20805	0.41609
GMDH	DAX	0.75591	0.24945	0.49890
KGate	DOW	0.00003	0.99997	0.00006
KGate	NIKKEI	0.94161	0.06054	0.12109
KGate	DAX	0.49327	0.51346	0.98654
RBF	DOW	0.00069	0.99936	0.00137
RBF	NIKKEI	0.97683	0.02421	0.04843
RBF	DAX	0.55369	0.45299	0.90598
SCNN	DOW	0.00000	0.99999	0.00001
SCNN	NIKKEI	0.76010	0.24545	0.49089
SCNN	DAX	0.61937	0.38708	0.77417

6   Discussion and Conclusions

Interesting observations for ablation experiments are based on the MAPE metric, giving a more universal measure of the overall accuracy for each prediction point across the multiple indexes. In contrast, RMSE, though more index-specific, penalizes outlier predictions more, even though few of them exist. ANN models with exponential and polynomial non-linearities are especially vulnerable to stock volatility especially observed in NIKKEI and DAX indexes behaviour, which is one of the faces of the Out-of-Distribution (OOD) problem. Among all ANN architectures, the most accurate and robust was LSTM sequence-to-vector architecture.

Table 7. Wilcoxon signed rank test p-values on accuracy distribution over 34 sessions for models trained on DOW data and tested on other indexes. P-values for alternative hypotheses of distribution shift for the not-cross-trained relative to cross-trained to the right, left, and two-sided.

Model	Index	Greater	Less	Two-side
Reg	NASDAQ	0.99988	0.00013	0.00027
Reg	NIKKEI	0.99993	0.00007	0.00015
Reg	DAX	0.99948	0.00056	0.00112
ANN	NASDAQ	0.99979	0.00023	0.00047
ANN	NIKKEI	0.99996	0.00005	0.00009
ANN	DAX	0.99952	0.00052	0.00104
ReLU	NASDAQ	0.99980	0.00022	0.00043
ReLU	NIKKEI	0.99995	0.00006	0.00012
ReLU	DAX	0.99977	0.00025	0.00050
Sig	NASDAQ	0.99940	0.00064	0.00128
Sig	NIKKEI	0.99854	0.00156	0.00311
Sig	DAX	0.99272	0.00766	0.01531
LSTMv	NASDAQ	0.99999	0.00001	0.00001
LSTMv	NIKKEI	0.99993	0.00008	0.00016
LSTMv	DAX	0.99999	0.00001	0.00002
LSTM	NASDAQ	0.99999	0.00001	0.00002
LSTM	NIKKEI	0.99952	0.00051	0.00103
LSTM	DAX	0.98755	0.01304	0.02609
GMDH	NASDAQ	0.99830	0.00181	0.00361
GMDH	NIKKEI	0.99956	0.00048	0.00095
GMDH	DAX	0.94564	0.05631	0.11262
KGate	NASDAQ	0.99992	0.00009	0.00018
KGate	NIKKEI	0.99995	0.00006	0.00012
KGate	DAX	0.99940	0.00064	0.00128
RBF	NASDAQ	0.99999	0.00001	0.00003
RBF	NIKKEI	0.99742	0.00274	0.00548
RBF	DAX	0.99272	0.00766	0.01531
SCNN	NASDAQ	0.99993	0.00008	0.00015
SCNN	NIKKEI	0.99969	0.00033	0.00065
SCNN	DAX	0.99019	0.01029	0.02058

Table 8. Wilcoxon signed rank test p-values on accuracy distribution over 34 sessions for models trained on NIKKEI data and tested on other indexes. P-values for alternative hypotheses of distribution shift for the not-cross-trained relative to cross-trained to the right, left, and two-sided.

Model	Index	Greater	Less	Two-side
Reg	NASDAQ	0.00137	0.99872	0.00273
Reg	DOW	0.00009	0.99991	0.00019
Reg	DAX	0.01536	0.98536	0.03071
ANN	NASDAQ	0.00128	0.99881	0.00255
ANN	DOW	0.00007	0.99993	0.00015
ANN	DAX	0.01851	0.98233	0.03701
ReLU	NASDAQ	0.00128	0.99881	0.00255
ReLU	DOW	0.00009	0.99992	0.00017
ReLU	DAX	0.00685	0.99351	0.01369
Sig	NASDAQ	0.00017	0.99984	0.00034
Sig	DOW	0.00001	0.99999	0.00001
Sig	DAX	0.00097	0.99909	0.00195
LSTMv	NASDAQ	0.04666	0.95510	0.09332
LSTMv	DOW	0.00009	0.99991	0.00019
LSTMv	DAX	0.15642	0.84783	0.31283
LSTM	NASDAQ	0.36879	0.63785	0.73757
LSTM	DOW	0.01148	0.98908	0.02295
LSTM	DAX	0.25674	0.74894	0.51348
GMDH	NASDAQ	0.00214	0.99799	0.00428
GMDH	DOW	0.00038	0.99965	0.00076
GMDH	DAX	0.01396	0.98670	0.02791
KGate	NASDAQ	0.00112	0.99896	0.00223
KGate	DOW	0.00001	0.99999	0.00002
KGate	DAX	0.00026	0.99976	0.00051
RBF	NASDAQ	0.00177	0.99834	0.00354
RBF	DOW	0.00003	0.99997	0.00007
RBF	DAX	0.00035	0.99967	0.00070
SCNN	NASDAQ	0.01536	0.98536	0.03071
SCNN	DOW	0.00000	1.00000	0.00000
SCNN	DAX	0.00097	0.99909	0.00195

It also should be noted that this study is not about transfer learning - the approach popular in image recognition, when an ANN model trained on as much as possible large data is slightly modified and rapidly retrained for recognition of images from other domains, because it has already learned common image micro-patterns. That approach could have been used for stock indexes when time series from other domains, for example, weather prediction, are retrained specifically for financial series. However, that is not the aim of the study - we principally do not allow any retraining, hoping to find synchronous information-shared invariants in the untouched cross-trained models.

Results of the Wilcoxon tests (Tables 6, 7 and 8) look mixed, with cross-trained accuracy being consistently better across all ANN models for NIKKEI index, consistently worse across all ANN models for Dow Jones, and mixed for NASDAQ and DAX. However, the results are shown for 0 expected shift. If the shift is bounded by 0.02 (which is less than the standard deviation of even the best and narrowest accuracy distributions), then all models and stocks demonstrate at least no worse accuracy of the cross-trained models compared to the ablation non-cross-trained ones.

As a conclusion for practical purposes, if the bounded to the circa standard deviation, degradation of the cross-trained modes is accepted, in such a Lipschitz sense, the study supports a weak form of EMH, indirectly supporting the idea of existing of the patterns common for global markets and different stock types, which is allegedly maps to information in EMH context, of course in the study limitations frames. Another observation arguing in favour of the information nature of EMH is the tendency of larger indexes to "explain" smaller indexes, i.e. NIKKEI and NASDAQ vs DOW and DAX.

From the general ML point of view, considering local stock indexes as observations of the global process (world market), this study offers experimental validation of the plausible, but still speculations, that ML model trained on some local observations may effectively predict other synchronous observations.

6.1   Limitations

The study has obvious limitations in the geographic coverage and representation of only developed economies' stock exchanges. An apparent technical limitation of the study is using ANN models only, skipping other ML and statistical approaches. Another technical one, imposed by the hardware limitations, is the relative simplicity of the ANNs being used, barely fitting in the theoretical minimum of the universal approximation. The more subject-related limitation is that accuracy was calculated uniformly along the whole time span, without concentrating on the times of perturbations when a lot of new information allegedly entered the market.

6.2   Future Work

As an area for future research in ablation experiments area, the common for many ML models, rare but catastrophic prediction failures call for relation-aware methods (for example, graph ANN or physics-aware, or rather "kinematic-aware" ANN) that would limit sudden changes in the stock prediction in a limited time period. Attention-aware methods, which would look for similar input period models in the past, may also be for future research.

Such an approach is also methodologically co-located with the abovementioned limitations - the study of the "disruption information"-rich time intervals, perhaps in contrast with new information-deprived periods. For example, the study of the models trained immediately before and tested immediately after the "disruption information" boundaries, such as the emergence of the COVID pandemic or the start of the war in Ukraine. "Information" here is used in its common sense meaning rather than in the information theory's.

Because such information may have prolonged inertia to be digested by markets (for example, intelligence or other expert communities may have that information months in advance), an extended temporal horizon "cross-training" may be a prospective area of research.

Extension of the research to emergent markets and behaviour of the individual stocks, especially those which do not follow the expected trend (cross-training of the successful company stocks on the failing companies' data and vice versa), is another area of future research.

References

1. Beheim, L., Zitouni, A., Belloir, F., de la Housse, C.D.M.: New RBF neural network classifier with optimized hidden neurons number. WSEAS Trans. Syst. (2), 467–472 (2004)
2. Broomhead, D.S., Lowe, D.: Radial basis functions, multi-variable functional interpolation and adaptive networks. Technical reprt, Royal Signals and Radar Establishment Malvern (United Kingdom) (1988)
3. Eun, C.S., Shim, S.: International transmission of stock market movements. J. Financ. Quant. Anal. 24(2), 241–256 (1989)
4. Fama, E.F.: Efficient capital markets: a review of theory and empirical work. J. Financ. 25(2), 383–417 (1970)
5. Girosi, F., Poggio, T.: Representation properties of networks: Kolmogorov's theorem is irrelevant. Neural Comput. 1(4), 465–469 (1989)
6. Hastie, T., Tibshirani, R., Friedman, J.: The Elements of Statistical Learning. Springer Series in Statistics. Springer, New York (2001)
7. He, Q.Q., Pang, P.C.I., Si, Y.W.: Multi-source transfer learning with ensemble for financial time series forecasting. In: 2020 IEEE/WIC/ACM International Joint Conference on Web Intelligence and Intelligent Agent Technology (WI-IAT), pp. 227–233. IEEE (2020)
8. Ivakhnenko, A.G.: Polynomial theory of complex systems. IEEE Trans. Syst. Man Cybern. 4, 364–378 (1971)
9. Jakaite, L., Schetinin, V., Maple, C., et al.: Bayesian assessment of newborn brain maturity from two-channel sleep electroencephalograms. Comput. Math. Methods Med. 2012 (2012)
10. Jensen, M.C.: The performance of mutual funds in the period 1945–1964. J. Financ. 23(2), 389–416 (1968)
11. Kolmogorov, A.N.: On the representation of continuous functions of several variables by superpositions of continuous functions of a smaller number of variables. Am. Math. Soc. (1961)

12. Kurkin, S.A., Pitsik, E.N., Musatov, V.Y., Runnova, A.E., Hramov, A.E.: Artificial neural networks as a tool for recognition of movements by electroencephalograms. In: ICINCO (1), pp. 176–181 (2018)
13. Kůrková, V.: Kolmogorov's Theorem is relevant. Neural Comput. **3**(4), 617–622 (1991)
14. Nabipour, M., Nayyeri, P., Jabani, H., Mosavi, A., Salwana, E.: Deep learning for stock market prediction. Entropy **22**(8), 840 (2020)
15. Nyah, N., Jakaite, L., Schetinin, V., Sant, P., Aggoun, A.: Evolving polynomial neural networks for detecting abnormal patterns. In: 2016 IEEE 8th International Conference on Intelligent Systems (IS), pp. 74–80. IEEE (2016)
16. Park, J., Sandberg, I.W.: Universal approximation using radial-basis-function networks. Neural Comput. **3**(2), 246–257 (1991)
17. Pinkus, A.: Approximation theory of the MLP model in neural networks. Acta Numer. **8**, 143–195 (1999)
18. Poterba, J.M., Summers, L.H.: Mean reversion in stock prices: evidence and implications. J. Financ. Econ. **22**(1), 27–59 (1988)
19. Ren, S., Sun, J., He, K., Zhang, X.: Deep residual learning for image recognition. In: CVPR, vol. 2, p. 4 (2016)
20. Roll, R.: Orange juice and weather. Am. Econ. Rev. **74**(5), 861–880 (1984)
21. Roll, R.: What every CFO should know about scientific progress in financial economics: what is known and what remains to be resolved. Financ. Manage. **23**(2), 69–75 (1994)
22. Schetinin, V., Jakaite, L., Schult, J.: Informativeness of sleep cycle features in Bayesian assessment of newborn electroencephalographic maturation. In: 2011 24th International Symposium on Computer-Based Medical Systems (CBMS), pp. 1–6. IEEE (2011)
23. Selitskaya, N., et al.: Deep learning for biometric face recognition: experimental study on benchmark data sets. Deep Biomet. 71–97 (2020)
24. Selitskiy, S.: Kolmogorov's gate non-linearity as a step toward much smaller artificial neural networks. In: Proceedings of the 24th International Conference on Enterprise Information Systems, vol. 1, pp. 492–499 (2022)
25. Selitskiy, S.: Elements of active continuous learning and uncertainty self-awareness: a narrow implementation for face and facial expression recognition. In: Goertzel, B., Iklé, M., Potapov, A., Ponomaryov, D. (eds.) AGI 2022. LNCS, vol. 13539, pp. 394–403. Springer, Cham (2023). https://doi.org/10.1007/978-3-031-19907-3_38
26. Selitskiy, S., Christou, N., Selitskaya, N.: Using statistical and artificial neural networks meta-learning approaches for uncertainty isolation in face recognition by the established convolutional models. In: Nicosia, G., et al. (eds.) LOD 2021. LNCS, vol. 13164, pp. 338–352. Springer, Cham (2022). https://doi.org/10.1007/978-3-030-95470-3_26
27. Selitsky, S.: Hybrid convolutional-multilayer perceptron artificial neural network for person recognition by high gamma EEG features. Medicinskiy Vest. Severnogo Kavkaza **17**(2), 192–196 (2022)
28. Sewell, M.: History of the efficient market hypothesis. Rn **11**(04), 04 (2011)
29. Shleifer, A.: Inefficient Markets: An Introduction to Behavioural Finance. OUP, Oxford (2000)
30. Stoean, C., Paja, W., Stoean, R., Sandita, A.: Deep architectures for long-term stock price prediction with a heuristic-based strategy for trading simulations. PLoS One **14**(10), e0223593 (2019)
31. Szegedy, C., et al.: Going deeper with convolutions. In: Proceedings of the IEEE Conference on Computer Vision and Pattern Recognition, pp. 1–9 (2015)

32. Tsang, G., Deng, J., Xie, X.: Recurrent neural networks for financial time-series modelling. In: 2018 24th International Conference on Pattern Recognition (ICPR), pp. 892–897. IEEE (2018)
33. Venugopal, V., Baets, W.: Neural networks and statistical techniques in marketing research: a conceptual comparison. Mark. Intell. Plann. (1994)
34. Wickstrøm, K., Kampffmeyer, M., Mikalsen, K.Ø., Jenssen, R.: Mixing up contrastive learning: self-supervised representation learning for time series. Pattern Recogn. Lett. **155**, 54–61 (2022)
35. Yen, G., Lee, C.F.: Efficient market hypothesis (EMH): past, present and future. Rev. Pac. Basin Financ. Mark. Policies **11**(02), 305–329 (2008)
36. Zhang, G.P.: A neural network ensemble method with jittered training data for time series forecasting. Inf. Sci. **177**(23), 5329–5346 (2007)

U-FLEX: Unsupervised Feature Learning with Evolutionary eXploration

Nicolo' Bellarmino$^{(\boxtimes)}$, Riccardo Cantoro , and Giovanni Squillero

CAD Research Group, Politecnico di Torino, Corso Duca degli Abruzzi 24,
10129 Turin, Italy
{nicolo.bellarmino,riccardo.cantoro,giovanni.squillero}@polito.it
http://www.cad.polito.it/

Abstract. Feature selection is an essential task in machine learning and data mining that involves identifying a subset of relevant features from a larger set. This paper proposes a novel technique for unsupervised feature selection based on a Neural Network in conjunction with an evolutionary algorithm. The proposed method aims to extract subsets of the most discriminative and relevant features from high-dimensional data, which can be eventually used for efficient and accurate machine learning. An evolutionary algorithm is employed to generate the feature subsets, and the goodness of a feature subset is evaluated through the ability of a neural network to reconstruct the whole original input space by mean squared error minimization (in an auto-encoder fashion). Experimental results demonstrate the effectiveness of the proposed approach in finding relevant feature subsets for successive learning tasks, achieving better classification and regression accuracy compared to state-of-the-art feature selection methods.

Keywords: Unsupervised Learning · Feature Selection · Genetic Algorithm · Deep AutoEncoder · Supervised Learning

1 Introduction

Feature selection is a crucial step in machine learning and data analysis, as it aims to identify the most informative subset of features from a large set of potential predictors [11]. This process can greatly improve the accuracy and efficiency of predictive machine learning models by reducing the dimensionality of the input space for successive supervised or unsupervised tasks, removing irrelevant or redundant features, and avoiding overfitting. It may also help to uncover underlying patterns and relationships in the data. However, traditional supervised feature-selection methods rely on labeled data, which can be a major limitation in real-world applications, where labels may be expensive or difficult to obtain. In recent years, unsupervised feature selection has emerged as an alternative active research area that aims to identify relevant feature subsets that can capture the underlying structure of the data without relying on labels [26].

© The Author(s), under exclusive license to Springer Nature Switzerland AG 2024
G. Nicosia et al. (Eds.): LOD 2023, LNCS 14505, pp. 364–378, 2024.
https://doi.org/10.1007/978-3-031-53969-5_27

Evolutionary Algorithms (EAs) represent a powerful, general-purpose, and well-known paradigm among evolutionary computation techniques that has been applied to a wide range of optimization problems, including feature selection [18]. EAs are particularly well-suited for feature selection tasks because they can explore a large search space of potential feature subsets and automatically generate candidate solutions using a combination of mutation, crossover, and selection operations [4]. In recent years, they have been used as effective techniques for unsupervised feature selection [1,7], allowing for the automatic discovery of non-redundant feature subsets.

This paper proposes a novel approach to unsupervised feature selection based on genetic programming. The proposed approach aims to identify a subset of relevant features that maximize the ability to reconstruct the whole input data features space from subsets of features, eliminating redundant or irrelevant features. Subsets of features are generated by a Genetic Algorithm, and the mean squared error (MSE) between the features space reconstructed by deep neural networks and the original input space is used to evaluate the goodness of feature subsets. We use the MSE in conjunction with a regularization term as the fitness of the Genetic Algorithm to guide the search towards simpler and more generalizable feature subsets, preferring lower-dimensional solutions. We used a Deep (asymmetric) Auto Encoder to reconstruct the input space, which takes as input a subset of the original features.

The experimental results demonstrate that our proposed method can effectively identify informative feature subsets in various real-world datasets. We evaluate the proposed approach on several benchmark datasets and compare its performance with two classical feature selection methods, both supervised and unsupervised. The experimental results show that the proposed approach outperforms the compared methods in terms of classification/regression accuracy and feature subset size. Our approach offers a promising direction for unsupervised feature selection, enabling more efficient and accurate predictive modeling in a wide range of applications.

The rest of the paper is organized as follows. Section 2 provides a review of related work on unsupervised feature selection and genetic programming. Section 3 presents background information that describes theory and concept useful for understanding successive experiments. Section 4 describes the proposed approach in detail, including the genetic programming framework, the fitness function, and the diversity preservation mechanism. Section 5 presents the experimental setup, the dataset used, the comparison method, and the results of the evaluation. Finally, Sect. 6 concludes the paper with the implications and limitations of the proposed approach, a summary of the main results, and future directions.

2   Related Work

Several surveys about feature selection have been published [11,20]: they provide basic information on feature selection approaches to better understand the underlying problem. Evolutionary Algorithms (EAs) have been successfully applied to feature selection problems in the past [18]. In [2], the problem of the high time needed to evaluate the fitness of any individual in the EA process has been faced: the computational time of the EAs may be quite high. Therefore, the combination of EA and a classification method maybe not be efficient. They faced the problem by fitness function approximation, reducing the time needed to evaluate each candidate solution. In [1], an unsupervised feature selection approach based on EA was developed for text-clustering, relying on the mean absolute difference in text embedding. A similar approach was developed in [7], where the fitness function relies on KMeans clustering metrics. In [5], an unsupervised, model-agnostic, wrapper method for feature selection has been proposed. The authors assumed that if a feature can be predicted using the others, it adds little information to the problem. Therefore, it could be removed without impairing the performance of whatever model will be eventually built. There, the proposed method iteratively identifies and removes predictable, or nearly-predictable, redundant features, allowing trade-off complexity with expected quality. The theory and the philosophy behind that approach are similar to the one proposed here, but our novel work is trying to reconstruct the whole feature set all at once by starting from feature subsets.

3   Background

3.1   Features Selection

Feature selection is the process of identifying and selecting the most relevant feature subset in a dataset from a large pool of potential input variables for building a predictive model, to identify the most informative features that capture the essential characteristics of the underlying data while discarding redundant or noisy features that may lead to overfitting or poor generalization. It is a critical pre-processing step in machine learning and a fundamental research topic.

The need for feature selection arises due to the increasing complexity and dimensionality of real-world datasets, which often contain numerous variables, many of which are irrelevant or redundant, and data analysts may have no or limited domain knowledge to pre-prune the data input space. These techniques may help reduce the data's dimensionality, improve the interpretability of the results, and enhance the performance of machine learning models.

Input space reduction can also be achieved with Feature Extraction [15]: it involves reducing the dimensionality of a dataset by extracting the most important features. Principal Component Analysis (PCA) is one of the most popular techniques for Feature Extraction [17]. The main difference between Feature Extraction and Feature Selection is that Feature Extraction creates new features by combining the original ones, while Feature Selection selects a subset of the original features.

Various methods of feature selection have been developed, and they can be grouped into three main groups: filter methods, wrapper methods, and embedded methods [11].

Filter methods rely on statistical measures or machine learning algorithms to rank the features based on their relevance and importance. Correlation-based metrics can be used for the selection of features that are less-correlated with each other, thereby discarding redundant information from the dataset in an unsupervised fashion. Alternatively, we can select only the features that are (highly) correlated with the target label we aim to predict in a supervised fashion [28]. A correlation threshold is chosen and the features that score below this threshold are discarded. Once a subset of features is selected, it can be used as an input to the chosen classifier algorithm. Unlike the other feature selection methods (wrapper and embedded), filter methods are independent/separated from the successive ML algorithm [25], and thus they can be considered an independent pre-processing step.

Wrapper methods use a machine learning model to evaluate the performance of different subsets of features and select the best one, by computing error measures on a selected test set, eventually relying on cross-validation. As an example, evolutionary algorithms can be used as wrapper methods for feature selection [18], by relying on an underlying ML model trained on possible candidate feature subsets. The ML model is then tested on a proper test set, usually in a cross-validation fashion. A widely-used wrapper method is the Recursive Feature Elimination (RFE) [12].

Finally, embedded methods incorporate feature selection into the process of model training, by optimizing the feature subset during the learning process, as it happens in tree-based models like Decision Trees or Random Forests.

The choice of feature selection method depends on the characteristics of the dataset, the type of machine learning model used, and the specific application. Each method has its strengths and weaknesses, and the selection of the best approach requires careful evaluation and experimentation. Furthermore, recent advances in deep learning and neural networks (NN) have opened up new opportunities for feature selection, by leveraging the representation learning capabilities of these models [10]: feature selection has traditionally been an important technique for shallow learning models, but for deep learning the need for explicitly selecting the features has been partially alleviated by the ability of the NNs to learn meaningful representations directly from raw data. In these types of models, the concept of feature selection is often replaced by the concept of representation learning, where the goal is to learn a set of features that capture the most relevant information from the input data. NNs are capable of automatically discovering and extracting high-level features from the raw data, without the need for manual feature engineering [10]. This is achieved by stacking multiple layers of nonlinear transformations, each layer learning increasingly abstract representations of the input data. In some cases, however, it may still be beneficial to apply feature selection techniques in conjunction with deep learning. For example, to reduce computational complexity and memory requirements of deep neural networks when dealing with high-dimensional data, while also improving their interpretability. Additionally, feature selection can be useful in situations where the dataset contains noisy or irrelevant features, which can negatively impact the performance of deep learning models.

3.2 Evolutionary Algorithm

EAs are adaptive search techniques, for which improvement over random and local search methods has been demonstrated [9,18]. An EA maintains a population of individuals representing potential solutions across the generations. Each individual is evaluated based on its fitness with respect to the objective function being minimized or maximized. The fittest individuals are selected, and through crossover and mutation, new individuals are generated, leading the population toward better solutions in the given domain. EAs have been proposed as a tool for identifying and selecting the optimal subset of features for subsequent machine learning algorithms, such as classification systems [9,18].

In a population of individuals, represented as chromosomes $(C_1, C_2 \cdots C_p)$, each chromosome is a potential solution of the underlying task (and in the feature selection domain, a set of relevant features). EAs are advantageous for classification processes due to their insensitivity to noise and independence from domain knowledge. EAs rely on selection, crossover, and mutation. The selection phase involves choosing the best feature subsets based on their fitness values. The crossover operator combines these selected subsets to generate offspring solutions, while the mutation operator introduces random changes to explore new areas of the search space. Various evolutionary operators have been developed in the past to drive the optimization process of EAs [16,21]. When utilizing EAs, several aspects need to be determined, including chromosome codification, chromosome length, population size, genetic operators, and fitness function. Typically, the population size is set to 50 or 100 [18]. Standard genetic operators, such as one-point crossover and standard mutation, are commonly employed [16,21]. The probabilities of crossover and mutation are hyper-parameters that should be optimized [22]. The choice of fitness function depends on the specific problem being optimized. In ML tasks, both supervised (based on labels, such as prediction performance in classification algorithms) and unsupervised (e.g., clustering capacity) fitness functions can be used, sometimes combined with penalties or additional factors.

4   Methods

The core idea behind the proposed approach is to have a model able to predict the whole feature space from a subset. This approach is similar to the one commonly used by autoencoders (AE), which project the input data into a latent space, and from this, they aim to reconstruct the original input space by minimizing the MSE between the inputs and their reconstruction. But while in traditional AE the inputs and the outputs of the networks are the same, in our models the inputs are feature subsets of the original one, thus making the model asymmetric. The chosen ML model is a fully-connected NN, that is a multi-layer perceptron, closely similar to an auto-encoder. It consists of an encoder and a decoder, where the encoder maps the input features to a lower-dimensional representation and the decoder reconstructs the original features from the encoded representation by MSE minimization. If the model can reconstruct the whole input space by a subset of the original feature set, the selected subset is representative of the whole feature space, and thus, redundant information from the dataset should have been removed. By starting from this idea, we propose to evaluate feature subsets based on how good is the model to reconstruct the original input space, generating feature subsets using an EA. The EA work on binary bit strings chromosomes, of length equal to the total number of features in the dataset. For each chromosome, if the gene in position i is zero, the ith feature was not selected. Each gene codes the presence or absence of a feature in the subset. The GA approach permits to evolve through feature subsets that both minimize the reconstruction MSE and try to keep the size of the feature subsets small. We added a regularization term to the MSE to prevent preferring high-dimensional feature subsets. This encourages simpler and better generalizing solutions while avoiding keeping the whole feature set as the best set. In formula, given X as the original input, $\tilde{X}$ the output of the autoencoder, Fs the chosen subset of features, our fitness function is defined as:

$$(\alpha \cdot MSE(X, \tilde{X}) + \beta \cdot length(Fs))^{-1} \tag{1}$$

Since we are solving a minimization problem (minimizing both the MSE and the length of the chosen feature subset) but we want to maximize the fitness function, we raised to the power of -1 to have that higher fitness function mean better solutions, to be compliant with classical EA framework. The same results can be obtained without raising to the power of -1, but directly minimizing the fitness function. We terminate the EA algorithm when a maximum number of generations g is reached (Algorithm 1). Then, the feature subsets in the final population are evaluated through supervised error metrics (R2, MSE, Accuracy, ...) on a proper validation set. The chromosome with the lowest error on the validation set is selected as the final solution. Then, we evaluate the selected feature subset by comparing its performance to the original model using all features, on a proper test set.

5 Experiments

The proposed method is implemented in Python using scikit-learn [24], pandas [19,27], and numpy [14] libraries for dataset preprocessing. The model was built and run on GPU using pytorch [23]. We conduct experiments on four benchmark datasets, to evaluate the effectiveness of the proposed method and compare it to state-of-the-art unsupervised feature selection methods. All experiments were conducted on a server equipped with an Intel ® Core™ i9-9900K CPU @ 3.60 GHz × 16, 32 GiB of RAM, and an Nvidia ® 2080 TI GPU with 12 GiB of reserved RAM.

Algorithm 1. U-FLEX Evolutionary Algorithm

$max_g \leftarrow generation$
$g \leftarrow 0$
$tournamentSize \leftarrow t$
$populationSize \leftarrow p$
$mutationRate \leftarrow m$
$population \leftarrow random(popSize)$
while $g \leq max_g$ **do**
    $fitnessValues = evaluate(population)$
    $parents = tournamentSelection(population, fitnessValues, tournamentSize)$
    $population \leftarrow []$
    for $i \in range(0, populationSize, 2)$ **do**   ▷ iterate over couples of parents
        $offspring1, offspring2 = crossover(parents[i], parents[i+1])$
        $population.append(mutate(offspring1, mutationRate))$
        $population.append(mutate(offspring2, mutationRate))$
    end for
    $g = g + 1$
end while

5.1 Datasets

Here, we describe the dataset used for validating our feature selection framework. The number of samples, features, and the task performed are described in Table 1.

MSD. The Million Song Dataset (MSD) [6] is a collection of audio features and metadata for over a million songs, created by The Echo Nest and released to the public in 2011. The year prediction task is one of several tasks that can be performed on the Million Song Dataset, and it has been used as a benchmark for evaluating machine learning models for music analysis. The goal of the task is to train a machine learning model able to predict the year of release for a set of test songs. The audio features used in the task include features like tempo, timbre, and spectral contrast, which are extracted from the audio signal using signal processing techniques. The metadata associated with each song, such as the artist name and album title, can also be used as input to the machine learning model. The year prediction task is typically framed as a regression problem, where the output is a continuous value representing the predicted year of release.

Artificial Dataset. A random artificial regression problem (ART). The input X was generated with a low-rank-fat tail singular profile. The dataset was created by means of `scikit-learn` package. Most of the variance can be explained by a bell-shaped curve of width 30, while we have 500 components. The low-rank part of the profile can be considered the structured signal part of the data while the tail can be considered the noisy part of the data that cannot be summarized by a low number of linear components. The output is generated by applying a random linear regression model with bias equal to 100 and with 30 informative and nonzero regressors to the previously generated input and Gaussian-centered noise with a standard deviation equal to 0.05. 30,000 samples were created.

Madelon. MADELON (MAD) is an artificial dataset for balanced binary classification, containing data points grouped in 32 clusters placed on the vertices of a five-dimensional hypercube and randomly labeled. The five dimensions constitute 5 informative features. 15 linear combinations of those features were added to form a set of 20 (redundant) informative features. An additional number of 480 distractor features called 'probes' having no predictive power were added, for a total of 500 features. It is a classical dataset used for feature selection challenges [13].

Table 1. Datasets information

Name	Pseudo	# Samples	# Features	Task	Reference
Million Song Dataset	MSD	515,345	90	Regression	[6]
Artificial	ART	30,000	100	Regression	[24]
Madelon	MAD	2600	500	Classification	[13]

5.2 Comparison

We compared our framework with two classical feature selection approaches: a Pearson Correlation-based univariate filter method, based on a threshold, and a wrapper method based on an Evolutionary Algorithm.

Univariate Feature Selection. Pearson correlation can be used as a filter method for unsupervised univariate feature selection by identifying the linear correlation between independent features in the dataset, without relying on the target label. The most informative features in a dataset are the less-correlated ones: if two pairs of features present high linear correlation, they may provide redundant or overlapping information and can be eliminated from the dataset without losing much information. It is important to note that Pearson correlation is only effective for identifying linear relationships between variables. If the relationship between variables is non-linear, other measures of correlation such as Spearman's rank correlation or Kendall's tau correlation may be more appropriate. By eliminating redundant or overlapping information, the resulting subset of features can provide a more efficient and informative representation of the data. An unsupervised correlation-based approach for feature selection can be performed by calculating the Pearson correlation between each pair of features in the dataset. Then, it is possible to identify pairs of highly correlated features by setting a threshold value for the correlation coefficient and removing any features that have a correlation value above the threshold. Perfectly correlated variables are truly redundant in the sense that no additional information is gained by adding them, but very high variable correlation (or anti-correlation) does not mean the absence of variable complementarity [11]. We used 80% as the correlation threshold.

Wrapper Supervised EA Feature Selection. We compared the obtained results with a supervised wrapper method for feature selection, based on EA (Supervised EA). The fitness function used is the error metric of a supervised model on a validation set, with the objective of maximizing the prediction performance of the underlying model in a 5-folds CV fashion. The chromosomes with the best fitness score are selected for reproduction, and the process continues until a stopping criterion is met (maximum number of generations). We used `scikit- learn genetic-opt` package [3]. It is a popular open-source Python package that provides a genetic algorithm-based approach for hyperparameter tuning and feature selection in `scikit-learn`. The genetic feature selection algorithm starts by randomly generating a set of chromosomes, each representing a different subset of features. It then applies a genetic algorithm to iteratively select the most relevant subset of features, optimizing the prediction performance (R2 score for regression, Accuracy for classification) of a specified estimator using k-fold cross-validation. The package implements the $\mu + \lambda$ evolutionary strategy, thus keeping both parent and children at each generation. μ represents the number of parent individuals in a population, while λ is the number of offspring individuals.

5.3   Parameter Setting

As written in Sect. 4, the fitness function we used is a combination of two terms that we aim to optimize: the ability of the network to reconstruct the original inputs and the dimensionality of the feature subsets found. The parameters α and β in Eq. 1 control the stability of the learning process, and they are dataset-dependent. In our settings, for a dataset with n total features, $\alpha = 10$ and $\beta = 1.5 \cdot \frac{1}{n}$ worked well in general, making the two terms of the objective functions do not prevail one over the other.

The EA was implemented by using very simple criteria for the creation of a new population at each generation. The selection was based on tournament selection with a tournament size of 5. First, the candidate with the best fitness function is added to the new population (*elitism*). The participants in each tournament are selected at random among the population, and the winner of each tournament is added to the parents, that will be subjected to crossover in the next phase (see Algorithm 1).

The crossover is a uniform crossover between each parent. The mutation was a single uniform bit-flip based on a mutation rate of 0.01, meaning that each gene has a probability of 0.01 of being flipped. The mutation was applied to each of the offspring. The population size was set to 30.

The deep autoencoder was trained for just 1 epoch: even if this number seems to be low or not enough to say something about the reconstruction ability of the autoencoder, we didn't find great improvement in the genetic learning process by increasing the number of epochs (we tried from units to tens of epochs): what we aim is having a rough idea of the ability of the network to reconstruct the original feature set from feature subset. Increasing the number of epochs of training leads for sure to more reliable measurements of the input reconstruction ability of the feature subset but at the cost of an increasing time in the evaluation of each individual of the population: the main bottleneck of our approach is precisely the evaluation step since evaluating each candidate requires the full training of a NN. Considering just one epoch of training is an approximation of the fitness function computation, but it is needed to reduce the fitness evaluation time [2]. More sophisticated ways of tackling the problem can be found. The architecture is a simple fully connected autoencoder. The encoder has 2 hidden layers of size 128 and 64, while the decoder is symmetric to the autoencoder, but the last layer has as output dimensionality the number of original features. The activation function is the Leaky-ReLU. We used ADAM optimizer with learning rate $lr = 1e - 3$ For fitting the successive supervised model, we used Linear Regression for regression tasks and SVM for the classification, with rbf kernel and parameter $C = 1$ (no hyperparameters tuning for the shallow model, we just used the default parameters provided by scikit-learn). We randomly split the dataset into train, validation, and test sets. We trained all the models (for the supervised and unsupervised evolutionary feature selection) on the same training set, and we tested them on the test set. We used 80% of the samples as the training set, and the remaining 20% as a blind test set, for each dataset. Additionally, the 15% of the training set was used as the validation set for the NN evaluation, for computing the fitness function, at each generation, and for choosing the best feature set.

The performances of the supervised model were tested on the test set. We used as error metrics the *accuracy*, in percentage, for the classification task and the *normalized Root Mean Squared Error* (nRMSE) for the regression task. The nRMSE is nothing more than the root mean squared error RMSE but normalized by the mean of continuous label y_true in the test set, i.e. $nRMSE = RMSE(y_{true}, y_{pred})/mean(y_{true})$. The lower this value, the better the performance of the model. These metrics were used to have error percentages rather than absolute values.

5.4   Results

The main results are presented in Table 2. Our framework successfully found relevant feature subsets, and the obtained results in the supervised task are comparable with the other methods considered, both for the number of features found, and prediction error reached (comparable with other research experiments on the same datasets, [6,8]). We let the algorithm run for a different number of generations, depending on the dataset. Since the MAD dataset contains a low number of samples, the evaluation of each candidate lasts a few seconds. In this case, to reach better performance (for both size and prediction error, thus more valuable feature subsets), we can let the algorithm run for more generations, and train the NN for more epochs (with a minor increase in time). With 50 steps, we were able to reach the 72.34% of accuracy on the test set with 136 features. With 150 generations, about 119 features are selected, with 77.50% of accuracy. With the Supervised EA, after 150 generations, for the MAD dataset 220 features were selected, with an accuracy of 74.23% (see Table 2). Figure 1 presents the trends of the mean fitness function and the max accuracy in the population for the MAD dataset. For the Supervised EA, we are directly optimizing the accuracy over the validation set (that increases until a plateau), while for U-FLEX not. But the fitness function correctly increases over the generation, meaning that we are pruning the original feature set. In MSD, after 50 generations, we found final feature subsets of similar size (just 9 more, 10% of the total) with respect to the supervised EA method, reaching the same prediction error (Table 2). For all the datasets explored, the Pearson Correlation method was not able to prune a relevant number of features, and in some cases (for example, the artificial dataset), no features were removed. Datasets with a high amount of correlation among features (MSD) are easily managed by our framework. Other types of datasets (such as MAD and ART), in which many features are practically noise, are not so easily tractable with reconstruction approaches but would require more knowledge on the type of data we must manage. These types of problems are more suitable to be solved with supervised approaches rather than unsupervised ones since the features are mutually independent, and thus we would need more features to reconstruct the original input space, even if they would be noise for a supervised task. But even with this type of problem, the difference between the number of features found by the supervised and unsupervised genetic algorithm in the ART dataset is not so pronounced (50 features, 10% of the total after 50 generations). A comparison between the fitness function of the supervised and unsupervised approach is shown in Fig. 2.[1].

Table 2. Error metrics (nRMSE for MSD and ART, Accuracy for MAD) on the benchmark datasets. 150 generations for the MAD dataset, 50 generations for MSD and ART datasets. The higher the accuracy (MAD) the better the result. The lower the nRMSE (ART, MSD), the better the result.

Dataset	Features Used/Error metric% All	Pearson	Supervised EA	U-FLEX (Our)
MSD	90/0.48%	87/0.49%	62/0.49%	71/0.51%
MAD	500/(68.85%)	490/(65.96%)	220/(74.23%)	119/(77.50%)
ART	500/(0.05%)	500/(0.05%)	232/(0.05%)	282/(0.13%)

---

[1] Promising results were also found on an industrial classified dataset for regression: the computed nRMSE on all the features was 1.83%. With Pearson filter methods, 2.5% of the features were selected, with a nRMSE of 1.86%. The wrapper-supervised approach selected the 33% of features, with an error of 1.37%. Our approach converged to 19% of the features, with 1.38% of nRMSE.

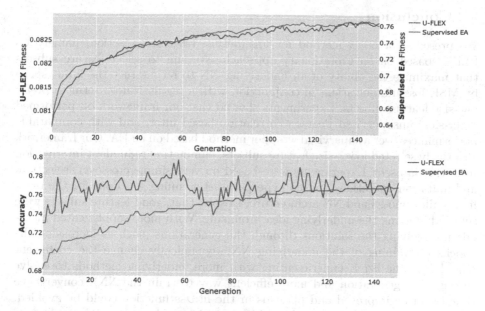

Fig. 1. Comparison of U-FLEX and Supervised EA mean fitness function (upper) and max accuracy (lower) over the population. MAD dataset, 150 generations. For the Supervised EA, we are directly optimizing the accuracy over the validation set (that increases until a plateau), while for U-FLEX not.

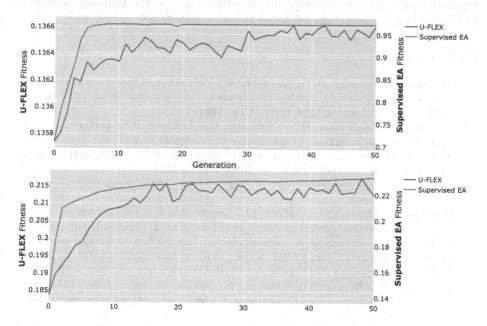

Fig. 2. Mean fitness function comparison between Supervised EA and U-FLEX in the population for each generation, computed on ART (upper) and MSD (lower) datasets.

6   Conclusion

We presented a novel method for unsupervised feature selection, namely U-FLEX, based on an evolutionary approach. We aim to select feature subsets that maximize the ability of an underlying NN to reconstruct the input space by MSE loss minimization, in conjunction with a regularization term to prefer low-size feature subsets. Experiments showed that our framework outperformed a classical unsupervised Pearson correlation-based filter, and with comparable performances over a supervised wrapper method based on an EA. Our framework effectively selected lower-size feature subsets, eliminating redundant information from the datasets. Experiments were based on very simple crossover, selection, and mutation operations and no hyperparameter tuning was done for the baseline shallow model and NN architecture. Despite that, good feature subsets were selected, making the underlying idea promising. With more sophisticated methods for evolving the solutions through the generations (such as increasing the epochs of training of the underlying NN, dynamically change the coefficient α and β in the Eq. 1, experimenting with more underlying methods to evolve through the generation and more efficient way to train the NN), convergence time could be improved and plateaus in the fitness function could be avoided, but these were not implemented for this work. Also, using more elaborated deep underlying NN could help in adapting the proposed techniques to other types of datasets (natural images, audio wavelengths, or spectra). More rigorous work should be done over a higher number of datasets and hyper-parameters, but the evaluation and the results make, in the authors' minds, the proposed approach promising in the unsupervised feature selection panorama.

References

1. Abualigah, L., Khader, A.T., Al-Betar, M.: Unsupervised feature selection technique based on genetic algorithm for improving the text clustering, pp. 1–6, July 2016. https://doi.org/10.1109/CSIT.2016.7549453
2. Altarabichi, M.G., Nowaczyk, S., Pashami, S., Mashhadi, P.S.: Fast genetic algorithm for feature selection - a qualitative approximation approach. Expert Syst. Appl. 118528 (2023). https://doi.org/10.1016/j.eswa.2022.118528. https://www.sciencedirect.com/science/article/pii/S0957417422016049
3. Arenas, R.: sklearn-genetic-opt (2023). https://github.com/rodrigo-arenas/Sklearn-genetic-opt
4. Barbiero, P., Lutton, E., Squillero, G., Tonda, A.: A novel outlook on feature selection as a multi-objective problem. In: Idoumghar, L., Legrand, P., Liefooghe, A., Lutton, E., Monmarché, N., Schoenauer, M. (eds.) EA 2019. LNCS, vol. 12052, pp. 68–81. Springer, Cham (2020). https://doi.org/10.1007/978-3-030-45715-0_6
5. Barbiero, P., Squillero, G., Tonda, A.: Predictable features elimination: an unsupervised approach to feature selection. In: Nicosia, G., et al. (eds.) LOD 2021. LNCS, vol. 13163, pp. 399–412. Springer, Cham (2022). https://doi.org/10.1007/978-3-030-95467-3_29
6. Bertin-Mahieux, T., Ellis, D.P., Whitman, B., Lamere, P.: The million song dataset. In: Proceedings of the 12th International Conference on Music Information Retrieval (ISMIR 2011) (2011)

7. Boutegrabet, W., Piot, O., Guenot, D., Gobinet, C.: Unsupervised feature selection by a genetic algorithm for mid-infrared spectral data. Anal. Chem. **94**(46), 16050–16059 (2022). https://doi.org/10.1021/acs.analchem.2c03118. pMID: 36346912

8. De Stefano, C., Fontanella, F., Scotto di Freca, A.: Feature selection in high dimensional data by a filter-based genetic algorithm. In: Squillero, G., Sim, K. (eds.) EvoApplications 2017. LNCS, vol. 10199, pp. 506–521. Springer, Cham (2017). https://doi.org/10.1007/978-3-319-55849-3_33

9. Eiben, A.E., Smith, J.E.: Introduction to Evolutionary Computing, 2nd edn. Springer, Heidelberg (2015). https://doi.org/10.1007/978-3-662-44874-8

10. Goodfellow, I., Bengio, Y., Courville, A.: Deep Learning. MIT Press (2016). http://www.deeplearningbook.org

11. Guyon, I., Elisseeff, A.: An introduction to variable and feature selection. J. Mach. Learn. Res. **3**, 1157–1182 (2003)

12. Guyon, I., Weston, J., Barnhill, S., Vapnik, V.: Gene selection for cancer classification using support vector machines. Mach. Learn. **46**, 389–422 (2002). https://doi.org/10.1023/A:1012487302797

13. Guyon, I.M.: Design of experiments for the NIPS 2003 variable selection benchmark (2003)

14. Harris, C.R., et al.: Array programming with NumPy. Nature **585**(7825), 357–362 (2020). https://doi.org/10.1038/s41586-020-2649-2

15. Hastie, T., Tibshirani, R., Friedman, J.: The Elements of Statistical Learning: Data Mining, Inference, and Prediction. Springer, New York (2009). https://doi.org/10.1007/978-0-387-84858-7

16. Heiss-Czedik, D.: An introduction to genetic algorithms. Artif. Life **3**, 63–65 (1997)

17. Jolliffe, I.T.: Principal Component Analysis. Springer, New York (2011)

18. Martin-Bautista, M., Vila, M.A.: A survey of genetic feature selection in mining issues. In: Proceedings of the 1999 Congress on Evolutionary Computation-CEC99 (Cat. No. 99TH8406), vol. 2, pp. 1314–1321 (1999). https://doi.org/10.1109/CEC.1999.782599

19. McKinney, W.: Data structures for statistical computing in Python. In: van der Walt, S., Millman, J. (eds.) Proceedings of the 9th Python in Science Conference, pp. 56–61 (2010). https://doi.org/10.25080/Majora-92bf1922-00a

20. Miao, J., Niu, L.: A survey on feature selection. Procedia Comput. Sci. **91**, 919–926 (2016). https://doi.org/10.1016/j.procs.2016.07.111. https://www.sciencedirect.com/science/article/pii/S1877050916313047. Promoting Business Analytics and Quantitative Management of Technology: 4th International Conference on Information Technology and Quantitative Management (ITQM 2016)

21. Mitchell, M.: An Introduction to Genetic Algorithms (1996)

22. Mitchell, M.: An Introduction to Genetic Algorithms. Complex Adaptive Systems, 7th edn. Cambridge (2001)

23. Paszke, A., et al.: Pytorch: an imperative style, high-performance deep learning library. In: Advances in Neural Information Processing Systems 32, pp. 8024–8035. Curran Associates, Inc. (2019). http://papers.neurips.cc/paper/9015-pytorch-an-imperative-style-high-performance-deep-learning-library.pdf

24. Pedregosa, F., et al.: Scikit-learn: machine learning in Python. J. Mach. Learn. Res. **12**, 2825–2830 (2011)

25. Pudjihartono, N., Fadason, T., Kempa-Liehr, A.W., O'Sullivan, J.M.: A review of feature selection methods for machine learning-based disease risk prediction. Front. Bioinform. (2022). https://doi.org/10.3389/fbinf.2022.927312. https://www.frontiersin.org/articles/10.3389/fbinf.2022.927312

26. Solorio-Fernández, S., Carrasco-Ochoa, J., Martínez-Trinidad, J.F.: A review of unsupervised feature selection methods. Artif. Intell. Rev. **53** (2020). https://doi.org/10.1007/s10462-019-09682-y
27. The Pandas Development Team: Pandas-dev/pandas: Pandas, February 2020. https://doi.org/10.5281/zenodo.3509134
28. Xie, J., Wang, M., Xu, S., Huang, Z., Grant, P.W.: The unsupervised feature selection algorithms based on standard deviation and cosine similarity for genomic data analysis. Front. Gen. **12** (2021). https://doi.org/10.3389/fgene.2021.684100. https://www.frontiersin.org/articles/10.3389/fgene.2021.684100

Improved Filter-Based Feature Selection Using Correlation and Clustering Techniques

Akhila Atmakuru[1]([✉]) [iD], Giuseppe Di Fatta[2] [iD], Giuseppe Nicosia[3] [iD], and Atta Badii[1] [iD]

[1] University of Reading, Reading, UK
`akhila.atmakurupt@gmail.com`, `atta.badii@reading.ac.uk`
[2] Free University of Bozen-Bolzano, Bolzano, Italy
`giuseppe.difatta@unibz.it`
[3] University of Catania, Catania, Italy

Abstract. Feature engineering and feature selection are essential techniques to most data science and machine learning applications, in which, respectively, raw data are transformed into features and features are selected to provide the most effective subset of features for the application. Feature selection techniques are particularly useful when dealing with high-dimensional datasets that contain noisy and redundant data. An optimised feature subset could enhance the performance as well as the interpretability of the model. There are three types of feature selection methods, namely filter, wrapper and embedded techniques. Amongst these methods, the filter method is more efficient than the others as it is computationally less expensive and more generalised. This work presents two improved filter-based feature selection methods based on a correlation coefficient and clustering techniques. The first approach is based on feature correlation where the feature subset consists of features above a similarity threshold to identify a kind of neighbourhood for each feature. The second method uses clustering analysis on the correlation data to identify features that can be used to represent the entire cluster. The obtained feature subsets have been applied as pre-processing step for logistic regression and artificial neural networks. The performance of the proposed methods has been compared against the popular ReliefF feature selection method. The experimental analysis shows that the proposed feature selection methods provide an observable improvement in accuracy by choosing the most effective features.

Keywords: Feature Selection · Correlation · Clustering · Principal Coordinate analysis · Neural Network and High Dimensional Dataset

1 Introduction

Machine learning models can identify patterns in data and make predictions accurately based on identified patterns. However, raw data are typically noisy and contain redundant and irrelevant information: machine learning models can be quite inaccurate and ineffective when applied directly to raw data.

G. Nicosia et al. (Eds.): LOD 2023, LNCS 14505, pp. 379–389, 2024.
https://doi.org/10.1007/978-3-031-53969-5_28

Feature selection is the process of choosing the most effective subset of features from raw data. The performance of a machine learning model is improved by minimising the number of the less effective, redundant, or noisy features in the dataset. Identifying and ranking the most effective features, enables researchers to have a better grasp of the relationships between the features and the target variable, making the output of a machine learning model more robust. Feature ranking could also aid in reducing overfitting, increasing model accuracy, reducing model complexity and training time. There are three different feature selection methods, namely filter, wrapper and embedded methods [1]. The filter-based methods rank each individual feature or an entire subset of features based on measures such as information, distance, consistency, similarity (correlation), and other statistical measures. Feature correlation is typically computed as a matrix representation of the relationship between each possible pair for all the variables and is useful to summarise a large dataset and visualise similarity patterns within the features. The performance of filter-based methods does not depend on to the data modelling approach chosen for the application. Thus, compared to the wrapper and embedded techniques, the filter-based approaches are more general and computationally less expensive since they are independent of the learning algorithm. ReliefF, Correlation based Feature Selection (CFS), and the max-relevancy min-redundancy (mRMR) feature selection algorithm are a few examples of filter-based techniques.

The wrapper methods use searching algorithms to identify the significant feature sub-sets, which are ranked based on their performance when applied to a specified modelling method. The creation of each subset is reliant on the search technique, and the evaluation is repeated for each subset. Since they are dependent on the resource requirements of the modelling process, wrappers are substantially slower than filters at identifying the more effective subsets. The feature subsets thus identified, are skewed in favour of the modelling approach that was used to rank them.

The embedded methods perform feature selection during the execution of the mod-elling algorithm. By maximising a performance parameter, such as accuracy or error rate, these techniques seek to identify and choose the more effective features during training. These methods are embedded in the algorithm either as its normal or extended functionality. Due to their iterative process, embedded-based approaches can be com-putationally costly. Moreover, they require more domain expertise than filter or wrapper approaches since they change an existing learning algorithm rather than just rank the efficacy of feature subsets independently of it, as is the case with other feature selection methods.

The ReliefF algorithm assumes that the variables are linear in nature and suitable for a binary classification problem. It calculates the significance of each feature relative to its performance target. Features with high correlation are retained. And so, some redun-dant features remain making the algorithm inefficient. The Correlation based Feature Selection (CFS) algorithm computes correlation for a subset of features. The method is then repeated for several feature subsets until a subset with a lower average of feature-to-feature correlation and a higher average of feature to target correlation is determined. The CFS algorithm based on iterative heuristic search strategies is computationally expen-sive as it requires repeated computation of correlation between features and target. The max-relevancy min-redundancy (mRMR) feature selection algorithm is based on mutual information. The feature relevancy, i.e., the relative effectiveness of a feature, is determined by mutual information computed between individual features and the target class. The redundant features are identified by applying mutually exclusive conditions. The algorithm is computationally expensive as many mutual information computations are required.

Another prominent strategy for identifying a subset of the relatively more significant features in large datasets is clustering-based feature selection. The characteristics features are clustered based on their closeness, and the best representative feature from each cluster is then selected. The disadvantage of the clustering methods arises from the loss of performance due to overlap of clusters and influence of outliers.

There are mainly four types of clustering techniques. The correlation-based clustering involves clustering features according to their correlation coefficient. The most representative feature from each cluster is chosen after the features with strong correlation are clustered together. K-Means clustering algorithm groups the features depending on their distance from the cluster centroid. The most representative feature of each cluster is chosen as a subset of features. Hierarchical clustering approaches group the features into hierarchical structures based on how similar these are. At various levels of the hierarchical structure, the most representative feature is chosen from each cluster. However, this approach, is inefficient while handling missing data or large datasets. The spectral clustering approach groups the features using a graph representation. The most representative feature of each cluster is chosen as a subset of features. Computing eigenvectors for a high dimensional dataset can cause bottlenecks in this approach.

This paper proposes two distinct strategies for implementing feature selection. These are based on a correlation matrix generated from a raw dataset. The first technique analyses the correlation matrix to find key attributes that will be used to represent the neighbourhood. The second technique utilises an unsupervised clustering method on the correlation matrix to identify the key features that represent the complete cluster. The findings showed that the features obtained from these techniques improved the performance interpretability by identifying key features and the corresponding performance.

The remainder of this paper is structured as follows: Sect. 2 presents an overview of the literature on current feature selection approaches. Section 3 provides a description of the dataset that was used in this work. Section 4 discusses the approach of the feature selection strategies that have been developed. Section 5 includes the Results and Discussion which: summarises the methodology outcomes utilising suggested feature selection methods. Section 6 concludes this paper by discussing the future trajectory of the work.

2 Literature Review

While dealing with a high dimensional dataset, it is beneficial and efficient to use filter-based feature selection methods as they do not need repeated training and evaluation and are independent of the model being used for training. The literature study will focus on methods based on correlation and clustering methods.

The widely used feature selection method known as "Relief" assumes the features are linear in nature and only works for two-class situations. This method is based on a distance-based metric function that weights each feature according to how relevant (correlation) it is to the target class [2]. The features with higher correlation to the target and lower correlation to other features were retained. The ReliefF [3], a Relief variant, can deal with multi-class problems and noisy datasets, but it is inefficient due to redundant features. ReliefF has been augmented by various other algorithms to handle the redundant features issue.

Another popular method for correlation-based feature selection (CFS) [4] is based on a heuristic best first search approach. In this algorithm, correlation between features, and features to the target is computed. The algorithm then determines which features could be used together in a subset in a heuristic manner. The algorithm repeats the process for several feature subsets until a subset containing a lower average of feature-to-feature correlation and higher average of feature-to-the-target correlation is identified. The algorithm is computationally expensive for a high dimensional dataset.

Another Fast Correlation-Based Filter (FCBF) [5] algorithm removes both irrelevant and redundant features by calculating the symmetrical uncertainty which is the measure of relevance between the feature and the target. Beginning with the whole feature set, FCBF adopts a heuristic backward selection approach combined with a sequential search strategy to eliminate unnecessary and duplicate features. The algorithm terminates once there are no more characteristics to discard. The algorithm is able to reduce the features, while still maintaining high accuracy. However, the algorithm is computationally expensive.

The algorithm maximum Relevancy Minimum Redundancy (mRMR) [6] is based on Mutual Information (MI), which is computed between the individual feature and target class for identifying feature relevancy. To handle the redundant features, mutually exclusive conditions are applied.

For clustering methods, one of the algorithms [7] is based on graph clustering to identify similar features for removing redundant features. Each cluster is ranked based on similarity measures, thereafter the top-ranked clusters are selected and the representative features retained. Another method [8] clusters the features based on the clustering chi-square statistical measure and then selects the most representative feature for each cluster. This approach utilises a modified K-Means clustering algorithm to form clusters. The method Multi Cluster feature selection [9] is based on a spectral clustering algorithm on the correlation similarity matrix between the features. The importance of each feature is assessed within each cluster using a non-parametric Wilcoxon rank-sum test, and the most relevant features are identified.

A clustering method based on mutual information [10] starts with splitting the dataset into training and testing sets and then groups the features in the training set according to how relevant they are to the classification task using a supervised clustering algorithm. The clustering technique uses the mutual information between each feature and the class label as the distance metric. The most representative characteristic from each cluster is then chosen as the final set of features after the clusters are sorted according to their mutual information with the class label. A classifier is then trained using the selected features on the training set, and its performance is assessed using the testing set.

A similar approach [11] uses a graph partitioning clustering algorithm. The data is represented as a graph and partitioned into clusters to select the most relevant feature from each cluster. Another method [12] is based on similarity measures to form clusters using hierarchical clustering algorithms. At each hierarchical level, both the intra-cluster and inter-cluster similarity score is computed based on which the features with each cluster are retained. This process continues until stopping criteria are met or the specified number of clusters are obtained.

An approach for arrhythmia classification using a feature selection schema for creating an ensemble of classifiers [13] improves accuracy in high dimensional datasets by identifying relevant feature sets that affect classification. Multiple feature subsets are extracted, and classification models are built based on each subset. The models are then combined using a voting approach to calculate each classifier score in the ensemble. In an experimental study, the proposed method generated three top distinct feature sets, and three classifiers were constructed based on these subsets. The classifier ensemble using the voting approach significantly improved classification accuracy in high dimensional datasets and led to a more stable classification model. The performance of each classifier and the ensemble was compared to a classifier using the entire feature space of the dataset.

Different feature selection methods have been compared [14] on the arrhythmia dataset to improve the accuracy of machine learning models for diagnosing heart disease. The two methods used are filters and wrappers, both of which perform well, but filters are faster. The paper finds that random forest with the wrapper method has the highest accuracy. The paper also suggests further investigation into the correlation of features and optimization for learning machines targeted on the arrhythmia dataset. A Gaussian naive Bayes classifier is also trained, and the paper finds that filters greatly improve the accuracy of this classifier. Support Vector Machines (SVMs) are also used, and the paper finds that an SVM classifier with feature selection methods scores higher on the test set than that on the train set.

3   Dataset Description

The Arrhythmia Dataset is a widely used dataset for research in the field of cardiac arrhythmia detection and classification. The dataset was initially created and made publicly available by the University of California, Irvine (UCI) Machine Learning Repository. The specific source reference for the dataset is "Arrhythmia Data Set" in the "UCI Machine Learning Repository" [15].

This dataset was collected to aid in the identification of various types of cardiac arrhythmias from regular 12-lead ECG recordings. It consists of records from 452 patients, with each record containing 280 features. The features encompass clinical measurements derived from ECG signals, such as QRS length, R-R interval, P-R interval, Q-T interval, and other demographic information including sex, age, weight, and a cardiologist's recommendation.

The primary objective of using this dataset is to classify patients into two categories: normal patients and those with one of the 15 different types of arrhythmias. The dataset poses challenges due to its high dimensionality and the presence of missing values, which require appropriate pre-processing techniques.

Researchers and machine learning practitioners often utilise this dataset as a benchmark for developing and evaluating arrhythmia detection algorithms, feature selection methods, and classification models.

The dataset contains features that need pre-processing, some of which have missing values. In order to address this, the missing values were replaced with 0. Furthermore, certain features were removed from the dataset due to their numerous NaN values. Despite this, there are still 262 features available for analysis. In this paper, target variables are divided into two classes: normal patients and arrhythmia patients.

4   Methodology

This section describes the feature selection methods developed using Python. The feature selection methods are based on the correlation between the input features. An Arrhythmia dataset was used to perform the feature selection. The developed methods were further tested on machine learning algorithms and neural network models to compare their performance.

4.1   CGN-FS: Correlation-Based Greedy Neighbourhood Feature Selection Method

The method involves computation of feature-to-feature correlation values, thresholding the values and selecting the features having the lowest correlation. This approach will hereafter be referred to as the Correlation-based Greedy Neighbourhood Feature Selection method (CGN-FS). The anticipated output of the algorithm is a subset of features from the main dataset.

To pick the features, first a correlation matrix is generated. The correlation matrix would include the correlation coefficient for each feature in relation to each of the other features in all possible pairs. The diagonal value of the correlation matrix, for each feature, represents the correlation coefficient value of the feature with itself which is equal to 1. The diagonal values are therefore excluded from the analysis. The absolute values of the element of the correlation matrix are then obtained. The correlation magnitude is represented by an absolute value in the matrix; the higher the value of the element, the more strongly the variables are related to one another.

The absolute correlation values of the matrix are utilised to calculate various evaluation metrics, such as 'sum' and 'count.' The cumulative total of the correlation values of each feature concerning all other features determines the 'sum' for each feature. The cumulative total for each feature is appended to the absolute correlation matrix under the column 'sum'.

Next, the count is determined by applying a threshold to the correlation values, which can range between 0 and 1.00. Further analysis excludes the correlation values of the feature with respect to all other features, while the features that exceed the threshold are identified as the neighbours of the feature. The count value, which corresponds to the number of these neighbours, is then added to a new column called 'count' in the absolute correlation matrix.

The above metrics are computed for different threshold values. At each threshold value the selected features are evaluated by two Machine Learning classifiers. The threshold value with best performance is finalised as the optimum threshold value. Starting with 0.50, the various threshold values used for testing are increase by 0.05. After deriving the necessary metrices, the matrix is organised in descending order using the primary and secondary columns of "Count" and "Sum". Additionally, a flag labelled "Keep" is initialised for each feature. The neighbouring features are then marked as "Remove," while the flag for each feature remains as 'Keep'. Ultimately, the list of features with the "Keep" flag exclusively assigned is retrieved. This final list represents the anticipated subset of features selected from the entire range. The process can be replicated on various datasets to obtain the chosen subset of features.

Two models have been evaluated to assess their performance for each threshold value. With the help of a 10-fold stratified cross-validation approach, these models are trained using a final subset of features that were generated using the CGN-FS methodology. For each of the two models, the accuracy mean and standard deviation were computed. The best performance is the one where the accuracy is maximum while the standard deviation of the accuracy is at a minimum.

After calculating the needed components, the matrix is sorted using the primary and secondary columns of "Count" and "Sum" in decreasing order. A flag labelled "Keep" is initialised for each feature as well. The neighbours of each feature are then marked as "Remove," while the flag for each feature itself remains as 'Keep'.

Finally, the list of features that only have the flag "Keep" assigned is retrieved. The expected subset of features chosen from all the features is represented through this final list. This process could be repeated on different datasets to obtain the selected subset of features.

The proposed method, Correlation-based Greedy Neighbourhood Feature Selection (CGN-FS), presents various advantages and limitations. On the positive side, CGN-FS efficiently detects pertinent features by means of feature-to-feature correlation computation. This results in an enhanced classification performance through the selection of informative features. The flexibility of the method enables customisation of the feature selection process by adjusting threshold values, making it applicable to various datasets and classification tasks. Nevertheless, the effectiveness of the approach is heavily dependent on the choice of the optimal threshold value, which can be challenging and reliant on the dataset. To maximise the benefits of CGN-FS, careful threshold selection and consideration of alternative evaluation metrics are crucial (Fig. 1).

4.2 RCH-FSC: Region and Correlation Based Heuristic Feature Selection with Clustering Analysis

The clustering analysis uses an arrhythmia dataset, which is a high-dimensional dataset, as input for the analysis. The objective of this method is to carry out clustering of the input features and feature selection by selecting the features that represent the complete clusters produced by the clustering algorithm. The analysis used K-medoids clustering, which is an improved version of the conventional K-means clustering and is more resistant to noise and outliers. A medoid indicates the centre of a cluster as a data-point in the cluster rather than the mean point. Medoid is the data-point in the cluster that is nearest to the

Algorithm 01: Correlation-based Greedy Neighbourhood Feature Selection Method (CGN-FS)
Input: High Dimensional Dataset
Threshold: To be chosen after thorough analysis
Output: Subset of features
Procedure:
1.   Calculate the feature correlation using Pearson Method.
2.   Obtain the absolute values of the feature correlation matrix.
3.   Calculate the sum of each features correlation values w.r.t all other features.
4.   For each feature i, count the number of correlation values above threshold w.r.t to all other features and identify these features as neighbours(i). The number of neighbours is 'count' value.
5.   Sort by decreasing order of count (primary) and sum (secondary).
6.   Initialize flag as 'Keep' for all features.
7.   For each feature i, if Flag(i) is "Keep" mark the features in the neighbours(i) as 'Removed".
8.   Return the features with flag "Keep".

Fig. 1. Algorithm for CGN-FS: Correlation-based Greedy Neighbourhood Feature Selection method

centre and has the least total distances from other locations. This developed method will be termed as Region and Correlation-based Heuristic Feature Selection with Clustering Analysis (RCH-FSC) method. This method is expected to produce a lower number of final subset features while not compromising on the performance. Initially, a correlation matrix is created for the dataset. The correlation matrix will include the correlation coefficients for each feature in relation to each other feature in all conceivable pairs. Next, the absolute correlation matrix is determined by taking absolute values.

The distance matrix for the absolute correlation matrix is calculated. Knowledge of the linear or non-linear dependency between the two variables might be gained from the distance matrix.

Next, the distance matrix is subjected to dimensionality reduction. For this purpose, a Principal Coordinate Analysis (PCoA) method, an improved version of the classical Principal Component Analysis, variant Multi-dimensional Scaling (MDS) is used. PCoA is a graphical depiction of the matrix in a low dimensional space. This approach emphasises the differences between the features by visualising similarity of features. In PCoA with MDS, the issues arising due to unstable eigen decomposition and non-linear dissimilarities are handled by the algorithm. The number of principal components is determined by using a cumulative explained variance with a threshold of 50%. The threshold used for an explanation was 50% and the resultant number of principal components were 219 as a result.

The K-medoids clustering method is applied to the resultant PCoA data that has been generated. The elbow technique and silhouette score are used to calculate the optimal number of clusters. The outcomes of the silhouette score depend on how well each feature fits into the cluster, which is crucial to the goal of the developing method. Thus, the silhouette score determines the K-value which is the number of clusters. K-medoids attempt to lower the total sum of dissimilarities between the data-points of a cluster and its medoid. The datapoint in the cluster with the most central location is a medoid. As the medoids reflect the complete cluster, these medoids represent the final subset of the selected features.

The RCH-FSC Method, which stands for Region and Correlation-based Heuristic Feature Selection with Clustering Analysis, presents various benefits for selecting features in datasets with high dimensions. Through the utilisation of the K-medoids clustering algorithm, it can identify features that are representative and form complete clusters. This makes for a robust selection process, particularly when dealing with noise and anomalies. To lower dimensionality and capture both linear and non-linear relationships among characteristics, the principal coordinate analysis (PCoA) using Multi-dimensional Scaling (MDS) is additionally implemented. By means of silhouette scores, this approach can efficiently determine the optimal number of clusters, thereby ensuring a more effective feature subset. Furthermore, the algorithm strives to attain a compromise between reducing the feature set and classification effectiveness, resulting in a smaller feature selection without any significant decrease in overall accuracy (Fig. 2).

Algorithm 02: Region and Correlation based Heuristic Feature Selection with Clustering Analysis Method (RCH-FSC)

Input: High Dimensional Dataset

Output: Subset of features

Procedure:

1. Calculate the feature correlation using the Pearson Method.

2. Obtain the distance matrix from the correlation matrix.

3. Perform Principal Co-Ordinate Analysis with Multi Dimensional Scaling.

4. Perform K-Medoids Clustering analysis.

5. Identify one feature to represent each and entire cluster.

6. Identified features are the final subset.

Fig. 2. Algorithm for RCH-FSC: Region and Correlation based Heuristic Feature Selection with Clustering Analysis

5   Results and Discussion

5.1   Correlation Method

For clustering-based method, the arrhythmia dataset was utilised to perform the analysis. The final number of features used for feature selection were 262.

Starting with 0.50, the various threshold values used for testing increased by 0.05. Three models have been evaluated to assess performance for each threshold value. With the help of a repeated 10-fold stratified cross-validation approach, these models were trained using a final subset of features that was generated using the CGN-FS methodology. For each of the two models, the accuracy mean, and standard deviation were computed. The best performance was taken as the one where the accuracy was maximum while the standard deviation of the accuracy was at minimum which occurred at 0.70 threshold. The number of keep values at the optimum threshold was 101 features.

The algorithm was evaluated on two different machine learning models namely, logistic regression, and neural networks. The threshold 0.80 was selected as optimum and the logistic regression accuracy was 68.57% with a standard deviation of 3.87. For the neural network the best threshold was at 0.85 and number of features selected were 200 features and accuracy was 92.10% with standard deviation of 2.77.

ReliefF was implemented for the same dataset used by the CGNFS algorithm. For ReliefF the model used for evaluation was logistic regression. The parameters required were the number of features to be selected which was 101 similar to the threshold value at .80. The obtained accuracy for the ReliefF method was 70.36%.

5.2 Clustering Method

The clustering-based method was used to carry out the analysis of the arrhythmia dataset. During the feature selection process, a total of 262 features were utilised. Furthermore, the PCoA with MDS method was implemented for dimensionality reduction. This resulted in a total of 219 dimensions for 50% cumulative explained variance.

Upon using the developed algorithm, 15 features were obtained for this particular dataset. Upon evaluating these 15 features with logistic regression, using the repeated (10 times) stratified 10 cross-validation method, an accuracy of 70.60% was achieved. The standard deviation was calculated to be 6.57. For comparison purposes, testing was conducted using all features with the repeated (10 times) stratified 10 cross-validation method. The accuracy achieved was 72.36%, with a standard deviation of 7.25. Additionally, using the neural network for 15 features with repeated (10 times) stratified 10 cross-validation method resulted in an accuracy of 67.39%, with a standard deviation of 3.90.

6 Conclusions

The application of correlation and clustering-based feature selection methods on the arrhythmia dataset has proven to be a valuable approach for enhancing the efficiency and interpretability of the predictive models. The correlation-based approach has successfully identified features with low inter-feature correlation, ultimately leading to a reduced feature set without compromising accuracy. This reduction in features has not only streamlined the modelling process but has also contributed to improved interpretability and robustness, consequently allowing for more meaningful insights into the underlying relationships between variables.

On the other hand, the clustering-based approach has demonstrated its efficacy in selecting a compact set of 15 relevant features, with minimal trade-offs in accuracy (approximately 1.5%). By leveraging the patterns of data points in a 2D space and determining cluster centroids, this method has effectively highlighted key features that capture essential information for classification tasks. Thus, the ability of the clustering algorithm to reduce feature dimensionality has led to a noticeable enhancement in model performance and interpretability.

The combination of these two feature selection techniques offers a promising avenue for optimising machine learning models in various domains, especially when faced with high-dimensional datasets such as the arrhythmia dataset. Not only do they contribute to improve predictive accuracy, but they also provide a clearer understanding of the underlying data patterns, leading to more reliable and transparent decision-making processes. As the field of feature selection continues to evolve, incorporating these methods can be significantly beneficial to practitioners seeking to develop efficient and effective machine learning models for real-world applications.

References

1. Gnana, D.A.A., Balamurugan, S.A.A., Leavline, E.J.: Literature review on feature selection methods for high-dimensional data. Int. J. Comput. Appl. **136**(1), 9–17 (2016)
2. Kira, K., Rendell, L.A.: A practical approach to feature selection. In: Machine Learning Proceedings, pp. 249–256. Morgan Kaufmann (1992)
3. Kononenko, I.: Estimating attributes: analysis and extensions of RELIEF. In: ECML, vol. 94, pp. 171–182, April 1994
4. Hall, M.A.: Correlation-based feature selection of discrete and numeric class machine learning (2000)
5. Yu, L., Liu, H.: Efficient feature selection via analysis of relevance and redundancy. J. Mach. Learn. Res. **5**, 1205–1224 (2004)
6. Peng, H., Long, F., Ding, C.: Feature selection based on mutual information criteria of max-dependency, max-relevance, and min-redundancy. IEEE Trans. Pattern Anal. Mach. Intell. **27**(8), 1226–1238 (2005)
7. Song, Q., Ni, J., Wang, G.: A fast clustering-based feature subset selection algorithm for high-dimensional data. IEEE Trans. Knowl. Data Eng. **25**(1), 1–14 (2011)
8. Li, Y., Luo, C., Chung, S.M.: Text clustering with feature selection by using statistical data. IEEE Trans. Knowl. Data Eng. **20**(5), 641–652 (2008)
9. Cai, D., Zhang, C., He, X.: Unsupervised feature selection for multi-cluster data. In: Proceedings of the 16th ACM SIGKDD International Conference on Knowledge Discovery and Data Mining, pp. 333–342, July 2010
10. Chow, T.W., Huang, D.: Estimating optimal feature subsets using efficient estimation of high-dimensional mutual information. IEEE Trans. Neural Netw. **16**(1), 213–224 (2005)
11. Mitra, S., Acharya, T., Luo, J.: Data mining: multimedia, soft computing, and bioinformatics. J. Electron. Imaging **15**(1), 019901 (2006)
12. Sotoca, J.M., Pla, F.: Supervised feature selection by clustering using conditional mutual information-based distances. Pattern Recogn. **43**(6), 2068–2081 (2010)
13. Namsrai, E., Munkhdalai, T., Li, M., Shin, J.H., Namsrai, O.E., Ryu, K.H.: A feature selection-based ensemble method for arrhythmia classification. J. Inf. Process. Syst. **9**(1), 31–40 (2013). https://doi.org/10.3745/JIPS.2013.9.1.031
14. Liu, Z.: Comparison of feature selection methods on arrhythmia dataset. In: Proceedings of the 2021 3rd International Conference on Image Processing and Machine Vision (IPMV 2021), New York, NY, USA, pp. 66–71. Association for Computing Machinery (2021). https://doi.org/10.1145/3469951.3469963
15. Guvenir, H., Acar, B., Haldun, M., Quinlan, R.: Arrhythmia. In: UCI Machine Learning Repository (1998). https://doi.org/10.24432/C5BS32

Deep Active Learning with Concept Drifts for Detection of Mercury's Bow Shock and Magnetopause Crossings

Sahib Julka[✉][ID], Rodion Ishmukhametov[ID], and Michael Granitzer[ID]

Data Science, University of Passau, Passau, Germany
{sahib.julka,michael.granitzer}@uni-passau.de

Abstract. Active learning has shown great potential for improving the efficiency of data annotation in scientific applications. In the field of planetary science, where large volumes of data are generated by remote sensing instruments, active learning can significantly reduce the cost and time required for data analysis. However, existing active learning (AL) methods for planetary science applications do not consider the potential for drifts in the data distribution, which can lead to model degradation over time. To address this issue, we propose a drift detection-based active learning approach for planetary science applications. The proposed approach uses a semi-supervised generative adversarial network (GAN) to detect concept drifts and an entropy-based AL sampling procedure to select the most informative orbits from each for training. We test this approach on the use case of detecting bow shock and magnetopause boundaries around Mercury, utilising data obtained from NASA's MESSENGER mission. Our key results indicate that by employing our approach, a near-maximal information gain can be obtained by training with less than 10% of the available data, surpassing the simpler entropy-based active learning. This approach has the potential to accelerate scientific discoveries in planetary science and other scientific domains that deal with large volumes of remote sensing data.

Keywords: Active Learning · Concept Drift Detection · Magnetospheres

1 Introduction

The magnetosphere of a planet is the surrounding region where its magnetic field is stronger than that of the interplanetary space. The magnetopause is the outer boundary of this region, while the magnetosheath lies above it and extends to the bow shock. The bow shock is a shock wave that decelerates the solar wind and deflects it around the planet's magnetospheric cavity (cf. Fig. 1(a)). The positions and features of these regions around a planet are chiefly determined by the varying solar wind conditions [7]. This is particularly the case for Mercury, the innermost planet in our solar system. Additionally, the magnetic field of

Mercury is relatively weak, with only about 1% of the strength of Earth's magnetic field, rendering its magnetic conditions even more dynamic and worthy of investigation. Analyzing such magnetospheres can provide valuable knowledge for comprehending more complex magnetospheres, such as the Earth's.

Studying the magnetic fields of planets has been a long-standing interest in planetary science. To this end, NASA launched a spacecraft called MESSENGER to orbit Mercury and conduct long-term empirical studies. During the four years of its voyage from 2011 to 2015, the spacecraft completed over 4000 orbits around the planet (cf. Fig. 1(b)).

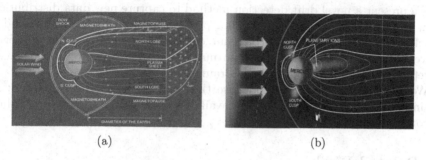

(a)   (b)

Fig. 1. (a) Schematic view of Mercury's magnetic conditions [24]. The bow shock slows down the approaching solar wind to subsonic speeds. The magnetopause further acts as an obstacle. (b) A typical MESSENGER orbit path: the spacecraft passed from the *interplanetary magnetic field* (IMF) through bow shock, magnetosheath, magnetopause and magnetosphere regions of Mercury and then through the same sequence in reverse [31].

Various studies have used data from the MESSENGER magnetometer to propose physical geometric models for Mercury's magnetosphere. However, these parametric models have been limited by their global and static nature, rendering them unable to generalise to all events. Data-driven statistical machine learning techniques, particularly deep neural networks, can capture more intricate signals and have been successfully applied to tasks such as rare event detection in audio signals [1]. Yet, these models assume that the underlying data distribution remains stationary over time. This highlights the need for methods that can detect and adapt to changes in the data distribution over time – also known as *concept drifts*. Although celestial orbits are cyclical in nature, environmental factors such as space weather, solar wind conditions etc. can induce significant variation. By detecting and adapting to changes in the data distribution, models can remain accurate and effective over long periods of time, making them more valuable for scientific study and exploration.

The planetary science community has recently witnessed a paradigm shift, gradually favouring data-centric to model-centric approaches [22]. In this vein, authors in [15] proposed to solve the problem of detection of bow shock and magnetopause crossings in a supervised deep learning setting, and further proposed an active learning approach to select the most informative orbits using an entropy criterion. The underlying notion here is that orbit similarity can be exploited to select the orbits with maximum unseen variation. We take a step further and introduce a more sophisticated active learning method that selects only the top uncertain samples from each identified concept drift. Our proposed approach further reduces the lower bound on required training samples. We validate this approach against the existing technique without concept drifts and random sampling. Preparing annotations in planetary science applications is a labour-intensive and costly endeavour – one that can be greatly ameliorated with active learning methods. In particular, our contributions can be listed as follows:

1. We present a novel drift detection method to capture new data distributions in the incoming stream of data and propose an active learning method to select the most informative orbit samples based on that. Using this approach, we demonstrate that the minimum number of orbits required to achieve a representative performance drops dramatically.
2. We provide a high-quality codebase that may be used as a framework for similar experiments. The code is available open source at: https://github.com/epn-ml/AL-Drift-Detection.

2   Related Work

2.1   Modelling Bow Shock and Magnetopause Boundaries

The task of modelling boundary crossings in the magnetosphere of planets has garnered increasing attention in recent years. Earth has been the subject of much related work [19,27], which has enabled subsequent studies to investigate various properties of the magnetopause [12]. However, the process of generating a consistent catalogue of boundary crossings from in-situ data has been recognised to be time-consuming, ambiguous, and poorly reproducible [15]. To address this, various models have been proposed, including threshold-based methods [14], and various geometric models [29,30]. However, all of these approaches share the drawback of applying static models that cannot capture variable conditions in the environment [21]. To overcome this limitation, the authors in [15] propose using a deep learning approach and introduce an entropy-based active learning strategy to select the most informative orbits. We leverage this and improve the technique by employing semi-supervised drift detection before the entropy-based sampling.

2.2   Concept Drift Detection

Concept drift detection refers to learning in non-stationary environments, wherein the underlying joint distribution of the data and labels $P(X, Y)$ evolves over time [28]. Broadly, it can be classified [18] as either *task drift* which refers to drift in the conditional distribution $P(Y|X)$ of the labels given the input data, or *domain drift* which refers to the drift in the marginal distribution i.e. $P(X)$ with the assumption that $P(Y|X)$ remains the same.

The overarching objective in this work is to detect the bow shock and magnetopause crossings during each orbit. However, the data distribution in such a dynamic environment is constantly changing. There have been a plethora of publications recently introducing techniques to handle drifts [17]. Despite the ever-increasing growth of concept drift detection, most popular approaches [4,6] and survey literature [13,20] focus on supervised settings. However, most real-world applications do not have annotations readily available. Such cases are also harder to model, as they require monitoring the evolving data features and distribution [13]. In comparison to supervised approaches, unsupervised and semi-supervised approaches are dramatically underexplored in the literature [9].

Early non-supervised approaches in drift detection include distributional comparison using K-S statistical test [16], linear regression to compare two adjacent windows [10], or linear SVMs [11]. While these methods are well suited to low-dimensional data, they are infeasible for high-dimensional data streams and are not automatic. To overcome this, ODIN [26] and a neural network-based approach [8] were recently proposed. While these approaches are a good foundation, they are only suited to standard drift detection settings and haven't been explored in active learning settings. Further, these may not be entirely suited to our task, where drifts are expected to recur in non-uniform cycles, and drift detection is to be used at the orbital level. We introduce a generative adversarial network (GAN) based method to detect the cyclical drifts and utilise these drifts to improve sampling in an active learning setup to reduce the need for labelled data for supervised classification.

3   Method

3.1   Dataset

The dataset we use for our experiments in this work contains magnetic field measurements from the MESSENGER magnetometer instrument. The data is already preprocessed with the removal of calibration signals, enrichment with Mercury position data, and split according to UTC day boundaries and Mercury apoapsis. The dataset introduced in [15] contains 2776 orbits and is ready for Machine Learning modelling. The final annotations that have been derived from [23] have finally a class distribution listed in Table 1. The boundary classes of interest are the Bow shock (SK) and the Magnetopause (MP).

For our drift detection algorithm, we use the *Extrema* values and cosine of the Mercury azimuth angle ($cos\alpha$) (cf. Sect. 3.2). Using these features, we also create an initial set of orbit groupings to train the drift detector GAN in a semi-supervised manner. For modelling the boundaries, we empirically select three three-dimensional features i.e. MSO position ($X_{MSO}, Y_{MSO}, Z_{MSO}$), flux density ($B_X, B_Y, B_Z$), and planetary dipole magnetic field values. Finally, we preprocess the data using Z-score standardisation.

Table 1. Class labels with their abbreviations and frequency of occurrence. The boundary classes (marked in bold) are highly underrepresented. Classification involves accurately detecting these relatively rare events.

label	magnetic region	share
0	interplanetary magnetic field (IMF)	64.8%
1	bow shock crossing (**SK**)	3.7%
2	magnetosheath (MSh)	14.8%
3	magnetopause crossing (**MP**)	2.3%
4	magnetosphere (MSp)	14.4%

3.2   Drift Detection

We follow the assumption that the orbits in the dataset can be grouped into sets of similar orbits based on their crossing locations and signatures. Visual inspection of orbits suggests that adjacent orbits are likely to be similar to one another with respect to bow shock and magnetopause crossing signatures, while orbits further apart tend to be more different. This pattern appears to have a cyclical nature, where patterns repeat periodically. We exploit this characteristic, present in the data, to identify groups with recurring patterns. The hypothesis here is that by identifying groups, we can constrain our sampling to include a fixed size of orbits from each during training to obtain a generalisable performance with a significantly lesser amount of training data. This would optimise the sampling over existing active learning methods. Orbit similarity can be based on various factors. The solar-wind – magnetosphere interaction is particularly sensitive to B_{z-IMF} [3], as it determines the state of the magnetosphere, specifically the amount of incoming energy to the magnetosphere and the magnetic flux magnitude on the open field lines. The cosine of Mercury azimuth angle, denoted as $cos\alpha$ [25], indicates whether the direction of shock normal is quasi-perpendicular or quasi-parallel to the direction of IMF. These shock conditions principally determine the magnetic signature during a bow shock boundary [25]. Thus, we use the similarity in these features to prepare a first set of groupings for use as labels to train the drift classifier. This classifier then classifies the next stream of orbital data into either of the previously seen groups or a new one. If an orbit is identified to have an unseen distribution, it is assigned a new label. This is achieved by training a semi-supervised GAN. From the 2776 orbits in the dataset, the model identified 21 drifts. The natural way to validate the goodness of the drift detection is to see if the training set with orbits sampled from each distinct group yields improvements over random sampling, discussed in Sect. 4.

We use a **Generator** $\theta_G(\mathcal{O})$ that accepts a sequence of orbits in a batch $\mathcal{O}_i = o_{k-n}, \ldots, o_k$ with a window size n, and outputs a vector x_i with the same dimensions, expressed as:

$$\theta_G(\mathcal{O}_i) := \{o_{n-k}, \ldots, o_n\} \longmapsto x_i.$$

The **Discriminator** $\theta_D(x)$ is a multi-class dynamic discriminator that predicts whether the input x_i generated by θ_G belongs to one of the distributions already discovered.

We define a mapping function $\theta_D(x)$, which maps the incoming feature vector to a set of known data distributions or an unidentified data distribution as:

$$\theta_D(x) := x_i \longmapsto \{\mathcal{D}_1, \ldots, \mathcal{D}_n, \mathcal{D}_{n+1}\},$$

where x_i is the vector generated by θ_G and, $\mathcal{D}_j$ is the j_{th} input distribution with $j \in [1, n]$ indicating the set of known distributions and $n + 1$ referring to the label reserved for a newly detected drift, which is added to the set j. The Discriminator is updated with a new output size $n + 1$ whenever a new drift is discovered.

We map all incoming orbit features to one of the distributions. A new drift is signalled when x_i is predicted to not belong to either of the already seen distributions $\mathcal{D}_j$, or the predicted label of the current input does not match the label of the previous input, formally expressed as:

$$\forall\, x_i \in \mathcal{O} : \theta_D(x_i) \neq \mathcal{D}_j \cup \forall(x_{i-1}, x_i) \in \mathcal{O} : \theta_D(x_{i-1}) \neq \theta_D(x_i),$$

where x is a generated input vector and, $\mathcal{O}$ is the input data batch to θ_D.

Initially, we start with a training dataset comprising arbitrarily 8 identified orbit groups with unique drifts. The size of the groups is a heuristic that defines the sensitivity of the drift detector. Through preliminary experimentation, we found a group size of 14, and 8 groups to be a good starting point for our experiments. With these, we train the GAN so that θ_D learns to predict the distribution of the data generated by θ_G as one seen before or a new one. For every newly detected distribution, an incremental integer is assigned as the label. The drift detection algorithm can be formally defined with parameters d_{gen} (generator label), $\mathcal{O}$ (sequential set of unlabelled orbits), $\mathcal{D}_{init}$ (hash table of initial orbit groupings with orbit numbers and drift labels) (cf. Algorithm 1).

3.3   Active Learning

We use the catalogue of detected drifts and sample orbits from them incrementally in an active learning setting. Within each drift group, we rank all the orbits according to an informativeness measure as used in [15]. As our classification model has a series of Multinoulli distributions for output, we measure the orbit uncertainty as a function of the output probabilities using *Shannon entropy*. Consider the training set $\mathcal{T} \subseteq \mathbb{R}^{d \times w} \times \{\text{IMF}, \text{SK}, \text{MSh}, \text{MP}, \text{MSp}\}^w$ with number of features $d \in \mathbb{N}$, and window size $w \in \mathbb{N}$. Given a model prediction $\hat{Y} = [\hat{y}^{(1)}, \ldots, \hat{y}^{(w)}] \in [0, 1]^{5 \times w}$, we define its uncertainty as:

$$\mathfrak{u}(\hat{Y}) := \max_j H(\hat{y}^{(j)}) = -\min_j \sum_{i=1}^{5} y_i^{(j)} \log(y_i^{(j)}),$$

where $H : \triangle^4 \to \mathbb{R}$ is the Shannon entropy on the standard 4-simplex.

Algorithm 1. detect_drifts($d_{gen}, \mathcal{O}, \mathcal{D}_{init} : \mathbb{R}^{d \times w} \to \mathbb{R}^{d_{gen}}$) :

1: $\mathcal{D} := \mathcal{D}_{init}$	▷ List of detected drifts.
2: $gen :=$ Generator() $: \mathbb{R}^{d \times w} \to \mathbb{R}^{d \times w}$	▷ Generator.
3: $dis :=$ Discriminator() $: \mathbb{R}^{d \times w} \to \mathbb{R}^{|\mathcal{D}|}$	▷ Discriminator.
4: $\mathcal{T} :=$ create_training_dataset$(\mathcal{O}, \mathcal{D})$	▷ Labelled training orbits.
5: $\mathcal{T}_{gen} :=$ equalise_and_concatenate$(\mathcal{T})$	▷ Training data for generator.
6: $dis :=$ train_discriminator$(\mathcal{T}, dis, gen, d_{gen})$	▷ Initial training.
7: $gen :=$ train_generator$(\mathcal{T}_{gen}, dis, gen, d_{gen})$	
8: $j := |\mathcal{D}|$	▷ Total detected drifts
9: $i := 0$	▷ Current orbit number.
10: **while** $i < |\mathcal{O}|$ **do**	
11: $d := dis(\mathcal{O}_i)$	▷ Get most probable drift label from discriminator.
12: **if** $d \in [\mathcal{D}_1 : \mathcal{D}_{|j|}]$ **then**	
13: reset_top_layer(dis)	▷ Reset the top **Linear** layer.
14: **else**	
15: update(dis) ▷ Reset the top **Linear** layer and increase its output size by 1.	
16: $d_{gen} = d_{gen} + 1$	
17: **end if**	
18: $n = 1$	▷ Number of adjacent orbits in a drift.
19: **if** $d \neq \mathcal{D}[i-1]$ **then**	
20: $n = 14$	
21: **end if**	
22: $k := 0$	
23: **while** $k < n$ **do**	
24: $[\mathcal{D}[i+j] := d$	▷ Assign drift label to orbits.
25: $k := k + 1$	
26: **end while**	
27: $\mathcal{T} :=$ create_training_dataset$(\mathcal{O}, \mathcal{D})$	▷ Dataset with new drift.
28: $\mathcal{T}_{gen} :=$ equalise_and_concatenate$(\mathcal{T})$	
29: $dis :=$ train_discriminator$(\mathcal{T}, dis, gen, d_{gen})$	
30: $gen :=$ train_generator$(\mathcal{T}_{gen}, dis, gen, d_{gen})$	
31: $i := i + n$	▷ Move to the next unlabelled orbit.
32: **end while**	
33: **return** $\mathcal{D}$	

To achieve this on the orbit level, we must reduce the individual window uncertainties to a single orbit score. In contrast to the approach in [15] who calculate the entropy for only the windows including the crossing labels, we remain agnostic to it and use the entire region to calculate the orbit uncertainty $\hat{f}_\theta : \mathbb{R}^{d \times w} \to \mathbb{R}^{d \times w}$ for our task by averaging over all the windows $\widetilde{\mathcal{D}}_o$ as:

$$\mathfrak{U}_{\hat{f}_\theta}(\widetilde{\mathcal{D}}_o) := \frac{1}{|\widetilde{\mathcal{D}}_o|} \sum_{(\boldsymbol{X}, \boldsymbol{y}) \in \widetilde{\mathcal{D}}_o} \mathfrak{u}(\hat{f}_\theta(\boldsymbol{X}))$$

Using this uncertainty measure, we select the top k orbits detected in each drift and incrementally add them to the training set. The algorithm is defined with parameters Ω (set of all orbits), $\mathcal{D}$ (set of all drift labels) and m (maximum number of training orbits per drift, hyperparameter) (cf. Algorithm 2).

Algorithm 2. `sample_orbits`$(\Omega, \mathcal{D}, k)$:

1: $\mathcal{T} := \varnothing$ ▷ Set of training orbits.
2: $\mathcal{S} := \varnothing$ ▷ Set of testing orbits.
3: $d = 1$ ▷ Current drift label.
4: **while** $d \leq |\mathcal{D}|$ **do**
5:       $\mathcal{S} := \mathcal{S} \cup$ `random.sample`$(\Omega_d, |\Omega_d|//5)$ ▷ Randomly sample 20% as testing data.
6:       $E := $ `hash_table()`
7:       **for** $\omega \in \Omega_d \backslash \mathcal{S}$ **do**   ▷ Get entropy for all non-testing orbits with current drift.
8:           $E[\omega] := $ entropy(ω)
9:       **end for**
10:      $\mathcal{T} := \mathcal{T} \cup$ `top_k`(E, k) ▷ Add k most uncertain orbits.
11:      $d := d + 1$ ▷ Move to the next drift.
12: **end while**
13: **return** $\mathcal{T}, \mathcal{S}$

3.4 Boundary Crossing Detection

To have an augmented set of fixed-shaped input vectors and extract aggregated features from them, we use a sliding window, with a stride of size one. This is done to ensure translation equivariance i.e. the position of the crossing should not matter to the model. Hence, the model's input is a window of $w \in \mathbb{N}$ successive time steps, each comprising $d \in \mathbb{N}$ scalar features. The input windows can thus be expressed as:

$$X := \begin{bmatrix} x^{(1)} & x^{(2)} & \cdots & x^{(w)} \end{bmatrix} \in \mathbb{R}^{d \times w}$$

Consequently, the target output matrix of one-hot vectors is $Y \in \mathbb{R}^{5 \times w}$.

Formally, the classification task is framed as a multidimensional multi-class classification: Given the window X, we predict a sequence of magnetic region probabilities, where each column sums up to one:

$$\hat{Y} := \begin{bmatrix} p_{1,1} & \cdots & p_{1,w} \\ \vdots & & \vdots \\ p_{5,1} & \cdots & p_{5,w} \end{bmatrix} \in [0,1]^{5 \times w}$$

We pass these normalised data features through a neural network comprised of a convolutional block and a recurrent block, referred to as CRNN [2], with the final activations represented as: $\gamma_\theta : p_{ij} = \gamma_\theta(X)$ where γ is the CRNN model, with θ as its parameters.

To measure the error between the prediction $\hat{Y}$ and the ground truth Y, we employ the standard categorical cross-entropy loss. The resulting weighted loss averaged across all time steps in a window thus can be expressed as:

$$\mathcal{L}(\hat{Y}, Y) := \frac{1}{w} \sum_{j=1}^{w} \mathcal{L}_j(\hat{Y}, Y) = -\frac{1}{w} \sum_{i=1}^{5} w_i \sum_{j=1}^{w} Y_{ij} \log(\hat{Y}_{ij})$$

The final activations are then passed through a smoothing function to obtain final continuous predictions. The smoothing function is applied to eliminate gaps in prediction or any pseudo-activations in the output signal. This procedure is defined as a function with parameters classification labels $(\hat{Y})$, gap threshold (g) (max size of fillable gap), and smoothing window size (c) in Algorithm 3.

Algorithm 3. smooth$(\hat{Y}, g, c)$:

1: $i = 1$ ▷ Index of the current label, first pass.
2: **while** $i \leq |\hat{Y}| - c$ **do**
3:     $\mathcal{W} := \varnothing$ ▷ Sliding window of labels.
4:     **for** $j \in [i : i + c]$ **do**
5:         $\mathcal{W} := \mathcal{W} \cup \hat{\mathcal{Y}}[j]$
6:     **end for**
7:     **if** $\mathcal{W}[1] = \mathcal{W}[|\mathcal{W}|]$ & $\mathcal{W}[1] \notin \{1, 3\}$ **then**   ▷ If both edges of the window are not a crossing label.
8:         **for** $j \in [i : i + c]$ **do**
9:             $\hat{\mathcal{Y}}[j] = \mathcal{W}[1]$ ▷ Fill the entire window with the same non-crossing label as the edge.
10:         **end for**
11:     **end if**
12:     $i := i + 1$
13: **end while**
14: $i = 1$ ▷ Second pass.
15: **while** $i \leq |\hat{Y}| - g$ **do**
16:     $\mathcal{W} := \varnothing$
17:     **for** $j \in [i : i + g]$ **do**
18:         $\mathcal{W} := \mathcal{W} \cup \hat{Y}[j]$
19:     **end for**
20:     **if** $\mathcal{W}[1] = \mathcal{W}[|\mathcal{W}|]$ & $\mathcal{W}[1] \in \{1, 3\}$ **then**   ▷ If both edges of the window are a crossing label.
21:         **for** $j \in [i : i + g]$ **do**
22:             $\hat{Y}[j] = \mathcal{W}[1]$ ▷ Fill the entire window with the same crossing label as the edge.
23:         **end for**
24:     **end if**
25:     $i := i + 1$
26: **end while**
27: **return** $\mathcal{D}$

---

4   Results

We evaluate the performance of our proposed approach by answering the following research questions:

1. **Effectiveness (RQ1):** Does our drift detection algorithm correctly identify the inherent drifts in the distribution of orbits?

2. **Efficiency (RQ2):** Is an active learning strategy with drift detection better than (a) sampling with entropy alone and, (b) random sampling? Consequently, can we reduce the lower bound on the minimum number of orbits to find a representative performance on classification?
3. **Effect of smoothing (RQ3):** Can our proposed smoothing algorithm help mitigate the gaps in prediction – a problem encountered in previous works, by yielding more consistent crossing timestamps?

4.1   Evaluation of Drift Detection (RQ1 + RQ2)

We evaluate our drift detection approach in a two-fold manner:

Qualitatively: First, we inspect the output of our GAN-based drift detection, in order to examine if the detected drifts seem reasonable based on the shapes and signatures of the crossings in predicted drifts. Figure 2 illustrates an example of two orbits 3 months apart that are assigned the same drift label. Likewise, the orbits with different detected labels appear to be visually distinct. Most of the groupings arguably have similar shapes and positions of the bow shock (SK) and magnetopause (MP) crossings. They vary at times in the duration, but it is likely due to inter-orbital variations in aircraft speed and environment. For a detailed catalogue of detected drifts, we refer the reader to the code repository.

Quantitatively: Next, we compare the performance of our approach (AD) against the entropy-based sampling (AL) introduced in [15], which serves as a benchmark for comparison. The only difference is that we sample the top k most uncertain orbits from each detected drift group. We formulate the hypothesis that by sampling the most informative orbits from the maximally distant orbits, i.e. distinct drifts, we would require fewer training samples to achieve comparable performance. We select 278 orbits from 13 drift groups and compare the results against a model trained with the same number of orbits sampled incrementally using entropy alone (cf. Table 2). The results show a marked improvement in the detection of bow shock (SK) crossing, while only a modest improvement in the magnetopause (MP) crossing. It should be noted that generally, SK is harder to detect than MP, as visually, MP is more distinct. Therefore, the marginal utility is higher with improvements in the detection of SK.

Additionally, we compare the performance to the model trained with data with the same number of orbits sampled randomly (RS). The results show an overall improved performance (cf. Table 2). We further analyse the development of the performance metrics as we increase the number of training samples using AD (cf. Fig. 3). We clearly observe the onset of saturation in learning capacity after about 250 orbits. This denotes a lower bound on the minimum number of orbits required for a representative model. Comparing it to the total number of MESSENGER orbits, it is less than 10% of the available data. This observation indicates that significantly fewer orbits actually need to be labelled in order to yield an optimal classification result.

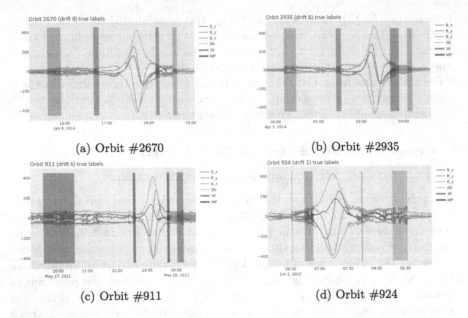

(a) Orbit #2670   (b) Orbit #2935

(c) Orbit #911   (d) Orbit #924

Fig. 2. An example of the same drift detected in two orbits spread 3 months apart (a & b). The shapes and locations of the boundary crossings are similar. (c & d) An example of variation between two different detected drifts in the data. The two boundaries of interest i.e.bow shock (BS) and magnetopause (MP) are marked in green and blue respectively. (Color figure online)

4.2   Effect of Smoothing

The previous work on automatic detection of the boundaries [15] observed that the crossing predictions had several gaps in them. In the case of SK crossing particularly, the similarity of flux density values to surrounding regions can lead the classifier to misfire and vote for the wrong classes mid-crossing. Overall, greater misfiring should produce an inferior result, both qualitatively and quantitatively. Figure 4 shows an example prediction in this ablative setting. The result without smoothing appears to incorrectly identify the total duration. The spurious gaps in prediction appear to have been smoothed out. We further evaluate the classification performance (cf. Table 3) on this orbit, to numerically corroborate the visual contrast.

Table 2. Classification performance of the CRNN using training data sampled with our approach vs benchmark. **AD** denotes Active Learning with Drift detection (our approach). **AL** denotes Active Learning with entropy alone and without drift detection [15]. **RS** denotes the model trained with training orbits sampled randomly. Best values are marked in bold.

	IMF			SK			MSh			MP			MSp		
	AD	AL	RS	AD	AL	RS	ALD	AL	RS	AD	AL	RS	AD	AL	RS
F1	**0.98**	**0.93**	0.72	**0.66**	0.5	0.39	**0.86**	**0.86**	0.51	0.63	0.64	0.46	**0.96**	0.93	0.78
Recall	**0.97**	**0.95**	0.59	**0.73**	0.71	0.44	**0.85**	0.84	0.53	0.65	**0.76**	0.49	**0.95**	0.94	0.9
Precision	**0.98**	0.92	0.91	**0.59**	0.39	0.36	**0.88**	**0.88**	0.50	**0.61**	0.56	0.42	**0.98**	0.93	0.66

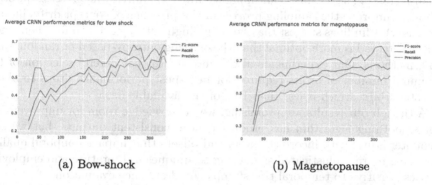

(a) Bow-shock   (b) Magnetopause

Fig. 3. Development of the performance metrics for the boundary crossings. The metrics start to plateau between the 250–300 epoch range. The X-axis depicts the number of training orbits, while the Y-axis depicts the absolute value for an individual metric.

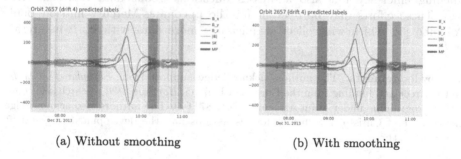

(a) Without smoothing   (b) With smoothing

Fig. 4. Smoothing filter applied to a sample output. Smoothing yields noticeably more consistent boundary predictions.

Table 3. Classification performance on an exemplary test orbit # 2657 with (**w**) and without (**w/o**) smoothing filter.

	IMF		SK		MSh		MP		MSp	
	w	w/o	w	w/o	w	w/o	w	w/o	w	w/o
F1	**0.96**	0.94	**0.41**	0.15	0.81	**0.83**	0.67	**0.69**	**0.97**	0.97
Recall	**0.95**	0.90	**0.63**	0.21	**0.68**	0.82	0.82	**0.95**	**0.99**	0.95
Precision	**0.97**	0.98	**0.30**	0.12	0.99	**0.84**	**0.58**	0.54	**0.95**	0.95

5   Conclusion

In this study, we devised a novel approach to detect drifts in orbital data around Mercury through the use of a generative (GAN) adversarial network-based method, with which we augmented an entropy-based active learning method to select the most informative orbits. Utilising the data obtained from NASA's MESSENGER mission, we demonstrated the efficacy of this technique as an active learning strategy. We then used this set of training data to train a convolutional recurrent neural network (CRNN) to detect the bow shock and magnetopause crossing signatures. Our approach aims to address the pressing question of how to improve the sampling process in the presence of concept drifts in data streams. Our findings suggest that sampling high entropy orbits only when a new drift is detected is more optimal than the existing entropy-based or random sampling methods. Our experiments also demonstrate that careful selection of the minimum number of orbits required for near-best performance drops to about 250–300 orbits, which is less than 10% of the available data.

Although our results are promising, we acknowledge room for improvement. Our smoothing algorithm has produced more consistent outputs, but we still encounter issues with incorrect onsets and offsets that impact temporal quality. Exploring more sophisticated smoothing techniques in literature and employing metrics sensitive to temporal timestamps could enhance evaluation.

In conclusion, our study reveals two significant insights. First, clever active learning strategies benefit planetary science applications with extensive data volumes and limited expert annotations. Second, our approach not only enhances labelling efficiency but also addresses questions about data representativeness. These efforts may be relevant for the upcoming Mercury mission Bepi-Colombo [5], which will collect significantly more data with its twin-aircraft probe.

Acknowledgements. The authors acknowledge support from *Europlanet 2024 RI* that has received funding from the European Union's *Horizon 2020* research and innovation programme under grant agreement No. 871149. The authors also acknowledge the support of Christofer Fellicious for discussions and the development of the drift detection method.

References

1. Amiriparian, S., et al.: Recognition of echolalic autistic child vocalisations utilising convolutional recurrent neural networks (2018)
2. Amiriparian, S., Cummins, N., Julka, S., Schuller, B.: Deep convolutional recurrent neural network for rare acoustic event detection. In: Proceedings of the DAGA, pp. 1522–1525 (2018)
3. Anderson, B., Johnson, C.: A magnetic disturbance index for mercury's magnetic field derived from messenger magnetometer data. Geochem. Geophys. Geosyst. **14**, 3875–3886 (2013). https://doi.org/10.1002/ggge.20242
4. Baena-García, M., Campo-Ávila, J., Fidalgo-Merino, R., Bifet, A., Gavald, R., Morales-Bueno, R.: Early drift detection method, January 2006

5. Benkhoff, J., et al.: BepiColombo-comprehensive exploration of mercury: mission overview and science goals. Planet. Space Sci. **58**(1–2), 2–20 (2010)
6. Bifet, A., Gavaldà, R.: Learning from time-changing data with adaptive windowing, vol. 7, April 2007. https://doi.org/10.1137/1.9781611972771.42
7. Fairfield, D.H.: Average and unusual locations of the earth's magnetopause and bow shock. J. Geophys. Res. **76**(28), 6700–6716 (1971)
8. Fellicious, C., Wendlinger, L., Granitzer, M.: Neural network based drift detection. In: Nicosia, G., et al. (eds.) LOD 2022. LNCS, vol. 13810, pp. 370–383. Springer, Cham (2023). https://doi.org/10.1007/978-3-031-25599-1_28
9. Gemaque, R.N., Costa, A.F.J., Giusti, R., Dos Santos, E.M.: An overview of unsupervised drift detection methods. Wiley Interdiscip. Rev. Data Min. Knowl. Discov. **10**(6), e1381 (2020)
10. Gözüaçık, Ö., Büyükçakır, A., Bonab, H., Can, F.: Unsupervised concept drift detection with a discriminative classifier. In: Proceedings of the 28th ACM International Conference on Information and Knowledge Management, pp. 2365–2368 (2019)
11. Gözüaçık, Ö., Can, F.: Concept learning using one-class classifiers for implicit drift detection in evolving data streams. Artif. Intell. Rev. **54**, 3725–3747 (2021)
12. Haaland, S., et al.: Characteristics of the flank magnetopause: MMS results. J. Geophys. Res. Space Phys. **125**(3), e2019JA027623 (2020)
13. Hu, H., Kantardzic, M., Sethi, T.S.: No free lunch theorem for concept drift detection in streaming data classification: a review. Wiley Interdiscip. Rev. Data Min. Knowl. Discov. **10**(2), e1327 (2020)
14. Jelínek, K., Němeček, Z., Šafránková, J.: A new approach to magnetopause and bow shock modeling based on automated region identification. J. Geophys. Res. Space Phys. **117**(A5) (2012)
15. Julka, S., Kirschstein, N., Granitzer, M., Lavrukhin, A., Amerstorfer, U.: Deep active learning for detection of mercury's bow shock and magnetopause crossings. In: Amini, M.R., Canu, S., Fischer, A., Guns, T., Kralj Novak, P., Tsoumakas, G. (eds.) ECML PKDD 2022. LNCS, vol. 13716. Springer, Cham (2022). https://doi.org/10.1007/978-3-031-26412-2_28
16. Justel, A., Peña, D., Zamar, R.: A multivariate Kolmogorov-Smirnov test of goodness of fit. Stat. Probab. Lett. **35**(3), 251–259 (1997)
17. Ksieniewicz, P., Zyblewski, P., Choraś, M., Kozik, R., Giełczyk, A., Woźniak, M.: Fake news detection from data streams. In: 2020 International Joint Conference on Neural Networks (IJCNN), pp. 1–8. IEEE (2020)
18. Lao, Q., Jiang, X., Havaei, M., Bengio, Y.: Continuous domain adaptation with variational domain-agnostic feature replay. arXiv preprint arXiv:2003.04382 (2020)
19. Lin, R., Zhang, X., Liu, S., Wang, Y., Gong, J.: A three-dimensional asymmetric magnetopause model. J. Geophys. Res. Space Phys. **115**(A4) (2010)
20. Lu, J., Liu, A., Dong, F., Gu, F., Gama, J., Zhang, G.: Learning under concept drift: a review. IEEE Trans. Knowl. Data Eng. **31**(12), 2346–2363 (2018)
21. Nguyen, G., Aunai, N., Michotte de Welle, B., Jeandet, A., Fontaine, D.: Automatic detection of the earth bow shock and magnetopause from in-situ data with machine learning. Ann. Geophys. Discuss. 1–22 (2019)
22. Nikolaou, N., et al.: Lessons learned from the 1st ariel machine learning challenge: correcting transiting exoplanet light curves for stellar spots. arXiv preprint arXiv:2010.15996 (2020)

23. Philpott, L.C., Johnson, C.L., Anderson, B.J., Winslow, R.M.: The shape of mercury's magnetopause: the picture from messenger magnetometer observations and future prospects for bepicolombo. J. Geophys. Res. Space Phys. **125**(5), e2019JA027544 (2020)
24. Slavin, J.A.: Mercury's magnetosphere. Adv. Space Res. **33**(11), 1859–1874 (2004)
25. Sundberg, T., et al.: Cyclic reformation of a quasi-parallel bow shock at mercury: MESSENGER observations. J. Geophys. Res. Space Phys. **118**, 6457–6464 (2013). https://doi.org/10.1002/jgra.50602
26. Suprem, A., Arulraj, J., Pu, C., Ferreira, J.: Odin: automated drift detection and recovery in video analytics. arXiv preprint arXiv:2009.05440 (2020)
27. Wang, Y., et al.: A new three-dimensional magnetopause model with a support vector regression machine and a large database of multiple spacecraft observations. J. Geophys. Res. Space Phys. **118**(5), 2173–2184 (2013)
28. Widmer, G., Kubat, M.: Learning in the presence of concept drift and hidden contexts. Mach. Learn. **23**(1), 69–101 (1996)
29. Winslow, R.M., et al.: Mercury's magnetopause and bow shock from messenger magnetometer observations. J. Geophys. Res. Space Phys. **118**(5), 2213–2227 (2013)
30. Zhong, J., et al.: Mercury's three-dimensional asymmetric magnetopause. J. Geophys. Res. Space Phys. **120**(9), 7658–7671 (2015)
31. Zurbuchen, T.H., et al.: Messenger observations of the spatial distribution of planetary ions near mercury. Science **333**(6051), 1862–1865 (2011)

Modeling Primacy, Recency, and Cued Recall in Serial Memory Task Using On-Center Off-Surround Recurrent Neural Network

Lakshmi Sree Vindhya[1] , R. Gnana Prasanna[2] , Rakesh Sengupta[3](✉) ,
and Anuj Shukla[4]

[1] School of CS & AI, SR University, Warangal 506371, Telangana, India
[2] Department of ECE, SR University, Warangal 506371, Telangana, India
[3] Center for Creative Cognition, SR University, Warangal 506371, Telangana, India
rakesh.sengupta@sru.edu.in
[4] Thapar School of Liberal Arts & Sciences, Thapar Institute of Engineering &
Technology, Patiala 147004, Punjab, India

Abstract. The serial recall paradigm has long been used to study short-term memory. Previous experiments have consistently revealed two key phenomena: the primacy effect and the recency effect. Essentially, it is easier to recall items at the beginning and end of a sequence compared to those in the middle. In this study, we present a single-layer fully connected recurrent neural network with self-excitation and mutual inhibition. By providing transient input to the network, we observed a dynamic steady-state output. We examined this output pattern to determine which inputs were ultimately "remembered" at the end of the simulation. Our results demonstrate that this network can replicate the serial recall curve observed in empirical studies when higher inhibition values are used. Additionally, it can account for the capacity limitation commonly observed in serial recall tasks. By varying the presentation duration in the model, we successfully explain both the primacy and recency effects within the same network. Furthermore, when we introduced cues to a single item in the sequence by elevating its input, we observed a decrease in the recall probability of neighboring items. In summary, our findings suggest that the dynamics of a single-layer recurrent on-center off-surround neural network can provide insights into the mechanisms underlying primacy, recency, and cued recall effects.

Keywords: Visual Working memory · Serial Recall · Recurrent neural networks · Recall probability

1 Introduction

Working memory has become a fundamental concept in cognitive psychology, thanks to the influential contributions of [1] and [2]. It is considered one of the

G. Nicosia et al. (Eds.): LOD 2023, LNCS 14505, pp. 405–414, 2024.
https://doi.org/10.1007/978-3-031-53969-5_30

key executive functions of the brain, explaining how the brain handles tasks that require temporary manipulation and storage of information beyond the immediate perceptual input. Visual working memory (VWM) specifically enables cognitive agents to maintain and modulate visual sensory information for a short duration, typically a few seconds. The presence of persistent delay period activity in higher cortical areas, even in the absence of a stimulus, is regarded as a characteristic feature of VWM [3,4].

Currently, several major theoretical frameworks exist for understanding working memory, and they are not necessarily mutually exclusive. These frameworks include:

- The multi-component model of working memory, which posits a central executive that governs the phonological loop, visuo-spatial sketchpad, and episodic buffer. These components collectively maintain and manipulate temporary task-related information [1,5].
- The view of working memory as the activation and maintenance of representations from long-term memory within the focus of attention [6].
- The perspective of working memory as a limited-capacity system that maintains discrete, fixed-resolution items necessary for task completion [7,8].
- The notion of working memory as a dynamically and flexibly distributed limited resource that encompasses all the items requiring maintenance [9].

The serial recall paradigm is a crucial experimental approach in the Visual Working Memory (VWM) literature. Typically, participants are presented with a stream of items at the center of the screen, followed by judgments on the primacy (earlier) or recency (later) of two randomly selected items from the stream. The recall probability of items usually follows a standard recency curve, where the most recent items are more easily recalled, while the recall probability for items presented 4 or 5 positions earlier than the end of the stream drops dramatically. In general, recency curves exhibit a concave sigmoid shape, and their steepness increases with shorter presentation durations of items [10]. Serial recall curves reveal that the memory tends to retain only the most recent 4 or so items with precision, highlighting the presence of capacity limitations. Some researchers [12,13] have developed models using center-surround mechanisms to simulate PFC (prefrontal cortex) neurons.

In this study, we are investigating two important aspects of how our memory works when we try to recall a series of items. These aspects are known as the primacy and recency effects. The primacy effect refers to our better ability to remember items at the beginning of the series, while the recency effect describes our tendency to recall items at the end of the series more easily. Understanding these effects gives us valuable insights into the functioning of short-term memory.

To uncover the underlying mechanisms of these effects, we propose a new approach using a single-layer fully connected recurrent neural network. This network is inspired by previous research [12,14] and it features self-excitation and mutual inhibition.

By employing this neural network model and studying its behavior with transient inputs, we can gain a deeper understanding of how these primacy and recency effects emerge during the process of serial recall. Our investigation aims to provide new insights into the workings of short-term memory and how our brain stores and retrieves information in everyday tasks. In this study, our main objective is to simulate the serial recall task using a simplified network comprising a single layer of center-surround neurons in the prefrontal cortex (PFC), which represents the abstract memory nodes (see Fig. 1). The subsequent sections provide a detailed description of our model.

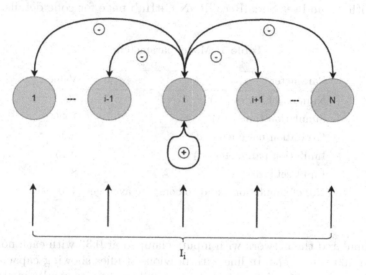

Fig. 1. Recurrent on-center off-surround architecture for modeling serial recall in Visual Working memory. Each node has self-excitation (represented by the +, and laterally inhibits all the other nodes.

2 Methods

In case of serial presentation the PFC nodes are activated one after another following that it tries to keep those nodes in persistent activity for recall.

Time evolution of the VWM network of N neurons is given by,

$$\frac{dx_i}{dt} = -x_i + \alpha F(x_i) - \beta \sum_{j=1, j \neq i}^{N} F(x_j) + I_i + \text{noise} \tag{1}$$

The activation function $F(x)$ is given by

$$F(x) = \begin{cases} 0 & \text{for } x \leq 0 \\ \frac{x}{1+x} & \text{for } x > 0 \end{cases} \tag{2}$$

Here α is the self-excitation parameter and β is the lateral inhibition parameter. $I_i \in [0,1]$ is the transient external input. Noise is sampled at every time step of the simulation from a Gaussian distribution with mean 0 and standard deviation 0.05. The simulation parameters are given in Table 1. The choice of parameters was prompted by our previous work [14,15]. For algorithm details see Algorithm 1 and see SerialRecallRNN GitHub page for code details.

Table 1. Model parameters

Parameter	Value
Number of nodes (N)	70
Simulation time (ms)	10000
Excitation parameter (α)	2.0
Inhibition parameter (β)	0.15
Input set size	8
No. of simulations used to compute averages	100

We simulated the network with inputs clamped at 0.33 with each node stimulated for 400 time steps. In line with previous studies showing capacity limits of 7 ± 2 items [2], activating an increasing number of units results in a decrease in the average activation for each node (Fig. 2).

3   Results

3.1   Serial Recall Curves

To obtain the serial recall curve, we conducted simulations for various input durations ranging from 50 ms to 1 s in 50 ms increments. For each simulation duration, we generated serial position curves by measuring the fraction of successful trials (out of 100 simulation runs) in which an item corresponding to a specific serial position remained active in memory at the end of the simulation. To estimate the proportion of correct recall, we made a simplified assumption: if an item was still active in memory when the recall cue for it was given, the recall would be considered successful. By averaging the recall curves across the presentation durations, we obtained the recall curve (see Fig. 3). Notably, the resulting recency curve exhibited several key characteristics consistent with experimental data [10].

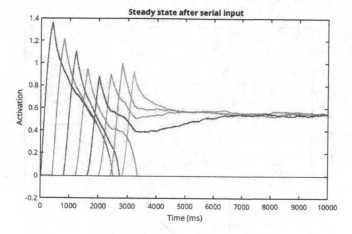

Fig. 2. Steady state dynamics of VWM network with 8 sequential inputs of 400 ms each.

The recall curve interestingly shows the effects of both primacy effect and recency effect - where subjects are able to recall the most recent and the earliest items in serial recall.

3.2   Primacy Effect

To provide a more compelling demonstration of the primacy effect, we utilized the same model, but this time, we computed the serial recall curve by averaging across presentation durations ranging from 50 ms to 300 ms (see Fig. 4). This approach aligns with the typical presentation time effect observed in the serial recall literature [10].

3.3   Capacity Limit

Furthermore, we investigated whether the predicted recall performance was affected by set size, as anticipated based on previous behavioral studies. Our findings aligned with these expectations, as the model predicted near-complete recall for set sizes up to 4 items but demonstrated poorer performance for larger set sizes (see Fig. 5).

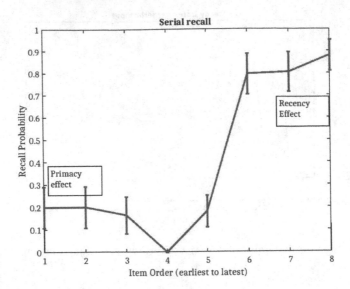

Fig. 3. Average predicted probability of recall for items. The recall probabilities are averaged across presentation durations 50 ms to 1 s (in steps of 50 ms). The corresponding error bars are also plotted. The recall curve shows properties of primacy and recency effect.

Algorithm 1. Serial Working Memory Model

Require: N (number of neurons), nt (total number of time steps), dt (time step size), $nstim$ (number of stimulus presentations per input), $ninputs$ (number of inputs), $alphabeta$ (parameter controlling the activation function), $params$ (structure of model parameters)

Ensure: $results$ (structure containing the final state of the network and the indices of the inputs)

1:   $a \leftarrow$ a random permutation of the integers 1 to N
2:   $inputs \leftarrow$ the first $ninputs$ elements of a
3:   $I \leftarrow$ a matrix of zeros with dimensions N by nt
4:   **for** $i = 1$ to $ninputs$ **do**
5:      $I(inputs(i), ((i - 1) * nstim + 1) : (i * nstim)) \leftarrow 0.33$
6:   **if** $nargin = 2$ **then**
7:      $k \leftarrow \{varargin1, 1\}$
8:      $I(inputs(k), ((k - 1) * nstim + 1) : (k * nstim)) \leftarrow 0.43$
9:   $x \leftarrow$ a matrix of zeros with dimensions N by nt
10:   $noise \leftarrow$ a matrix of random numbers drawn from a normal distribution with mean 0 and standard deviation equal to $params.noise$ and dimensions N by nt
11:   **for** $i = 1$ to $nt - 1$ **do**
12:      $b \leftarrow$ the indices of the elements of $x(:, i)$ that are less than 0
13:      **if** $isempty(b) = 0$ **then**
14:         $x(b, i) \leftarrow 0$
15:      $fx \leftarrow x(:, i)./(1 + x(:, i))$
16:      $x(:, i + 1) \leftarrow (1 - dt) \cdot x(:, i) + dt \cdot (alphabeta \cdot fx + I(:, i) + noise(:, i))$
17:   $results.x \leftarrow x$
18:   $results.b \leftarrow inputs$

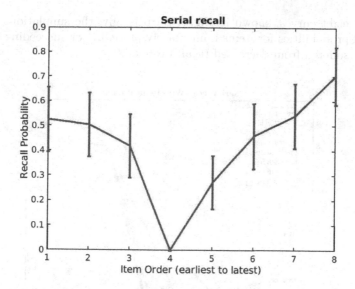

Fig. 4. Average predicted probability of recall for items. The recall probabilities are averaged across presentation durations 50 ms to 300 ms (in steps of 50 ms). The corresponding error bars are also plotted. The recall curve shows properties of primacy and recency effect.

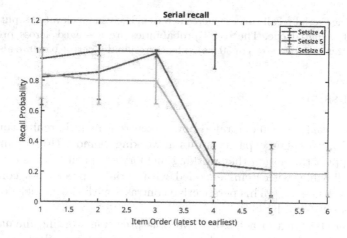

Fig. 5. Average serial recall curves for setsizes 4, 5, and 6. It shows that up to 4 items we almost have perfect recall, and performance deteriorates towards standard recency behavior for higher set sizes.

3.4   Serial Recall with Spatial Cue

In our investigation, we simulated the effect of cues on a single item within the sequence by introducing elevated input (0.43) to that particular item while calculating the recall probabilities of all the items in the sequence. The results were plotted as eight curves, shown in Fig. 6. Surprisingly, the simulations revealed that recall probabilities for items immediately preceding or succeeding the cued item often suffered from decreased recall probability.

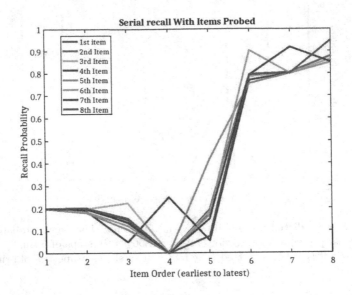

Fig. 6. Average serial recall curves where each item is given an elevated input imitating the effect of a spatial cue. The recall probabilities are averaged across presentation durations 50 ms to 1 s (in steps of 50 ms). The corresponding error bars are also plotted.

4   Discussion

In this study, we have demonstrated how a recurrent neural architecture (Fig. 1) can dynamically explain capacity limits in working memory (Figs. 2 and 5). Our findings support the notion that working memory's capacity limits are not fixed but rather flexible, with items encoded with variable precision-a concept also proposed by others [11]. This perspective contrasts with slot models for working memory, as put forth by [8].

Moreover, the idea of a flexible, variable precision working memory allows for a characterization of the serial recall curve and recency effect, as evidenced by our model's predictions (Fig. 3). Moreover, we have shown how altering the presentation durations to averaged over, we can also recover the primacy curves as shown in VWM literature (Fig. 4). Notably, the model suggests that increasing the saliency of an item through spatial cues should boost its recall probability while suppressing temporally adjacent items (Fig. 6).

The computational model presented in this work exhibits significant explanatory and predictive power and could be readily integrated with visual processing models meeting the constraints outlined in the computational section, such as the cognitive programs architecture proposed by [16]).

In light of recent advances in neural networks, we draw attention to the work of [17], which delves into the encoding of information in working memory (WM). Their study proposes that WM may not solely rely on persistent neuronal activity but can also involve 'activity-silent' hidden states, such as short-term synaptic plasticity and synaptic efficacies. However, in tasks that require actively manipulating information, persistent activity naturally emerges from learning, and its extent correlates with the degree of manipulation required. This suggests that the level of persistent activity in WM can vary significantly depending on the cognitive demands of short-term memory tasks. Interestingly some recent works [18] have used activity-silent scheme using spike-timing-dependent plasticity (STDP) modeling to explore working memory. However, their simulations also were comparable to behavioral data in persistent-activity state.

In our future research, we plan to explore how the implementation of such advanced working memory models, along with knowledge from circuit-level studies, can contribute to a comprehensive understanding of the neural processes involved in cognitive functions. By building on these connections, we aim to make significant strides in the field of cognitive neuroscience and its practical applications.

5   Conclusion

In this study, our recurrent neural architecture (Fig. 1) has provided dynamic insights into the flexible nature of working memory capacity, challenging the fixed limits proposed by slot models [8]. The variable precision encoding of items aligns with a concept supported by others [11], offering a nuanced perspective on working memory function. Our model not only characterizes the serial recall curve and recency effect but also predicts primacy curves when presentation durations are averaged over (Figs. 3 and 4). Intriguingly, the model suggests that enhancing an item's saliency through spatial cues boosts recall probability while suppressing adjacent items (Fig. 6). The presented computational model demonstrates both explanatory and predictive prowess, with potential integration into visual processing models meeting outlined constraints [16]. As we acknowledge the recent insights of [17] regarding hidden states in working memory, our future research aims to explore advanced models and insights from circuit-level studies as well as explore the predictions from the model using behavioral experiments.

References

1. Baddeley, A.: Working memory: looking back and looking forward. Nat. Rev. Neurosci. **4**(10), 829–839 (2003)
2. Miller, A.G.: The magical number seven, plus or minus two: some limits on our capacity for processing information. Psychol. Rev. **63**, 81–97 (1956)
3. Sreenivasan, K.K., Curtis, C.E., D'Esposito, M.: Revisiting the role of persistent neural activity during working memory. Trends Cogn. Sci. **18**(2), 82–89 (2014)
4. Goldman-Rakic, P.: Cellular basis of working memory. Neuron **14**(3), 477–485 (1995)

5. Baddeley, A.: Working memory: theories, models, and controversies. Annu. Rev. Psychol. **63**, 1–29 (2012)
6. Cowan, N.: The magical number 4 in short-term memory: a reconsideration of mental storage capacity. Behav. Brain Sci. **24**, 87–185 (2001)
7. Luck, S.J., Vogel, E.K.: The capacity of visual working memory for features and conjunctions. Nature **390**, 279 (1997)
8. Zhang, W., Luck, S.J.: Discrete fixed-resolution representations in visual working memory. Nature **453**(7192), 233–235 (2008)
9. Ma, W.J., Husain, M., Bays, P.M.: Changing concepts of working memory. Nat. Neurosci. **17**(3), 347–356 (2014)
10. Waugh, N.C., Norman, D.A.: Primary memory. Psychol. Rev. **72**, 89–104 (1965)
11. van den Berg, R., Shin, H., Chou, W., George, R., Ma, W.J.: Variability in encoding precision accounts for visual short-term memory limitations. Proc. Natl. Acad. Sci. **109**(22), 8780–8785 (2012)
12. Usher, M., Cohen, J.D.: Short term memory and selection processes in a frontal-lobe model. In: Heinke, D., Humphreys, G.W., Olson, A. (eds.) Connectionist Models in Cognitive Neuroscience, pp. 78–91 (1999)
13. Grossberg, S.: Contour enhancement, short term memory, and constancies in reverberating neural networks. In: Studies in Applied Mathematics, vol. LII, pp. 213–257 (1973)
14. Sengupta, R., Surampudi, B.R., Melcher, D.: A visual sense of number emerges from the dynamics of a recurrent on-center off-surround neural network. Brain Res. **1582**, 114–224 (2014)
15. Knops, A., Piazza, M., Sengupta, R., Eger, E., Melcher, D.: A shared, flexible neural map architecture reflects capacity limits in both visual short term memory and enumeration. J. Neurosci. **34**(30), 9857–9866 (2014)
16. Tsotsos, J.K., Kruijne, W.: Cognitive programs: software for attention's executive. Front. Psychol. **5**, 1260 (2014)
17. Masse, N.Y., Yang, G.R., Song, H.F., Wang, X.J., Freedman, D.J.: Circuit mechanisms for the maintenance and manipulation of information in working memory. Nat. Neurosci. **22**(7), 1159–1167 (2019)
18. Huang, Q.-S., Wei, H.: A computational model of working memory based on spike-timing-dependent plasticity. Front. Comput. Neurosci. **15**, 630999 (2021)

Joining Emission Data from Diverse Economic Activity Taxonomies with Evolution Strategies

Michael Hellwig[✉][ID] and Steffen Finck[ID]

Josef Ressel Center for Robust Decision Making, Research Center Business Informatics, FH Vorarlberg University of Applied Sciences, Dornbirn, Austria
{michael.hellwig,steffen.finck}@fhv.at

Abstract. In this paper, we consider the question of data aggregation using the practical example of emissions data for economic activities for the sustainability assessment of regional bank clients. Given the current scarcity of company-specific emission data, an approximation relies on using available public data. These data are reported in different standards in different sources. To determine a mapping between the different standards, an adaptation to the Covariance Matrix Self-Adaptation Evolution Strategy is proposed. The obtained results show that high-quality mappings are found. Nevertheless, our approach is transferable to other data compatibility problems. These can be found in the merging of emissions data for other countries, or in bridging the gap between completely different data sets.

Keywords: Evolution Strategies · Application · Sustainability · Constrained Optimization · Sustainable Finance · Self-Adaptation

1 Introduction

The paper addresses the problem of data aggregation using the practical example of emissions data for economic activities for the sustainability assessment of regional bank clients. To assess the impact of climate change European financial institutions are required to perform analyses and risk investigations considering the Ecologic, Social & Governance (ESG) regulations. Part of these regulations is to determine and monitor the risk exposure of companies that are part of a bank's portfolio. Beyond the use case, the paper presents a convenient modeling of the problem that can be answered by a modified Evolution Strategy.

Many practical problems are associated with resource or capacity limits, which must be ensured by introducing constraints to the optimization objective. The search for the optimal solution of an objective function subject to such

---

We would like to thank Hypo Vorarlberg Bank AG for providing the problem and for inspiring discussions on the topic.

G. Nicosia et al. (Eds.): LOD 2023, LNCS 14505, pp. 415–429, 2024.
https://doi.org/10.1007/978-3-031-53969-5_31

restrictions on the parameter vectors is called constrained optimization. Particularly, in black-box optimization, the involvement of constraints in an optimization task increases the complexity, as Evolutionary Algorithms (EA) need to learn how to navigate through the restricted search space without violating the constraints of the problem. In recent years, many EAs have been designed for solving constrained black-box optimization problems. Some of these algorithms are developed for competitiveness on state-of-the-art benchmark functions [3,10,11]. While algorithm variants that turn out to be successful in such benchmark environments are thought to be applicable to a broad range of constrained problems, the highly individual character of real-world applications usually renders it necessary to tailor the algorithm design to the particular use case at hand. That is, very specific problem representations must be selected and novel operators may be developed to ensure constraint satisfaction [1,8,9,14,16,18].

In this paper, we identify a use case for the application of EAs for constrained optimization. In the broadest sense, the use case comes from the field of sustainability assessment. In this context, it is common to group companies and their key figures (e.g., various emissions) into industrial sectors, since individual data are still scarce. If one intends to combine different data sources, one encounters the problem of different industrial sector taxonomies whose mutual correlation has a surprisingly large degree of freedom. Although rough guidelines exist that regulate the transition, and these can be modeled by respective constraints, they do not provide clear conversion keys. Hence, it is not trivial to merge public data, often gathered by national authorities, if they are reported in different taxonomies. Furthermore, the conversion keys for the same taxonomies differ at the country level, which also complicates the comparability of the data on a country level. Consequently, it is desirable to find a reasonable mapping for transferring data from one economic activity taxonomy to another. In this work two specific taxonomies are considered, one based on the European reporting format, the other on a national reporting format.

It is important to note that the nature of the constraints prevents us from formulating the problem as a system of linear equations and simply generating the best-fit solution by computation of the associated Moore-Penrose inverse. However, the problem can be formulated as a constrained optimization problem that must be solved for the most appropriate mapping matrix that satisfies all restrictions and realizes a minimal total deviation between the data reported in different taxonomies. To this end, we propose a specifically designed Evolution Strategy (ES) that is based on the well-known Covariance Matrix Self-Adaptation Evolution Strategy (CMSA-ES) [2]. The newly proposed CMSA-ES variant relies on a matrix representation of the individuals and introduces a problem-specific repair step that ensures feasibility during the search process. The variation operators maintain the matrix form of the decision variables. The selection operator of the ES is depending on the definition of the fitness function designed to measure the approximation quality of a mapping.

The results of the optimization are validated by comparison to available historical data. In particular, in the economic sectors most relevant for the specific use case, the solution found is a good match. It can therefore be used as an approximation for data merging and in further steps for the sustainability assessment of companies. In this regard, the approach provides a contribution towards closing the data gap that exists in this context [12]. The methodology further allows for the identification of potential differences at a country level and to determine reasonable country-specific conversion keys.

This paper is organized in the following way: Sect. 2 provides a description of the real-world use case and derives the corresponding problem formulation. In Sect. 3 the ES for the proposed constrained optimization problem is explained. After that, the empirical algorithm results are shown in Sect. 4 and discussed in Sect. 5. Section 6 concludes the paper with a brief outlook on future extensions and possible implications.

2   Problem Description

This section motivates the real-world use case that is the origin of the mapping task considered. Further, it presents the mathematical representation as a constrained non-linear optimization problem.

2.1   Practical Context

With the increasing impact of climate change, there is a growing need to incorporate sustainability data into future-oriented corporate decision-making. In the face of the ESG regulations imposed by the European Union (EU) [4], this applies in particular to the European financial industry. Authorities demand capital market participants to monitor and manage their ESG-related risk exposure as detailed as possible by the year 2026.

The use case originates from the requirement of our use case partner, the Austrian banking company Hypo Vorarlberg Bank AG, to allocate the associated Green House Gas (GHG) emissions to each credit customer as precisely as possible. The desired data are often only available for a few market participants (large companies with disclosure obligations) and for relatively short time horizons. In particular, for small and medium-sized companies (SMEs), one solution is to rely on sector-specific sustainability and company data. These are thought to suffice as a first approximation of the company-specific indicators. However, assigning the same value to all companies in an economic activity class does not appear to be very useful, as this paints a distorted picture that does not take company sizes and regional differences into account. For this reason, the aim is to use regional GHG data and combine them with other company key indicators.

Unfortunately, already searching only for GHG emissions, the data sources use different taxonomies for economic activities. In Austria, the following three data sets on GHG emissions per economic activity are openly available: Air Emission Accountants (AEA) on a European reporting basis (Eurostat) [7], Austria's National Inventory Reports (NIR) [17], and the Bundesländer Luftschadstoff-Inventur (BLI), Air Pollutant Inventory, reports provided by the Federal States

Table 1. CRF and NACE Rev. 2 classifications of economic activities

CRF		NACE	
Code	**Economic activity description**	**Code**	**Economic activity description**
1A1a	Public electricity and heat production	A	Agriculture, forestry and fishing
1A1b	Petroleum Refining	B	Mining and quarrying
1A1c	Manufacture of solid fuels and energy industries	C	Manufacturing
		D	Electricity, gas, and steam supply
1A2	Manufacturing industries and construction	E	Water supply; waste management
1A3a	Domestic aviation	F	Construction
1A3b	Road transportation	G	Wholesale and retail trade
1A3c	Railways	H	Transporting and storage
1A3d	Domestic navigation	I	Accommodation and food service activities
1A3e	Other transportation	J	Information and communication
1A4a	Commercial/Institutional	K	Financial and insurance activities
1A4b	Residential	L	Real estate activities
1A4c	Agriculture/Forestry/Fisheries	M	Professional, scientific and techn. activities
1A5	Other	N	Administr. and support service activities
1B	Fugitive emissions from fuels	O	Public administr. & defense; social security
2	Total industrial processes	P	Education
3	Total agriculture	Q	Human health and social work activities
5	Total waste	R	Arts, entertainment and recreation
		S	Other services activities
		HH	Households

of Austria. The AEA dataset applies the obligatory taxonomy for statistical reporting in economic activities within the European Statistical System, i.e. the Nomenclature of Economic Activities (NACE) Rev. 2 definitions from 2008 [6]. Yet, the two national data sets (NIR and BLI) use the so-called Common Reporting Format (CRF) according to the recommendations for inventories in the Intergovernmental Panel on Climate Change (IPCC) guidelines from 2006 [15].

As the bank Hypo Vorarlberg Bank AG holds an international customer portfolio, the bank intends to gather the available data in the most current and comprehensive economic sector taxonomy covering all countries of interest. That is, it is necessary to transfer the regional GHG emission data from CRF into NACE classification. In order to be able to map such a transfer in a meaningful way, it is necessary that the annually reported total GHG values match each other in both classifications. In that case, one searches for the most appropriate percentages of the CRF classes that contribute to each NACE class. To the best of our knowledge, there exist no literature dealing with this specific problem. In this paper, we restrict ourselves to finding a mapping between the reported CRF classes within the Austrian NIR reports and the zeroth level of the NACE classification presented in Table 1.

A consideration of the total GHG emissions shows that the BLI and NIR reports (both using CRF) disclose equal GHG emission values. Using the NACE taxonomy, the AEA reports do deviate slightly. Over the years, one observes a mean deviation of about 2.5% between AEA and NIR. Still, such rather small deviations allow for the determination of an approximate mapping between the CRF and the NACE taxonomy.

There are, of course, certain dependencies and references between the classes of both taxonomies. These are specified in the correspondence table [5], in which valid transmissions between both classifications are specified. However, these guidelines provide only ground rules. That is, they roughly suggest which sectors of CRF contribute to a certain NACE sector without disclosing any shares. For example, the tables indicate that, among others, CRF sectors 1A1b (*Petroleum Refining*) and 1B (*Fugitive emissions from fuels*) may contribute to NACE sector C (*Manufacturing*), but they do not indicate any magnitudes. On the other hand, the tables rule out infeasible correspondences like establishing a connection from CRF 1A3c (*Railways*) to NACE P (*Education*). Hence, the correspondence table represents a valuable set of constraints to be respected in determining the approximate shares.

Furthermore, the correspondence table contains many country-specific exceptions that cannot be specified in detail. Since the country-specific peculiarities are not tangible in the Austrian case, they are not taken into account in this paper.

2.2 Problem Formulation

The source data consists of the reported GHG emissions in all sectors for the years 2008 to 2019. These are available for Austria in both taxonomies (CRF and NACE).

The tabulated CRF data can be regarded as a matrix $A \in \mathbb{R}^{k \times m}$ where k and m denote the number of years, and CRF sectors, respectively. Each column of the matrix A then describes a time series for one of the $m = 17$ economic CRF sectors reported in the NIR, e.g. CRF sector 1A1a over the $k = 12$ years from 2008 to 2019. Each row provides information about the reported sectoral emissions in a specific year. Analogously, the tabulated NACE data can be understood as matrix $B \in \mathbb{R}^{k \times n}$ with n being the number of distinct NACE sectors considered. Note, that this paper considers only the $n = 20$ zeroth level NACE categories as a first step, c.f. Table 1.

The task of finding the most appropriate shares for mapping data from CRF to NACE taxonomy can then be rephrased as searching for a matrix $X \in \mathbb{R}^{m \times n}$ that satisfies $AX = B$. Encoding the decision variable in matrix form allows a relatively simple problem formulation despite the inclusion of constraints. Due to the constraints from the correspondence table, and due to the small deviations within the total GHG emission reporting, it will not be possible to find a matrix X that satisfies this equality. Hence, we reformulate the question accordingly and search for the matrix X that minimizes the deviation between the components of AX and B. The deviation can be measured in multiple ways, e.g. as absolute, as relative, or as quadratic deviation. With a focus on the economic activities that show the largest magnitude of GHG emissions, we consider minimizing the quadratic deviation in the first step. This motivates the choice of the objective function

$$f(X) = \sqrt{\sum_{i=1}^{k} \sum_{j=1}^{n} (AX - B)_{ij}^2}. \tag{1}$$

At the same time, the components X_{ij} should satisfy the necessary constraints. Each entry in X represents the relative contribution of a CRF class to a specific NACE class. Consequently, the X_{ij} should be non-negative and less than or equal to 1, i.e.

$$0 \leq X_{ij} \leq 1 \quad \forall i \in \{1, \ldots, m\}, j \in \{1, \ldots, n\}. \tag{2}$$

In the correspondence table [5], the acceptable sectoral linkages between both taxonomies are modeled by a binary matrix $F \in \mathbb{B}^{m \times n}$ whose entries are 1 if there is a feasible relationship between a CRF and a NACE sector and 0 otherwise.

$$F = (F_{ij})_{\substack{i = 1, \ldots, m \\ j = 1, \ldots, n}} \tag{3}$$

with

$$F_{ij} = \begin{cases} 1, & \text{CRF } i \text{ feasible contribution to NACE } j \\ 0, & \text{otherwise} \end{cases}. \tag{4}$$

Considering the component-wise multiplication of X and F, the matrix F can be regarded as a kind of filter that sets the irrelevant (or infeasible) entries in X to zero. Note, that the component-wise multiplication of two matrices of the same type is indicated by $X \circ F$. Further, the total emissions of each single CRF sector should be distributed across all NACE sectors. This implies that the row sums of the matrix $X \circ F$ must always be equal to 1,

$$\sum_{j=1}^{n} X_{ij} F_{ij} = 1, \qquad \forall i \in \{1, \ldots, m\}. \tag{5}$$

Considering (1), (2), and (5), leads to a constraint optimization problem in matrix form

$$\min f(X)$$
$$s.t. \sum_{j=1}^{n} X_{ij} F_{ij} = 1, \quad \forall i \tag{6}$$
$$0 \leq X_{ij} \leq 1 \quad \forall i, j.$$

All constraints but the non-negativity constraint can be handled by a specifically designed mutation operator within the proposed ES. For that reason, the non-negativity is going to be enforced by the inclusion of the static penalty function

$$p(X) = P \cdot \sum_{i=1}^{m} \sum_{j=1}^{n} \mathbb{1}_{X_{ij} < 0} \cdot |X_{ij}| \tag{7}$$

with penalty parameter $P \in \mathbb{R}^{\geq 0}$ and indicator function $\mathbb{1}_{X_{ij} < 0}$ into the objective function. Noticing, that the upper bound on the matrix elements of X is satisfied by enforcing the equality constraint (5), and incorporating the penalty function (7), problem (6) can be reformulated as

$$\min \ f(\boldsymbol{X}) + p(\boldsymbol{X})$$
$$s.t. \ \sum_{j=1}^{n} X_{ij} F_{ij} = 1 \qquad \forall i. \tag{8}$$

This problem is going to be dealt with by a specifically designed CMSA-ES variant that makes use of a matrix encoding for the candidate solutions evolved during the search process. By construction of the mutation operator, the constraints in problem (8) are always satisfied. That is, the proposed CMSA-ES can be regarded as an interior-point method that always operates within the feasible region of problem (8). The penalized objective function of (8) serves as the fitness function of the ES. The algorithmic peculiarities of the proposed CMSA-ES are presented next, in Sec. 3. Afterward, Sec. 4 illustrates the experimental results of the approach.

3   Algorithm Design

The applied CMSA-ES is based on [2]. It is adapted for using a matrix representation of the solutions and includes a basic repair mechanism to generate mutations yielding feasible solutions respecting the constraint in Eq. (8). The pseudo-code for the algorithm is given in Algorithm 1.

The CMSA-ES variant requires the specification of the offspring and parent population sizes (λ and μ, respectively), the available computational budget, the size (m and n) of the desired matrix $\boldsymbol{X}$ that represents the mapping and the filter matrix $\boldsymbol{F}$ from Eq. (3). Each solution is represented as an ($m \times n$) matrix, $\boldsymbol{X}$ for the recombinant and $\boldsymbol{Y}_l$ for the offspring. The initial recombinant is derived from the filter matrix by normalizing the entries in $\boldsymbol{F}$ with the respective row sums, cf. line 5. Thus, the recombinant satisfies constraint (5). The design of the algorithm is based on the intention that this constraint remains satisfied for each offspring and each updated recombinant. The initial recombinant is used in line 6 to initialize the best solution obtained ($\boldsymbol{X}_b$) and its respective fitness value (f_b) which is obtained by evaluation of the recombinant at the objective function $ObjFun$. The objective function of (8) is composed of Eq. (1) and Eq. (7).

The CMSA-ES learns the covariance matrix $\boldsymbol{C}$ of the multivariate normal distribution for mutation during the evolutionary process. It is initialized as an ($mn \times mn$) matrix in line 5. The shape of $\boldsymbol{C}$ allows to model correlations between the different entries X_{ij}. Instead of the covariance matrix itself, its Choleksy decomposition, i.e. matrix $\tilde{\boldsymbol{C}}$, is used in the mutation creation process. Using the Cholesky decomposition is computationally more efficient than using the spectral decomposition.

As long as parts of the budget are available, offspring are created and evaluated (lines 9 to 15), and the recombinant and strategy parameter updates are calculated (lines 17 to 21). By employing σ-self-adaptation [13], each offspring l obtains a mutation strength s_l by variation of the current mutation strength

Algorithm 1. Pseudo code of the CMSA-ES for sectoral mappings.

1: **Initialize:** μ, λ, σ, $budget_{max}$, m, n, F
2: $\tau \leftarrow \frac{1}{\sqrt{2mn}}$, $c_\mu \leftarrow 1 + mn\frac{mn+1}{2\mu}$
3: $C \leftarrow I_{|mn \times mn|}$, $\tilde{C} \leftarrow C$
4: $budget \leftarrow 0$
5: $X_{ij} \leftarrow \frac{F_{ij}}{\sum_{j=1}^{n} F_{ij}}$ $\forall i \in [1, m] \wedge j \in [1, n]$
6: $X_b \leftarrow X, f_b = ObjFun(X)$
7: **while** $budget < budget_{max}$ **do**
8:     **for** $l = 1$ to λ **do**
9:         $s_l \leftarrow \sigma e^{\tau\mathcal{N}(0,1)}$
10:        $r \leftarrow \tilde{C}^T \mathcal{N}(0, I)$
11:        R from reshaping r to $|m \times n|$ matrix
12:        $Z \leftarrow R \circ F$
13:        $\tilde{Z}, z_l$ from repair Alg. 2
14:        $Y_l \leftarrow X + s_l\tilde{Z}$
15:        $f_l \leftarrow ObjFun(Y_l)$
16:    **end for**
17:    $X \leftarrow \mu^{-1}\sum_{k=1}^{\mu} Y_{k:\lambda}$
18:    $\sigma \leftarrow \mu^{-1}\sum_{k=1}^{\mu} s_{k:\lambda}$
19:    $C \leftarrow (1 - c_\mu^{-1})C + (\mu c_\mu)^{-1}\sum_{k=1}^{\mu} z_{k:\lambda}z_{k:\lambda}^T$
20:    $\tilde{C} \leftarrow$ Cholesky(C)
21:    $f \leftarrow ObjFun(X)$
22:    **if** $f < f_b$ **then**
23:        $f_b \leftarrow f$
24:        $X_b \leftarrow X$
25:    **end if**
26:    $budget = budget + \lambda + 1$
27: **end while**
28: **return** X_b, f_b

σ (cf. line 9). The mutation vector r is drawn from the multivariate normal distribution with zero mean and the current covariance matrix. The mutation vector needs to be reshaped into an $(m \times n)$ matrix, R in line 11, for the creation of the offspring (line 14). Since R is a fully populated matrix, adding it to the current recombinant yields with a high probability to a change in the row sums for the offspring. To keep the row sums constant at one, two steps are applied: a) Setting all entries R_{ij} to zero where the filter matrix is also zero. To this end, a filtering of R through a component-wise multiplication of R and F is performed in line 12. b) The obtained matrix, Z, from the previous step still allows for a deviation of the row sums. Hence a repair mechanism was designed which ensures that all row sums in Z are equal to zero. This repair mechanism is provided in Algorithm 2. The repair mechanism builds a mean mutation matrix where each entry is specified by the respective row sums of Z and F, cf. line 2 in Alg. 2. This mean mutation matrix is subsequently subtracted from Z and

Algorithm 2. Pseudo code of the repair mechanism for CMSA-ES for sectoral mappings.

1: **Input:** $, m, n, \boldsymbol{ZF}$
2: $\hat{\boldsymbol{Z}}_{ij} \leftarrow \dfrac{\sum_{j=1}^{n} Z_{ij}}{\sum_{j=1}^{n} F_{ij}}$
3: $\tilde{\boldsymbol{Z}} \leftarrow (\boldsymbol{Z} - \hat{\boldsymbol{Z}}) \circ \boldsymbol{F}$
4: $\boldsymbol{z}$ from reshaping $\boldsymbol{Z}$ to $|mn \times 1|$ vector
5: **return** $\tilde{\boldsymbol{Z}}, \boldsymbol{z}$

---

the result obtained is filtered with $\boldsymbol{F}$ in line 3. The repaired mutation matrix, $\tilde{\boldsymbol{Z}}$, has the following properties

$$\tilde{Z}_{ij} = 0, \text{ if } F_{ij} = 0 \qquad \text{and} \qquad \sum_{j=1}^{n} Z_{ij} = 0 \quad \forall i.$$

Therefore the constraint in Eq. (8) is satisfied by each offspring. Next to the repaired mutation matrix, the repair algorithm provides the repaired mutation vector $\boldsymbol{z}$ that is required for the update of the covariance matrix. An offspring is created and evaluated by using the current recombinant, the mutation strength s_l, and the repaired mutation matrix, cf. lines 14 and 15 in Alg. 1.

Recombination is applied to the selected parents (i.e., the μ best offspring and their respective mutation strengths) to obtain the new recombinant $\boldsymbol{X}$ and the updated mutation strength σ. Selection is indicated by the order notation[1] in the index of the respective terms. In lines 19 to 21 in Algorithm 1 the update of the covariance matrix is performed. Then, the new recombinant is evaluated and compared to the current best solution. If it is better, the respective values are updated. In the end, the budget counter is increased. In the pseudo-code, the number of function evaluations is used as a measurement unit for the budget.

4   Experiments

This section describes the experiments performed and presents the related results of the mapping determination.

The paper is restricted to the derivation of the mapping between CRF and NACE sectors which is based on the emission data published in 2019 exclusively. The resulting shares of the contribution of the individual CRF sectors to the NACE classes are then validated against the previous years.

In fact, the problem formulation is composed in a more general way allowing to determine mappings that balance the CRF data reported for multiple years. For example, one might search for a mapping that relies on the data relationships over the years from 2013 to 2019. The interest in such a mapping would be justified by a change in the CRF reporting guidelines resolved at the 2012 UN Climate Change Conference in Qatar. However, the mapping found for the 2019

---
[1] $k : \lambda$ refers to the k-best, i.e. k-smallest fitness value, out of λ possible solutions.

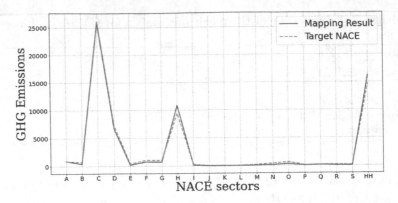

Fig. 1. Illustration of the GHG emission allocation based on the mapping determined by the CMSA-ES. The dashed orange line represents the target distribution of the reported total GHG emissions over the NACE categories in the year 2019. The solid blue line displays the desired allocation of CRF emissions to the NACE classes. (Color figure online)

data alone provides a good solution that can also be used for the GHG allocation of previous years and yields a mean error in the range of the anticipated reporting error Since a detailed description of further experiments would exceed the page limit, we will only briefly address such aspects in the discussion in Sec. 5.

Using the information provided in [5], the filter matrix F is constructed accordingly, see Eqs. (3), and (4). The matrix contains 52 non-zero entries out 340 possible entries. There exist at least one entry and at most 8 entries for each NACE sector.

For the presented experiments, the CMSA-ES variant makes use of an offspring population size of $\lambda = 50$ and a parental population size of $\mu = 3$. The mutation strength is initialized at $\sigma = 1$ and a fixed penalty parameter of $P = 10^4$ is considered. The algorithm uses a budget of $5 \cdot 10^4$ function evaluations. According to Table 1, the dimensions of the desired mapping matrix X are $m = 17$ and $n = 20$. The configuration of the ES is based on a large number of experiments performed so far and uses the common algorithmic parameters [2]. However, it is conceivable that targeted parameter tuning might still lead to improvements, but this step must be postponed to future investigations.

The results of a single run of the CMSA-ES variant for sectoral mappings on problem (8) are illustrated in Fig. 1 to Fig. 3. Figure 1 displays the GHG emissions per economic activity in the NACE taxonomy. The dashed orange line represents the data provided by the AEA report for 2019 that serve as target values for the determination of an appropriate mapping (c.f. the matrix B in Sec. 2.2). The solid blue line displays the CRF data (originating from the NIR 2019 report) that have been transferred to the NACE taxonomy by application of the derived mapping matrix X_b, i.e. AX_b. The overall agreement of both curves indicates that the matrix X_b represents a reasonable mapping between both taxonomies.

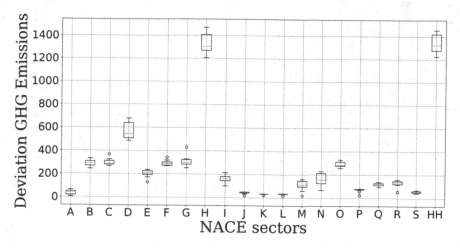

Fig. 2. Illustration of the GHG emission allocation based on the mapping determined by the CMSA-ES in 10 consecutive runs. Each boxplot shows the distribution of the absolute deviation between the obtained results and the reported GHG emissions in NACE taxonomy.

The CMSA-ES variant is able to ensure feasible mappings of consistently good quality. This assessment is supported by the experimental results displayed in Fig. 2. For 10 independent runs a mean value of 3434.15 with standard deviation of 26.65 was obtained for the objective function. This corresponds to about 5.38% of the total GHG emissions reported in the NACE taxonomy. The sum of the mean deviations for each NACE class correspond to about 0.72% of the total reported emissions. The given boxplots for each NACE class provide the dispersion of the absolute deviation between the approximation and the reported values. They show consistent results, larger spreads and larger absolute errors are found for the classes D, H and HH. These classes correspond to the classes with larger emissions (cf. Fig. 1). However, class C, the one with the largest GHG emission, is approximated quite well in all runs as expected due to the use of quadratic deviations as objective function.

Figure 3 provides a more detailed view of the agreement of the mapping results to specific NACE sectors (C, D, I, and HH). To this end, the mapping X_b obtained for the 2019 GHG emission data is applied to the rest of the CRF data. The resulting data points are compared to the original NACE data over time. The displayed illustrations present the commonly observed dynamics and serve as representatives for the classes not shown.

For all NACE sectors, one observes a very good qualitative agreement with the target curves. This supports the idea to allow for the allocation of the CRF-reported data in appropriate proportions to the NACE classes. Noticeably, for the years after 2013, the agreement significantly improves yet again. This observation is in line with the adjustment of the CRF reporting guidelines after the 2012 UN Climate Change Conference in Qatar and with the determination of

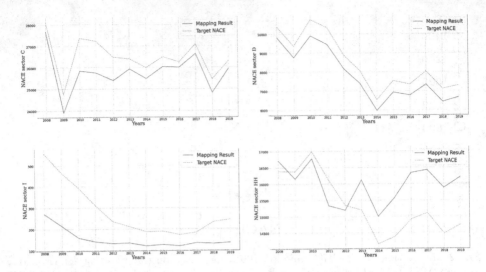

Fig. 3. Illustration of the dynamic agreement over the past years for the NACE sectors C, D, I, and HH (upper left to lower right). The solid blue line depicts the results of the mapping of CRF emissions to the considered NACE sector while the dashed orange line displays the target values reported in NACE taxonomy. (Color figure online)

the mapping X_b based on the data reported in this revised CRF manner for 2019.

Over these seven years (2013–2019), a mean relative deviation between 1% and 8% between AX_b and B is measured for those NACE activities that contribute to about 90% of all GHG emissions in 2019. With reference to the basic deviation of both data sets, the observed quantitative difference between the curves is at least tolerable.

The mapping determined by the CMSA-ES variant can therefore be considered an acceptable conversion tool between the taxonomies.

5   Discussion

The mapping obtained can be considered a reasonable tool for the practical and, above all, quantitative conversion of CRF-classified GHG emissions data into the NACE classification. In this way, regionally published data can be embedded in the international context and thus structural differences can be included in the sustainability considerations.

Yet, a closer look at the resulting plots reveals still some deviations in individual NACE sectors. One part of such deviations must be explained with the difference in total emissions disclosed (mean deviation of about 2.5% over the years). Still, sectors with relatively low GHG emission exposure appear to be given lower priority in the optimization process. Regarding the objective function definition in problem (8), this observation makes sense as the algorithm is

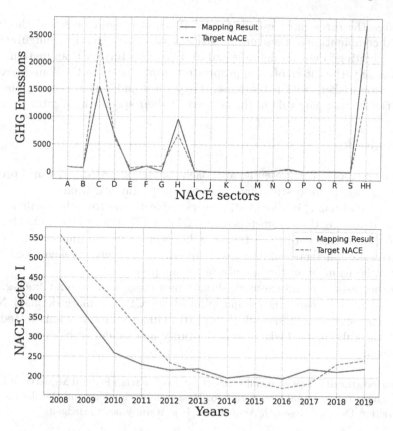

Fig. 4. Illustration of the overall agreement and the dynamic agreement over the past years for the NACE sectors I. The solid blue line depicts the results of the mapping of CRF emissions to the considered NACE sector while the dashed orange line displays the target values reported in NACE taxonomy. (Color figure online)

predominantly searching to close the gap between AX and B for the sectors with the largest emission exposure. Changing the objective function definition can reverse this effect. By minimizing the relative error, the focus of the optimization process is placed on the classes with a small share of the total emissions.

For this case, Fig. 4 displays the results of a single CMSA-ES run optimizing the relative error. In the upper figure, one immediately sees that the deviations in the peaks of the figure increase. A closer look at individual NACE sectors (here Sector I in the lower figure) shows that the agreement in the classes of smaller GHG emission contributions is improving. This is indicated by the blue solid line which has moved closer to the target curve (c.f. Fig. 3 lower left). In this turn, the relative deviation of the classes of low GHG contributions is also significantly reduced. As this comes at the expense of degradation of the deviation related to the large NACE classes, one faces a balancing problem that depends on the definition of the objective function. A balanced mapping that distributes the error more evenly across all classes could be found through a weighted combination of these optimization goals. However, the configuration of such a weighting is not straightforward and requires further examination.

Ultimately, the choice of an appropriate mapping is up to the domain experts and decision-makers within the bank. They must judge the meaningfulness of the particular mapping rule in the context of their specific application.

6   Outlook

While the results of the presented approach for the proposed real-world problem look promising, there is still some room for future investigations.

On the one hand, it should still be possible to improve the quality of the mapping by tuning the parameters of the CMSA-ES appropriately. On the other hand, the investigation of the trade-off between, and the impact of, different objective function representations (e.g. quadratic vs. relative deviation), offers the potential to increase the balance of the mapping.

Another aspect of future work should be taking into account finer-grained baseline data, i.e. searching for mappings from CRF to more detailed NACE sectors (e.g. level 1 or 2 instead of level 0). This way, more sophisticated data sets for the subsequent tasks of sustainability assessment of companies can be established.

Acknowledgment. The financial support by the Austrian Federal Ministry of Labour and Economy, the National Foundation for Research, Technology and Development and the Christian Doppler Research Association is gratefully acknowledged.

References

1. Beyer, H.G., Finck, S.: On the design of constraint covariance matrix self-adaptation evolution strategies including a cardinality constraint. IEEE Trans. Evol. Comput. **16**(4), 578–596 (2012). https://doi.org/10.1109/TEVC.2011. 2169967
2. Beyer, H.G., Sendhoff, B.: Covariance matrix adaptation revisited - the CMSA evolution strategy. In: Rudolph, G., Jansen, T., Beume, N., Lucas, S., Poloni, C. (eds.) PPSN 2008. LNCS, vol. 5199, pp. 123–132. Springer, Heidelberg (2008). https://doi.org/10.1007/978-3-540-87700-4_13
3. Brockhoff, D., Hansen, N., Tušar, T., Mersmann, O., Sampaio, P.R., Auger, A., Atamna et al., A.: COCO documentation repository. http://github.com/numbbo/coco-doc
4. European Banking Authority (EBA): Report on management and supvervision of ESG risks for credit institutions and investment firms EBA/REP/2021/18 (2021). https://www.eba.europa.eu/eba-publishes-its-report-management-and-supervision-esg-risks-credit-institutions-and-investment
5. Eurostat: Annual air emissions accounts (AEA) - Annex I (Correspondence between CRF/NFR - NACE Rev. 2) to Manual for Air Emissions Accounts, (2015). https://ec.europa.eu/eurostat/web/environment/methodology. see: Emissions of greenhouse gases and air pollutants (air emission accounts)

6. Eurostat: NACE Rev. 2 - Statistical classification of economic activities (2008). https://ec.europa.eu/eurostat/web/products-manuals-and-guidelines/-/ks-ra-07-015

7. Eurostat: Luftemissionsrechnungen nach NACE Rev. 2 Tätigkeit (2022). https://ec.europa.eu/eurostat/databrowser/view/ENV_AC_AINAH_R2__custom_4682524/default/table

8. Finck, S.: Worst case search over a set of forecasting scenarios applied to financial stress-testing. In: Proceedings of the Genetic and Evolutionary Computation Conference Companion, pp. 1722–1730. ACM, Prague Czech Republic (2019). https://doi.org/10.1145/3319619.3326835

9. Hellwig, M., Arnold, D.V.: Comparison of constraint-handling mechanisms for the $(1,\lambda)$-ES on a simple constrained problem. Evol. Comput. **24**, 1–23 (2016)

10. Hellwig, M., Spettel, P., Beyer, H.G.: Comparison of contemporary evolutionary algorithms on the rotated klee-minty problem. In: Proceedings of the Genetic and Evolutionary Computation Conference, GECCO 2019. ACM, New York, NY, USA (2019). https://doi.org/10.1145/3319619.3326805

11. Kumar, A., Wu, G., Ali, M.Z., Mallipeddi, R., Suganthan, P.N., Das, S.: A test-suite of non-convex constrained optimization problems from the real-world and some baseline results. Swarm Evol. Comput. **56**, 100693 (2019)

12. Network for Greening the Financial System (NGFS): Final report on bridging data gaps. Tech. rep. (2022). https://www.ngfs.net/en/final-report-bridging-data-gaps

13. Schwefel, H.P.: Adaptive mechanismen in der biologischen evolution und ihr einfluss auf die evolutionsgeschwindigkeit. Interner Bericht der Arbeitsgruppe Bionik und Evolutionstechnik am Institut für Mess-und Regelungstechnik Re 215(3) (1974)

14. Sickel, J.H.V., Lee, K.Y., Heo, J.: Differential evolution and its applications to power plant control. In: 2007 International Conference on Intelligent Systems Applications to Power Systems. IEEE (2007)

15. Simon, E., Buendia, L., Miwa, K., Ngara, T., Tanabe, K.: 2006 IPCC Guidelines for National Greenhouse Gas Inventories – IPCC (2006). https://www.ipcc.ch/report/2006-ipcc-guidelines-for-national-greenhouse-gas-inventories/

16. Spettel, P., Beyer, H.G., Hellwig, M.: A Covariance Matrix Self-Adaptation Evolution Strategy for Optimization under Linear Constraints. IEEE Trans. Evol. Comput. (2018). https://ieeexplore.ieee.org/document/8470948/

17. Umweltbundesamt GmbH: Österreichische Emissionen von Treibhausgasen im Zeitraum 1990 bis 2021 nach CRF-Sektoren (2023). https://www.data.gv.at/katalog/dataset/78bd7b69-c1a7-456b-8698-fac3b24f7aa5

18. Zavoianu, A.C., et al.: Multi-objective optimal design of variably constrained 2D path network layouts with application to ascent assembly engineering. J. Mech. Des. (2018). http://dx.doi.org/10.1115/1.4039009

GRAN Is Superior to GraphRNN: Node Orderings, Kernel- and Graph Embeddings-Based Metrics for Graph Generators

Ousmane Touat[2] , Julian Stier[1]([✉]) , Pierre-Edouard Portier[2] ,
and Michael Granitzer[1]

[1] University of Passau, Passau, Germany
julian.stier@uni-passau.de
[2] INSA Lyon, Villeurbanne, France

Abstract. A wide variety of generative models for graphs have been proposed. They are used in drug discovery, road networks, neural architecture search, and program synthesis. Generating graphs has theoretical challenges, such as isomorphic representations – evaluating how well a generative model performs is difficult. Which model to choose depending on the application domain?

We extensively study *kernel-based metrics* on distributions of graph invariants and *manifold-based* and *kernel-based metrics* in *graph embedding space*. Manifold-based metrics outperform kernel-based metrics in embedding space. We use these metrics to compare GraphRNN and GRAN, two well-known generative models for graphs, and unveil the influence of node orderings. It shows the superiority of GRAN over GraphRNN - further, our proposed adaptation of GraphRNN with a depth-first search ordering is effective for small-sized graphs.

A guideline on good practices regarding dataset selection and node feature initialization is provided. Our work is accompanied by open-source code and reproducible experiments.

Keywords: Graph Generative Models · Graph Neural Network · Graph Manifolds Metrics · Geometric Deep Learning

1 Introduction

For a few years now, graph generation via deep generation models has been a subject of particular attention, notably because of its promising applications, especially in pharmacy. Beyond these specified applications, graphs are also central to many fields, such as physics or sociology.

The attention on graph generation has been accentuated with the arrival of autoregressive generation models [13,14,27], able to build a graph with a sequence of elementary steps (e.g., the addition of a node and its connection to the current graph). These models are among the most scalable graph generator while still being able to approach graph distribution better. The main challenge of these generators is that since they must learn to generate graphs in sequence, these models deal with the problem of non-unique representation of graphs, where the node orderings dictate how to represent the graph in sequence. As there can be numerous node orderings, and consequently various ways to sequentially represent graphs, the problem of sequential generation quickly becomes challenging to model. One approach is to model the problem by estimating a lower bound on the maximum likelihood, considering only a subset of the set of node orderings of a graph.

Another challenge is the evaluation of the performance of these models. Until now, the primary method of evaluation consists in using metrics based on the computation of kernels from statistical measures related to the studied graphs, computing, for example, the Maximum Mean Discrepancy of the node degree distribution [8, 14, 16, 27]. With advances in graph neural networks (GNN), evaluation methods using data embeddings, originally applied to image generation models, have recently been transposed to the domain of generative models for graphs [4, 22]. Nevertheless, these techniques still need to be sufficiently studied, and their use is just beginning to spread.

Our **contributions** comprise 1st/ extensive studies on recent evaluation techniques based on graph embeddings which are learned through pre-trained graph classifiers, on several factors such as the pre-training dataset or the node feature initialization technique, giving insights on good practices for graph evaluation accompanied with a *reproducible code repository*. 2nd/ *independent* and *new empirical studies* on GraphRNN and GRAN compared using several node orderings, adding experiments with a *newly proposed node ordering* (depth-first search) on GraphRNN, using both graph-embeddings- and kernel-based evaluation techniques. These contributions show new insights on graph embeddings-based metrics for generative models of graphs and motivate using depth-first-search as a node ordering when learning GraphRNN on small graph datasets.

2   Metrics for Generative Models of Graphs

Generative models of graphs are trained on graph datasets and are then used to sample new graphs. Good evaluation metrics should be able to rank graph generative models on how well they capture the characteristics of the underlying graph distribution P_G from the training set samples $X_G^{train} \sim P_G$. The graph edit distance is a standard similarity measure. However, in general, its computation is NP-hard, [28]. First papers on deep graph generation used graph statistics such as node degree and clustering coefficients to measure the performance of generative models [3].

2.1   Kernel-Based Evaluation Metrics

In their work, You et al. employ Maximum Mean Discrepancy **MMD** as a measure of distance between graph invariants, such as node degree or the number of 4-orbits [27]. This allow them to assess how effectively the generated graphs from their trained model capture properties found in reference graphs. MMD between two drawn distribution x and y based on a kernel k is defined as:

$$\text{MMD}^2(p\|q) = \mathbb{E}_{x,y\sim p}[k(x,y)] + \mathbb{E}_{x,y\sim q}[k(x,y)] - 2\mathbb{E}_{x\sim p,y\sim q}[k(x,y)] \quad (1)$$

For graphs, p and q refer to the distributions of computed graph invariants, such as the node degree distribution from the generated graphs and the set of reference graphs. Those metrics are currently widely used in the graph generation community [16].

2.2 Graph Embeddings-Based Evaluation Metrics

We introduce the graph embeddings-based metrics used in our study: *Fréchet Distance, Precision, Recall, Density, Coverage* and *F1-Score* use graph representations learned by the *Graph Isomorphism Network* model.

Graph Isomorphism Network. Graph Isomorphism Network (GIN) is a robust message-passing neural network architecture, usually designed for graph classification tasks [26]. We define $G = (V, E)$ as an undirected graph defined by its set of n nodes denoted V, and a set of e edges E. A node $v_i \in V$ have its d-dimensional node feature $X \in \mathbb{R}^{e \times d}$. The neighborhood $\mathcal{N}(v_i)$ is a set of nodes connected to v_i. The hidden representation at the k-th layer $h^{(k)}(v_i)$, with $\phi^{(k)}$ being the node aggregation function, is modelized using a multilayer perceptron (MLP):

$$h^{(k)}(v_i) = \text{MLP}^{(k)}((1 + \epsilon^{(k)})h^{(k-1)}(v_i), \ \phi^{(k)}(\{h^{(k)}(u) \mid u \in \mathcal{N}(v_i)\})) \quad (2)$$

The graph level readout ξ_{rd}, obtaining graph-level representations, "can be a simple permutation invariant function such as summation or a more sophisticated graph-level pooling function" [26] that can be either a sum or an average over graph representations of each layer. We get the graph's embeddings by performing such sum "$\sum$" or concatenation "$\|$" over pooled node representations at each layer of the GIN [26]:

$$x_{\|} = \| \left(\xi_{rd}(\{h^{(k)}(u) \mid u \in N(v_i)\}) \mid l \in \{1, 2, .., K\} \right) \quad (3)$$

$$x_{\sum} = \sum_{l=1}^{K} \xi_{rd}(\{h^{(k)}(u) \mid u \in N(v_i)\}) \quad (4)$$

For the first expression, Xu et al. proves that GIN is equivalent to the Weisfeiler-Lehman Isomorphism Test (1-WL) [25].

Pre-training GIN for Graph Feature Extraction. To evaluate the quality of generative image models, metrics such as the *Fréchet Inception Distance* (FID) [10] have been used and are computed based on representations from image embedding space. Image embeddings are obtained from learned image classification

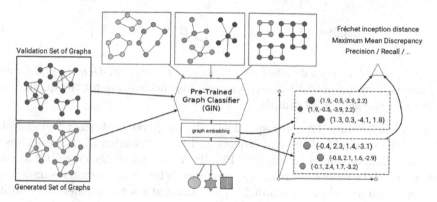

Fig. 1. Embedding-based metrics are computed in embedding spaces of a GNN such as the GIN [26]. After training this GNN on classifying types of graphs, we compare two sets of graphs by computing metrics based on their graph representations – a $N \times d$ matrix, with N the number of graphs and d the graph manifold size. The reference set of graphs is compared to a trained model generated set of graphs, to assert the generative model performance.

models such as Inception v3 on ImageNet [21]. This approach was first explored on small graphs using GIN [4]. Similarly, we use metrics in *graph* embedding space as sketched in Fig. 1. *Graph* embeddings are extracted from a graph neural network trained on a multi-class classification task, where each class represents one type of graph, which has also been used in training for the generative tasks (i.e., grid graphs, community graphs).

Graph Embeddings-Based Metrics. Let $(X_1, \cdots, X_N) \in \mathbb{R}^{N \times d}$ be vector representations of size d in the graph embedding space of the pre-trained GIN obtained from graphs $(G_1, \cdots, G_N)$ through Eqs. 3 or 4.

The first studied metric, called Fréchet Distance *FD*, is a popular metric derived from *FID* [10] used in the image generation domain. This metric compares both generated and authentic samples of graphs in the graph embedding space, taken as Gaussian distributions with means and covariances $p_{gen} = (m_{gen}, C_{gen})$ and $p_{real} = (m_{real}, C_{real})$. Fréchet Distance, accounting for the quality and diversity of generated samples, is computed as:

$$d^2\left(p_{gen}, p_{real}\right) = \|m_{gen} - m_{real}\|_2^2 + \text{Tr}\left(C_{gen} + C_{real} - 2\left(C_{gen}C_{real}\right)^{1/2}\right) \quad (5)$$

We also use two pairs of metrics, denoted Data-Manifold metrics, based on multi-dimensional representation manifolds [12,15], meaning sets of close neighbors forming a hypersphere. On graphs, with $B(x, r)$ a ball of center x and radius r, and $\text{NND}_k(X_i)$ the distance between X_i, the representation of one graph extracted from the GNN, and the kth nearest neighbor in the embedding space, the manifold is defined as

$$\text{manifold}(X_1, \cdots, X_N) := \bigcup_{i=1}^{N} B(X_i, \text{NND}_k(X_i)) \tag{6}$$

The two pairs of metrics denoted *Precision and Recall*, and *Density and Coverage*, are used to compute *F1 Score*, a metric also accounting for the quality and diversity of the generated sample.

Precision and Recall [12] are respectively the proportion of generated samples being in the manifold of the real objects and the proportion of the real objects in the manifold of generated samples. Finally, *Density and Coverage* [15] are the proportion of the number of real object manifolds for each generated sample and the proportion of real object manifolds containing at least one generated sample.

Node Feature Engineering on Unattributed Graphs. Generally, GNNs are very powerful when the graphs are attributed, where the nodes are associated with natural feature vectors corresponding to known attributes. With unattributed graphs, we have no information other than the graph's topology given by its adjacency matrix. In order to compute *FD* using GIN, [4] used a one-hot vector encoding of node degrees as input node feature vector, as in [26] on social network graphs. Such feature vectors have been studied and shown to be among the feature initialization that gives overall good performance in graph classification [5,7]. Recently, other feature engineering has been explored, using random node features initialization [1,17]. [17] notably takes up GIN, adding the ability to generate a random component at each pass through GIN, demonstrating its power on some problems that GIN alone cannot solve. We also implemented this approach alongside the existing node feature initialization approaches explored for generative graph evaluation.

Related Work. First attempts of importing GAN-related metrics such as the Fréchet Inception Distance *FID* [10] *Improved Precision and Recall* [15] were implemented using recent Graph Neural Network techniques such as Graph Isomorphism Network *GIN* [26], but were only marginally used to implement graph version of existing metrics to work on [4,22]. Obray and al. [16] investigated kernel-based evaluation metrics using graph invariants, especially MMD on graph statistics. They show how the kernel choice and fine-tuning of parameters are critical as the metric may become unreliable, giving an unstable ranking of the compared generative model. Thompson et al. [23] recently investigated graph embeddings-based metrics, justifying the potential of using untrained GNN to extract graph embeddings, using similar experiments as in [16]. We took [16] approach for systematically evaluating those graph embeddings-based metrics. Nevertheless, we developed our paper independently from [23] as we differentiate on studying different hyperparameter settings and their impact on the behavior of the graph embeddings metrics, such as the choice of node feature initialization and pre-training dataset choice.

3   GraphRNN and GRAN

Both Graph Recurrent Neural Networks **GraphRNN** [27], and Graph Recurrent Attention Networks **GRAN** [14] are autoregressive graph generators. We will then present the autoregressive graph generation problem before detailing how GRAN and GraphRNN work.

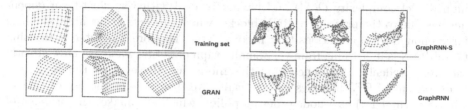

Fig. 2. Visualisation of generated 2d grid graphs from GRAN, GraphRNN-S and GraphRNN-RNN compared to originally generated grid graphs used as *training set*.

Autoregressive Graph Generation Problem. We define $G = (V, E)$ as an undirected graph defined by its set of nodes V and edges between nodes E. With the node ordering π of G, and $V^\pi = (v_1, \ldots, v_n)$ the ordered set of nodes, we have A^π the adjacency matrix of G ordered by π, a $n \times n$ boolean matrix representing G. Graph generative models learn a probability distribution P_G by learning the sequence $S^\pi = (S_n^\pi, \ldots, S_1^\pi)$ with S_i^π corresponding to the i-th row in the adjacency matrix A^π. We usually solve the problem of graph generation by maximizing the log-likelihood over $log(P_G)$. Learning to generate rows of adjacency matrices in sequences is difficult because the search space of different adjacency matrices associated to one graph can be huge, up to $n!$, due to the factorial nature of node permutations. Both **GRAN** and **GraphRNN** precisely use chosen node orderings to alleviate this sampling complexity.

Graph Recurrent Neural Networks **GraphRNN** are based on two RNNs to model the sequence S^π. The first RNN is called graph-level RNN, used to save the current state of the generated graph. The second RNN is called node-level RNN, generating the rows of the adjacency matrix from the current state of the graph given by the graph-level RNN. Breadth-First-Search (BFS) restricts the maximum node permutations for one graph. BFS ordering is also used to improve scalability by allowing the model to work on adjacency matrices presenting a known lower graph bandwidth, reducing the model complexity from n^2 to $n \cdot m$ with $m \leq n$.

Graph Recurrent Attention Networks **GRAN** use attention-based Graph Neural Networks to generate a graph in, at best $O(n)$ autoregressive steps. Each step generates rows of the adjacency matrix representation of G. GRAN allows for generating multiple rows at once for a quality/time efficiency tradeoff. Compared to GraphRNN, this model does not save all the previous graph representations and only works on the current graph representation for the autoregressive step. This model also allows using several sets of node orderings (called canonical node orderings) that could improve generation by having a higher lower bound of the maximum likelihood $\log(P_G)$. But this comes with a tradeoff between training time and sample quality.

Using Several Node Ordering on GraphRNN and GRAN. GraphRNN uses BFS to improve its scalability, but this model still suffers from having a relatively high model complexity $O(n^2)$ and having to deal with very long-term dependencies due to the presence of RNN blocks, which prevents achieving scalability large enough for applications that need it [14]. Therefore, the interest in using GraphRNN is in learning from very small graph datasets, where we have already seen an excellent performance from this model [6]. However, at this scale, the main benefit of BFS becomes less relevant. We then present the use of Depth-First-Search (DFS) as a node ordering policy, which also makes the problem of graph generation easier by reducing the search space of the graph representation. However, how much do the performance and quality change when DFS as the node ordering?

GRAN already allows for the possibility of using different node orderings to train the model. We compare the use of these node orderings on GRAN to accompany the comparative study on GraphRNN. The main difference between node ordering in both models is that GRAN uses fixed matrix representations of the graphs, as the node ordering function is only called at the beginning of the training. In GraphRNN, graphs matrix representations change through training because the stochastic node ordering function (e.g., GraphRNN BFS function is computed with a random starting node) is called each time a graph is picked from the dataset during the training loop to give a node ordering that can be different from the one computed previously for the same graph.

Related Graph Generative Models. Early graph generative models are low parameterized and not powerful enough to model complex dependencies in real-world networks. Graph generators use deep learning techniques to mimic graphs from real-world data. Models such as variational auto-encoders [11,19] or generative adversarial networks [3] have been studied for graphs. Breakthroughs in quality and scalability happened with autoregressive models like GraphRNN [27]. Due to their sequential nature, auto-regressive formulations must restrict the space of studied node orderings to make the problem tractable. This problem has been studied e.g., in DGMG, and GraphRNN [13,20,27].

4   Experiments

Our study has two concerns: 1/Studying the behavior of graph embeddings-based metrics using two experiments where we vary several hyperparameters. 2/See whether GRAN or GraphRNN generates better graphs in terms of quality and how node orderings can impact both models' performances. All experiments were done using the same Python environment (Python 3.7 + Pytorch 1.8.1).

Table 1. Sum Graph Manifolds metrics on Gr dataset from 10 independently trained GIN. All of the metrics show consistent ranking fidelity with the base ranking hypothesis and stability. F1 Score metrics are more stable in regards to the dataset configuration, with low variance compared to FD.

	Base dataset configuration			
	Mean Rank ↓	FD ↓	F1 PR ↑	F1 DC ↑
Re-sampled	1	-0.0010 ± 0.0007	1.0 ± 0.0	0.994 ± 0.003
GRAN	2	41004 ± 14957	0.845 ± 0.006	0.79 ± 0.030
GraphRNN-RNN	3	91512 ± 24215	0.040 ± 0.010	0.019 ± 0.009
GraphRNN-S	4	319669 ± 94417	0.0 ± 0.0	0.0 ± 0.0
	Ladder dataset configuration			
	Mean Rank ↓	FD ↓	F1 PR ↑	F1 DC ↑
Re-sampled	1	-0.0010 ± 0.0008	1.0 ± 0.0	0.992 ± 0.005
GRAN	2	30773 ± 23098	0.870 ± 0.041	0.81 ± 0.10
GraphRNN-RNN	3	127572 ± 29497	0.042 ± 0.015	0.02 ± 0.01
GraphRNN-S	4	362265 ± 149052	0.001 ± 0.006	0.003 ± 0.001
	Full dataset configuration			
	Mean Rank ↓	FD ↓	F1 PR ↑	F1 DC ↑
Re-sampled	1	$\mathbf{-0.0002 \pm 0.0001}$	$\mathbf{1.0 \pm 0.0}$	$\mathbf{0.990 \pm 0.006}$
GRAN	2	737 ± 340	0.880 ± 0.016	0.847 ± 0.039
GraphRNN-RNN	3	45270 ± 16467	0.046 ± 0.006	0.027 ± 0.007
GraphRNN-S	4	75501 ± 29895	0.0 ± 0.0	0.0 ± 0.0

Datasets. For both parts, we focus on synthetic datasets for two reasons: First, some synthetic graphs datasets are based on mathematical models such as Barabasi-Albert **BA** and Watts-Strogatz **WS** models or have directly identifiable topologies such as community datasets. Finally, using such synthetic datasets enable us to directly compute benchmarks for a *perfect* generation being the synthetic graph generator itself, which is further explained in the Results section.

The following datasets are used for our experiment: (1) **BA:** 500 Barabasi-Albert graphs [2] with $100 \leq n \leq 200$ using the B-A model, using the integrated networkx.barabasi_albert_graph function [9], with parameter $k = 4$.

(2) **WS:** 500 Watts-Strogatz graphs [24] with $100 \leq n \leq 200$ using the W-S model, where each node is linked to its four nearest neighbor and the probability of rewiring is 0.1.

(3) **C2L:** 500 2-community graphs using the synthetic community generator from [27], where communities are formed with Erdos-Renyi graphs of $6 \leq n_{community} \leq 10$ and $p = 0.7$ from which $0.1 \times n$ edges between communities are added uniformly at random.

(4) **C2S:** 500 2-community graphs using the synthetic community generator from [27], where communities are formed with Erdos-Renyi graphs of $30 \leq n_{community} \leq 80$ and $p = 0.3$ from which $0.05 \times n$ edges between communities are added uniformly at random.

Table 2. Concatenation Graph Manifolds metrics on Gr dataset from 10 independently trained GIN. All of the metrics show consistent ranking fidelity with the base ranking hypothesis and stability. F1 Score metrics are more stable in regards to the dataset configuration, with low variance compared to FD.

	Base dataset configuration			
	Mean Rank ↓	FD ↓	F1 PR ↑	F1 DC ↑
Re-sampled	1	-0.001 ± 0.001	1.0 ± 0.0	0.994 ± 0.003
GRAN	2	26639 ± 14495	0.852 ± 0.010	0.789 ± 0.016
GraphRNN-RNN	3	65917 ± 24978	0.042 ± 0.005	0.026 ± 0.006
GraphRNN-S	4	201977 ± 89750	0.0 ± 0.0	0.0 ± 0.0
	Ladder dataset configuration			
	Mean Rank ↓	FD ↓	F1 PR ↑	F1 DC ↑
Re-sampled	1	-0.002 ± 0.002	1.0 ± 0.0	0.990 ± 0.003
GRAN	2	10228 ± 11339	0.879 ± 0.029	0.832 ± 0.069
GraphRNN-RNN	3	91113 ± 24230	0.047 ± 0.005	0.034 ± 0.011
GraphRNN-S	4	212305 ± 78070	0.003 ± 0.008	0.0006 ± 0.001
	Full dataset configuration			
	Mean Rank ↓	FD ↓	F1 PR ↑	F1 DC ↑
Re-sampled	1	$-6e\text{-}4 \pm -6e\text{-}4$	1.0 ± 0.0	0.990 ± 0.007
GRAN	2	1264 ± 360	0.885 ± 0.0147	0.862 ± 0.029
GraphRNN-RNN	3	59471 ± 9750	0.048 ± 0.001	0.033 ± 0.003
GraphRNN-S	4	91310 ± 19924	0.001 ± 0.006	0.0003 ± 0.001

(5) **Gr:** 100 2D grid graphs with $100 \le n \le 400$ using the integrated networkx.2d_grid_graph function.

(6) **Ld:** 500 ladder graphs with $100 \le n \le 200$ using the integrated networkx.ladder_graph function (Table 2).

(7) **ER:** we create this dataset with one of the previously mentioned graph datasets, where an ER graph is created with $n = max(n_g, \text{label}(g) = i)$ and parameter $p = \frac{e_g}{n_g^2}$, this dataset contains the same amount of graph as the linked graph dataset (e.g., we would generate 500 ER graphs "equivalent" to the **BA** dataset, or 100 ER graphs "equivalent" to the **Gr** dataset).

4.1   Testing Evaluation Metrics with Graph Embeddings

We evaluate the behavior of graph embeddings-based metrics computed using Graph Neural Networks trained on a graph classification task. We want to observe how changes in the training process, such as the initialization of node feature vectors or the choice of the training dataset, would impact the behavior of the resulting metrics and gain insights into optimizing the training process to get more expressive metrics.

Table 3. Graph manifolds metrics of Gr dataset on 10 independentely trained GIN ($\boxed{F}$ull dataset), using rGIN and constant features. While the metrics were able to keep consistent results through GIN training repetition, they kept unconsistent ranking results in regards to the base hypothesis, ranking the GraphRNN-S better than the GraphRNN-RNN generated grid graphs.

With constant feature				
	Mean Rank	FD ↓	F1 PR ↑	F1 DC ↑
Re-sampled	1	$-$**8e-5** $\pm$ **5e**$-$**5**	**1.0** $\pm$ **0.0**	**0.975** $\pm$ **0.010**
GRAN	2	211 ± 29	0.973 ± 0.015	0.973 ± 0.015
GraphRNN-RNN	4	30904 ± 2275	0.007 ± 0.010	0.052 ± 0.0001
GraphRNN-S	3	26335 ± 2016	0.60 ± 0.027	0.056 ± 0.010
With random feature (rGIN)				
	Mean Rank	FD ↓	F1 PR ↑	F1 DC ↑
Re-sampled	1	**0.87** $\pm$ **0.82**	**0.988** $\pm$ **0.006**	**1.013** $\pm$ **0.021**
GRAN	2	61 ± 13	0.967 ± 0.013	0.989 ± 0.036
GraphRNN-RNN	4	8989 ± 1076	0.017 ± 0.006	0.058 ± 0.005
GraphRNN-S	3	8062 ± 925	0.535 ± 0.074	0.065 ± 0.021

Experiment Description. In their work, Obray and al. discussed criteria for evaluating comparison metrics, among which we find expressiveness, where the ideal comparison metric can rank graph samples effectively [16]. To evaluate expressiveness, we use two experiments: First, in the **Perturbation experiment**, we compare a set of graphs to a copy set but perturbed with interpolation between those graphs and rewired ER-graphs as in [16,27]. We expect the metrics to react monotonically to those perturbations, keeping the similarity rank of each perturbed graph set with the reference set. We also look for metric stability if those metrics keep the low variance regarding experiment repetition, which includes pre-training and metrics computation. Second, in the **Grid experiment**, we compare Gr graphs from the real data distribution with ones sampled from GRAN and GraphRNN, which gives a closer outlook on using those metrics in an actual research workflow. We expect the metrics based on graph embeddings to rank these graph sets consistently with our visual inspection.

Experiment Parameters. In both experiments, we varied training training hyperparameters of the graph classifier: *1st/*Node feature vector initialization for pretraining and evaluation. We compare three feature initializations: One-hot degree encoding, Constant node encoding, and GIN-random [17], where each node feature is generated by sampling a random integer each time the model is called during training and inference time. *2nd/*The graph dataset used for pre-training. We concatenated the graph data to work with three graph classification datasets: (1) The **base dataset** $\boxed{B}$ consists of the graph datasets BA, WS, C2L, C2S and Gr. (2) The **ladder dataset** $\boxed{L}$ takes the base dataset and add Ld graphs set. (3) The **full dataset** $\boxed{F}$ use the ladder dataset and add ER graph sets for each already existing graph set. *3rd/*The GIN graph embedding function, whether summation or concatenation.

Table 4. Ablation study on Node Ordering of GRAN W and GraphRNN ♣ on BA and WS graphs, using reported MMD values. Lower values yields better generation quality. Overall GRAN gets lower MMD metrics vazlues on both WS and BA datasets, while the choice of node ordering clearly influence results. On GraphRNN, DFS ordering give comparable results to the vanilla BFS orderings.

		WS				BA			
		Deg	Clust	4-orbits	Spec	Deg	Clust	4-orbits	Spec
Re-sampled		**2.718e−6**	**0.0013**	**9.780e−6**	**0.0003**	3.822e−5	0.0035	0.0047	0.0001
E-R		0.3412	1.3495	0.1904	0.1436	0.1074	0.6967	0.6889	0.0564
GRAN	BFS	0.0038	0.4425	0.0468	0.0282	**0.0199**	**0.0534**	**0.0360**	0.0023
	DFS	0.0054	0.1818	0.0043	0.0035	0.0339	0.0751	0.0436	**0.0012**
	Degree Descent	0.0180	0.5491	0.0511	0.0090	0.0706	0.1045	0.0553	0.0015
	Kcore	0.0208	0.4999	0.0232	0.0091	0.0587	0.0992	0.0379	0.0018
	Default	**0.0001**	**0.0752**	**0.0046**	**0.0023**	0.1017	0.0967	0.0367	0.0031
GraphRNN	BFS	0.0986	**0.6891**	0.1641	0.0250	0.1083	**0.3152**	0.1525	0.0184
	BFS-Max	0.0873	0.7388	0.2242	0.0277	0.1280	0.3378	**0.1377**	0.0150
	DFS	**0.0780**	0.7094	**0.1508**	**0.0203**	**0.0342**	0.3234	0.1690	**0.0090**
	Uniform	0.1801	1.1016	1.7263	0.0711	0.0937	0.3775	0.1531	0.0105

Experimental Setup. We follow [4] by using GIN [26] as our primary graph classifier. The official implementation of GIN by the authors of the paper is available on github , using the GIN-0 version. We use the recommended hyperparameters, i.e., five layers of GIN with two layers of MLP for each, a hidden size of 64, and summation as the graph- and node-pooling method. For the pre-training of the models, we set the number of epochs to 64, and we used the Adam optimizer with a learning rate of 0.01. Then, we train our model using an 80-20 train-test split on all our datasets. We use the PRDC library by [15] and the Frechet distance implementation from pytorch-fid [18]. For the perturbation experiment, we use the absolute Spearman rank correlation coefficient (see [29]) to assess the ranking ability of the given metric, with good ranking ability evaluated with an absolute value close to one and lousy ranking ability evaluated close to zero.

Grid Experiment. In Table 1, we first observe that all metrics, computed using the degree node feature initialization, are consistently ranking the resampled grid graphs first when using the node, followed by GRAN generated grid graphs and the grid graph generated by both GraphRNN variants, which is consistent with the base assumption (see Fig. 2). We also observe how The F1-Score metrics (PR and DC) show way less variance than the FD through independent runs. FD variance between independent runs is affected by the training dataset configuration, with the full dataset heavily reducing the variance for the resampled and GRAN-generated grid graphs. F1-Score is less affected by the training dataset configuration. The choice in the embedding function inside GIN (summation x_Σ or concatenation $x_\parallel$) does not change the behavior of all used metrics. In Table 3, we see that using other node feature initialization techniques, such as constant and random node features, gives inconsistent metrics results, with all tested graph embeddings-based metrics, in terms of ranking in the grid experiment.

Table 5. Ablation study on Node Ordering of GRAN and GraphRNN on C2L and C2S graphs, using reported MMD values. Lower values yields better generation quality. While on C3L graphs GRAN gets lower MMD values, GraphRNN gets similar results except for the Uniform ordering. On C2S graphs both model also gave similar results, with a slight advantage for GRAN. GraphRNN-DFS gave comparable results to the other GraphRNN node ordering variants.

		C2L				C2S			
		Deg	Clust	4-orbits	Spec	Deg	Clust	4-orbits	Spec
Re-sampled		**0.0004**	**0.0036**	**0.0039**	**0.0001**	**0.0005**	**0.0037**	**0.0006**	**0.0011**
E-R		0.1481	0.2562	0.1042	0.0452	0.1715	0.1603	0.2597	0.0730
GRAN	BFS	0.0052	0.0189	0.0070	0.0017	**0.0009**	0.0441	**0.0074**	0.0223
	DFS	**0.0012**	0.0190	0.0057	**0.0007**	0.0022	**0.0374**	0.0087	**0.0108**
	Degree Descent	0.0262	0.0236	0.0073	0.0029	0.0038	0.0417	0.0138	0.0215
	Kcore	0.0082	0.0215	0.0070	0.0011	0.0038	0.0483	0.0405	0.0263
	Default	0.0042	**0.0115**	**0.0054**	0.0010	0.0080	0.0473	0.0801	0.0248
GraphRNN	BFS	0.0136	0.0400	**0.0068**	0.0092	0.0085	0.0486	0.0088	0.0411
	BFS-Max	**0.0075**	**0.0346**	0.0075	**0.0050**	0.0041	**0.0472**	0.0103	0.0400
	DFS	0.0171	0.0524	0.0083	0.0068	**0.0027**	**0.0472**	**0.0051**	**0.0196**
	Uniform	0.1594	0.2463	0.0103	0.0079	0.0431	0.0503	1.193	0.0478

Key Takeaways. Like kernel-based measurement techniques, graph-embedding metrics behavior is highly sensitive to underlying hyperparameters [16]. More precisely, graph embeddings-based metrics that rely on trained graph classifiers are highly sensitive to more factors related to the training process, leading to reduced reproducibility and limited applicability of these metrics. Manifold-based metrics in graph embedding space can alleviate the sensitivity issue, making the selection of such metrics more interesting. Graph embeddings-based metrics use is more exciting in benchmarking graph generators compared to kernel-based metrics as they can capture global information of the graph. For instance, in the last section, two well-known graph generators were also compared using graph embeddings-based metrics.

4.2 Comparing GRAN and GraphRNN

We evaluate two state-of-the-art graph generation methods using the kernel-based MMD on distributions of graph invariants and previously explored metrics in graph embedding space. For both models, we used the official github source code at JiaxuanYou/graph-generation and lrjconan/GRAN with recommended hyperparameters.

Experiment Parameters. In this experiment comparing GRAN and GraphRNN, we also varied the used node ordering policy used to reduce the complexity graph adjacency matrices space to learn from. For GRAN, we use BFS and DFS that is computed using the node with the highest degree, and other node ordering functions such as the default node ordering, degree-descent, and their denoted K-core ordering [14]. For GraphRNN, we use the vanilla BFS version, the uniform

Table 6. Graph embedding metrics for re-sampled graphs of a graph type are denoted in the first row with V. Reported pretrained Graph manifolds-based metrics on GRAN (W) and GraphRNN (♣) for BA and WS dataset. Those graph manifolds metrics give similar observation with the evaluation using MMD-based metrics.

		BA			WS			C2S			C2L		
		FD ↓	F1 PR ↑	F1 DC ↑	FD ↓	F1 PR ↑	F1 DC ↑	FD ↓	F1 PR ↑	F1 DC ↑	FD ↓	F1 PR ↑	F1 DC ↑
V		13.89	0.9540	0.9567	4.450	0.9799	1.001	6.80	0.424	0.375	1487	0.9799	1.011
W	BFS	**2399**	**0.2132**	**0.0771**	1332	0.4297	0.1344	16.29	**0.8207**	0.3466	47782	0.8881	0.6984
	DFS	3323	0.1431	0.0315	1708	0.4546	0.2341	26.59	0.7951	**0.3557**	21005	0.9201	0.7523
	DegDes	17275	0.0366	0.0065	2189	0.2535	0.0646	31.41	0.7335	0.2813	54148	0.8300	0.5670
	Kcore	7057	0.0514	0.0101	1843	0.2535	0.0771	25.18	0.7415	0.2764	22293	0.7822	0.5559
	Default	14364	0.0068	0.0011	**308**	**0.6964**	**0.4212**	74.07	0.6491	0.2679	45217	0.8819	0.7630
♣	BFS	20077	0.0	0.0	44503	0.0	0.0	44.69	0.2208	0.0549	12751	0.9588	0.7883
	BFS-Max	30684	0.0	0.0	37641	0.0	0.0	61.37	0.2305	0.0552	**5312**	**0.9678**	**0.8590**
	DFS	9762	0.0	0.0	22886	0.0	0.0	**13.94**	0.8060	0.3336	12946	0.9242	0.7411
	Uniform	28248	0.0	0.0	72402	0.0	0.0	182.1	0.5762	0.1847	794711	0.1463	0.0976

random ordering, our introduced DFS ordering, and a BFS variant where we do not reduce the graph matrix representation.

Experimental Setup. For GraphRNN, we use 4-layer GRU cells with a 128-dimensional hidden state to encode the graph information and a 16-dimensional hidden state to encode the edge information. The output of the edge-level RNN is mapped from 16 to 8 dimensions using an MLP, then passed through a ReLU to be mapped to a scalar to pass through a sigmoid activation function finally. The model is trained on each dataset in a minibatch of size 32 using the Adam optimizer for 3000 epochs each, with a learning rate of 0.001 reduced by 0.3 at the 400th and 1000th epoch. For GRAN, we keep the same hyperparameters for the model, using GNN with 7 layers and the block size and stride fixed at 1 because this will assure the highest generation quality possible. The model also uses an Adam optimizer for the training phase and is trained with a learning rate of 0.0001 and no decay. We train both models on the **BA, WS, C2L**, and **C2S** graph datasets using several node ordering, using 80-20 train-test splits.

We compared GRAN and GraphRNN evaluation results and also added two reference measurements for each dataset in order to give a sense of "scale", as recommended by O'Bray et al. [16]. First, the evaluation compares the reference set of graphs with a resampled set of graphs from the same generator denoted with V or "Resampled", a very similar set of graphs. Furthermore, we use the **ER** graph model as a benchmarking reference [14], computing metrics for a random graph generative model compared to the test graph sets. We resample an instance of BA, WS, C2L, and C2S graph datasets using their base generation function, which will be compared to the generated graph datasets. We train GIN on the full dataset configuration with one-hot node degree feature initialization and concatenation graph embedding function for Graph-Embeddings-based measurements. We then use the BA, WS, C2L, and C2S sets from the training dataset, which will be compared to generated graph sets.

Results. Measurements based on kernel computation and those from GNN training give consistent comparisons between GraphRNN and GRAN (compare 5 and 6). Through our experiments and measurements, we find that GRAN generates graph lists that are more faithful to the base dataset, especially on medium-sized datasets like BA or WS in Table 4. However, GraphRNN shines when trained on small graph datasets like on C2S (compare 5). We found discrepancies in generation quality across several tested node orderings for both GraphRNN and especially GRAN. Our proposed DFS ordering for GraphRNN offers good performance, competing with the BFS ordering on our datasets, and even offers better performances on small datasets.

5   Conclusion and Future Work

This work compares GraphRNN and GRAN on different node orderings. We notice cases where GRAN wins, especially on medium-sized graphs, while GraphRNN has an advantage on smaller graphs. The node ordering influences these models' performance, especially the generated graphs' quality. Through extensive experiments, we have traced several insights into using GNNs to evaluate graph generators in expressiveness and stability. The first insight is with changing pre-training datasets, where we show that augmenting pre-training datasets with Erdős-Rényi-graphs can improve the performance of the GNN for graph generator evaluation. Also, we recommend using one-hot degree vectors to initialize the features of the nodes, which have shown the best results in terms of expressiveness and stability of the metrics. Finally, we find that using manifold-based metrics also reduces the influence of other hyperparameters on the evaluation expressiveness and stability, in line with an observation made by [23], thus giving our recommendation to favor these metrics as reference metrics for future studies of GGMs. For future work, this work raises the interest in looking at other node orderings for using GraphRNN on small graphs. Elaborate feature engineering approaches for graph nodes could improve the behavior of graph-embedding-based metrics.

References

1. Abboud, R., Ceylan, İ.İ., Grohe, M., Lukasiewicz, T.: The surprising power of graph neural networks with random node initialization (2021)
2. Barabasi, A.L., Albert, R.: Emergence of scaling in random networks. Science **286**(5439), 509–512 (1999)
3. Bojchevski, A., Shchur, O., Zügner, D., Günnemann, S.: NetGAN: generating graphs via random walks (2018)
4. Chia-Cheng Liu, H.C., Luk, K.: Auto-regressive graph generation modeling with improved evaluation methods (2019)
5. Cui, H., Lu, Z., Li, P., Yang, C.: On positional and structural node features for graph neural networks on non-attributed graphs (2021)
6. Du, Y., et al.: GraphGT: machine learning datasets for graph generation and transformation. In: NeurIPS 2021 (2021)

7. Errica, F., Podda, M., Bacciu, D., Micheli, A.: A fair comparison of graph neural networks for graph classification (2020)
8. Goyal, N., Jain, H.V., Ranu, S.: GraphGen: a scalable approach to domain-agnostic labeled graph generation. In: Proceedings of The Web Conference 2020, pp. 1253–1263 (2020)
9. Hagberg, A., Swart, P., S Chult, D.: Exploring network structure, dynamics, and function using networkx. Technical report, Los Alamos National Lab. (LANL), Los Alamos, NM (United States) (2008)
10. Heusel, M., Ramsauer, H., Unterthiner, T., Nessler, B., Hochreiter, S.: GANs trained by a two time-scale update rule converge to a local nash equilibrium (2018)
11. Kipf, T.N., Welling, M.: Variational graph auto-encoders (2016)
12. Kynkäänniemi, T., Karras, T., Laine, S., Lehtinen, J., Aila, T.: Improved precision and recall metric for assessing generative models (2019)
13. Li, Y., Vinyals, O., Dyer, C., Pascanu, R., Battaglia, P.: Learning deep generative models of graphs (2018)
14. Liao, R., et al.: Efficient graph generation with graph recurrent attention networks. CoRR abs/1910.00760 (2019). http://arxiv.org/abs/1910.00760
15. Naeem, M.F., Oh, S.J., Uh, Y., Choi, Y., Yoo, J.: Reliable fidelity and diversity metrics for generative models (2020)
16. O'Bray, L., Horn, M., Rieck, B., Borgwardt, K.: Evaluation metrics for graph generative models: Problems, pitfalls, and practical solutions (2021)
17. Sato, R., Yamada, M., Kashima, H.: Random features strengthen graph neural networks (2021)
18. Seitzer, M.: pytorch-fid: FID Score for PyTorch (2020). http://github.com/mseitzer/pytorch-fid. version 0.2.1
19. Simonovsky, M., Komodakis, N.: GraphVAE: towards generation of small graphs using variational autoencoders (2018)
20. Stier, J., Granitzer, M.: DeepGG: a deep graph generator. In: Abreu, P.H., Rodrigues, P.P., Fernández, A., Gama, J. (eds.) IDA 2021. LNCS, vol. 12695, pp. 313–324. Springer, Cham (2021). https://doi.org/10.1007/978-3-030-74251-5_25
21. Szegedy, C., Vanhoucke, V., Ioffe, S., Shlens, J., Wojna, Z.: Rethinking the inception architecture for computer vision. CoRR abs/1512.00567 (2015). http://arxiv.org/abs/1512.00567
22. Thompson, R., Ghalebi, E., Devries, T., Taylor, G.W.: Building LEGO using deep generative models of graphs. ArXiv abs/2012.11543 (2020)
23. Thompson, R., Knyazev, B., Ghalebi, E., Kim, J., Taylor, G.W.: On evaluation metrics for graph generative models. In: International Conference on Learning Representations (2022). http://openreview.net/forum?id=EnwCZixjSh
24. Watts, D.J., Strogatz, S.H.: Collective dynamics of 'small-world' networks. Nature 393(6684), 440–442 (1998)
25. Weisfeiler, B., Lehman, A.: A reduction of a graph to a canonical form and an algebra arising during this reduction. Nauchno-Tech. Inform. 2(9), 12–16 (1968)
26. Xu, K., Hu, W., Leskovec, J., Jegelka, S.: How powerful are graph neural networks? (2019)
27. You, J., Ying, R., Ren, X., Hamilton, W.L., Leskovec, J.: GraphRNN: a deep generative model for graphs. CoRR abs/1802.08773 (2018). http://arxiv.org/abs/1802.08773
28. Zeng, Z., Tung, A.K., Wang, J., Feng, J., Zhou, L.: Comparing stars: on approximating graph edit distance. Proc. VLDB Endow. 2(1), 25–36 (2009)
29. Zwillinger, D., Kokoska, S.: CRC Standard Probability and Statistics Tables and Formulae Sect.14.7. Chapman & Hall/CRC, Boca Raton (2000)

Can Complexity Measures and Instance Hardness Measures Reflect the Actual Complexity of Microarray Data?

Omaimah Al Hosni and Andrew Starkey(✉) iD

School of Engineering, University of Aberdeen, Aberdeen, Scotland, UK
{o.alhosni.19,a.starkey}@abdn.ac.uk

Abstract. Despite the significant contribution of the research community in the context of the Microarray data analysis, little attention has been made in understanding the Microarray dataset characteristics using Complexity Measures and Instance Hardness Measures; thus, this study aims to examine the performance of both datasets with Microarray properties. The study assumes that since these measures are data dependent, they might also be negatively affected by complex data characteristics -like the classification algorithm- and provide values that do not reflect the actual data complexity. To investigate this, we have adopted a different experiment strategy than other works undertaken in this context by using a controlled environment with synthetic data that match Microarray properties to assess the effect of each data challenge individually without relying on the classification algorithm performance. The study argues that the experiment strategy adopted by others in correlating the classification algorithm performance to the performance of the measures is not a good independent indicator for validating the measures performance in estimating the actual data difficulty nor for showing the causes of the poor prediction of the learning algorithm's performance as both are data dependant. The experiment outcomes indicate that among 35 measures covered in this study the measures responded differently to each data challenge due to the different assumptions they adopted and their sensitivity to the different data challenges. Thus, the study has confirmed that complex data characteristics result in the measures not reflecting the actual data complexity.

Keywords: Complexity Measures · Instance Hardness Measures · Small Sample size · High Dimensionality · Imbalanced Classes

1 Introduction

The information in the Microarray dataset is collected from tissue and cell samples that contain many genes. According to the difference in gene expression, the diseases are diagnosed, or specific tumours are distinguished [1]. Due to the complexity of Microarray datasets characteristics, and to the severity of the consequence in misdiagnosing a particular disease through "false positives" or "false negatives", a new line of research has formed in the research community that combines both bioinformatics and machine learning researchers. The complexity of the Microarray data analysis task comes from the fact that this type of dataset is mainly associated with a small sample size (often less than 100 samples) but with a high dimensionality (larger than 1000 features) [2], which limits the generalisation ability of the classifier. However, the situation worsens with the existence of other data properties such as class imbalance, non-linearity, overlapping between classes and data sparsity which add more complexity to the classification task of the Microarray dataset [3]. Having such complex data characteristics will create a high likelihood of misdiagnosing a particular disease through false positives (incorrectly diagnosing existence of disease) or false negatives (incorrectly diagnosing absence of disease). In such cases, knowing the causes of poor prediction accuracy - such as those associated with the Microarray dataset - is a nontrivial task due to the interactive effects of different data challenges. Thus, several.

tools have been proposed in data analysis to quantify the dataset's complexity level; however, the most widely used in the research community are Complexity Measures proposed by [4]. These measures are considered as a pre-processing data task used to assess the difficulty of the geometrical descriptions of the given dataset and have a variety of uses for pre-processing data tasks, such as noise identification [5], extracting the geometrical descriptions of the shape of the decision boundaries, data separability, and data distribution. [6, 7]. Therefore, these measures can play an essential role in identifying the specific data characteristic(s) that significantly impacts the classification performance. This is useful when identifying what machine learning or statistical model will be appropriate for the classification problem in hand. Therefore, it can be an essential step in identifying the specific data characteristic that significantly impacts the classification performance.

Despite the significant contribution of the research community in the context of the Microarray data analysis, still, the performance of Complexity Measures to the type of data problems found in Microarray – and other types of problems – has not been properly examined. Most of the works undertaken in this context are limited to examining the correlation between the values of the measures with the learning algorithms' prediction accuracy performance. Indeed, the above comments are true for other types of real-world problem beyond Microarray datasets. Another problem is that the relevant feature(s) in those datasets is also unknown, making it challenging to identify the actual effect that degrades the classification accuracy, especially with the interactive effect of different data challenges. The motivation behind this work is therefore evaluate the performance of Complexity measures to different data problems commonly found with Microarray datasets. The contribution of this paper can therefore be summarised as follows:

- To the best of our knowledge, this is the first study that examines the performance of Instance Hardness Measures in the context of the Microarray dataset or other similar problems
- Most of the works undertaken in context of Complexity Measures are limited to examining the correlation between the values of the measures with the learning algorithms' prediction accuracy performance. However, since both are mainly data dependent [8, 9], our study argues that examining the correlation between them is not a good independent indicator to validate the Complexity Measure performance in estimating the actual data difficulty nor for showing the causes of the poor prediction of the

learning algorithm's performance, since these measures might also be affected by the same complex data characteristic that affects the classification performance. This work therefore aims to highlight this gap by investigating the performance of these measures using a controlled environment through creating synthetic data that match Microarray dataset properties. The purpose of using synthetic data in this study is to enables the assessment of the effect of each data challenge individually without relying on the classification algorithm performance. This will help us to understand in depth which Complexity measures are affected or unaffected by these data challenges, and therefore are most suitable for use with Microarray datasets, or any other datasets that possess similar data challenges.

- A further motivation for the study is that most of the works conducted in this context are restricted to real-world datasets, where the properties of such datasets are primarily unknown, limiting the understanding of the impact of different data challenges in the classification algorithm and/or Complexity measure. Another problem is that the relevant feature(s) in those datasets is also unknown, making it challenging to identify the actual effect that degrades the classification accuracy, especially with the interactive effect of different data challenges. Therefore, the use of the synthetic data will overcome these limitations.

In general, following the above experiment strategy will help to precisely identify the best-performing measure(s) that suit Microarray or similar data analysis tasks, which will provide meaningful insights for the practitioners and researchers in exploring the domain competence of different machine learning algorithms, which can help in hyperparameters optimisation, reducing the data complexity and eventually improving the classification performance. In addition, it will guide practitioners and researchers in choosing the correct measures that are more appropriate for a particular dataset.

The paper is organized into five sections; Sect. 2 represents work that has applied these measures in the context of Microarray data analysis. Section 3 discuss the study methodology explaining the experimental setup designed for this work and the experimental results are shown and discussed in Sect. 4. Finally, the conclusion and future works will be presented in Sect. 5.

2 Related Work

Complexity Measures (global/data-level) were proposed by [4] and have been widely used for pre-processing data tasks, such as noise identification [5], extracting the geometrical descriptions of the shape of the decision boundaries, data separability, and data distribution [6, 7]. Further enhancement has been added to these measures by [5] to understand the difficulty of the imbalanced classification by considering each class individually, including the minority class, which was neglected by the original Complexity Measure as per their study.

Instance Hardness Measures (local/instance-level) have been proposed by [10] to overcome the limitations of the Complexity Measures (global-level measures). The authors argue that the original data Complexity Measures focus on characterising the overall complexity of the entire dataset and fail to provide information at the local/instances level, and that obtaining such information is essential to identify the misclassified instances and to understand why they have been misclassified. Based on the literature, Instance Hardness Measures have received less attention in the Microarray dataset context than the global level (Complexity Measure), which might be due to the latter's seniority.

However, extended with the same study motivation of [10, 11] have decomposed some of the Complexity Measures (global level) into sample/local-level measures, so the analysis is conducted based on the individual contribution of each instance instead of global complexity of the entire dataset. However, in their study, [11] compared the performance of the proposed decomposed sample/local-level measures against the global one. They concluded that the former provided better performance than the latter.

In the context of the Microarray data analysis, little attention has been given to understand in depth the Microarray dataset characteristics using Complexity Measures. However, a recent study conducted by [1] proposed a two-stage framework for binary classification that aims to use Complexity Measures (global level) to tackle the curse of dimensionality in microarray datasets. In the first stage, data complexity measures are used to reduce the feature space search in order to create less complex data. Then, in the second stage, Complexity Measures are used to produce the static belief space in an evolutionary cultural algorithm. According to their experimental results, the proposed method has effectively reduced the features while increasing the classification performance. Another study by [2] reviewed the most frequently used microarray databases in the literature while listing the problematic data characteristics in the Microarray domain. This work demonstrated the importance of using the Complexity Measures (global level) to enhance the classification performance of Microarray datasets. [3] used Complexity Measures to analyse the intrinsic complexity of several microarray datasets. Their study investigated the measure performance under two conditions: datasets with only a training set and datasets divided initially into training and test sets. Then, they explored the connection between the complexity measures and the actual error rates, proving that there is a correlation between them. Another study by [4] analysed the intrinsic complexity of several microarray datasets with and without feature selection and then explored the connection with the classification performance in both conditions. Their experimental results indicated a correlation between microarray data complexity and the classification error rates. [12] investigated the suitability of Complexity measures in deciding the optimal number of relevant features over 27 publicly available Microarray datasets. Their study findings indicated that the feature selection method's classification performance is highly correlated with data complexity measures, specifically those that estimate the complexity based on neighbourhood learning.

In summary, it can clearly be seen that most of the works undertaken in the literature are limited to examining the correlation between the values of the measures with the learning algorithms' prediction accuracy performance using only real-world datasets. However, using real-world datasets limits the analysis from validating the actual effect of different data challenges in the classification performance since many data properties are unknown. In addition, examining the correlation between them is not a good independent indicator to validate the complexity measure performance in estimating the actual data difficulty nor for showing the causes of the poor prediction of the learning algorithm's performance since the complexity measures might also be affected by the complex data characteristic. Thus, this work aims to highlight this gap by investigating the performance of these Measures using a controlled environment that enables us to assess the effect of each data challenge individually.

In addition, this study has found that most of the works conducted in Microarray datasets have only investigated the performance at the global level measures (Complexity Measures). To the best of our knowledge, the local level measures (Instance Hardness Measure) still have not been covered in this context of Microarray datasets which might be due to the former's seniority. Thus, a comparative study of the global/data-level (Complexity Measures) against local/instance-level (Instance Hardness Measures) is undertaken in order to assess the actual difficulty of the complex data characteristics using these approaches.

In conclusion, although previous studies show how these measures can lead to an improved classification performance, they do not give an understanding as to when the measures perform well or what data characteristics in particular affect their performance. This is a significant issue when the data characteristics for new real-world data is unknown – i.e. measures that are affected in their performance by a data characteristic will give an unrealistic estimation of the complexity of real-world data. This study aims to fill this gap in knowledge to give a comprehensive understanding of the measures' performance under different data characteristics controlled using synthetic datasets.

3 Methodology

This section describes the measures being studied and also the experimental strategy designed to cover the common data issues that Microarray datasets suffer from, namely: small sample size, high dimensionality, and class imbalance. As noted, and unlike other studies in the literature, the classification performance analysis will be skipped in this study since it is assumed that examining the correlation between the measure performance and the prediction accuracy is not necessarily a good indicator of the measure's ability to characterise the data challenges. In addition, since the datasets used in this study are synthetically generated, and therefore the underlying data characteristics are known, the focus is to examine the effect of the scenarios described below on the performance of the measures; therefore, the classification learning algorithm's performance is not necessary.

3.1 Measures Used in This Study

This study has covered 35 measures (from both global and local perspectives) as shown in Table 1. The measures are categorised according to the problem they focus on, with a reference to the original work given. These measures are chosen as they are the most used in literature, with all measures having a range of 0 to 1 for their output.

3.2 Synthetic Data Generation and Description of Scenarios for Study

In order to investigate the performance of the measures against specific data problems, three scenarios are created to accomplish this task for the problems of small vs large sample size, small vs large feature size, and finally balanced vs unbalanced classes. The synthetic data was generated using the scikit-learn library in Python, which includes a set of data generation functions. In this study, a four multiclass dataset was generated using the make_blobs() function to generate isotropic Gaussian blobs for clustering. Noise is added to the data using a standard Gaussian probability function, which allows the control of the degree of class overlap.

Scenario 1: Small-Sample Size vs Large-Sample Size. As a common issue in the Microarray dataset is sample size limitation, synthetic datasets are generated to compare the measures' ability to differentiate between the difficulty of a small-sample size against a large-sample size. The data characteristics for this scenario are given in Table 2.

Table 1. List of measures used in this study

Measure Categories		Measures description
Overlapping of Individual Features Values	Global- Level Measures [4]	F1: Maximum fisher's discriminant ratio
		F2: Volume of overlap region
		F3: Maximum feature efficiency
		F4: Collective feature efficiency
	Local -Level Measures [11]	$F1_{HD}$: Frac. Feature values overlapping
		$F2_{HD}$: Volume of overlap region
		$F3_{HD}$: Maximum feature efficiency
		$F4_{HD}$: Collective feature efficiency
Separability of Classes	Global-Level Measures [4]	N1: Fraction of points on the class boundary
		N2: Ratio of inter/intra class nearest neighbour distance
		N3: Leave-one-out error rate of the 1NN classifier
		L1: Sum of the error distance by linear programming
		L2: Error rate of linear classifier
	Local -Level Measures [11]	$N1_{HD}$: Fraction of points on the class boundary
		$N2_{HD}$: Ratio of Inter/Intra class nearest neighbor distance
Geometry, Topology and Density of Manifolds	Global-Level Measures [4]	N4: Nonlinearity of a 1-NN classifier
		N5/T1: Fraction of hyperspheres covering data
		L3: Nonlinearity of the linear classifier

(continued)

Table 1. (*continued*)

Measure Categories		Measures description
	Local-Level Measures [13, 14]	LSC: Local set cardinality
		LSR: Local set radius
		H: Harmfulness
		U: Usefulness
Data Sparsity and Dimensionality	Global-Level Measures [4, 15, 16]	T2: Average number of features per points
		T3: Average number of PCA dimensions per points
		T4: Ratio of the PCA dimension to the original dimension
Structural Representation		Density: Average density of network
		ClsCoef: Clustering coefficient
		Hubs: Average hub score
Instance Hardness Measures	Local -Level Measures [10]	kDN: k-disagreeing neighbors
		DS: Disjunct size
		DCP: Disjunct class percentage
		TDP: Tree depth pruned
		TDU: Tree depth unpruned
		CL: Class likelihood
		CLD: Class likelihood difference

Table 2. Synthetic datasets characteristics for Scenario 1

Synthetic Datasets	Sample Size	No. of Relevant Feature	No. of Irrelevant Feature	Class Ratio
Dataset_1	1000	6	0	25:25:25:25
Dataset_2	100			

Scenario 2: Small-Feature Size vs Large-Feature Size. Another issue that harms the Microarray's prediction accuracy is the data sparsity caused by high dimensionality. Therefore, synthetic datasets are generated to compare the measures' ability to differentiate the complexity of a small-feature size against a large-feature size. The data characteristics for this scenario are given in Table 3. It is worth noting that Dataset_3 in Table 3 has the same relationship as Dataset_2, but has 994 irrelevant features added.

Scenario 3: Balanced vs Imbalanced. The imbalanced class occurs in the Microarray dataset because healthy patients (majority class) are typically more than the unhealthy patients (minority class). Therefore, this scenario examines the measure's ability to estimate the complexity of the imbalanced issue by comparing the measure's performance in the balanced against imbalanced classes. Note that the overall sample size for Dataset 4 has reduced due to the reduction in size of Classes 2, 3 and 4 as shown in Table 4.

Table 3. Synthetic datasets characteristics for Scenario 2

Synthetic Datasets	Sample Size	No. of Relevant Feature	No. of Irrelevant Feature	Class Ratio
Dataset_2	100	6	0	25:25:25:25
Dataset_3	100	6	994	

Table 4. Synthetic datasets characteristics for Scenario 3

Synthetic Datasets	Sample Size	No. of Relevant Features	No. of Irrelevant Features	Class Ratio
Dataset_2	100	6	0	25:25:25:25
Dataset_4	60			25:20:10:5

4 Results and Discussion

The experimental results will be discussed for each scenario in the subsections below.

4.1 Experiment Results of Scenario 1: Small-Sample vs Large-Sample Size

As a small-sample size is more complex than the large-sample size due to the limited representation, unstable performance, increased bias, and higher risk of overfitting [17–19], the values of the measures should increase in the small-sample size dataset (i.e. Dataset_2). The study results will be presented in the below subsections in line with the measure categories given in Table 1.

Measures of Overlapping of Individual Feature Values. These measures have been proposed to describe the data complexity caused by class overlap by evaluating the features' discriminative power in estimating each feature's individual efficiency in separating the classes [20]. The experiment results indicate that they could not capture the complexity of a small-sample size since most of the measures at both global and local perspectives produced lower values than the large-sample size dataset. According to [20], these measures are influenced by the number of features, the noise, and the sample size, which this study's results have also proven.

Measures of Separability of Classes. The measures in this category are proposed to quantify the complexity according to the class separability by analysing the distance between the examples [16]. All the Separability of Classes Measures at both global and local level can distinguish the complexity of a small-sample size dataset, returning higher values in the small-sample size data as shown in Table 5.

Measures of Geometry, Topology and Density of Manifolds. The measures in this category capture the geometry of the manifolds covering each class by extracting the information from the geometry (local) and topology (global) structure of the data to measure the class separability [4]. In this scenario, N4 has produced identical values in both datasets, whereas N5/T1 and L3 have shown better performance distinguishing the complexity of small-sample size dataset by producing higher values in this dataset. The local measures levels have produced lower values in the small-sample size dataset, demonstrating that they cannot estimate the complexity of a small-sample size, except for the H measure whose value slightly increased. However, [13] have stated in their study that these local-level measures are seriously affected by noise and class overlapping, which this study results have also proven since Class 1, Class 3, and Class 4 in the examined dataset are overlapped.

Table 5. Experiment Results of the Scenario 1

Measure Categories		Measures	Dataset_1	Dataset_2
Overlapping of Individual Features Values	Global- Level	F1	0.53	0.54
		F2	0.57	0.46
		F3	0.83	0.67
		F4	0.83	0.67
	Local -Level	F1_HD	0.81	0.6
		F2_HD	0.3	0.24
		F3_HD	0.39	0.34
		F4_HD	0.45	0.43
Separability of Classes	Global-Level	N1	0.23	0.32
		N2	0.39	0.42
		N3	0.15	0.2
		L1	0.13	0.15
		L2	0.23	0.3
	Local -Level	N1_HD	0.15	0.22
		N2_HD	0.39	0.42
Geometry, Topology and Density of Manifolds	Global-Level	N4	0.03	0.03
		N5/T1	0.45	0.63

(continued)

Table 5. (*continued*)

Measure Categories		Measures	Dataset_1	Dataset_2
		L3	0.22	0.3
	Local-Level	LSC	0.92	0.79
		LSR	0.7	0.54
		H	0.00	0.04
		U	0.92	0.80
Data Sparsity& Dimensionality	Global-Level	T2	0.00	0.06
		T3	0.00	0.06
		T4	1	1
Structural Representation	Global-Level	Density	0.87	0.86
		CIsCoef	0.33	0.30
		Hubs	0.81	0.80
Instance Hardness Measures	Local -Level	kDN	0.17	0.29
		DS	0.60	0.32
		DCP	0.13	0.18
		TDP	0.34	0.16
		TDU	0.49	0.45
		CL	0.13	0.19
		CLD	0.12	0.16

Measures of Data Sparsity and Dimensionality. These measures have been proposed to measure the data sparsity caused by the high dimensionality, which leads to difficulty extracting meaningful information because of the low-density areas imposed by data sparsity. The results reveal that T2 and T3 show similar performance giving a slight increase in the small-sample size. On the other hand, T4 has produced identical values at both small and large-sample size datasets incorrectly representing an extremely hard-set complexity.

Measures of Structural Representation. These measures have been proposed by [16] to characterise the data complexity according to the structural representation of the data set using graphs. Table 5 indicates that the measure values incorrectly decrease in the small-sample size dataset.

Instance Hardness Measures. As mentioned earlier, Instance Hardness Measures were proposed by [10] to overcome the limitation of the global Complexity Measures in which they estimate the data complexity at the sample level by identifying which samples are frequently misclassified in a dataset using various learning algorithms. As shown in Table 5, kDN, DCP, CL, and CLD have correctly shown an increase in complexity. In contrast, DS, TDP and TDU have produced lower values in the small-sample size, indicating that they cannot capture the complexity of the small-sample size.

Table 6. The Experiment Results of the Scenario 2

Measure Categories		Measures	Dataset_2	Dataset_3
Overlapping of Individual Features Values	Global- Level	F1	0.54	0.96
		F2	0.46	0
		F3	0.67	0.59
		F4	0.67	0.59
	Local -Level	F1_HD	0.6	0.88
		F2_HD	0.24	0.22
		F3_HD	0.34	0.40
		F4_HD	0.43	0.49
Separability of Classes	Global-Level	N1	0.32	0.73
		N2	0.42	0.50
		N3	0.2	0.68
		L1	0.15	0
		L2	0.3	0
	Local -Level	N1_HD	0.22	0.63
		N2_HD	0.42	0.50
Geometry, Topology and Density of Manifolds	Global-Level	N4	0.03	0
		N5/T1	0.63	1
		L3	0.3	0
	Local-Level	LSC	0.79	0.97
		LSR	0.54	0.08
		H	0.04	0.04
		U	0.80	0.98
Data Sparsity& Dimensionality	Global-Level	T2	0.06	10
		T3	0.06	0.89
		T4	1	0.089
Structural Representation	Global-Level	Density	0.86	0.93
		ClsCoef	0.30	0.68
		Hubs	0.80	0.90
Instance Hardness Measures	Local -Level	kDN	0.29	0.71
		DS	0.32	0.24
		DCP	0.18	0.13
		TDP	0.16	0.30
		TDU	0.45	0.36
		CL	0.19	0.55
		CLD	0.16	0.39

Scenario 1 Summary. The results indicate that the measures that correctly demonstrated an increasing complexity value are from the Geometry category: N5/T1, L3, H; from Data Sparsity and Dimensionality category: T2 and T3; and from instance Hardness Measures: kDN, DCP, CL, and CLD. The]remaining measures did not correctly describe the change in complexity. The variation in the values generated is also of note:

ranging from 0 to 0.63 for the methods giving an increase in complexity, and since the underlying relationship is not so difficult higher values in complexity overestimate the difficulty of the problem.

4.2 Experiment Results of Scenario 2: Small-Feature vs Large-Feature Size

In Scenario 2, the measure's performance is examined by comparing their output from a small-sample size dataset with a small-feature size against a large-feature size to identify their applicability for use. In both datasets, the classes are balanced with identical sample sizes; the data characteristics of this scenario are shown in Table 3.

Measures of Overlapping of Individual Feature Values. The results in Table 6 show that F1, F1_HD, F3_HD and F4_HD have correctly shown an increase in complexity. The remaining measures failed to correctly capture the increased complexity of the datasets.

Measures of Separability of Classes. In Table 6, L1 and L2 values have incorrectly decreased in the large-feature size dataset. The reason for such performance is overfitting since these measures use Support Vector Machine (SVM) as a linear classifier to estimate the data complexity; thus, in high dimensional space, the data piling phenomenon occurs, which adversely affects the generalisation ability of SVM [21]. In contrast, the remaining measures at both global and local levels have performed well.

Measures of Geometry, Topology and Density of Manifolds. N5/T1, LSC and U have shown reasonable performance in describing the problem of this scenario. However, the remaining measures have failed to describe the problem. The reason L3 does not capture the data sparsity issue is due to the overfitting issue because of the effect of irrelevant features - L3 uses SVM to estimate the complexity, so it shares the identical drawback of L1 and L2.

Measures of Data Sparsity and Dimensionality. The results indicate that T2 and T3 have correctly described the problem as they produced higher values in a large-feature size dataset. However, T4 has shown a similar performance to Scenario 1 in the inability to describe the problem of high sparsity.

Measures of Structural Representation. Table 6 indicates that the value of the measures has correctly increased for Dataset_3.

Instance Hardness Measures. The measures DS, DCP and TDU failed to correctly identify the increased complexity, due to overfitting since they use C4.5 as a classifier to estimate the data complexity and inherit the classifier's limitation in high dimensional space, where the overfitting occurs due to the large scale of the created decision tree [22]. However, the rest of the measures correctly described the problem by producing higher values in a large-feature dataset.

Scenario 2 Summary. In this scenario, the measures' ability to capture the interactive effect of the high dimensionality with the small sample size problem was examined. The study outcomes demonstrate that the local-level measures that failed to describe the problem are F2_HD, LSR, H, DS, DCP and TDU. In contrast, the global level measures

Table 7. The Experiment Results of the Scenario 3

Measure Categories		Measures	Dataset_5	Dataset_6
Overlapping of Individual Features Values	Global-Level	F1	0.54	0.58
		F2	1/0.2/0.09/0.5 Average=0.46	0.73/0.18/0.04/0.87 Average= 0.46
		F3	1/0.6/0.2/0.8 Average=0.67	0.92/0.65/0.2/0.8 Average=0.64
		F4	1/0.6/0.2/0.8 Average=0.67	0.92/0.65/0.2/0.8 Average= 0.64
	Local - Level	F1_HD	0.6	0.51
		F2_HD	0.24	0.23
		F3_HD	0.34	0.33
		F4_HD	0.43	0.42
Separability of Classes	Global-Level	N1	0.52/0.08/0.52/0.16 Average=0.32	0.52/0.35/0.3/1 Average= 0.54
		N2	0.47/0.39/0.34/0.48 Average=0.42	0.43/0.42/0.39/0.59 Average= 0.46
		N3	0.28/0.12/0.04/0.36 Average= 0.2	0.2/0.25/0.1/1 Average= 0.38
		L1	0.42/0.07/0/0.13 Average= 0.15	0.14/0.029/1/0.31 Average= 0.12
		L2	0.92/0.08/0/0.2 Average=0.3	0.32/0.05/0/0.4 Average= 0.19
	Local - Level	N1_HD	0.22	0.3
		N2_HD	0.42	0.44
Geometry, Topology and Density of Manifolds	Global-Level	N4	0.08/0.01/0/0.04 Average= 0.03	0.01/0.03/0/0.18 Average= 0.05
		N5/T1	0.92/0.4/0.32/0.88 Average= 0.63	0.76/0.7/0.5/1 Average= 0.74
		L3	0.97/0.09/0/0.17 Average= 0.3	0.18/0.01/0/0.45 Average= 0.16
	Local-Level	LSC	0.79	0.79
		LSR	0.54	0.53
Data Sparsity and Dimensionality	Global-Level	H	0.04	0.09
		U	0.80	0.81
		T2	0.06	0.1
		T3	0.06	0.1
		T4	1	1
Structural Representation	Global-Level	Density	0.86	0.86
		ClsCoef	0.3	0.41
		Hubs	0.80	0.78
Instance Hardness Measures	Local - Level	kDN	0.29	0.35
		DS	0.32	0.32
		DCP	0.18	0
		TDP	0.16	0.37
		TDU	0.45	0.37
		CL	0.19	0.23
		CLD	0.16	0.18

that could not reflect this data issue are F2, F3, F4, L1, L2, L3N4 and T4. As previously discussed, the range of complexity values produced by these measures is also of note: from 0.06 to 1 which represents almost the entire range of complexity output.

4.3   Experiment Results of Scenario 3: Imbalanced vs Balanced Classes

The datasets in this scenario have the same feature size (six features) where all features are relevant; the data characteristics of this scenario are presented in Table 4, with the under-sampling technique applied to generate unbalanced classes with the imbalance ratio for the classes as follows: Class 1 (41%), Class 2 (33%) Class 3 (16%) and Class 4 (8%). Table 7 shows the outputs of this scenario; as can be seen, in some global-level measures, the values are decomposed according to the classes' complexity. Therefore, the class value is highlighted with different colours indicating the class under consideration in order to assess the measure's ability to recognise the complexity of the minority class in which the measure should assign a higher value than the majority classes (red indicates Class 1, cyan is Class 2, magenta is Class 3 and blue is Class 4).

Measures of Overlapping of Individual Feature Values. None of the measures except F1 in this category at both global and local levels recognise the interactive effect of the small sample size with imbalanced classes challenge.

Measures of Separability of Classes. The L1 and L2 measures do not identify the complexity of the imbalanced classes, but the rest of the measures in this category show good performance in estimating the complexity of the imbalanced problem by assigning a higher value to the minority classes.

Measures of Geometry, Topology and Density of Manifolds. The experiment indicates that N4, N5/T1 (global level) and H, U (local level) correctly assigned higher values in the imbalanced dataset. However, the increases in N4, H and U are small. In contrast, the remaining measures in this category failed to describe the complexity of the imbalance dataset.

Measures of Data Sparsity and Dimensionality. Table 7 indicates that T2 and T3 correctly described the imbalance problem as they assigned a higher value in the imbalance dataset.

Measures of Structural Representation. The only measure in this category that correctly describes the complexity of the imbalanced class is CIsCoef, as shown in Table 7.

Instance Hardness Measures. The measures kDN, TDP, CL and CLD correctly give increases in their value for the imbalanced datasets. In contrast, the rest of the measures in this category have produced lower values in the imbalance dataset.

Scenario 3 Summary. In this scenario, the ability of each measure to estimate the interactive effect of the small sample size with the imbalanced class issue was investigated. The results indicate that the measures which have shown good performance in describing the data complexity from the global-level measures are: F1, N1, N2, N3, N5/T1, T2 and T3, and the local level measures are: N1_HdD, N2_HD, kDN, TDP, CL and CLD.

Table 8. Summary Outcomes of the overall performance of the Measures

Measure Categories		Measures	Small Sample Size	High Dimensionality	Imbalanced Classes	The Good Performance Measures
Overlapping of Individual Features Values	Global-Level	F1	✓	✓	✓	F1
		F2	✗	✗	✗	-
		F3	✗	✗	✗	-
		F4	✗	✗	✗	-
	Local-Level	F1_HD	✗	✓	✗	-
		F2_HD	✗	✗	✗	-
		F3_HD	✗	✓	✗	-
		F4_HD	✗	✓	✓	-
Separability of Classes	Global-Level	N1	✓	✓	✓	N1
		N2	✓	✓	✓	N2
		N3	✓	✓	✗	-
		L1	✓	✗	✗	-
		L2	✓	✗	✗	-
	Local-Level	N1_HD	✓	✓	✓	N1_HD
		N2_HD	✓	✓	✓	N2_HD
Geometry, Topology and Density of Manifolds	Global-Level	N4	✗	✗	✓	-
		N5/T1	✓	✓	✓	N5/T1
		L3	✓	✗	✓	-
	Local-Level	LSC	✗	✓	✗	-
		LSR	✗	✗	✗	-
		H	✓	✗	✗	-
		U	✗	✓	✓	-
Data Sparsity and Dimensionality	Global-Level	T2	✓	✓	✓	T2
		T3	✓	✓	✓	T3
		T4	✗	✗	✗	-
Structural Representation	Global-Level	Density	✗	✓	✗	-
		ClsCoef	✗	✓	✓	-
		Hubs	✗	✓	✗	-
Instance Hardness Measures	Local-Level	kDN	✓	✓	✓	kDN
		DS	✗	✗	✗	-
		DCP	✓	✗	✗	-
		TDP	✗	✓	✓	-
		TDU	✗	✗	✗	-
		CL	✓	✓	✓	CL
		CLD	✓	✓	✓	CLD

As in other Scenarios, the range of complexity is also of note: from 0.06 to 0.81. This is also a very wide range of output values, for a data problem which does not merit an exceptionally high score.

4.4 Summary of Results

A summary of results from all three scenarios can be seen in Table 8. As can be seen, there are a number of measures that are affected by the data problems investigated in this study. There are a number of measures that appear to capture the correct relationship (i.e. increasing complexity) that is created by the synthetic datasets for each scenario, namely from the global level: F1, N1, N2, N5/T1, T2 and T3, and from the local level measures: N1_HD, N2_HD, kDN, CL and CLD. All other measures did not identify the change in data characteristics and therefore give a misleading output in terms of their measure of complexity. It should be noted that although the increasing complexity has been identified by the above identified measures, the underlying complexity has often been either under- or over-stated by the measure. The synthetic datasets created are based on relatively simply blob relationships with equal variance for each feature, and so it would not be expected that the actual complexity for these datasets is measured at a high level.

5 Conclusion

In this study, the aim is to investigate each complexity measure's performance in reflecting the complexity of the Microarray dataset by simulating the common issues of this type of dataset. Having a controlled environment provided by synthetic datasets enables the identification of which measures have any issues with the specific data problem, and which are unable to capture the increased data complexity in each scenario. The study results show that the measures have responded differently to each scenario due to the different assumptions that the measures adopt and their sensitivity to the different data challenges. Thus, for many measures it has been demonstrated that complex data characteristics have negatively affected these measures from capturing the actual difficulty, as summarised in Table 8. The table also shows the measures that exhibit good performance in reflecting the change in data complexity across all scenarios from the global level: F1, N1, N2, N5/T1, T2 and T3, and from the local level measures: N1_HD, N2_HD, kDN, CL and CLD. This indicates that only these measures should be used if the data is suspected as possessing any of the characteristics studied in this paper. However, the outcomes of the comparative analysis of the global against the local level indicated that the measure's outputs did not show a clear indication to rely on since the measures' behaviour corresponds differently in both perspectives. In addition, the range of values produced by the measures is also to be noted, demonstrating that the output of the complexity measures may not reflect the difficulty of the underlying data relationship but can be negatively affected by the dataset attributes. This implies that the Complexity measure value may under- or over-state the difficulty of the problem, which may result in a more complicated model being used than is necessary. This is problematic when Green AI considerations are to be implemented – i.e. when the computational footprint of the overall analysis and model is to be reduced, which is increasingly a priority. This work clearly demonstrates the difficulty of using complexity measures in real world datasets since the dataset characteristics are largely unknown, and even using a combination of complexity measures may lead to a confusing overall picture as to the underlying complexity of the data problem. This work identifies that improvements in these or other measures are required in order to give a consistent and reliable output which describes the difficulty of any given dataset.

References

1. Sarbazi-Azad, S., Saniee Abadeh, M. Mowlaei, M.E.: Using data complexity measures and an evolutionary cultural algorithm for gene selection in Microarray Data. Soft Comput. Lett. **3**, 100007 (2021). https://doi.org/10.1016/j.socl.2020.100007

2. Alonso-Betanzos, A. et al.: A review of microarray datasets: where to find them and specific characteristics. Methods Mol. Biol. 65–85 (2019). https://doi.org/10.1007/978-1-4939-944 2-7_4

3. Bolan-Canedo, V., Moran-Fernandez, L., Alonso-Betanzos, A.: An insight on complexity measures and classification in Microarray Data. In: 2015 International Joint Conference on Neural Networks (IJCNN) [Preprint] (205). https://doi.org/10.1109/ijcnn.2015.7280302

4. Ho, T.K., Basu, M.: Complexity measures of supervised classification problems. IEEE Trans. Pattern Anal. Mach. Intell. **24**(3), 289–300 (2002)

5. Barella, V.H., Garcia, L.P.F., de Souto, M.P., Lorena, A.C., de Carvalho, A.: Data Complexity Measures for Imbalanced Classification Tasks. In: 2018 International Joint Conference on Neural Networks (IJCNN), pp. 1–8 (2018). https://doi.org/10.1109/IJCNN.2018.8489661

6. Garcia, L.P.F., Lorena, A.C., de Souto, M.C.P., Ho, T.K.: Classifier recommendation using data complexity measures. In: 2018 24th International Conference on Pattern Recognition (ICPR), pp. 874–879 (2018). https://doi.org/10.1109/ICPR.2018.8545110

7. Shah, R., Khemani, V., Azarian, M., Pecht, M., Su, Y.: Analyzing data complexity using metafeatures for classification algorithm selection. In: 2018 Prognostics and System Health Management Conference (PHM-Chongqing), pp. 1280–1284 (2018). https://doi.org/10.1109/PHM-Chongqing.2018.00224

8. Morán-Fernández, L., Bolón-Canedo, V. Alonso-Betanzos, A.: CAN classification performance be predicted by complexity measures? A study using microarray data. Knowl. Inf. Syst. **51**(3), 1067–1090 (2016). https://doi.org/10.1007/s10115-016-1003-3

9. Al Hosni, O., Starkey, A.: Assesing the stability and selection performance of feature selection methods under different data complexity. Int. Arab J. Inf. Technol. **19**(3A) (2022). https://doi.org/10.34028/iajit/19/3a/4

10. Smith, M., Martinez, T., Giraud-Carrier, C.: An instance level analysis of data complexity. Mach. Learn. **95**(2), 225–256 (2013)

11. Arruda, J.L.M., Prudêncio, R.B.C., Lorena, A.C.: Measuring instance hardness using data complexity measures. In: Cerri, R., Prati, R.C. (eds.) BRACIS 2020. LNCS (LNAI), vol. 12320, pp. 483–497. Springer, Cham (2020). https://doi.org/10.1007/978-3-030-61380-8_33

12. Dong, N.T., Khosla, M.: Revisiting feature selection with Data Complexity (2019). https://doi.org/10.1101/754630

13. Leyva, E., Gonzalez, A., Perez, R.: A set of complexity measures designed for applying meta-learning to instance selection. IEEE Trans. Knowl. Data Eng. **27**(2), 354–367 (2015). https://doi.org/10.1109/tkde.2014.2327034

14. Leyva, E., Gonzalez, A., Perez, R.: A set of complexity measures designed for applying meta-learning to instance selection. IEEE Trans. Knowl. Data Eng. **27**(2), 354–367 (2014). https://doi.org/10.1109/tkde.2014.2327034

15. Lorena, A.C., et al.: Analysis of complexity indices for classification problems: cancer gene expression data. Neurocomputing, **75**(1), 33–42 (2012). https://doi.org/10.1016/j.neucom.2011.03.054

16. Garcia, L.P.F., de Carvalho, A.C.P.L.F., Lorena, A.C.: Effect of label noise in the complexity of classification problems. Neurocomputing, **160**, 108119 (2015). https://doi.org/10.1016/j.neucom.2014.10.085

17. Wang, Z., Zou, Q., Li, X.: Small sample size problem in machine learning. In: International Conference on Bioinformatics and Biomedical Engineering, pp. 373–381. Springer, Cham (2019)

18. García, S., Fernández, A., Luengo, J., Herrera, F.: Advanced nonparametric tests for multiple comparisons in the design of experiments in computational intelligence and data mining: experimental analysis of power. Inf. Sci. **180**(10), 2044–2064 (2010)

19. Zhang, M.L., Zhou, Z.H.: A review on multi-label learning algorithms. IEEE Trans. Knowl. Data Eng. **28**(1), 10–24 (2016)

20. Lorena, A., Garcia, L., Lehmann, J., Souto, M., Ho, T.: How complex is your classification problem? ACM Comput. Surv. **52**(5), 1–34 (2020)

21. Du, N.D., Huy, N.H., Hoai, N.X.: The impact of high dimensionality on SVM when classifying ERP data - a solution from LDA. In: Proceedings of the Sixth International Symposium on Information and Communication Technology [Preprint] (2015). https://doi.org/10.1145/2833258.2833290

22. Nasution, M.Z., Sitompul, O.S., Ramli, M.: PCA based feature reduction to improve the accuracy of Decision Tree C4.5 classification. J. Phys. Conf. Ser. **978**, 012058 (2018). https://doi.org/10.1088/1742-6596/978/1/012058

Two Steps Forward and One Behind: Rethinking Time Series Forecasting with Deep Learning

Riccardo Ughi⬤, Eugenio Lomurno(✉)⬤, and Matteo Matteucci⬤

Politecnico di Milano, Milano, Italy
riccardo.ughi@mail.polimi.it,
{eugenio.lomurno,matteo.matteucci}@polimi.it

Abstract. The Transformer is a highly successful deep learning model that has revolutionised the world of artificial neural networks, first in natural language processing and later in computer vision. This model is based on the attention mechanism and is able to capture complex semantic relationships between a variety of patterns present in the input data. Precisely because of these characteristics, the Transformer has recently been exploited for time series forecasting problems, assuming a natural adaptability to the domain of continuous numerical series. Despite the acclaimed results in the literature, some works have raised doubts about the robustness and effectiveness of this approach. In this paper, we further investigate the effectiveness of Transformer-based models applied to the domain of time series forecasting, demonstrate their limitations, and propose a set of alternative models that are better performing and significantly less complex. In particular, we empirically show how simplifying Transformer-based forecasting models almost always leads to an improvement, reaching state of the art performance. We also propose shallow models without the attention mechanism, which compete with the overall state of the art in long time series forecasting, and demonstrate their ability to accurately predict time series over extremely long windows. From a methodological perspective, we show how it is always necessary to use a simple baseline to verify the effectiveness of proposed models, and finally, we conclude the paper with a reflection on recent research paths and the opportunity to follow trends and hypes even where it may not be necessary.

Keywords: Transformer · Time Series · Forecasting · Shallow Models · SLP · Sencoder · Sinformer · Baseline · Persistence

1 Introduction

Time series forecasting has often attracted the attention of researchers in fields as diverse as bioengineering, finance, climatology, and mechanics. With the advent

This paper is supported by the FAIR (Future Artificial Intelligence Research) project, funded by the NextGenerationEU program within the PNRR-PE-AI scheme (M4C2, investment 1.3, line on Artificial Intelligence).

G. Nicosia et al. (Eds.): LOD 2023, LNCS 14505, pp. 463–478, 2024.
https://doi.org/10.1007/978-3-031-53969-5_34

of widespread data availability, it has become increasingly common to use computational models that can analyse long historical data series to identify patterns and use this information to make accurate predictions about future. Early efforts in the field of time series forecasting were based on autoregressive statistical models such as ARIMA and SARIMA. These models were particularly effective for their predictive power, especially when used in conjunction with domain specific knowledge about the signals being predicted [1]. At the same time, the scientific community began to explore the potential of artificial neural networks to further reduce estimation error via their adaptive capabilities. Recurrent neural networks have emerged as a natural choice for forecasting due to their ability to extract valuable information from both feature and time domains [9,16,21]. Another popular deep learning approach to forecasting have been temporal convolutional neural networks. These networks adapt the successful paradigm of convolutional neural networks for image analysis to extract hierarchical patterns from temporal sequences and make accurate predictions [3,10,19].

A major turning point in the world of deep learning has been the introduction of the Transformer model [17]. This model set a new benchmark for performance in a wide range of applications, including natural language processing [2,4], computer vision [6,12], and speech analysis [5,15]. The key to the Transformer success lies in its attention mechanism which uses a sophisticated representation of the input and a large amount of training data to identify complex spatial and temporal correlations. These correlations are then used during the learning process to improve prediction quality. The Transformer model has since been widely adopted and has inspired numerous variations and extensions. More recently, the Informer model has emerged as a leading alternative to traditional forecasting techniques [25]. Drawing heavily on the structure of the Transformer model, the Informer has been designed to be computationally efficient while maintaining high levels of performance. The authors believe that this model is well suited to predicting very long sequences, and, since its introduction, numerous Transformer-based techniques have been proposed and continue to set new standards for time series forecasting.

Despite the widespread adoption of these models and the interest they have generated, there are critical issues that leave room for questions and doubts. Firstly, these models are often compared with each other without a baseline or a common reference, so there is no way of knowing whether they actually work well or badly. Secondly, as Zeng *et al.* points out, there are cases where embarrassingly simple models can not only compete with, but even outperform, the Transformer-based model of the moment [22]. For these reasons, and in order to shed some light on this extremely important area of research, in this article we examine the effectiveness of the most popular time series forecasting techniques presented in the literature as state of the art over the last three years. First, we demonstrate the importance of comparing one's own new model at least against an extremely trivial baseline such as the Persistence model, show that there are models based on Transformer that perform worse than this model still being considered in the literature to be very good. We then introduce two models derived from simplifying the Transformer, called respectively Sinformer and Sincoder, which are capable of outperforming current Transformer-based models deemed as the state of the art in forecasting. Finally, we present two shallow models, the novel Sinusoidal Layered Perceptron (SLP) and a conventional Multi-Layer Perceptron (MLP), show how both are able to outperform any Transformer-based model, in some cases even by a large margin. We also analyse the behaviour of these models in the presence of extremely long forecasting windows - which would be prohibitive for Transformer-based models due to memory constraints - and demonstrate their stability and robustness as predictive techniques. We conclude our contribution with a reflection on recent research trends, which are often too focused on chasing the deep learning model of the moment, and sometimes pay little attention to promising alternatives being trapped in sort of evolutionary niches.

The rest of the paper is divided into the following sections. Section 2 contains a description of the latest time series forecasting techniques and provides a background to the models that will be compared in the experiments. Section 3 describes our proposed models in details focusing on the simplification path that led us to extremely shallow networks. Section 4 provides information on the datasets and hyperparameters used in the experiments. Section 5 describes in details the experiments conducted, comparing and commenting on the results of the proposed models against those selected from the literature. Reflections on current trends and empirically more promising alternatives are shared. Section 6 concludes the article by summarising what has been discussed and laying the foundations for future research.

2   Related Works

The Vanilla Transformer model is a sequence-to-sequence architecture consisting of an encoder and a decoder. Both the encoder and the decoder consist of multiple identical L blocks. Each encoder block consists primarily of a multi-head self-attention module and a position-wise feed-forward network (FFN). To construct a deeper model, residual connections around each module are used, followed by layer normalisation. Compared to the encoder blocks, the decoder blocks contain additional cross-attention modules inserted between the multi-head self-attention modules and the position-wise FFNs. In addition, the self-attention modules in the decoder are modified to prevent each position from attending to subsequent positions [17]. The overall Vanilla Transformer architecture is shown in Fig. 1.

The earliest efforts to apply the Transformer architecture to time series forecasting were aimed at adapting exactly the vanilla model to a continuous domain rather than a dictionary-generated embedding, and at reducing its spatial complexity to enable longer predictions. Shiyang Li *et al.* identified two major weaknesses in the canonical Transformer architecture: first, the point-wise dot product self-attention is insensitive to local context, which can make the model vulnerable to anomalies in time series; second, the spatial complexity of the canonical

Transformer grows quadratically with sequence length L, making it unfeasible to directly model long time series. To address these issues, the authors proposed a convolutional self-attention mechanism that generates queries and keys with causal convolution to better incorporate local context into the attention mechanism. They also proposed a LogSparse Transformer with a memory cost of only $O(L(\log L)^2)$, where L is the length of the sequence, which also improves prediction accuracy for time series with fine granularity and strong long-term dependencies under a constrained memory budget [11].

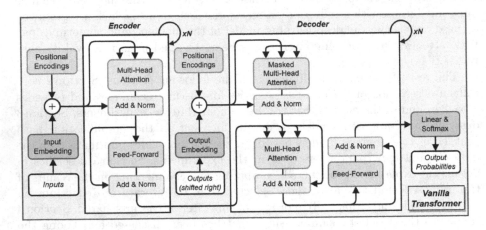

Fig. 1. The Vanilla Transformer architecture [17].

Haoyi Zhou *et al.* proposed an efficient Transformer-based model for forecasting long time series called Informer. The model has three key features: first, a ProbSparse self-attention mechanism that achieves $O(L(\log L))$ time complexity and memory usage while maintaining comparable performance on sequence dependency alignment; second, a self-attention distillation that highlights dominant attention by halving the cascading layer input and efficiently handles extremely long input sequences; and third, a generative style decoder that predicts long time series sequences in a single forward operation rather than step-by-step, significantly improving the inference speed of long sequence prediction [25]. Haixu Wu *et al.* subsequently proposed a novel decomposition-based architecture with an auto-correlation mechanism called Autoformer. This model departs from the preprocessing convention for series decomposition and instead incorporates it as a fundamental inner block of deep models, providing Autoformer with progressive decomposition capabilities for complex time series. Drawing on stochastic process theory, the authors developed an auto-correlation mechanism based on series periodicity that detects dependencies and aggregates representations at the sub-series level. This auto-correlation block outperformed the self-attention one in terms of both efficiency and accuracy [20].

Tian Zhou *et al.* recently introduced a novel method that combines the Transformer architecture with the seasonal trend decomposition technique. The decomposition method captures the global profile of the time series, while the Transformers capture more detailed structures. The resulting method, known as the Frequency Enhanced Decomposed Transformer or FEDformer, is more efficient than standard Transformers with linear complexity with respect to sequence length [26]. In recent months, Yunhao Zhang et al. published a novel method that combines the Transformer architecture with the seasonal trend decomposition method from multivariate time series forecasting. The model, called Crossformer, exploits cross-dimensional dependency and embeds the input into a 2D vector array through Dimension-Segment-Wise embedding to preserve time and dimensional information. The Two-Stage Attention layer is then used to efficiently capture both cross-time and cross-dimension dependencies. Extensive experimental results demonstrate the effectiveness of Crossformer over previous state-of-the-art methods [23].

Despite the recent tendency to reward the latest Transformer-based model, recent studies suggest that there may be alternatives to this paradigm. Zeng et al. argued that while Transformers are successful at extracting semantic correlations between elements in a long sequence, they may not be as effective at extracting temporal relations in an ordered set of continuous points. The authors claimed that permutation-invariant attention mechanisms, which are present in many of the aforementioned approaches, may result in a loss of temporal information. To validate their claim, the authors introduced a set of simple one-layer linear models into the comparison, namely Linear, NLinear, and DLinear. Experimental results on nine real-world datasets showed that these simple approaches surprisingly outperformed existing sophisticated Transformer-based models in all cases, often by a large margin. The authors hope that this surprising finding opens up new research directions for long time series forecasting, and advocate re-examining the validity of Transformer-based solutions [22].

3   Method

The purpose of this work is to advance the study of the long-term sequence prediction problem and to deepen the behavior of Transformer-based models advocating the use of much less complex alternatives. In this section, we will present in detail the models and techniques that we have developed and utilized in our experiments.

3.1   Baseline

The starting point for this work was to determine an appropriate baseline for a predictive model. While recurrent neural networks or Transformer-based models have been widely used in the literature and were previously considered the gold standard, they have limitations. These limitations include non-deterministic learning, dependence on the choice of input window and other hyperparameters, and thus the inability to be considered a robust and replicable baseline. To address these limitations, this paper proposes the use of a non parametric model, known as the **Persistence** model, as reference baseline for comparison in any forecasting problem. This model asserts that future predictions of length L will be identical to their previous L samples. Experiments in Sect. 5 will demonstrate how even this approach can compete with, and sometimes outperform, state of the art models recently presented at prestigious venues, highlighting the need for a more precise evaluation in time series forecasting.

3.2   Embedding

Having defined a robust comparison baseline, we sought to determine a suitable latent representation for temporal sequences to be used as embeddings in Transformer-based models. We assumed that the presence of periodic patterns and a sufficiently large amount of data were the only two conditions necessary to extract useful information for both Transformer-based models and shallow neural networks. In this perspective, we were inspired by the Time2Vec model presented by Kazemi *et al.* [8]. This technique is intended to replace the positional encoding used in linguistic Transformers and is characterised by two linearly learnable tensors, one with periodic activation, which receive the raw temporal sequence as input and concatenate the output as a latent representation to be fed into the predictive model. The aim of Time2Vec is to extract and isolate high-level periodic patterns. The main drawback of this technique is the increase in spatial complexity due to the concatenation operation, which can be particularly limiting in problems with very long prediction windows. To address this issue, we used a revised version of the original architecture called Additive Time To Vector (**AddT2V**) in our experiments. This variant uses the additive operator instead of concatenation, the sine function as the periodic function, and has tensors equal in size to the prediction window to map the periodicities in the input directly onto the prediction space. From the ablation studies we have carried out, the performance of the Time2Vec and AddT2V models appears to be equivalent.

3.3   Transformer-Based Models

In contrast to the literature, where Transformer-based models typically have deep and complex architectures, the new models presented in this paper aim to simplify the original Transformer architecture and analyse its behaviour in the context of long time series prediction. With reference to Fig. 1, the most complex model, which we have named **Sinformer**, is exactly the smallest Vanilla Transformer, i.e. with $N = 1$ for both the encoder and the decoder. The position encodings are removed, while the embeddings of this model are the previously described AddT2V, both for the encoder and the decoder inputs. This simultaneously solves the problem of the intrinsic permutation invariance of the self-attention operation by injecting the order information into the input sequence and the embedded representation of the input features. The output of the model is a dense layer with the dimension of the prediction window. The activation function is sinusoidal rather than linear. A further rationale for these choices is that in an end-to-end learning context, AddT2V aims to map periodic patterns from the input to the prediction space, taking into account permutation invariant reworking and the presence of periodic components in the model output. The second Transformer-based model is a simplification of Sinformer where the decoder is completely removed. This model, which thus consists of a Vanilla Transformer encoder with a number of blocks $N = 1$, an input AddT2V embedding and an output with sinusoidal activation, has been named **Sencoder**.

3.4 Shallow Models

Seeking for further simplification, the logical progression involves the complete removal of all components associated with the Transformer architecture, in particular the attention operators. Consequently, we have developed a model consisting of AddT2V followed by a single dense layer with the same dimensionality as the predicted window and a sinusoidal activation function. This model has been called the Sinusoidal Layered Perceptron (**SLP**). The aim of this architectural choice is to evaluate whether a less sophisticated operator such as a simple internal linear combination can result in a reduction in prediction error compared to the Sencoder model. As a final model, we have constructed a Multi-Layer Perceptron (**MLP**) that is completely independent from the blocks previously presented and at the same time different from the simple models proposed by Zeng et al. [22]. This MLP consists of three dense layers with ReLU activation and the same dimensionality as the predicted window. The MLP model is elastically regularised with $l_1 = 10^{-5}$ and $l_2 = 10^{-4}$, and the last layer has a linear activation function.

4 Experiments Setting

To facilitate the interpretation of the results, this section provides a detailed description of the datasets and hyperparameters used in the experiments. In addition, the models under comparison, the evaluation criteria, and the data pre-processing methods prior to the learning phase are presented.

4.1 Datasets

In order to gain a detailed insight into the effectiveness of the forecasting techniques under consideration, in this research we identified and selected six datasets for their heterogeneous characteristics, which are summarised in Table 1. More specifically, the selected datasets are:

- **ETTh1 and ETTm1** [24]: The Electricity Transformer Temperature (ETT) datasets contain information relevant to the long-term distribution of electricity. The data were collected over a period of two years from two different counties in China. The ETTh1 dataset, which refers to the first station, consists of 17520 hourly samples. The ETTm1 dataset, also related to the first

Table 1. This table provides a summary of the datasets utilized in the experiments conducted within this study.

Dataset	Length	Features	Frequency	Batch Size	Training/Validation/Test
ETTh1	17520	7	1 h	64	60/20/20
ETTm1	70080	7	15 m	64	60/20/20
Electricity	26304	321	1 h	32	70/10/20
Milan T°	7360	1	24 h	32	66/17/17
Venice	330000	1	1 h	256	60/20/20
Weather	52696	21	10 m	16	70/10/20

station, consists of 70080 samples taken at 15 min intervals. Both datasets consist of eight features and were used in all proposed experiments with a training/validation/test split of approximately 14/5/5 months. Each model was trained on these datasets with a batch size of 64. These datasets represent two of the most important benchmarks for time series forecasting due to their multivariate nature and different sampling frequencies.

- **Electricity** [7]: The Electricity dataset contains several features on the electricity consumption of 321 customers from 2012 to 2014. It consists of 26304 hourly samples, each with 322 features. The data was divided into training, validation and test sets of approximately 24, 4 and 8 months respectively. Each model was trained on this dataset with a batch size of 32. This dataset was chosen to represent a medium-sized multivariate dataset with a significant number of features.

- **Temperature of Milan (Milan T°)** [13]: This dataset covers the daily mean temperature history of Milan from 2001 to 2021. It is the smallest dataset analysed in this study, consisting of 7360 points sampled at 24-hour intervals. The data was divided into training, validation and test sets with durations of approximately 160, 40 and 40 months respectively. Each model was trained on this dataset with a batch size of 32. This dataset was chosen to investigate the behaviour of models dealing with a small scale univariate forecasting task.

- **High Water of Venice (Venice)** [18]: This dataset contains the historical series of sea level values recorded in Venice from 1983 to the present. It is the largest dataset analysed in this study, consisting of 330000 hourly samples. The data were divided into training, validation and test sets with durations of 24, 8 and 8 years respectively. Due to the large amount of data, a batch size of 256 was chosen for the experiments. This dataset was chosen for its high cardinality, which allows the analysis of the behaviour of different models in a univariate context for very long forecasts.

- **Weather** [14]: This dataset, which contains 21 meteorological features such as air temperature and humidity, was collected in Germany in 2020. It is a medium-sized dataset consisting of 52696 samples taken at 10-min intervals. Following the literature, the data was divided into training, validation and test sets with proportions of 70%, 10% and 20% respectively. A batch size of 16 was used for the experiments. This multivariate dataset was included due to its average length and number of features relative to the other datasets.

The experiments were designed to provide a comprehensive comparison of the models' capabilities by running multivariate predictions on each multivariate dataset and univariate predictions on all others.

4.2   Models and Setup

The models selected for comparison on the above datasets are grouped into three categories. The first category, consisting only in the Persistence model, represents the non parametric model used as a baseline. The second category consists of shallow models, including the SLP and MLP models proposed by us and the Linear, NLinear and DLinear models proposed by Zeng *et al.* [22]. The third and final category consists of Transformer-based models, including the Sencoder and Sinformer models we propose in this paper, and the FEDformer [26], Autoformer [20], Informer [25], and Crossformer [23] models.

Among the various metrics used in the literature to evaluate forecasting quality, this paper uses the Mean Absolute Error (MAE) to allow a more direct comparison, independent from the magnitude of individual errors, as opposed to quadratic metrics. In terms of the learning details of the models, the best epoch was selected according to the validation error from a training of 50 epochs performed with the Adam optimiser and an exponentially decaying learning rate starting from 10^{-3} and shrinking to 10^{-6}. The optimal portion of the training set and the input window were determined by tuning. Data standardisation was the only form of pre-processing used. All experiments were run on a single A6000 GPU with 48 GB of memory.

5   Results and Discussion

In the first set of experiments, we conduct a comparative analysis between the selected Transformer-based and shallow models from the literature and the newly proposed models. The Persistence model is used as a baseline for comparison. The analysis is performed on six datasets, as described in the previous section, and it considers four different forecasting windows.

The results presented in Table 2 reveal a remarkable finding: the Informer, previously considered to be the best forecasting model up to 2021 and superior to all other Transformer-based models, performs worse than the Persistence model in almost half of the cases, regardless of the forecasting window considered. A similar observation can be made for the Autoformer, although to a lesser extent, and the same holds partially also for the FEDformer. The Crossformer is the only exception within this family of models; it consistently outperforms the non parametric model, as would be desirable, in some cases with a lot of room for improvement. The Sencoder and Sinformer models show remarkable performance in the forecasting task, consistently outperforming the baseline across all

Table 2. This table presents the results of the experiments conducted on the test sets in the form of Mean Absolute Error. The evaluations were carried out over four different forecasting windows for each dataset. Results for models not presented in this paper were obtained from their respective original papers, with the exception of the results for the Milan T° and Venice datasets, which were computed as part of this paper. Crossformer results were all recalculated with the authors code. Results that outperform the baseline are coloured green, while those that underperform it are coloured red. The best results are highlighted in **bold**, while the best results considering only the Transformer-based models are underlined.

Dataset		Persistence (baseline)	SLP (ours)	MLP (ours)	Linear AAAI2023	NLinear AAAI2023	DLinear AAAI2023	Sencoder (ours)	Sinformer (ours)	FEDformer ICML2022	Autoformer NeurIPS2021	Informer AAAI2021	Crossformer ICLR2023
ETTh1	96	0.480	0.392	**0.388**	0.397	0.394	0.399	0.390	0.415	0.419	0.459	0.713	0.426
	192	0.530	0.422	0.424	0.429	**0.415**	0.416	0.433	0.444	0.448	0.482	0.792	0.440
	336	0.571	0.456	0.464	0.476	**0.427**	0.443	0.463	0.457	0.465	0.496	0.809	0.459
	720	0.718	0.528	0.540	0.592	**0.453**	0.490	0.542	0.537	0.507	0.512	0.865	0.519
ETTm1	96	0.389	0.335	0.334	0.352	0.348	0.343	0.333	0.345	0.419	0.475	0.571	0.449
	192	0.421	0.358	0.366	0.369	0.375	0.365	0.389	0.382	0.441	0.496	0.669	0.413
	336	0.845	0.380	0.403	0.393	0.388	0.386	0.415	0.433	0.459	0.537	0.871	0.455
	720	0.851	0.414	0.419	0.435	0.422	0.421	0.425	0.455	0.490	0.561	0.823	0.528
Electricity	96	0.477	0.217	0.210	0.237	0.237	0.237	0.218	0.221	0.308	0.317	0.368	0.285
	192	0.372	0.242	0.233	0.250	0.248	0.249	0.236	0.242	0.315	0.334	0.386	0.313
	336	0.435	0.277	0.297	0.268	0.265	0.267	0.296	0.311	0.329	0.338	0.394	0.353
	720	0.561	0.383	0.371	0.301	0.297	0.301	0.363	0.397	0.355	0.361	0.439	0.449
Milan T°	96	1.337	0.328	0.318	0.304	0.303	0.299	0.360	0.282	0.503	0.722	0.305	0.283
	192	1.718	0.326	0.321	0.304	0.295	0.302	0.306	0.303	0.384	0.644	0.293	0.303
	336	0.543	0.312	0.328	0.314	0.295	0.310	0.309	0.296	0.584	0.602	0.298	0.286
	720	0.496	0.382	0.356	0.332	0.295	0.328	0.380	0.363	0.271	0.904	0.321	0.308
Venezia	96	0.936	0.223	0.274	0.275	0.279	0.274	0.273	0.270	0.357	0.665	0.302	0.267
	192	1.190	0.242	0.364	0.325	0.330	0.322	0.312	0.313	0.421	0.633	0.327	0.308
	336	0.525	0.252	0.394	0.359	0.372	0.357	0.342	0.343	0.492	0.853	0.766	0.331
	720	0.540	0.262	0.413	0.396	0.410	0.398	0.360	0.360	0.503	0.814	0.749	0.355
Weather	96	0.889	0.321	0.486	0.236	0.232	0.237	0.501	0.509	0.296	0.336	0.384	0.233
	192	0.956	0.394	0.541	0.276	0.269	0.282	0.572	0.544	0.336	0.367	0.544	0.263
	336	1.090	0.452	0.566	0.312	0.301	0.319	0.571	0.566	0.380	0.395	0.523	0.308
	720	0.714	0.479	0.551	0.365	0.348	0.362	0.541	0.559	0.428	0.428	0.741	0.349

datasets and forecasting windows. In particular, the two proposed models turn out to be the best Transformer-based choice, rivalling the Crossformer model and outscoring it for half of the datasets considered. Despite being significantly simpler than their counterparts, they consistently outperform the state of the art architectures, in some cases by a considerable margin. When evaluating the performance of the Sencoder and Sinformer models, it is clear that neither model consistently outperforms the other. On average, the Sencoder performs slightly better than the Sinformer, although differences are negligible. As the Sencoder model is the first half of the Sinformer, and therefore has fewer parameters and operations, it should be considered the preferred choice by Occam's razor.

The analysis of the results obtained by shallow models shows that none of them performs worse than the Persistence model across datasets and forecast windows considered. Further analysis of the data presented in Table 2 shows that the shallow models consistently produce the best overall results. Furthermore, with the exception of the MLP model - which is the most complex of all the shallow models and at the same time the less performing one - no single model emerges as the preferred choice. Rather, they are all interchangeable and equally competitive, consistently producing a very low MAE. This confirms the hypothesis that shallow models are highly effective in generating accurate forecasts, outperforming their Transformer-based counterparts. In addition, it is notewor-

Table 3. This table displays the results of experiments conducted on test sets, calculated as the average percentage improvement per forecasting window between the Mean Absolute Error of the selected models and the Mean Absolute Error of the baseline, represented by the Persistence model. Results that outperform the baseline are coloured green, while those that underperform it are coloured red. The best results are highlighted in **bold**, while the best results among the Transformer-based models are underlined.

Window	SLP (ours)	MLP (ours)	Linear AAAI2023	NLinear AAAI2023	DLinear AAAI2023	Sencoder (ours)	Sinformer (ours)	FEDformer ICML2022	Autoformer NeurIPS2021	Informer AAAI2021	Crossformer ICLR2023
96	0.512	0.464	0.511	**0.518**	0.511	0.460	0.464	0.414	0.304	-0.134	0.434
192	0.494	0.439	0.495	**0.512**	0.497	0.439	0.331	0.419	0.336	-0.223	0.439
336	**0.403**	0.279	0.363	0.401	0.373	0.308	0.322	0.200	0.065	-0.348	0.401
720	0.331	0.222	0.309	**0.393**	0.359	0.230	0.212	0.315	-0.006	-0.240	0.348

thy that the MLP model, although not specifically designed for forecasting tasks and often considered unsuitable in the literature, also performs exceptionally well, achieving excellent levels of MAE, particularly for small forecast windows, and competing effectively with any current variant of the Transformer (Fig. 2).

To gain a comprehensive understanding of the performance of the models with respect to the size of the forecasting window and to assess their advantage over the baseline, we calculated the percentage improvement between the MAE of the models considered and that of the Persistence model. The average of each forecast window is then calculated over the different data sets. The results are summarised in Table 3. A first look at the results shows that, in the context of a time series forecasting problem, it is generally preferable to use the Persistence model over the Informer, regardless of the forecasting window considered. Furthermore, it is evident that the Sencoder and Sinformer models are the best Transformer-based options for medium and small windows, reaching the state of the art in this model family, while the Crossformer is preferable for longer forecasts. It is also fair to say that SLP and NLinear are the optimal choices due to their simplicity and superior performance. Overall, the results convey a clear message: on average, any shallow model is preferable to any Transformer-based model.

Figure 1 shows a graphical comparison between SLP and Sencoder on a subset of the test set for each dataset analysed. Looking at all the plots, it is clear that neither SLP nor Sencoder emerges as the clear winner in every context. SLP produces predictions that are accurate but noisy, while Sencoder produces smoother ones. Both models are successful in learning patterns and periodicity, but struggle to make good predictions when these features are less obvious or absent. For the ETTh1, ETTh2 and Venice datasets, the predictions are close to the ground truth in many cases. However, for the Electricity, Milan T° and Weather datasets, the models are less successful in approximating non-periodicity, but still accurately capture the trend and quota of the time series.

Thanks to the availability of a particularly large dataset such as Venice, the last set of experiments concerns the analysis of forecasting when the prediction window becomes extremely long. Table 4 shows the results of this study.

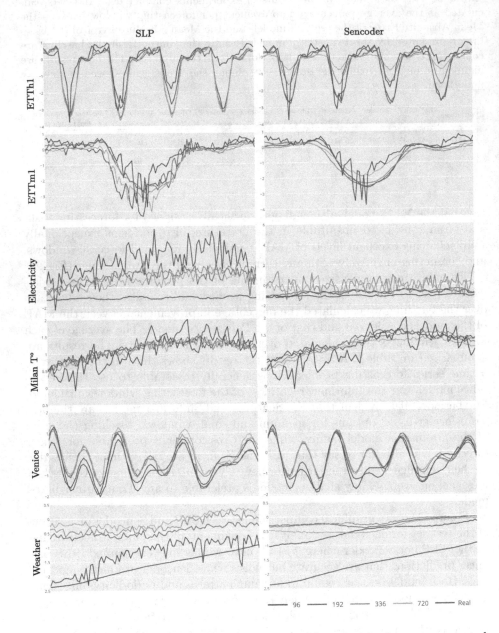

Fig. 2. This figure depicts a qualitative comparison between the forecasts generated by the SLP model and those produced by the Sencoder model. All predictions were calculated with respect to the datasets test sets and were made across four distinct forecasting windows.

Table 4. This table presents the results of the experiments carried out on the test sets of the Venice dataset in the form of Mean Absolute Error. Evaluations were performed over five different forecasting windows, distinguished by their increased length relative to the standard windows used in most other papers. Results that outperform the baseline are coloured green, while those that underperform it are coloured red. The best results are highlighted in **bold**, while the second best results are underlined.

Window	Persistence (baseline)	SLP (ours)	MLP (ours)	Linear AAAI2023	NLinear AAAI2023	DLinear AAAI2023
1440	0.680	**0.395**	0.402	0.432	0.447	0.432
2880	0.966	**0.420**	0.426	0.477	0.486	0.483
5760	1.125	**0.410**	0.418	0.480	0.501	0.479
11520	1.364	**0.407**	0.418	0.482	0.488	0.481
23040	1.018	**0.421**	0.466	0.472	0.492	0.472

The first relevant result concerns the absence of the Transformer-based models which, on average, saturate the available memory and thus became intractable already with a window between 1440 and 2880 points and were therefore omitted. This is certainly another feature to be taken into account; despite much work has been done to optimise the spatial complexity of Transformer-based models, it is still unthinkable to use them for extremely long forecasts. As far as the shallow models are concerned, many of the considerations made for windows shorter than 720 samples still apply, such as the fact that each of them manages to achieve a smaller error than Persistence. Although other shallow models achieve good performance, the SLP model proved to be the best among the alternatives analysed. This suggests the effectiveness of the AddT2V embedding and sinusoidal activations to exploit periodicities during the training phase and achieve to good generalisation on unseen data. Another general result, but particularly evident from the SLP model results, is the lack of a clear and linear proportionality between prediction length and prediction MAE. For example, a 240-fold increase in prediction size from 96 to 23040 points resulted in only an 88.8% increase in error, as shown in Table 2. This demonstrates the potential of a simple yet powerful model for forecasting medium, long and extremely long time series.

The final message from the analyses conducted in this study is quite clear: at the current state of research, Transformer-based models, while theoretically superior, are not, on average, the best choice for solving time series forecasting problems. By this we do not mean that it will not be possible to disprove the following statement as research progresses. What we are saying is that this type of analysis requires a thoroughness that has been lacking in almost all of the work presented in the literature in recent years, and which has led to probably overestimate the Informer, which, in comparison, is not even able to compete with the Persistence model. Finally, the results obtained with the simplest models, such as the SLP model, suggest that perhaps the best strategy in the field of time series forecasting is to start with simple techniques and use them as a basis for applying methods and strategies currently used in complex models.

6 Conclusion

In this article we discussed the effectiveness of applying Transformer-based techniques in the context of time series prediction. The results of the experiments showed that the key to improving these architectures is simplification, and that currently the best performing models are simplified to the point where they are no longer Transformers, but even shallow neural networks. We discussed the importance of a baseline and showed how the Persistence model is able to outperform models that have been awarded state of the art techniques in recent years. Finally, we showed how shallow models are able to make accurate predictions of extremely long time series, which are computationally prohibitive for current Transformer-based models due to their polynomial complexity.

Nevertheless, the main objective of this research was not to present new forecasting models, but rather to share with the reader a reflection supported by experimental results. It seems that in recent years, at least in the time series forecasting field, research has focused more on the desire to apply a particular fashionable technique at all costs than on finding a solution to the problem to be solved. Having shown that very simple or even non parametric models are able to outperform algorithms with millions of parameters, is it fair to ask whether part of the scientific community has reached a dead end and talks about it as if it were a highway? We leave it to the reader to answer this question. Our hope is that, having shown that there are better techniques from a performance point of view, and much better ones from a performance-complexity perspective, it will be possible to continue down the path of finding better solutions by looking forward while keeping an eye on the past. We hope that in this area, as in others, we can move on by taking two steps forward and one behind.

References

1. Ariyo, A.A., Adewumi, A.O., Ayo, C.K.: Stock price prediction using the ARIMA model. In: 2014 UKSim-AMSS 16th International Conference on Computer Modelling and Simulation, pp. 106–112. IEEE (2014)
2. Brown, T., et al.: Language models are few-shot learners. In: Advances in Neural Information Processing Systems, vol. 33, pp. 1877–1901 (2020)
3. Chen, Y., Kang, Y., Chen, Y., Wang, Z.: Probabilistic forecasting with temporal convolutional neural network. Neurocomputing **399**, 491–501 (2020)
4. Devlin, J., Chang, M.W., Lee, K., Toutanova, K.: BERT: pre-training of deep bidirectional transformers for language understanding. arXiv preprint arXiv:1810.04805 (2018)
5. Dong, L., Xu, S., Xu, B.: Speech-transformer: a no-recurrence sequence-to-sequence model for speech recognition. In: 2018 IEEE International Conference on Acoustics, Speech and Signal Processing (ICASSP), pp. 5884–5888. IEEE (2018)

6. Dosovitskiy, A., et al.: An image is worth 16x16 words: transformers for image recognition at scale. arXiv preprint arXiv:2010.11929 (2020)
7. Dua, D., Graff, C.: UCI machine learning repository. https://archive.ics.uci.edu/ml/datasets/ElectricityLoadDiagrams20112014 (2017)
8. Kazemi, S.M., et al.: Time2vec: learning a vector representation of time. arXiv preprint arXiv:1907.05321 (2019)
9. Kong, W., Dong, Z.Y., Jia, Y., Hill, D.J., Xu, Y., Zhang, Y.: Short-term residential load forecasting based on LSTM recurrent neural network. IEEE Trans. on Smart Grid **10**(1), 841–851 (2017)
10. Lara-Benítez, P., Carranza-García, M., Luna-Romera, J.M., Riquelme, J.C.: Temporal convolutional networks applied to energy-related time series forecasting. Appl. Sci. **10**(7), 2322 (2020)
11. Li, S., et al.: Enhancing the locality and breaking the memory bottleneck of transformer on time series forecasting. In: Advances in Neural Information Processing Systems, vol. 32 (2019)
12. Liu, Z., et al.: Swin transformer: hierarchical vision transformer using shifted windows. In: Proceedings of the IEEE/CVF International Conference on Computer Vision, pp. 10012–10022 (2021)
13. Lombardia: Milan Temperature arpa lombardia. https://www.arpalombardia.it/Pages/Aria/Richiesta-Dati.aspx
14. NCEI: Weather national centers for environmental information. https://www.ncei.noaa.gov/data/local-climatological-data/
15. Pham, N.Q., Nguyen, T.S., Niehues, J., Müller, M., Stüker, S., Waibel, A.: Very deep self-attention networks for end-to-end speech recognition. arXiv preprint arXiv:1904.13377 (2019)
16. Siami-Namini, S., Tavakoli, N., Namin, A.S.: The performance of LSTM and BiLSTM in forecasting time series. In: 2019 IEEE International Conference on Big Data (Big Data), pp. 3285–3292. IEEE (2019)
17. Vaswani, A., et al.: Attention is all you need. In: Advances in Neural Information Processing Systems, vol. 30 (2017)
18. Venice: Sea Level of Venice city of Venice. https://www.comune.venezia.it/it/content/archivio-storico-livello-marea-venezia-1
19. Wan, R., Mei, S., Wang, J., Liu, M., Yang, F.: Multivariate temporal convolutional network: a deep neural networks approach for multivariate time series forecasting. Electronics **8**(8), 876 (2019)
20. Wu, H., Xu, J., Wang, J., Long, M.: Autoformer: decomposition transformers with auto-correlation for long-term series forecasting. In: Advances in Neural Information Processing Systems, vol. 34, pp. 22419–22430 (2021)
21. Yan, K., Li, W., Ji, Z., Qi, M., Du, Y.: A hybrid LSTM neural network for energy consumption forecasting of individual households. IEEE Access **7**, 157633–157642 (2019)
22. Zeng, A., Chen, M., Zhang, L., Xu, Q.: Are transformers effective for time series forecasting? arXiv preprint arXiv:2205.13504 (2022)
23. Zhang, Y., Yan, J.: Crossformer: transformer utilizing cross-dimension dependency for multivariate time series forecasting. In: The Eleventh International Conference on Learning Representations (2023)
24. Zhou, H., et al.: Electricity Transformer Dataset etdataset. https://github.com/zhouhaoyi/ETDataset (2021)

25. Zhou, H., et al.: Informer: Beyond efficient transformer for long sequence time-series forecasting. In: Proceedings of the AAAI Conference on Artificial Intelligence, vol. 35, pp. 11106–11115 (2021)
26. Zhou, T., Ma, Z., Wen, Q., Wang, X., Sun, L., Jin, R.: FEDformer: frequency enhanced decomposed transformer for long-term series forecasting. In: International Conference on Machine Learning, pp. 27268–27286. PMLR (2022)

Real-Time Emotion Recognition in Online Video Conferences for Medical Consultations

Dennis Maier$^{(\boxtimes)}$ (iD), Matthias Hemmje(iD), Zoran Kikic, and Frank Wefers

University of Hagen, Universitätsstraße 11, 58097 Hagen, Germany
dennis.maier@fernuni-hagen.de

Abstract. In this paper, we present a novel solution for real-time emotion analysis in video conferencing, aiming to enhance therapies for both patients and therapists. Building upon the SenseCare project, we extend its capabilities to include a video conferencing platform with scalable real-time emotion analysis using widely available software and frameworks avoiding critical vendor locks. Our architecture focuses on ease of adaptation and further development, connecting a WebRTC conferencing platform to a scalable Kubernetes backend for emotion analysis. Emphasizing low latency, we implement the producer-consumer pattern and utilize a message broker. For emotion analysis, we use convolution neural networks. We propose a methodology for identifying an optimal batch size that maximizes backend efficiency while maintaining low latency. Our approach exhibits scalability, allowing for seamless adaptation during periods of high system utilization. Our findings demonstrate the feasibility of employing CNNs for sub-second emotion analysis on an affordable Kubernetes cluster, enabling multiple users to effectively engage in the system as patients and therapists.

Keywords: Machine learning · Telehealth · Realtime

1 Introduction

Emotions are a fundamental aspect of human life, influencing various areas ranging from communication and social interaction to decision-making and mental well-being. This highlights the importance of mental health in today's world, with psychotherapy as a proven method of maintaining it. Traditionally, psychotherapies have been conducted in person, but the ongoing COVID-19 pandemic has necessitated a shift towards online therapies as a means of continuing treatment [12]. As a result, videoconferencing has gained popularity as a platform for online therapy, with some studies suggesting potential advantages over face-to-face therapy [18]. Our work is positioned within the EU-funded project Sensor Enabled Affective Computing for Enhancing Medical Care (SenseCare) [9]. SenseCare uses affective computing to improve healthcare processes and systems by analyzing emotions from facial landmarks of video streams, working asynchronously on a back-end system using Support Vector Machines (SVM) [14]. A synchronous analysis prototype has also been implemented in the client browser, but this requires robust hardware for emotion analysis, often resulting in disruptive fan noise. The shift to consumer hardware introduces several technical constraints. Not all home computers have sufficient processing power to run such intensive processes smoothly. To make matters worse, recent studies [1] indicate that internet usage via smartphones, accounting for 60% of total usage in 2023, has surpassed that on home computers, positioning smartphones as a potential platform for telemedicine. However, they lack the processing power required for such tasks. These limitations, combined with the lack of a scalable platform for real-time emotion analysis during video conferencing between patients and therapists, contribute to the challenges of improving the effectiveness of online psychotherapy. Furthermore, while external web services could potentially perform machine learning tasks, the lack of an open-source solution that can be tailored to the specific needs of a medical practice is an obstacle, as commercially available systems typically restrict access to their underlying processes and algorithms. Therefore, our research addresses these critical technical and scalability issues by investigating the feasibility of developing a scalable and efficient real-time emotion recognition system for online therapy sessions. As an extension of the SenseCare project, our research aims to fill the existing gap in freely available conferencing software with emotion recognition capabilities. Importantly, our proposed solution does not require high-performance hardware on the client side. This allows for its widespread use, including on mobile devices, without generating disruptive fan noise. Additionally, it supports self-hosting and promotes further research, thereby broadening its potential applications.

2   Methods

Real-time emotion recognition has numerous applications in healthcare, education, and entertainment. However, integrating emotion recognition into videoconferencing systems is challenging due to architectural and communication issues, leading to latency and degraded performance. In this chapter, we explain an implementation for our proposed solution, discussing the system's architecture, integration with a videoconferencing platform, and emotion recognition system. We provide strategies for optimal and scalable usage of the server infrastructure. While other approaches have been proposed, they lack a specific emphasis on scalability [11] or near real-time video conferences [10].

2.1   Overview

Our proposed methodology builds upon the WebRTC [2] video conferencing solution Edumeet [3], which serves as the foundation for our approach. Edumeet, an open-source video conferencing platform, is based on the mediasoup framework and operates as a Selective Forwarding Unit (SFU) [13], allowing centralized access to all relevant data within a video conference. This data is subsequently relayed from the SFU to an emotion analysis backend, and the resulting output is disseminated throughout the conference room, as shown in Fig. 1. The SFU architecture is designed in such a way that both patients and therapists transmit their video signals to a designated SFU server instance, which in turn distributes the streams to other users in the same virtual conference room. We opted for the SFU architecture over alternative approaches, including peer-to-peer and Multi-point Conference Unit (MCU) configurations. The mesh architecture establishes peer-to-peer connections between all participants, resulting in increased band-width consumption as the number of streams and participants grows. Conversely, the MCU aggregates all incoming streams and distributes the video signals as a single stream to the participants, conserving bandwidth but imposing substantial computational demands on the server side. Thus, the SFU architecture offers an optimal solution for our specific setup.

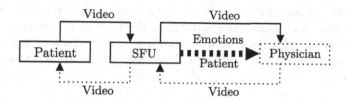

Fig. 1. A server exchanges the video signals between the patient and the physician. It also sends a copy to an analysis backend and forwards the results to the physician.

2.2 Infrastructure

Our proposed framework encompasses three integral components: the client interface, virtual machine (VM) infrastructure, and Kubernetes deployment (Fig. 2). Firstly, the client component corresponds to browser instances executing on the user's local device, facilitating real-time interaction. Secondly, the VM is responsible for operating the video conferencing server, specifically the Selective Forwarding Unit (SFU). Lastly, the Kubernetes cluster is employed to manage the emotion analysis functionality.

The Client program of Edumeet was augmented with additional modules for face detection and visualization. Before sending the image to an emotion classifier, it is necessary to crop the image to the face, as it was trained on faces only. To achieve this, Google's Mediapipe [5] Face Detection was used. This face detection is based on BlazeFace [8] and is able to detect faces in sub-milliseconds. It was created for mobile devices, thus it aims to be fast and efficient. The face detection system is built to function on the user's hardware, thereby utilizing edge computing to minimize the load on the server. This process can alternatively be switched to run on the server backend, making it feasible to use less powerful mobile clients. However, this approach leads to an increase in server load. For this reason, the standard approach is client-side face detection, as this places only minimal demands on the hardware and does not generate any client-side noise pollution due to computations in our experiments. Additionally, the client was enhanced with a feature for visualizing emotions using a line graph representation.

The SFU Server is responsible for managing not only video streams but also the facial detection data obtained from the client-side analysis. The extracted facial features, known as portrait excerpts, are forwarded to the emotion analysis backend. Accompanying these image data packets are essential metadata, including client ID, room ID, and a timestamp. This information makes it possible to assign emotion analysis results to the right rooms and people, without needing a synchronous connection. Furthermore, the SFU has been enhanced with an OAuth-based Keycloak [4] authorization service, enabling role-based access within a video conference setting. In the context of our proposed solution, two distinct roles have been established, namely, patient and therapist. This role-based setup allows customized and safe access to the videoconferencing platform.

The Emotion Backend runs the computational heavy emotion analysis. To address this challenge, we adopted the message broker RabbitMQ [6] and the producer-consumer pattern, which facilitates the dynamic scaling of consumers by decoupling them from producers. During the conceptualization phase, we avoided utilizing inferencing servers to maintain technology agnosticism. The Celery and Pika frameworks were selected for implementing the producer-consumer pattern in tandem with RabbitMQ. By utilizing Celery, it becomes possible to construct future pipelines (or ensembles) through the implementation of Directed Acyclic Graphs. These graphs provide task flows that allow the execution of various analysis models in parallel, with the ability to merge or process their results. This comes in handy when audio and text analysis will be added to the solution. In the context of real-time analysis, latency was prioritized over the comprehensive processing of all requests. Meeting all requests might cause issues during times of heavy load, leading to increased latency and memory use. As a result, we chose to remove messages older than 400ms to prevent a buildup of outdated requests and maintain the best system performance. Each worker pod in the system utilizes a convolution neural network [16] (MobileNetV3 [15], Small) that is trained to classify emotions. The input dimension of this model is $224 \times 224 \times 1$ and it produces a result of 8 dimensions. The underlying framework used is Tensorflow [7]. Depending on the setup, worker pods are also able to run face detection in case a weak client cannot run the face detection by itself.

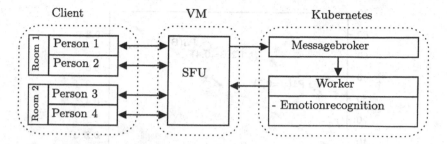

Fig. 2. Structure of outsourced emotion recognition. Participants send a video signal to the SFU, which forwards the signal to other participants. The SFU taps the video signal and sends it to a message broker. Worker pods process the tasks from the broker and send the emotion data to the SFU. The SFU sends the results to the corresponding participants.

2.3   Latency and Throughput

The challenge of balancing high throughput with low latency is a well-documented problem in data processing systems. Existing solutions often fall into two categories: batch processing and stream processing. Batch processing enables high throughput by optimising computation, but the process is discontinuous, resulting in potential latency. Stream processing, on the other hand, processes data as it comes in, one piece at a time. With this in mind, our solution uses a strategy known as micro-batching. Micro-batching offers a compromise between these two processing methods, combining the high throughput of batch processing with the low latency of stream processing. This is mainly because neural networks, which we use in our solution, tend to achieve higher throughput with larger batch sizes. However, this advantage has to be balanced with the need to limit latency. In pursuit of this balance, we have established a benchmarking system to determine the optimal batch size for our specific hardware configuration. Our benchmarking system operates within a Kubernetes pod on an IBM $b \times 2.8 \times 32$ server, which is equipped with 1 virtual CPU and 3072 Mbytes of memory for each worker pod. This hardware configuration allowed us to effectively balance the need for efficient computation with the constraints of our hardware capabilities. To calculate the optimal batch size, we used the formula shown in Eq. 1, which represents a trade-off between computation time and overall system throughput. This formula, represented by $f(x)$, calculates the average batch time for a given value x, within a range of possible values from 0 to y. Given a latency limit L, the optimal or "sweet spot" batch size is the maximum number of images per second that can be processed within the given constraints. Our benchmarking results, displayed in Fig. 3 and 4, confirm the effectiveness of our approach. Through the application of micro-batching, we have successfully developed a solution that balances the need for high throughput with the constraints of latency.

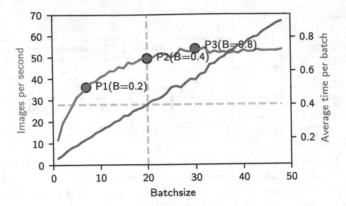

Fig. 3. The graph illustrates a benchmark of throughput within a designated timeframe, with 'B' denoting the maximum batch time. We've set the boundary at 400 ms, which equates to identifying the peak of the red graph within the horizontal range between 0 and the point where the blue line intersects the 0.4 mark on the y-axis. This gives us an optimal batch size of 20. It can be observed that the average time needed for a specific batch size increases proportionally with the batch size. (Color figure online)

$$\max_{0 \le x \le y} \frac{x}{f(x)} \quad \text{subject to} \quad f(x) \le L \tag{1}$$

2.4 Configuration

Several parameters can be set in the context of message processing to optimize the procedure. The Message Broker framework utilizes the Time To Live (TTL) parameter to manage the lifetime of messages. When messages exceed the TTL limit, they are deleted from the queue as they are deemed irrelevant to our use case. In addition, the batch-size parameter is employed to determine the optimal dimensions for concurrent image processing. A larger batch size can enhance throughput but also leads to longer computation times. This can result in a less desirable outcome for real-time analysis. Furthermore, the flush-time parameter is specifically designed to streamline the processing of task batches before reaching the specified batch size. This reduces the waiting time for a calculation to commence, which results in lower latency. Lastly, the prefetch-count parameter is utilized to effectively retrieve tasks from the message broker and cache them locally for faster access. The prefetch memory acts as a buffer, and a buffer that is too large may lead to significant latency. These parameters have a notable impact on message loss, throughput, and system latency. Detailed results on the effects of these parameters are presented in the Results chapter.

3 Results

The findings presented in this study were derived from an IBM b × 2.8 × 32 configuration, where each worker pod was allocated 1 virtual CPU and 3072

MBytes of memory. A total of two worker pods were employed in the experiment. To emulate the presence of connected clients, a testing tool was created. Each simulated client transmitted a set of 1000 images at a rate of 10 images per second. The experiment entailed a successive incrementation of clients, ranging from one to ten. The following parameters were utilized for our analysis backend: TTL: 400 ms, flush-time: 400 ms, prefetch-count: 100, and a variable micro-batch size. As shown in Fig. 4 there is a wide range of possible outcomes using the same hardware. The batch size parameter has a significant influence on latency and task loss. We use the calculated batch size to limit the latency based on our system hardware. By using the scaling function of Kubernetes, the system itself will work near this maximum utilization. Our experiment shows, that it is possible (depending on the neural network used) to analyze emotions on an external server in sub-second range.

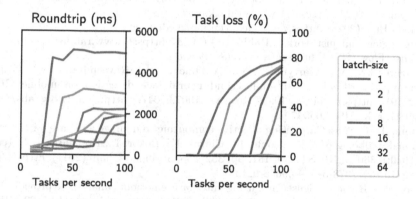

Fig. 4. Left graph shows the time between sending a message and receiving a result for a given batch size. Messages lost due to TTL are shown on the right graph.

4   Discussion

In conclusion, this study demonstrates the effectiveness of our scalable system for real-time emotion recognition in videoconferencing. This approach success-fully analyzes and transmits emotions to participants in a video conference with minimal latency. Our findings confirm the feasibility of outsourcing a real-time emotion recognition system to reduce client load. Thanks to this architecture, our solution eliminates the need for additional emotion analysis steps on the client side and makes face detection optional. The backend can easily be scaled using Kubernetes. Under optimal system configuration, emotion recognition can be carried out within a few milliseconds even under high load, minimizing the delay due to outsourced analysis. Overall, our results provide valuable insights for the development of real-time emotion recognition systems in video conferenc-ing. Our system provides a platform to test machine learning models in real-time in combination with a video conference system for further research.

Future developments include the integration of audio and text analysis into the system and the transfer of live recordings to the SenseCare project for subsequent analysis of video files. Corresponding changes to the infrastructure are to be examined. Micro-batching is supported by additional inferencing software. Whether this software increases or simplifies complexity will be investigated in future experiments. The source code for this project can be found in our repository [17].

References

1. Mobile Vs. Desktop Internet Usage (2023). https://www.broadbandsearch.net/blog/mobile-desktop-internet-usage-statistics
2. WebRTC (2021). https://webrtc.org/
3. eduMEET - web-based videoconferencing platform (2023). https://edumeet.org/
4. Keycloak (2023). https://www.keycloak.org/
5. MediaPipe (2023). https://google.github.io/mediapipe/
6. Messaging that just works - RabbitMQ (2023). https://www.rabbitmq.com/
7. TensorFlow (2023). https://www.tensorflow.org/
8. Bazarevsky, V., Kartynnik, Y., Vakunov, A., Raveendran, K., Grundmann, M.: BlazeFace: sub-millisecond neural face detection on mobile GPUs (2019). https://doi.org/10.48550/arXiv.1907.05047. http://arxiv.org/abs/1907.05047. arXiv:1907.05047 [cs]
9. Bond, R.R.: SenseCare: using affective computing to manage and care for the emotional wellbeing of older people. In: Giokas, K., Bokor, L., Hopfgartner, F. (eds.) eHealth 360°. LNICST, vol. 181, pp. 352–356. Springer, Cham (2017). https://doi.org/10.1007/978-3-319-49655-9_42
10. Donovan, R., et al.: SenseCare: using automatic emotional analysis to provide effective tools for supporting, pp. 2682–2687 (2018). https://doi.org/10.1109/BIBM.2018.8621250
11. Franzen, M., Gresser, M.S., Müller, T., Mauser, P.D.S.: Developing emotion recognition for video conference software to support people with autism. arXiv:2101.10785 [cs] (2021)
12. Glueckauf, R.L., et al.: Survey of psychologists' telebehavioral health practices: technology use, ethical issues, and training needs. Prof. Psychol. Res. Pract. 49(3), 205 (2018). https://doi.org/10.1037/pro0000188. https://psycnet.apa.org/fulltext/2018-28691-004.pdf. publisher: US: American Psychological Association
13. Grozev, B.: Efficient and scalable video conferences with selective forwarding units and WebRTC. Ph.D. thesis, Université de Strasbourg (2019). https://doi.org/10.13140/RG.2.2.11791.20645
14. Healy, M., Donovan, R., Walsh, P., Zheng, H.: A machine learning emotion detection platform to support affective well being. In: 2018 IEEE International Conference on Bioinformatics and Biomedicine (BIBM), pp. 2694–2700 (2018). https://doi.org/10.1109/BIBM.2018.8621562
15. Howard, A., et al.: Searching for MobileNetV3 (2019). http://arxiv.org/abs/1905.02244. arXiv:1905.02244 [cs]
16. LeCun, Y., Kavukcuoglu, K., Farabet, C.: Convolutional networks and applications in vision. In: Proceedings of 2010 IEEE International Symposium on Circuits and Systems, pp. 253–256. IEEE, Paris, France (2010). https://doi.org/10.1109/ISCAS.2010.5537907. http://ieeexplore.ieee.org/document/5537907/

17. Maier, D.: SourceCode Paper-Realtime-Emotionrecognition. https://github.com/DennisMaier/Paper-Realtime-Emotionrecognition
18. Sander, J., et al.: Online therapy: an added value for inpatient routine care? Perspectives from mental health care professionals. Euro. Arch. Psychiatry Clin. Neurosci. **272**, 107–118 (2021). https://doi.org/10.1007/s00406-021-01251-1

Attentive Perturbation: Extending Prefix Tuning to Large Language Models Inner Representations

Louis Falissard[(✉)], Séverine Affeldt, and Mohamed Nadif

Centre Borelli UMR 9010, Université Paris Cité, 75006 Paris, France
{louis.falissard,severine.affeldt,mohamed.nadif}@u-paris.fr

Abstract. From adapters to prefix-tuning, parameter efficient fine-tuning (PEFT) has been a well investigated research field in the past few years, which has led to an entire family of alternative approaches for large language model fine-tuning. All these methods rely on the fundamental idea of introducing additional learnable parameters to the model, while freezing all pre-trained representations during training. This fine-tuning process is generally done through refitting all model parameters to the new, supervised objective function. This process, however, still requires a considerable amount of computing power, which might not be readily available to everyone. In addition, even with the use of transfer learning, this method requires substantial amounts of data. In this article, we propose a novel and fairly straightforward extension of the prefix-tuning approach to modify both the model's attention weight and its internal representations. Our proposal introduces a "token-tuning" method relying on soft lookup based embeddings derived using attention mechanisms. We call this efficient extension "attentive perturbation", and empirically show that it outperforms other PEFT methods on most natural language understanding tasks in the few-shot learning setting.

Keywords: Large language models · Parameter efficient fine-tuning · Adapters · Prefix-tuning · Natural language processing · Natural Language Understanding

1 Introduction

The use of pre-trained large language models, such as BERT [3], T5 [11] or GPT [10] have been ubiquitous in a wide range of natural language processing tasks. These models are pre-trained in a self-supervised fashion on massive textual datasets and are made readily available online. The machine learning practitioner can then use these models' internal representations as a starting point – either as an initialization scheme or a feature extraction tool – to build powerful models for NLP related tasks (eg., summarization, machine translation, language generation).

The fine-tuning of pre-trained language models to perform downstream tasks requires storing and modifying a copy of all the model's parameters for each task.

G. Nicosia et al. (Eds.): LOD 2023, LNCS 14505, pp. 488–496, 2024.
https://doi.org/10.1007/978-3-031-53969-5_36

Yet, some powerful large language models involve a prohibitively high number of parameters (eg. 774M parameters for GPT-2 [10], 175B parameters for GPT-3 [2]). In such cases, fine-tuning can be readily unfeasible, in particular for areas where collected datasets are of small sample sizes. Indeed, while the smallest dataset used in the experiments presented in BERT models' seminal article contains $2.5K$ observations, such amount of data is generally impractical to obtain, for instance in clinical studies involving a cohort of patients.

Adapters and prefix-tuning approaches have motivated numerous investigations, as they enable lightweight fine-tuning by introducing additional learnable parameters to the model, while freezing all pre-trained representations during training [4,12]. Hence, such approaches are interesting ways to leverage large pre-trained language models for downstream NLP tasks. Over the last few years, they have been at the origin of a wide family of alternative approaches for large language model fine-tuning.

Although each individual approach has its specificity, they can be gathered into two main categories: *(i)* adapter-like approaches, that modify the model's internal representations with the use of bottleneck perceptrons introduced at strategic points in every transformer stack, and *(ii)* prefix or prompt based approaches, that introduce learnable virtual tokens in the input sequence, which allows to modify each transformer stack's attention weights. Some attempts have been made to unify these two families of methods, but mostly consist of combining them independently in a mosaic of adapter-like methods, which both considerably raises the number of model hyperparameters and doesn't necessarily yields the expected gains in predictive performances.

In this work, we demonstrate that a straightforward extension of the prefix-tuning approach, named `Attentive perturbation`, can efficiently outperforms the *parameter efficient fine-tuning* (PEFT) methods in the few-shot learning setting. In the following, we introduce in Sect. 2 the family of PEFT methods. Then, Sect. 3 provides detailed information on our novel prefix tuning extension. Section 4 describes our specific experimental settings, based on the GLUE benchmark [13]. Finally, we discuss our results in Sect. 5, before the conclusion in Sect. 6.

2   PEFT Methods

The field of *parameter efficient fine-tuning*, which has been an active area of research for a few years now, proposes alternatives to this standard fine-tuning scheme that allows practitioners to leverage large language models' powerful learned representations, even with a few labeled instances. Although these approaches tend to differ in their implementation, most of them rely on the same idea, which is to freeze all the language model's parameters during training, and instead train additional, trainable ones, typically injected in the model in an additive fashion. These additional parameters are typically orders of magnitude less numerous than the actual model, leading to better performances in the few-shot learning setting, and less computational requirement for model fitting.

Since the introduction of Adapters methods [4], an entire family of PEFT methods has been introduced in the academic literature. These methods have varying degrees of performances, both in terms of parameter efficiency and predictive power. This family of methods embeds:

- `Adapters` and their variants [4,9], which use a 2 layers bottleneck perceptron to additively corrupt the transformer stack's inner representations at one or several points after its attention module.
- `Bitfit` [1], which have the particularity to not introduce any additional parameter to the model, but instead limits itself to fine-tuning solely the bias terms in all of the model's layers.
- `Low Rank Adaptation` (`LoRA`) [5], which additively reparameterize the queries and values projection matrices with a linear bottleneck multilayer perceptron.
- `Prefix-tuning` [7], which, instead of modifying the transformer's stack's signal itself, preprends to the query and value sequences additional learnable tokens, in an approach similar to continuous prompt learning.
- `Prompt fine-tuning` [6], a simpler variant of prefix tuning where learnable virtual tokens are prepended to the input sequence. This method, however, tends to work only for larger models, and is as such discarded from the scope of this article.
- `Fusion methods` such as *Unified Framework for Parameter Efficient Language model Tuning* (UniPELT) [8], which leverages a combination of several adapter modules at once associated with gating mechanisms, in order to federate each approach's advantages while alleviating their respective drawbacks.

3   Proposed Approach: Attentive Perturbation

As aforementioned, apart from prefix (or *prompt*) tuning, most parameter efficient fine-tuning methods are based on the additive introduction to the model of a bottleneck two layer perceptron, with varying activation function, at varying points in the transformer stack. Prefix-tuning, however, does not directly modify the model's signal, but instead focuses on adapting each transformer stack's attention weights, and is based on learnable embedding.

Following this latter observation, we propose a method to additively modify the model's internal representation based on embeddings, rather than bottleneck perceptrons, and choose to apply these perturbations to the attention layer's queries and values, in an approach conceptually similar to LoRA.

Applying these embeddings in an additive fashion to the attention layer's input sequence, which is continuous and of varying length by essence, is not as straightforward as the perceptron approach used in LoRA or adapter methods. We propose to do so using a soft lookup mechanism on the embeddings, derived in a similar fashion to attention mechanisms.

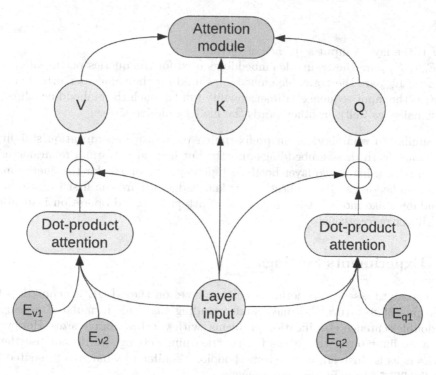

Fig. 1. Overview of the attentive perturbation module. The bottleneck adapter used for the reparameterization trick is hidden for simplicity

In other words, for each perturbation module (one for the transformer stack's queries, and another for its values), we introduce two distinct embeddings E_q and E_v, and combine them with the transformer layer's input in a scaled dot-product attention module as keys, values and queries, respectively. The module's output is then additively reinjected into the layer's input sequence, prior to the transformer stack's multi-head attention module, as can be seen in Fig. 1. We call these modules "attentive perturbators" and formally define them as follows for a given transformer layer:

$$P_q(I) = I + \mathbf{softmax}(\frac{IE_{q,1}^\top}{\sqrt{d_I}})E_{q,2} \qquad (1)$$

$$P_v(I) = I + \mathbf{softmax}(\frac{IE_{v,1}^\top}{\sqrt{d_I}})E_{v,2} \qquad (2)$$

where

- I is the layer's input sequence of vectors
- $E_{q,1}$, $E_{q,2}$ are the trainable embeddings used for the queries perturbation
- $E_{v,1}$, $E_{v,2}$ are the trainable embeddings used for the values perturbation
- d_I is the input sequence's dimensionality (and as such the embeddings dimensionality as well), in other words the model's hidden size

Similar to what is done in prefix-tuning to improve optimization stability, we do not learn these embeddings directly, but instead perform a reparameterization trick using a two layer bottleneck perceptron, shared for the queries and values embeddings. To the best of our knowledge, we are the first to introduce an adapter-like module that makes use of attention based operation instead of multilayer perceptrons.

4   Experiments Settings

We choose to assess our method's performances on the GLUE benchmark [13], a corpus of 8 natural language understanding tasks all formulated as single or double sentence classification problems, with varying sample sizes. However, since we limit ourselves to the few-shot learning setting, we do not use these datasets as is. Instead, we use a methodology similar to what was presented in the UniPELT paper for our experiments.

Basically, we sample observations from the GLUE benchmark in order to build a series of small datasets with 50, 100, 200 and 500 observations. Our method for building those datasets, however, differs from theirs on a few key points.

First, in [13] the authors chose to build validation sets of 1,000 observations for all sampled training datasets, which we feel is unrealistic in a few-shot learning settings (validation datasets do not typically contain more than ten times the number of training examples). Second, they use the GLUE benchmark validation datasets as test datasets. However, some of these validations datasets have considerably small sample sizes (277 for RTE, 408 for MRPC), which might lead to noisy test metric evaluation. As a consequence, for each of the glue benchmark dataset, a dataset of sample size K is built as follows:

1. The original training and validation sets were concatenated into one sampling dataset.
2. A test dataset was built from the observations, using random sampling with a sample size of half the total sample size (capped at $5K$ observations). For a given task, the same test dataset is used for all experiments (including all sampled training sizes).
3. From the remaining observations, K were sampled randomly, and divided with a 70/30 split into a training and validation set, respectively

Ten distinct training datasets were sampled from all glue benchmarks and for all sample sizes, so as to allow for performance metric averaging and statistical testing of performance differences.

For all these datasets, we compare our method with standard fine-tuning and 4 other parameter efficient methods, namely standard `Adapters`, `LoRA`, `Prefix-tuning` and `UniPELT`. For all methods, we follow the same training procedure and hyperparameter search that is done in the `UniPELT` paper. All models were fit for 50 epochs using $AdamW$ with linear weight decay and an early stopping mechanism with a patience of 10 non increasing epochs. The batch size was fixed to 16, and all method's specific hyperparameters were set up as follows:

- `Standard fine-tuning`: Learning rate $\in [2 \times 10^{-5}, 1 \times 10^{-5}]$.
- `Adapters`: Learning rate of 1×10^{-4} and a reduction rate taken from $\{12, 6, 3\}$.
- `LoRA`: rank and α values of 8, and a learning rate $\in [5 \times 10^{-4}, 1 \times 10^{-4}]$.
- `Prefix-Tuning`: The prefix length was fixed to 50. The reparameterization trick was applied to the method with a bottleneck size of 512, and the learning rate taken from $\{5 \times 10^{-4}, 2 \times 10^{-4}, 1 \times 10^{-4}\}$.
- `UniPELT`: Prefix-tuning with 10 prefixes, adapter with a reduction factor of 16 and standard `LoRA`.
- `Attentive perturbation (our approach)`: Same exact hyperparameter configuration as prefix-tuning. Embedding size fixed to 50, reparameterization trick applied with a bottleneck size of 512 and learning rate taken from $\{5 \times 10^{-4}, 2 \times 10^{-4}, 1 \times 10^{-4}\}$.

The proposed approach was implemented using the adapter-hub library, which already proposes implementations for all chosen baseline, as to use the same code base for all experiments to ensure fairness across all experiments.

5   Results

The performances of all selected PEFT methods as well as the proposed approach on the GLUE benchmark for varying sample sizes can be seen in Table 1.

Overall `Attentive perturbation` outperforms all other baselines on average for all sample sizes. This gain in performance of more than 1 point compared to prefix-tuning, the second highest baseline, on all sample sizes except for 500 observations, where the performance gain starts to decrease, strongly supporting that the proposed approach is better suited to the few shot learning setting. The fact that the best performance gains for sample sizes of 50 and 100 is associated with the COLA dataset, which is strongly imbalanced, tends to bring further evidence to confirm this observation.

In addition, the proposed approach yields the best and second best performances on a given task for a given sample size 18 and 9 times out of 32, respectively. The second best method, `Prefix-tuning`, only yields the best and second best performances 5 and 16 times, respectively. In other words, our proposed approach is not in the top 2 best methods only 5 times in the 32 experiments presented in this paper.

Surprisingly, UniPELT, which fuses LoRA, Prefix-tuning and Adapter methods, does not outperform Prefix-tuning. This might be due to our experiment set up, with considerably smaller validation sets than those used in their experimental setup.

Table 1. Experiment results. F1 scores are reported for QQP and MRPC. Spearman correlations are reported for STS-B. Matthews correlation are reported for CoLA. Accuracy measurements are reported for all other datasets. Results in bold and underlined correspond to the best and second best results for the selected dataset and sample size, respectively.

Method	MNLI	QNLI	SST-2	QQP	CoLA	STS-B	MRPC	RTE	Avg.
[K = 50]									
Fine-tuning	35.5	**65.9**	57.57	45.6	<u>3.5</u>	45.1	<u>81.1</u>	50.6	48.1
Adapter	35.6	62.6	64.7	35.3	0.0	59.7	80.2	**53.1**	48.9
LoRA	35.7	63.9	68.4	47	1.0	56.5	**81.4**	<u>52.8</u>	50.8
UniPELT	35.3	62.4	73.1	42.3	1.1	64.3	80.7	51.8	51.4
Prefix-tuning	**37.8**	63.5	<u>74.9</u>	**53.1**	1.8	59.2	80.4	52.6	<u>52.9</u>
Attentive perturbation	<u>37.7</u>	<u>65.0</u>	**76.9**	<u>52.2</u>	**5.0**	<u>63.6</u>	80.3	52.4	**54.1**
[K = 100]									
Fine-tuning	35.5	68.9	73.9	52.6	3.0	64.1	**81.3**	52.1	53.9
Adapter	36.3	66.7	72.8	54.0	7.2	63.8	80.5	<u>53.0</u>	54.3
LoRA	37.3	64.9	73.2	54.2	7.3	60.4	**81.3**	52.9	53.9
UniPELT	37.7	66.9	79.1	53.6	5.1	<u>68.4</u>	79.7	52.0	55.3
Prefix-tuning	<u>38.3</u>	<u>69.4</u>	<u>80.8</u>	<u>57.2</u>	<u>8.1</u>	66.6	<u>81.1</u>	54.2	<u>57.0</u>
Attentive perturbation	**40.4**	**70.7**	**81.25**	**58.5**	**12.9**	**69.3**	80.7	<u>53.0</u>	**58.3**
[K = 200]									
Fine-tuning	42.3	**71.9**	80.8	<u>63.0</u>	20.2	69.0	80.8	54.6	60.3
Adapter	42.7	69.1	83.1	59.5	**26.5**	70.3	80.7	**56.2**	61.0
LoRA	41.0	67.1	82.2	61.2	19.8	67.8	80.1	54.5	59.2
UniPELT	41.6	70.2	82.8	58.7	16.4	<u>72.8</u>	**81.7**	54.9	59.9
Prefix-tuning	<u>44.9</u>	<u>71.4</u>	**84.2**	<u>63.0</u>	22.2	71.3	79.6	<u>56.0</u>	<u>61.6</u>
Attentive perturbation	**45.8**	**71.9**	<u>83.7</u>	**64.2**	<u>24.9</u>	**73.7**	<u>81.1</u>	**56.2**	**62.7**
[K = 500]									
Fine-tuning	52.7	74.3	<u>85.4</u>	<u>66.8</u>	32.2	**78.0**	<u>82.5</u>	59.8	66.5
Adapter	51.1	72.4	<u>85.4</u>	65.7	**38.9**	76.1	81.9	59.8	66.4
LoRA	50.1	73.6	84.6	66.5	35.3	75.6	82.3	58.3	65.8
UniPELT	50.7	74.2	<u>85.4</u>	63.4	34.2	77.2	82.1	57.8	65.6
Prefix-tuning	<u>54.0</u>	<u>74.7</u>	**85.6**	66.2	35.7	<u>77.8</u>	82	<u>60</u>	<u>67.0</u>
Attentive perturbation	**54.7**	**75.0**	84.6	**67.0**	<u>37.4</u>	77.7	**83.0**	**61.4**	**67.6**

6   Conclusion

In this paper we introduced a new parameter efficient fine-tuning method, the attentive perturbator. This method is fairly straightforward and easy to use; it requires a comparable amount of computation to prefix-tuning, which is considerably lower than most adapter-like methods at inference time. In addition, we empirically showed that this method behaves better on average than all other selected baseline PEFT methods, as well as traditional fine-tuning, on a variant of the GLUE benchmark specifically tailored to assess model performances in the few-shot learning setting.

Acknowledgments. This work was supported by a grant overseen by the French National Research Agency (ANR) (ANR-19-CE23-0002). It also received the labelling of *Cap Digital* and *EuroBiomed* competitiveness clusters.

References

1. Ben Zaken, E., Goldberg, Y., Ravfogel, S.: BitFit: simple parameter-efficient fine-tuning for transformer-based masked language-models. In: Proceedings of the 60th Annual Meeting of the Association for Computational Linguistics, pp. 1–9. Association for Computational Linguistics, Dublin, Ireland (2022)
2. Brown, T., et al.: Language models are few-shot learners. In: Advances in Neural Information Processing Systems, vol. 33, pp. 1877–1901 (2020)
3. Devlin, J., Chang, M.W., Lee, K., Toutanova, K.: BERT: pre-training of deep bidirectional transformers for language understanding. arXiv preprint arXiv:1810.04805 (2018)
4. Houlsby, N., et al.: Parameter-efficient transfer learning for NLP. In: International Conference on Machine Learning, pp. 2790–2799. PMLR (2019)
5. Hu, E.J., et al.: LoRA: low-rank adaptation of large language models (2021). https://doi.org/10.48550/ARXIV.2106.09685
6. Lester, B., Al-Rfou, R., Constant, N.: The power of scale for parameter-efficient prompt tuning. In: Proceedings of the 2021 Conference on Empirical Methods in Natural Language Processing, pp. 3045–3059. Association for Computational Linguistics, Online and Punta Cana, Dominican Republic (2021)
7. Liu, X., et al.: P-tuning: prompt tuning can be comparable to fine-tuning across scales and tasks. In: Proceedings of the 60th Annual Meeting of the Association for Computational Linguistics (vol. 2: Short Papers), pp. 61–68. Association for Computational Linguistics, Dublin, Ireland (2022)
8. Mao, Y., et al.: UniPELT: A unified framework for parameter-efficient language model tuning. In: Proceedings of the 60th Annual Meeting of the Association for Computational Linguistics (vol. 1: Long Papers), pp. 6253–6264. Association for Computational Linguistics, Dublin, Ireland (2022)
9. Pfeiffer, J., Kamath, A., Rücklé, A., Cho, K., Gurevych, I.: AdapterFusion: non-destructive task composition for transfer learning. In: Proceedings of the 16th Conference of the European Chapter of the Association for Computational Linguistics, pp. 487–503. Association for Computational Linguistics (2021)
10. Radford, A., Wu, J., Child, R., Luan, D., Amodei, D., Sutskever, I.: Language Models are Unsupervised Multitask Learners (2019)

11. Raffel, C., et al.: Exploring the limits of transfer learning with a unified text-to-text transformer (2019). https://doi.org/10.48550/ARXIV.1910.10683
12. Rebuffi, S.A., Bilen, H., Vedaldi, A.: Learning multiple visual domains with residual adapters. In: Advances in Neural Information Processing Systems, vol. 30 (2017)
13. Wang, A., Singh, A., Michael, J., Hill, F., Levy, O., Bowman, S.: GLUE: a multi-task benchmark and analysis platform for natural language understanding. In: Proceedings of the 2018 EMNLP Workshop BlackboxNLP: Analyzing and Interpreting Neural Networks for NLP, pp. 353–355. Association for Computational Linguistics, Brussels, Belgium (2018)

SoftCut: A Fully Differentiable Relaxed Graph Cut Approach for Deep Learning Image Segmentation

Alessio Bonfiglio[(✉)] , Marco Cannici , and Matteo Matteucci

Politecnico di Milano, 20133 Milan, Italy
alessio.bonfiglio@mail.polimi.it,
{marco.cannici,matteo.matteucci}@polimi.it

Abstract. Graph cut algorithms can produce consistent high-quality image segmentation masks by minimizing a predefined energy function over pixels. However, defining such a function is often impracticable, especially when it comes to semantic segmentation where pixel values must convey information about the class of a pixel. On the other hand, convolutional neural networks, like U-Net, can learn to implicitly extract meaningful information from an image, but they lack explicit constraints, leading to potential rugged boundaries in the produced masks. In recent years, many solutions have been proposed to implement graph-cut algorithms into a neural network layer, and thus combine the best of both worlds, but all lack in speed or quality of the results. SoftCut, the approach proposed in this work, is a differentiable relaxation of the graph cut problem, equivalent to an intuitive electric circuit, that, used as an output activation function, is shown to outperform both U-Net and submodular optimization in terms of IoU on real-world images taken from Cityscapes, while being faster than the latter.

Keywords: Graph cut · Differentiable · Image segmentation · Artificial neural network · Deep learning · Machine learning

1 Introduction

Image segmentation is the process of partitioning a given image into regions of interest, by assigning a label to each pixel. It has a crucial role in tasks like medical image analysis [15] and autonomous driving [6], where the relevant elements are cut out from the background. Due to the ability of artificial neural networks to learn how to extract information from their input, architectures like U-Net [15] are used to produce the segmentation masks from the images. Such networks though cannot enforce any constraint on their outputs to guarantee smooth and consistent output masks. Traditionally, this task has been instead formally modeled as a graph cut problem that finds a binary partition S^* isolating pixels of the foreground from that in the background [8]:

$$S^* = \underset{S \subseteq \mathcal{V}}{\text{argmin}} \sum_{\{i,j\} \in \mathcal{E}} w_{ij} [\![|S \cap \{i,j\}| = 1]\!] - \sum_{i \in S} p_i \qquad (1)$$

G. Nicosia et al. (Eds.): LOD 2023, LNCS 14505, pp. 497–511, 2024.
https://doi.org/10.1007/978-3-031-53969-5_37

where p_i defines the data score, i.e., to which degree a pixel i should be considered part of the foreground and background partitions, $\mathcal{E}$ is the set of edges between adjacent pixels, and $w_{i,j}$ is a smoothness term representing the correlation between two pixels i and j.

By taking into account the correlation between each neighbor pixel, such formulation can produce precise segmentation masks, provided that meaningful pixel features p_i and $w_{i,j}$ are available. Because those values are not trivially obtainable from the image, one may like to train an artificial neural network to compute them. Training such a network would require examples of the correct values of p_i together with their corresponding image, making the problem circular. On the other hand, an end-to-end approach, where a model consists of a neural network followed by a graph cut solver, only requires examples of images and their segmentation masks for the training. The network would extract the needed features from the images and compute p_i, that the solver can thus to produce a consistent segmentation mask. Unfortunately, as (1) is not differentiable, such a direct approach is not trivially applicable.

The solution proposed in this work, named SoftCut, is an output activation layer that solves a differentiable relaxation of the graph cut problem (1) in the form of a sparse linear system of equations which allows for an intuitive interpretation as an electric circuit. The proposed solution simplifies the network task, allowing the network to only predict coarse segmentations which are later refined by solving the graph cut problem. By applying the conjugate gradient method [10], SoftCut is able to surpass similar techniques (such as [8]) in terms of both computation time and results quality.

1.1   Related Work

Over the years, there has been a shift towards neural network-based solutions for semantic segmentation, which have shown remarkable improvements over classical techniques like graph-cut algorithms thanks to their ability to autonomously extract meaningful features, without human intervention. Starting from simple yet effective encode-decoder architectures, such as the U-Net [15], the field has lately progressed toward more articulated network designs involving complex feature aggregation and training schemes [5,18] and, more recently, transformer-based solutions [14].

Despite the impressive level of accuracy achieved by these architectures, an alternative research direction is exploring how to combine learning-based solutions together with traditional optimization-based techniques to exploit the benefits of both worlds. Indeed, while optimization-based methods alone might not achieve the same level of accuracy as recent solutions, they produce explainable, certifiable, and constrained solutions, which are instead lacking in the latter. Nevertheless, designing such a hybrid solution is not trivial, as it requires backpropagating through the graph-cut solver. In [3], the authors build upon stochastic perturbations methods and propose to derive an approximate gradient by stochastically perturbing the problem's inputs. In OptNet [2], the implicit function theorem is instead employed over the Karush-Kuhn-Tucker conditions to derive a correct gradient formulation for quadratic optimization problems, and therefore also relaxed graph-cut problems. An extension has later been proposed in [1], where the authors demonstrate that a similar approach can also be applied to the class of log-log convex programs. More efficient, yet similarly accurate, solutions have also been proposed, such as the one in [16] where an alternating differentiation scheme is proposed to optimize computation time and memory efficiency of the Jacobian matrix computation. Finally, the solution proposed in [8], instead, exploits the submodular nature of graph-cut problems to derive a principled relaxation of the graph-cut problem as a total variation minimization. In this paper, we follow this research direction and propose a simple yet effective solution that linearizes the graph-cut into a sparse linear system of equations which is computationally faster and easier to integrate into end-to-end training schemes.

2   Method

In this section, we present a novel approach for bridging the gap between deep neural networks and graph-cut algorithms for image segmentation. We propose to achieve this by designing a hybrid architecture composed of a neural network followed by a graph-cut solver. We design a differentiable output layer, named SoftCut, which seamlessly integrates the two allowing the architecture to leverage the expressive power of neural networks for feature extraction while preserving the advantages of graph cut algorithms in producing consistent and high-quality image segmentations. The layer implements a relaxation of the graph cut problem which allows gradients to flow through the solution of the problem, thus enabling end-to-end optimization of the neural network, without intermediate supervision.

We design the SoftCut layer around the graph cut problem introduced in (1). In particular, we consider graph-cut problems defined on a lattice graph representing pixels of an image and indicate pixel scores, and correlations between them, by their i, j position in the grid. Formally, the SoftCut layer takes as input a pixel scores matrix $P \in \mathbb{R}^{h \times w}$ whose elements $p_{i,j}$ tend to $\pm\infty$ proportionally to how much the pixel $\{i, j\}$ is certain to belong to the foreground or the background, and two matrices, $C \in \mathbb{R}_+^{h-1 \times w}$ and $R \in \mathbb{R}_+^{h \times w-1}$, whose elements define the correlation between adjacent pixels vertically and horizontally, respectively. We assume all these matrices to be produced by a neural network that, without any direct supervision, automatically learns to produce them.

The output of the SoftCut activation layer is defined as the solution $x \in \mathbb{R}^{1 \times hw}$ of a linearized version of the problem in (1) in the form $Ax = b$, with $A \in \mathbb{R}^{hw \times hw}$ and $b \in \mathbb{R}^{hw \times 1}$, whose elements, in the same way as those of P, represent how much a pixel i, j belongs to the foreground or the background.

A and b are defined as

$$A_{k,l} = \begin{cases} \begin{aligned} & 1 + c_{i,j}^u + c_{i,j}^d \\ & + r_{i,j}^l + r_{i,j}^r \end{aligned} & \text{if } k = l = f(i,j) \\ -c_{i,j}^u & \text{if } k = f(i,j), l = f(i-1,j) \\ -c_{i,j}^d & \text{if } k = f(i,j), l = f(i+1,j) \\ -r_{i,j}^l & \text{if } k = f(i,j), l = f(i,j-1) \\ -r_{i,j}^r & \text{if } k = f(i,j), l = f(i,j+1) \\ 0 & \text{otherwise} \end{cases} \qquad b = \begin{bmatrix} p_{0,0} \\ \vdots \\ p_{m,n} \end{bmatrix} \quad (2)$$

with $k, l \in \{1, 2, \ldots, hw\}$, $f(i,j) = (i-1)w + j$ maps the 2d indices $\{i,j\}$ of the pixels of the input image to the single dimension used in A and b, $c_{i,j}^u \in C^u = \begin{bmatrix} 0 \cdots 0 \\ C \end{bmatrix}$, $c_{i,j}^d \in C^d = \begin{bmatrix} C \\ 0 \cdots 0 \end{bmatrix}$, $r_{i,j}^l \in R^l = \begin{bmatrix} 0 \\ \vdots R \\ 0 \end{bmatrix}$ and $r_{i,j}^r \in R^r = \begin{bmatrix} 0 \\ R \vdots \\ 0 \end{bmatrix}$.

Note that A, while having a size of $hw \times hw$, is very sparse, with at most 5 entries per row, hence the system can be solved efficiently with a sparse linear solver as the one presented in [13]. In order to allow seamless backpropagation of the gradients during training and enable the network to learn discriminative features, gradients $\frac{\partial L}{\partial P} = \frac{\partial L}{\partial b} \cdot \frac{\partial b}{\partial P}$, $\frac{\partial L}{\partial C} = \frac{\partial L}{\partial A} \cdot \frac{\partial A}{\partial C}$, and $\frac{\partial L}{\partial R} = \frac{\partial L}{\partial A} \cdot \frac{\partial A}{\partial R}$ must be properly defined.

It can be shown that $\frac{\partial L}{\partial b} = (A^T)^{-1} \frac{\partial L}{\partial x} = solve(A^T, \frac{\partial L}{\partial x})$ and $\frac{\partial L}{\partial A} = -\frac{\partial L}{\partial b} \otimes x$ (where $solve(A, b)$ is the solution x of a generic system $Ax = b$ and $\otimes$ is the outer product), while the $\frac{\partial b}{\partial P}$, $\frac{\partial A}{\partial C}$ and $\frac{\partial A}{\partial R}$ terms can be trivially obtained by differentiating (2). Note that, being A^T still sparse, the gradient can reuse the same solver used in the forward pass.

Notice that, it is possible to generalize the SoftCut formulation to segment higher-dimensional volumes than 2D images. Given a d-dimension volume of size $n_1 \times n_2 \times \cdots \times n_d = N$, the input of the layer now become $d+1$ tensors. The pixels scores tensor $P \in \mathbb{R}^{n_1 \times n_2 \times \cdots \times n_d}$ has rank d. Every pixel of this volume has at most (when not on the border of the volume) $2 \cdot d$ edges that are stored in d tensors $E^1, E^2, \ldots, E^d$ with sizes $[n_1 - 1 \times n_2 \times \cdots \times n_d], [n_1 \times n_2 - 1 \times \cdots \times n_d], \ldots, [n_1 \times n_2 \times \cdots \times n_d - 1]$ and elements $e_{i_1, i_2, \ldots, i_d}^m \in \mathbb{R}_+, m \in \{1, 2, \ldots, d\}$. In this case, $A \in \mathbb{R}^{N \times N}$ and $b \in \mathbb{R}^{N \times 1}$ are defined as

$$A_{k,l} = \begin{cases} \begin{aligned} & 1 + e_{i_1, \ldots, i_d}^{1,b} + e_{i_1, \ldots, i_d}^{1,a} \\ & + \ldots \\ & + e_{i_1, \ldots, i_d}^{d,b} + e_{i_1, \ldots, i_d}^{d,a} \end{aligned} & \text{if } k = l = f(i_1, \ldots, i_d) \\ -e_{i_1, \ldots, i_d}^{1,b} & \text{if } l = f(i_1 - 1, \ldots, i_d) \\ -e_{i_1, \ldots, i_d}^{1,a} & \text{if } l = f(i_1 + 1, \ldots, i_d) \\ \vdots & \\ -e_{i_1, \ldots, i_d}^{d,b} & \text{if } l = f(i_1, \ldots, i_d - 1) \\ -e_{i_1, \ldots, i_d}^{d,a} & \text{if } l = f(i_1, \ldots, i_d + 1) \\ 0 & \text{otherwise} \end{cases} \qquad b = \begin{bmatrix} p_{0,0,\ldots,0} \\ \vdots \\ p_{n_1, n_2, \ldots, n_d} \end{bmatrix}$$

$$(3)$$

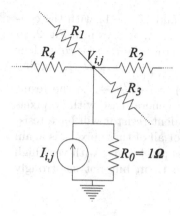

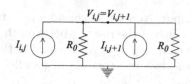

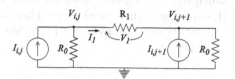

Fig. 1. The module of the circuit for each pixel.

Fig. 2. SoftCut in the case of two pixels connected by a hard edge.

Fig. 3. SoftCut in the case of two pixels connected by a soft edge.

with $k, l \in \{1, 2, \ldots, N\}$, $f(i_1, i_2 \ldots, i_d) = \sum_{i=1}^{d-1}(i_i - 1)n_i + i_d$ mapping the d-dimensional indices $\{d_1, d_2, \ldots, d_n\}$ of the pixels of the input volume to the single dimension used in A and b, $e_{i_1, i_2 \ldots, i_d}^{m,b} \in E^{m,a} = padbefore_m(E^m)$ and $e_{i_1, i_2 \ldots, i_d}^{m,a} \in E^{m,b} = padafter_m(E^m)$, where $padbefore_k(T)$ and $padafter_k(T)$ pad the tensor T along the dimension k before and after the other elements respectively. In this formulation, A is still sparse with at most $1 + 2d$ elements per row, and so both the system and the gradient can be solved in the same way as in the 2D case.

2.1   Equivalence to an Electrical Circuit

We now demonstrate that the proposed formulation is equivalent to an electric circuit. While this equivalence has no practical consequences in this work, we use the parallelism to give some intuition on the layer's functioning as well as to provide an intuitive derivation of the equations provided in (2). In particular, the circuit is built by repeating the module shown in Fig. 1 once for each pixel of the input, keeping their grid alignment as in the lattice structure presented in Sect. 2.

The resistors R_1, R_2, R_3 and R_4 are shared by every two adjacent modules. The generator's current $I_{i,j}$ is set to the score $p_{i,j}$ of said pixel. The conductance of R_1 and R_3 are provided by the edge weights $c_{i-1,j}$ and $c_{i,j}$ from matrix C, while the ones of R_2 and R_4 came from $r_{i,j-1}$ and $r_{i,j}$ of matrix R. The output of the SoftCut activation layer are the voltages $V_{i,j}$ of each module.

Proceeding from simple to complex pixel arrangements, let us start by considering the case of an image made up of a single pixel. By Ohm's law, $V_{i,j} = R_0 \cdot I_{i,j} = I_{i,j}$, thus the result, i.e., $V_{i,j}$, is the same pixel score received as input. Now let us consider only two pixels $\{i,j\}$ and $\{i,j+1\}$ with the conductance of the resistor between them equal to $+\infty\ \Omega^{-1}$, as shown in Fig. 2. In this case, $V_{i,j} = V_{i,j+1}$ is equal to the voltage drop across the resistor equivalent to the two resistors R_0 in parallel, that is $R_{eq} = \frac{1}{\frac{1}{1\,\Omega}+\frac{1}{1\,\Omega}} = \frac{1}{2}\ \Omega$. According to Ohm's law, $V_{i,j} = V_{i,j+1} = R_{eq} \cdot (I_{i,j} + I_{i,j+1}) = \frac{I_{i,j}+I_{i,j+1}}{2}$, so the result is the average between the two pixel scores. In the general case with N pixels connected with zero resistance as before, the equivalent resistor will have resistance $R_{eq} = \frac{1}{N}\ \Omega$, and hence the resulting voltage, for all of these pixels, is again the average of their pixel score. This property makes pixel scores with very high absolute values compensate for those connected to them but not as strongly (Fig. 4).

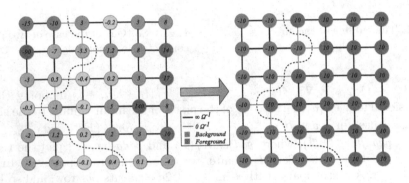

Fig. 4. Example of input and output of SoftCut in the case of hard edges ($0\ \Omega^{-1}$ and $+\infty\ \Omega^{-1}$) that partition the graph exactly around the dashed line.

If, instead, the resistor between $\{i,j\}$ and $\{i,j+1\}$ has a conductance $\leq +\infty\ \Omega^{-1}$, the circuit becomes as shown in Fig. 3. In this case, $V_{i,j} \neq V_{i,j+1}$, but it is still possible to compute them by applying Kirchhoff's voltage and current laws on the voltage nodes as follows:

$$V_{i,j} = V_1 + V_{i,j+1}, \qquad I_{i,j} = I_1 + \frac{V_{i,j}}{R_0}, \qquad I_{i,j+1} + I_1 = \frac{V_{i,j+1}}{R_0} \qquad (4)$$

which results in

$$V_{i,j} = \frac{G_1(I_{i,j} + I_{i,j+1}) + I_{i,j}}{2G_1 + 1}, \qquad V_{i,j+1} = \frac{G_1(I_{i,j} + I_{i,j+1}) + I_{i,j+1}}{2G_1 + 1}. \qquad (5)$$

Here, $G_1 = \frac{1}{R_1}$, shows how $V_{i,j}$ and $V_{i,j+1}$ gradually go from $I_{i,j}$ and $I_{i,j+1}$ to $\frac{I_{i,j}+I_{i,j+1}}{2}$ when G_1 goes from $0\ \Omega^{-1}$ to $+\infty\ \Omega^{-1}$. Doing this, SoftCut is able to propagate the influence of a pixel score to its neighbors in a controlled way, using the information provided by the weights of the edges (Fig. 5).

These intuitive examples show how the weights of the edges provide the general shape of how the image should be partitioned, while the pixel score values diffuse in these shapes. Therefore, the network before the SoftCut layer only needs to compute how much two adjacent pixels are correlated to each other and an estimate measure of the belonging of each pixel to the background or the foreground class, instead of precisely segmenting them. The SoftCut layer, leveraging the underlying graph-cut structure, is then capable, without any learning, to refine the final score of each pixel by smoothing out uncertainties or disagreements of the pixel scores while also taking into account their correlation.

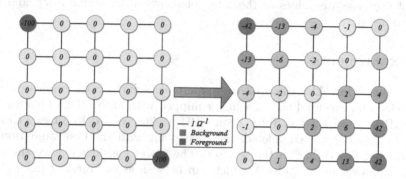

Fig. 5. Example of input and output of SoftCut in the case of soft edges.

2.2 Derivation of SoftCut Equations

Given the electric circuit just introduced, is now trivial to define the system of equations introduced in (2) by just applying Kirchhoff's current law on every pixel:

$$
\begin{aligned}
I_{i,j} &= \frac{V_{i,j}}{R_0} + \frac{V_{i,j} - V_{i+1,j}}{R_3} + \frac{V_{i,j} - V_{i,j+1}}{R_2} \\
&\quad - \left(\frac{V_{i-1,j} - V_{i,j}}{R_1} + \frac{V_{i,j-1} - V_{i,j}}{R_4}\right) \\
&= (1 + G_1 + G_2 + G_3 + G_4)V_{i,j} \\
&\quad - G_1 V_{i-1,j} - G_2 V_{i,j+1} - G_3 V_{i+1,j} - G_4 V_{i,j-1} \\
&= (1 + c_{i-1,j} + r_{i,j} + c_{i,j} + r_{i,j-1})x_{i,j} \\
&\quad - c_{i-1,j}x_{i-1,j} - r_{i,j}x_{i,j+1} - c_{i,j}x_{i+1,j} - r_{i,j-1}x_{i,j-1} \\
&= (1 + c_{i,j}^u + r_{i,j}^r + c_{i,j}^d + r_{i,j-1}^l)x_{i,j} \\
&\quad - c_{i,j}^u x_{i-1,j} - r_{i,j}^r x_{i,j+1} - c_{i,j}^d x_{i+1,j} - r_{i,j}^l x_{i,j-1} \\
&= p_{i,j}
\end{aligned}
\tag{6}
$$

In the case of the segmentation of d-dimensional volume, the pixels would have $2 \cdot d$ edges. Modifying the circuit in Fig. 1 by adding a resistor for each one of the neighbors of pixel $\{i,j\}$ it is possible to obtain back the formulations in (3) in the same way as done for the 2d formulation in (6).

2.3 Approximate Solution

While algorithms like KLU [13] are very efficient at solving sparse linear systems, they have the downside of being sequential, due to the very nature of the problem. Luckily, it is also possible to solve the linear systems using an iterative algorithm, like the conjugate gradient method [10], directly on the GPU. This approach is possible because matrix A is symmetric (by its definition (2)) and strictly diagonally dominant (the diagonal elements are always 1 more than the sum of the absolute values of the other elements in the same row), and thus positive definite.

3 Experiments

Implementation Details. All the experiments have been implemented in PyTorchs[1] and executed on a machine equipped with an NVIDIA Quadro RTX 6000 GPU paired with an Intel Xeon CPU E5-2687W v4. In all experiments, after the training was completed, the checkpoint with the best validation loss was selected for testing on the test set. The scores considered were the mean intersection over union (mIoU), which can be seen as the ratio of the TP pixels over $TP + FP + FN$ pixels, and the average number of distinct figures in the segmentation mask, that gives an idea of the spatial consistency of those masks. The scores are shown together with their standard deviation over the test set. The exact solver for the sparse linear system comes from [11] which has been modified in order to parallelize the computation of the mini-batches across multiple CPU cores. The approximate solution of SoftCut is computed directly on the GPU by an implementation of the conjugated gradient method [10] in a custom PyTorch layer. The algorithm has been set up in order to stop when $\sum_{r \in r} |r| < 10^{-6} bhw$ (where r is the residual vector and b is the batch size, h and w the height and width of the images) and the initialization value of the result x has been set to b, as the pixel scores are a first approximation of the actual solution.

The architecture used in the experiments consists of a base network, that computes the segmentation mask when trained alone or the pixel scores when a graph cut is applied at the end of it, and some other smaller decoders for predicting the edge weights that are used only by the graph cuts algorithms (details are provided in the architectures section). Being SoftCut a differentiable graph cut algorithm, we directly compare it with the submodular approach from [8] using the official PyTorch layer implementation [7] which we parallelized to run on multiple CPU cores.

The networks have been trained to solve *Binary Segmentation*, *Multiclass Segmentation* and *Specific Object Segmentation* (a form of binary segmentation where an additional mask is provided to indicate the target object to segment).

Architectures and Evaluation Setups. We first consider the setup proposed in [8]. The network architecture, shown in Fig. 6, consists of 3 convolutions for

---

[1] Code available at https://github.com/alessiobonfiglio/softcut-lod.

the base network, which produces the segmentation mask, when the network is used alone, or the pixel scores, when used in combination with a graph cut layer, together with other four convolutions that compute the edge weights. Note that this architecture does not use any kind of pooling, so the receptive field of each pixel is particularly limited. As in [8], we test this architecture to only solve *Binary Segmentation* on the *Weizmann Horse* [4] dataset, with a split of 180, 50 and 98 for the training, validation and test set. All the models have been trained for 100 epochs, with a batch size of 1, by minimizing the binary cross-entropy on each of the 0.1% of revealed pixels, using both *Adam* and *Adagrad* as optimizers with a learning rate of 10^{-2} and 10^{-3}. Moreover, as [7] supports and enables by default refining the results of the total variation algorithm with an isotonic regression, the effects of this option have also been tested.

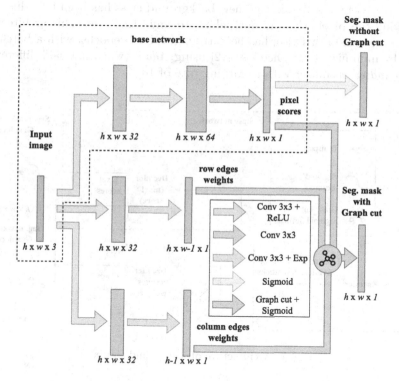

Fig. 6. Architecture of the setup from [8].

In the second evaluation, we test the effect of adding the proposed SoftCut layer on more complex architectures and datasets. We use a U-Net architecture from Segmentation Models [17]. In Fig. 7, the encoder shown is the contracting path of the U-Net architecture, which outputs the lowest resolution embeddings plus all the previous skip connections. The decoders are the expansive path of U-Net, which takes as input the various resolution features and returns a full-resolution output. Moreover, instead of the exponential function used in the previous architecture to assure that the edge weights were ≥ 0, the Soft-Plus function has been applied for numerical stability. Two different encoders have been tested: *resnet18* (18 layers, 11M parameters) and *resnet34* (34 layers, 21M parameters), from [9]. When not said otherwise, *resnet18* was used. Two decoders have been used: one for computing the pixel scores and another (not used in the base network alone) for the edge weights. The edge weights decoder has been reused for both the column and row weights, exploiting the fact that the transposed row edges of a transposed image are its column edges. Note that the decoders use much fewer parameters (3M) compared to the encoder, so a second decoder for the edge weight does not give an unfair advantage to the models with graph cut. For the multiclass segmentation task, the graph cut layer is repeated for each channel and the final sigmoid activation function has been replaced by a softmax, so a new background class has been introduced for classifying the pixels that do not belong to any other one. This architecture, except when said otherwise, has been trained for 100 epochs, with a batch size of 32, by minimizing the Dice loss [12] using *Adam* (when not said differently) or *Adagrad* as optimizer with a learning rate of 10^{-3}.

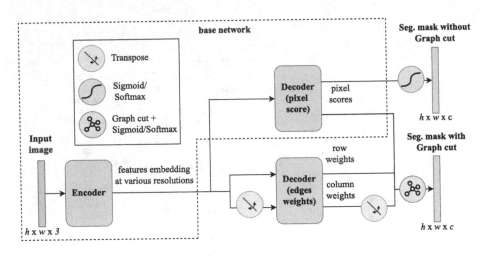

Fig. 7. Architecture of the second setup.

The datasets used in this setup are derived from *Cityscapes* [6]. We first divide images into tiles of size 256 × 256 and then sample the tiles so that the vast majority of them contain the considered classes. We split the original *Cityscapes* training set into two sets that we use for training and validation, while we use the original validation as the test set.

- *Custom Cityscapes for Binary segmentation*. It features only the *traffic signs* class, 13078 train samples, 1024 validation samples, and 2344 test samples.

- *Custom Cityscapes for Specific Object Detection.* it features *traffic signs* and *cars*, 21644 train samples, 609 validation samples, and 3724 test samples. In addition to the three channels of the input image, we also add an additional fourth channel containing all zeros except for a single value set to one in the middle of the object to be segmented.
- *Custom Cityscapes for Multiclass Segmentation.* It consists of *traffic signs* and *cars* classes, 11983 train samples, 877 validation samples, and 4833 test samples.

We finally test a dataset (*Cityscapes Full Resolution*) made of all the training and validation samples of *Cityscapes* at their original full resolution. In order to train the networks efficiently, however, we crop samples into overlapping tiles (of size $256 + 32 \times 256 + 32$), thus preserving the surrounding context for prediction, and then recombine back the output segmentation masks into a full-resolution one. During training, we compute the loss function also on the overlapping parts, while we discard them at inference time when we recombine the tiled masks back to full-resolution. Because of the limited computation power available for the training and size of the networks, only the *traffic signs, cars* and *persons* classes have been considered.

Table 1. Results of experiments in the first setup from [8]

Model	mIoU	loss	#distinct figs.	train time
Base Network	0.4720 (0.0243)	0.3936 (0.0300)	72.336 (5.5184)	**5 min**
Submod	0.5304 (0.0282)	0.3485 (0.0279)	2.0204 (0.1450)	50 min
Submod (iso)	**0.6832 (0.0220)**	**0.2637 (0.0176)**	**1.8571 (0.1547)**	50 min
SoftCut (cpu)	0.5937 (0.0218)	0.3024 (0.0306)	7.4693 (0.8242)	11 h 35 min
SoftCut (gpu)	0.6022 (0.0211)	0.3041 (0.0285)	8.7346 (0.8209)	35 min

Table 2. Results of experiments in the second setup trained over *Custom Cityscapes for Binary segmentation*

Model	Enc.	Opt.	mIoU	loss	#distinct figs.
Base Network	rn18	Adam	0.4324 (0.0074)	0.4918 (0.0079)	1.5059 (0.0289)
Submod (iso)	rn18	Adam	0.4141 (0.0076)	0.5188 (0.0081)	**0.9914 (0.017)**
SoftCut (cpu)	rn18	Adam	0.4240 (0.0075)	0.5050 (0.0081)	1.1335 (0.0198)
SoftCut (gpu)	rn18	Adam	**0.4790 (0.0075)**	**0.4468 (0.0078)**	1.2636 (0.0212)
Base Network	rn34	Adam	0.4512 (0.0074)	0.4724 (0.0079)	2.1467 (0.0420)
Submod (iso)	rn34	Adam	0.4252 (0.0076)	0.5049 (0.0081)	**1.0230 (0.0173)**
SoftCut (cpu)	rn34	Adam	0.4357 (0.0076)	0.4933 (0.0081)	1.1766 (0.0211)
SoftCut (gpu)	rn34	Adam	**0.4784 (0.0077)**	**0.4512 (0.0081)**	1.1139 (0.0188)
Base Network	rn18	Adagrad	0.3452 (0.0069)	0.5806 (0.0076)	1.6301 (0.0288)
Submod (iso)	rn18	Adagrad	0.3563 (0.0073)	0.5795 (0.0079)	**0.9142 (0.0168)**
SoftCut (gpu)	rn18	Adagrad	**0.3647 (0.0070)**	**0.5640 (0.0076)**	1.2585 (0.0220)

3.1   Results

Table 1 shows the results of the experiments in the first setup from [8] (*Binary Segmentation* on the *Weizmann Horse* dataset). The results shown are the best in terms of mIoU for each combination of optimizers and learning rates for the same model. For all the models this combination ends up being Adam with a learning rate of 0.001. As already seen in the results of [8], the submodular approach greatly improves the results of the network. Moreover, while it did not reach the same score as the submodular approach with isotonic refinement, SoftCut is able to surpass both the base network and the basic submodular model. Finally, it can be noticed how the approximate version of SoftCut performs better than the exact solution, while being faster than the latter and the submodular approach. The great improvements that the graph cut techniques provided in this setup may be attributed to the increased receptive field that these approaches implicitly provide by considering all the pixels and their relationship together.

Table 2 shows the results of experiments in the second setup trained over *Custom Cityscapes for Binary segmentation*. The submodular approach has been tested only with the isotonic regression refinement as it appears to always outperform the same method without it. In this new harder setup, the approximate solution of SoftCut is always able to outperform the base network, but also its exact solution and the submodular approach, in all the cases tested. Interestingly, a bigger encoder does not improve its scores, differently from the other techniques. The models trained with *Adam* greatly outperforms those trained with *Adagrad*.

Table 3. Results of experiments in the second setup trained over *Custom Cityscapes for Specific Object Detection*

Model	Enc.	mIoU	loss	#distinct figs.
Base Network	rn18	0.8454 (0.0022)	0.0930 (0.0016)	1.3053 (0.0140)
Submod (iso)	rn18	0.8138 (0.0025)	0.1168 (0.0019)	**1.0048 (0.0011)**
SoftCut (gpu)	rn18	**0.8540 (0.0021)**	**0.0886 (0.0015)**	1.0208 (0.0024)
Base Network	rn34	0.8434 (0.0022)	0.0942 (0.0016)	1.3943 (0.0180)
Submod (iso)	rn34	0.8025 (0.0027)	0.1270 (0.0022)	**1.0022 (0.0016)**
SoftCut (gpu)	rn34	**0.8517 (0.0021)**	**0.0907 (0.0016)**	1.0173 (0.0024)

Table 3, Table 4 and Table 5 show the results for the *Specific* and *Multiclass Segmentation* as well as the *Full Resolution* setups. Due to the slow training time of the exact solution of SoftCut and given that the approximate version achieved better scores in the previous experiments, we decide not to include them in this evaluation. In the case of multiclass segmentation, we report scores computed by averaging all the values of each class, including the background. Again, the approximate solution of SoftCut reaches better scores than both the base network and the submodular approach in all the experiments. The submodular approach [8], on the other hand, seems to not be able to reach the same results of U-Net

Table 4. Results of experiments in the second setup trained over *Custom Cityscapes for Multiclass Segmentation*

Model	mIoU	loss	#distinct figs.
Base Network	0.7505 (0.0027)	0.1673 (0.0022)	5.2375 (0.0672)
Submod (iso)	0.7417 (0.0027)	0.1936 (0.0023)	**2.3482 (0.0164)**
SoftCut (gpu)	**0.7788 (0.0026)**	**0.1612 (0.0022)**	2.7159 (0.0202)

Table 5. Results of experiments in the second setup trained over *Cityscapes Full Resolution*

Model	mIoU	loss	#distinct figs.	train time
Base Network	0.5910 (0.0064)	0.2660 (0.0059)	62.106 (1.6043)	**2 h 53 min**
Submod (iso)	0.5641 (0.0054)	0.2903 (0.0055)	54.020 (1.2437)	94 h 12 min
SoftCut (gpu)	**0.6202 (0.0059)**	**0.2493 (0.0053)**	**29.512 (0.7691)**	35 h 30 min

(a) The input image example (left) and the relative ground truth (right).

(b) The output of the base network without graph cut.

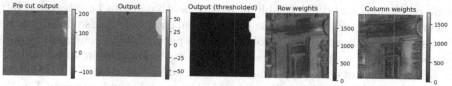

(c) The output of the base network with the submodular approach.

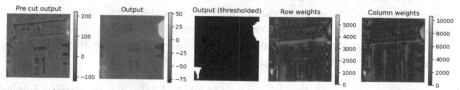

(d) The output of the base network with SoftCut computed exactly.

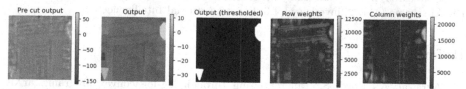

(e) The output of the base network with SoftCut computed approximately.

Fig. 8. Example of the results on a sample of the test set from the experiments of Table 2. The pre-cut outputs are the pixel scores.

alone when used in these harder scenarios. Moreover, the results show that all the graph cut techniques end up reducing the number of spurious predictions, as the number of distinct figures in the mask is significantly lower than in the case of the base network alone. We attribute the facts that both the submodular approach and the exact solution of SoftCut perform worse than the base network in these scenarios to the nature of the graph cut problem. Indeed, its binary nature could produce a more complex loss landscape that makes its minimization more difficult compared to that resulting from the approximated solution.

Finally, Fig. 8 shows a comparison of the results obtained using the techniques from Table 2. Both the base network and the submodular approach fail to predict the mask of the bottom left sign. Moreover, the differences between the *pre cut output* and the *output* show the properties of SoftCut described in Fig. 4 and Fig. 5. SoftCut is able to take coarse predictions of the class of the pixels and diffuse them in the shapes given by the edge weights, and thus produce a smooth, continuous, but also sharp and well-defined prediction. As a result, the mask produced by the approximate solution of SoftCut is qualitatively superior, featuring sharper edges and overall more correct segmentation masks.

4   Conclusion

This work presented a novel approach to image segmentation in deep learning. SoftCut is able to provide the same advantages of a graph cut algorithm, while being continuous and differentiable in nature, so that it can be integrated into a neural network architecture. Due to its formulation as a sparse linear system of equations, it allows different solving techniques, like the conjugated gradient method, that is shown to greatly improve the computation speed and the quality of the results. The experiments conducted show that SoftCut is able to outperform similar techniques like [8] and U-Net in real-world scenarios. Finally, it has a nice interpretation in the form of an electric circuit, which provides an easier way to understand its working and opens to the possibility of designing an analog hardware acceleration module for the task. Future works could focus on improving the computational speed of SoftCut, as it appears to currently be the main downside of this technique.

Acknowledgements. This paper is supported by the FAIR (Future Artificial Intelligence Research) project, funded by the NextGenerationEU program within the PNRR-PE-AI scheme (M4C2, investment 1.3, line on Artificial Intelligence).

References

1. Agrawal, A., Boyd, S.: Differentiating through log-log convex programs. arXiv (2020)
2. Amos, B., Kolter, J.Z.: OptNet: differentiable optimization as a layer in neural networks. In: Proceedings of the 34th International Conference on Machine Learning. Proceedings of Machine Learning Research, vol. 70, pp. 136–145. PMLR (2017)

3. Berthet, Q., Blondel, M., Teboul, O., Cuturi, M., Vert, J.P., Bach, F.: Learning with differentiable pertubed optimizers. In: Advances in Neural Information Processing Systems, vol. 33, pp. 9508–9519 (2020)

4. Borenstein, E., Ullman, S.: Class-specific, top-down segmentation. In: Heyden, A., Sparr, G., Nielsen, M., Johansen, P. (eds.) ECCV 2002. LNCS, vol. 2351, pp. 109–122. Springer, Heidelberg (2002). https://doi.org/10.1007/3-540-47967-8_8

5. Borse, S., Cai, H., Zhang, Y., Porikli, F.: HS3: learning with proper task complexity in hierarchically supervised semantic segmentation. In: 32nd British Machine Vision Conference 2021, BMVC 2021, Online, 22–25 November 2021, p. 175. BMVA Press (2021)

6. Cordts, M., et al.: The cityscapes dataset for semantic urban scene understanding. In: 2016 IEEE Conference on Computer Vision and Pattern Recognition (CVPR). IEEE (2016). https://doi.org/10.1109/cvpr.2016.350

7. Djolonga, J.: torch-submod (2017). https://github.com/josipd/torch-submod

8. Djolonga, J., Krause, A.: Differentiable learning of submodular models. In: Advances in Neural Information Processing Systems, vol. 30. Curran Associates, Inc. (2017)

9. He, K., Zhang, X., Ren, S., Sun, J.: Deep residual learning for image recognition. In: 2016 IEEE Conference on Computer Vision and Pattern Recognition (CVPR), pp. 770–778. IEEE (2016). https://doi.org/10.1109/cvpr.2016.90

10. Hestenes, M.R., Stiefel, E.: Methods of conjugate gradients for solving. J. Res. Natl. Bur. Stand. **49**(6), 409 (1952)

11. Laporte, F.: Torch sparse solve (2020). https://github.com/flaport/torch_sparse_solve

12. Milletari, F., Navab, N., Ahmadi, S.A.: V-net: fully convolutional neural networks for volumetric medical image segmentation. In: 2016 Fourth International Conference on 3D Vision (3DV), pp. 565–571. IEEE (2016). https://doi.org/10.1109/3dv.2016.79

13. Natarajan, E.P.: KLU-A high performance sparse linear solver for circuit simulation problems. Ph.D. thesis, University of Florida (2005)

14. Ranftl, R., Bochkovskiy, A., Koltun, V.: Vision transformers for dense prediction. In: 2021 IEEE/CVF International Conference on Computer Vision (ICCV), pp. 12179–12188. IEEE (2021). https://doi.org/10.1109/iccv48922.2021.01196

15. Ronneberger, O., Fischer, P., Brox, T.: U-net: convolutional networks for biomedical image segmentation. In: Navab, N., Hornegger, J., Wells, W.M., Frangi, A.F. (eds.) MICCAI 2015. LNCS, vol. 9351, pp. 234–241. Springer, Cham (2015). https://doi.org/10.1007/978-3-319-24574-4_28

16. Sun, H., Shi, Y., Wang, J., Tuan, H.D., Poor, H.V., Tao, D.: Alternating differentiation for optimization layers. In: The Eleventh International Conference on Learning Representations (2023)

17. Yakubovskiy, P.: Segmentation models pytorch (2020). https://github.com/qubvel/segmentation_models.pytorch

18. Zhang, X., et al.: DCNAS: densely connected neural architecture search for semantic image segmentation. In: 2021 IEEE/CVF Conference on Computer Vision and Pattern Recognition (CVPR), pp. 13956–13967. IEEE (2021). https://doi.org/10.1109/cvpr46437.2021.01374

Author Index

G. Nicosia et al. (Eds.): LOD 2023, LNCS 14505, pp. 513–515, 2024.
https://doi.org/10.1007/978-3-031-53969-5